GAS LIFT
MANUAL

GÁBOR TAKÁCS, PhD

Petroleum Engineering Department
University of Miskolc, Hungary

Copyright ©2005
PennWell Corporation
1421 South Sheridan Road
Tulsa, Oklahoma 74112

1-800-752-9764
sales@pennwell.com
www.pennwell.com
www.pennwell-store.com

Managing Editor: Marla Patterson
Production Editor: Sue Rhodes Dodd
Cover Designer: Shanon Moore
Book Designer: Clark Bell
Book Layout: Brigitte Pumford-Coffman

Library of Congress Cataloging-in-Publication Data Available on Request

Takács, Gábor. Gas Lift Manual / Gábor Takács.
    p. cm. Includes index.
ISBN 0-87814-805-1

Printed in the United States of America

1 2 3 4 5   09 08 07 06 05

To my wife Beata,
my son Zsolt,
and my daughter Reka.

Contents

Preface .... **xv**

1. Introduction to Gas Lifting 1
   1.1 Artificial Lifting 1
   1.2 Short History of Gas Lifting 4
   1.3 Basic Features of Gas Lifting 5
   References 8

2. Production Engineering Fundamentals 9
   2.1 Introduction 9
   2.2 Properties of Oilfield Fluids 9
       2.2.1 Introduction 9
       2.2.2 Basic thermodynamic properties 10
       2.2.3 Liquid property correlations 12
           2.2.3.1 Water 12
           2.2.3.2 Crude oil 13
       2.2.4 Properties of natural gases 15
           2.2.4.1 Behavior of gases 15
           2.2.4.2 Gas property correlations 18
   2.3 Inflow Performance of Oil Wells 21
       2.3.1 Introduction 21
       2.3.2 Basic concepts 21
       2.3.3 The productivity index concept 23
       2.3.4 Inflow performance relationships 24
           2.3.4.1 Introduction 24
           2.3.4.2 Vogel's IPR correlation 24
           2.3.4.3 Fetkovich's method 26
   2.4 Single-phase Flow 26
       2.4.1 Introduction 26
       2.4.2 Basic principles 27
       2.4.3 Pressure drop calculations 30
           2.4.3.1 Single-phase liquid flow 30
           2.4.3.2 Single-phase gas flow 32

2.4.4   Gas gradient in annulus 35
    2.4.4.1   Theoretical background 35
    2.4.4.2   Computer solution 37
    2.4.4.3   Universal gas gradient chart 39
2.4.5   Choke flow 39
    2.4.5.1   Calculation of gas throughput of chokes 40
    2.4.5.2   Gas capacity charts 42
2.5   Two-phase Flow 44
  2.5.1   Introduction 44
  2.5.2   Basic principles 44
    2.5.2.1   Two-phase flow concepts 45
      2.5.2.1.1   Flow patterns 45
      2.5.2.1.2   Superficial velocities 46
      2.5.2.1.3   Gas slippage 48
      2.5.2.1.4   Liquid holdup 49
      2.5.2.1.5   Mixture properties 50
    2.5.2.2   Multiphase flow 52
    2.5.2.3   Pressure gradient equations 53
  2.5.3   Two-phase flow in oil wells 54
    2.5.3.1   Background theories 54
      2.5.3.1.1   Vertical flow patterns 54
      2.5.3.1.2   Liquid holdup calculations 58
      2.5.3.1.3   Frictional and kinetic losses 59
    2.5.3.2   Inclined and annulus flow 60
    2.5.3.3   Empirical correlations 60
      2.5.3.3.1   Introduction 60
      2.5.3.3.2   Poettmann-Carpenter and related correlations 62
      2.5.3.3.3   Duns-Ros correlation 64
      2.5.3.3.4   Hagedorn-Brown correlation 73
      2.5.3.3.5   Orkiszewski correlation 78
      2.5.3.3.6   Chierici et al. correlation 85
      2.5.3.3.7   Beggs-Brill correlation 85
      2.5.3.3.8   Cornish correlation 92
      2.5.3.3.9   Mukherjee-Brill correlation 93
    2.5.3.4   Mechanistic models 97
      2.5.3.4.1   Introduction 97
      2.5.3.4.2   Aziz-Govier-Fogarasi model 98
      2.5.3.4.3   Hasan-Kabir model 103
      2.5.3.4.4   Further models 110
    2.5.3.5   Calculation of pressure traverses 111
    2.5.3.6   Accuracy and selection of pressure drop calculation models 114
      2.5.3.6.1   Introduction 114
      2.5.3.6.2   Possible sources of prediction errors 115
      2.5.3.6.3   Results of published evaluations 117
      2.5.3.6.4   Selection of an optimum model 120
      2.5.3.6.5   Conclusions 121
    2.5.3.7   Gradient curves 121
      2.5.3.7.1   Gilbert's gradient curves 121
      2.5.3.7.2   Other collections of gradient curves 123
  2.5.4   Horizontal and inclined flow 127
    2.5.4.1   Introduction 127
    2.5.4.2   Horizontal flow patterns 128

	2.5.4.3	Empirical correlations	130
		2.5.4.3.1 Lockhart-Martinelli correlation	130
		2.5.4.3.2 Dukler correlation	134
	2.5.4.4	Calculation of pressure traverses	137
2.5.5	Flow through restrictions		139
	2.5.5.1	Critical flow correlations	139
	2.5.5.2	Critical flow criteria	142
	2.5.5.3	General calculation models	144

2.6 Well Temperature 146
   2.6.1 Introduction 146
   2.6.2 Ramey's model 147
      2.6.2.1 Modification by Hasan-Kabir 151
   2.6.3 The Shiu-Beggs correlation 151
   2.6.4 The model of Sagar-Doty-Schmidt 153
   2.6.5 Gas lifted wells 155

2.7 Systems Analysis Basics 158
   2.7.1 Introduction 158
   2.7.2 The production system 159
   2.7.3 Basic principles 160
   2.7.4 Application to gas lifting 161

References 163

3. Gas Lift Valves **169**

3.1 Introduction 169
   3.1.1 Downhole gas injection controls 170
   3.1.2 Evolution of gas lift valves 171
   3.1.3 Overview of valve types 172
   3.1.4 Supporting calculations 174
      3.1.4.1 Dome charge pressure calculations 174
      3.1.4.2 Injection pressure vs. depth 177

3.2 Pressure-operated Gas Lift Valves 180
   3.2.1 Introduction 180
      3.2.1.1 Valve parts terminology 180
      3.2.1.2 Valve construction details 181
         3.2.1.2.1 Core valve and tail plug 181
         3.2.1.2.2 Gas charge 182
         3.2.1.2.3 Bellows assembly 182
         3.2.1.2.4 Spring 185
         3.2.1.2.5 Ball and seat 186
         3.2.1.2.6 Check valve 187
   3.2.2 Valve mechanics 188
      3.2.2.1 Introduction 188
      3.2.2.2 Unbalanced valves with spread 189
         3.2.2.2.1 IPO valves 189
         3.2.2.2.2 PPO valves 197
      3.2.2.3 Unbalanced valves without spread 200
         3.2.2.3.1 IPO valves 200
         3.2.2.3.2 PPO valves 203
      3.2.2.4 Balanced valves 204
         3.2.2.4.1 IPO valves 204
         3.2.2.4.2 PPO valves 206

3.2.2.5　Pilot valves 206
　　　　3.2.2.5.1　IPO pilot valves 206
　　　　3.2.2.5.2　Combination valves 210
3.2.3　Dynamic performance of gas lift valves 215
　3.2.3.1　Introduction 215
　3.2.3.2　Early models 216
　3.2.3.3　General models 219
　　　　3.2.3.3.1　The API RP 11V2 procedure 220
　　　　3.2.3.3.2　Valve performance curves 226
3.2.4　Setting of gas lift valves 228
　3.2.4.1　Introduction 228
　3.2.4.2　Sleeve-type valve tester 228
　3.2.4.3　Encapsulated valve tester 229
3.2.5　Application of gas lift valves 230
　3.2.5.1　Valve requirements for different services 230
　　　　3.2.5.1.1　Continuous flow 230
　　　　3.2.5.1.2　Intermittent lift 231
　　　　3.2.5.1.3　Well unloading 232
　3.2.5.2　Advantages and limitations of different valves 232
3.3　Running and Retrieving of Gas Lift Valves 234
　3.3.1　Gas lift valve mandrel types 234
　　3.3.1.1　Conventional mandrels 235
　　3.3.1.2　Wireline retrievable mandrels 236
　3.3.2　Running and retrieving operations 237
3.4　Other Valve Types 238
　3.4.1　Differential gas lift valve 238
　3.4.2　Orifice valve 239
　3.4.3　Nozzle-Venturi valve 239
3.5　API Designation of Gas Lift Valves and Mandrels 240
　3.5.1　Gas lift valves 240
　3.5.2　Gas lift mandrels 241
References 243

4.　Gas Lift Installation Types 245
4.1　Introduction 245
4.2　Tubing Flow Installations 245
　4.2.1　The open installation 245
　4.2.2　The semi-closed installation 247
　4.2.3　The closed installation 248
　4.2.4　Chamber installations 248
　　4.2.4.1　Two-packer chamber installations 249
　　4.2.4.2　Insert chamber installations 249
　　4.2.4.3　Special chamber types 250
　4.2.5　Multiple Installations 251
　4.2.6　Miscellaneous Installation Types 251
　　4.2.6.1　Pack-off installations 251
　　4.2.6.2　Macaroni installations 252
　　4.2.6.3　CT installations 253
4.3　Casing Flow Installations 253
References 254

5. Continuous Flow Gas Lift 255

5.1 Introduction 255
5.2 Basic Features 256
    5.2.1 The mechanism of continuous flow gas lift 256
    5.2.2 Applications, advantages, limitations 256
5.3 Principles of Continuous Flow Gas Lifting 257
    5.3.1 Dead wells vs. gas lifting 257
    5.3.2 Basic design of a continuous flow installation 258
    5.3.3 Basic considerations 261
        5.3.3.1 The effect of injection depth 261
        5.3.3.2 Multipoint vs. single-point gas injection 262
    5.3.4 The effects of operational parameters 263
        5.3.4.1 Wellhead pressure 263
        5.3.4.2 Gas injection pressure 264
        5.3.4.3 Tubing size 265
5.4 Description of System Performance 266
    5.4.1 Introduction 266
    5.4.2 Constant WHP cases 267
        5.4.2.1 Injection pressure given 267
        5.4.2.2 The Equilibrium Curve method 272
    5.4.3 Variable WHP cases 277
        5.4.3.1 Solution at the well bottom 277
        5.4.3.2 Solution at the wellhead 280
    5.4.4 System stability 284
    5.4.5 Conclusions 287
5.5 Optimization of Continuous Flow Installations 288
    5.5.1 Introduction 288
    5.5.2 Optimization of a single well 289
        5.5.2.1 Prescribed liquid rate 289
            5.5.2.1.1 Existing compressor 290
            5.5.2.1.2 Compressor to be selected 293
        5.5.2.2 Unlimited liquid rate 298
            5.5.2.2.1 Economic considerations 298
            5.5.2.2.2 Existing compressor 300
            5.5.2.2.3 Compressor to be selected 303
    5.5.3 Allocation of lift gas to a group of wells 308
        5.5.3.1 Introduction 308
        5.5.3.2 Conventional calculation models 310
        5.5.3.3 Field-wide optimization 314
5.6 Unloading of Continuous Flow Installations 315
    5.6.1 Introduction 315
    5.6.2 The unloading process 316
    5.6.3 General design considerations 318
    5.6.4 Unloading valve string design procedures 320
        5.6.4.1 IPO valves 321
            5.6.4.1.1 Variable pressure drop per valve 322
            5.6.4.1.2 Constant pressure drop per valve 327
            5.6.4.1.3 Constant surface opening pressure 329
        5.6.4.2 Balanced valves 332
        5.6.4.3 PPO valves 334
        5.6.4.4 Throttling valves 335
            5.6.4.4.1 Injection transfer 335
            5.6.4.4.2 Valve spacing and setting calculations 336

5.6.5   Practical considerations 342
5.6.6   Conclusions 343
5.7   Surface Gas Injection Control 344
5.7.1   Choke control 344
5.7.2   Choke and regulator control 345
5.7.3   Other control methods 346
References 347

6. Intermittent Gas Lift **351**
6.1   Introduction 351
6.2   Basic Features 351
6.2.1   Mechanism of operation 351
6.2.2   Applications, advantages, limitations 353
6.3   Surface Gas Injection Control 353
6.3.1   Introduction 353
6.3.2   Choke control 354
6.3.3   Choke and regulator control 356
6.3.4   Intermitter control 357
6.3.4.1   Simple intermitter control 358
6.3.4.2   Use of an intermitter and a choke 359
6.3.4.3   Use of an intermitter and a regulator 360
6.3.4.4   Casing pressure control 360
6.3.4.5   Other controls 361
6.3.5   Control devices 361
6.3.5.1   Fixed and adjustable chokes 361
6.3.5.2   Pressure regulators 361
6.3.5.3   Time cycle controllers 362
6.4   Intermittent Lift Performance 363
6.4.1   The intermittent cycle 363
6.4.2   Calculation of operational parameters 366
6.4.2.1   Rules of thumb 366
6.4.2.2   Empirical correlations 367
6.5   Design of Intermittent Installations 371
6.5.1   General considerations 371
6.5.2   Constant surface closing pressure 371
6.5.3   The Opti-Flow design procedure 376
6.5.4   Unloading procedure 380
6.6   Chamber Lift 381
6.6.1   Basic features 381
6.6.2   Equipment selection considerations 382
6.6.3   Design of chamber installations 383
6.6.3.1   Determination of chamber length 383
6.6.3.2   Installation design 384
6.7   Optimization of Intermittent Installations 388
6.7.1   Intermitter control 388
6.7.2   Choke control 389
References 390

7. Plunger-assisted Intermittent Lift **393**
7.1   Introduction 393
7.2   Equipment Considerations 394

7.2.1 Installation types 394
7.2.2 Surface equipment 395
7.2.3 Subsurface equipment 395
7.3 Operating Conditions 397
7.3.1 The intermittent cycle 397
7.3.2 Calculation of operating parameters 398
References 399

8. Dual Gas Lift **401**
8.1 Introduction 401
8.2 General Considerations 402
8.3 Installation Design Principles 403
8.3.1 Both zones on continuous flow 404
8.3.2 One intermittent and one continuous zone 404
8.3.3 Both zones on intermittent lift 405
References 406

9. The Gas Lift System **407**
9.1 Introduction 407
9.1.1 Functions of the gas lift system 407
9.1.2 Types of gas lift systems 408
9.1.3 Gas lifting vs. field depletion 409
9.2 Operation of Gas Lift Systems 409
9.2.1 System components 409
9.2.2 System operation 411
9.3 System Design 413
9.3.1 Introduction 413
9.3.2 Factors to consider 413
9.3.3 Design procedure 415
References 420

10. Analysis and Troubleshooting **421**
10.1 Introduction 421
10.2 Troubleshooting Tools and Their Use 422
10.2.1 Two-pen pressure recordings 422
10.2.1.1 Pressure recorders 422
10.2.1.2 Chart interpretation 422
10.2.2 Gas volume measurements 426
10.2.3 Downhole pressure and temperature surveys 426
10.2.3.1 Introduction 426
10.2.3.2 Running procedures for downhole surveys 427
10.2.3.3 Flowing pressure surveys 428
10.2.3.4 Flowing temperature surveys .428
10.2.4 Acoustic surveys 429
10.3 Common Gas Lift Malfunctions 430
10.3.1 Downhole problems 430
10.3.2 Problems in the distribution system 432
10.3.3 Problems in the gathering system 433
References 434

11. Appendices . . . . **.435**
    Appendix A 435
    Appendix B 437
    Appendix C 439
    Appendix D 443
    Appendix E 447
    Appendix F 455

Index . . . . **.461**

Preface

The first use of air to lift water to the surface was reported in the mines of Chemnitz, Hungary around the middle of the 18th century. First in the wells of Pennsylvania, at around 1864, oil was lifted by compressed air that was later displaced by natural gas as the lifting medium. Throughout its almost 150-year long history, gas lifting proved to be one of the most popular methods of lifting liquids from wells. It can be applied in oil wells with high gas production rates where other artificial methods are plagued with frequent failures or cannot be used at all. It is especially suited to the offshore environment where extremely high liquid volumes have to be lifted.

Gas lifting can be used throughout the whole lifespan of an oil well: from the time it dies until its abandonment. At early production times, higher liquid rates are achieved by continuous flow gas lift. As the field depletes and formation pressure and liquid rates decrease, easy conversion to intermittent gas lift ensures that production goals are met. Close to well abandonment, another version of gas lifting—chamber lift—can be applied. Because of these features, gas lifting is probably the most flexible means of artificial lifting available today.

I wrote *Gas Lift Handbook* to become a handbook of up-to-date gas lift theories and practices and to cover the latest developments in this important field of artificial lifting technology. Since I present a complete review of gas lift technology and include references to all of the important literature sources, the practicing engineer can use the book as a reference on the subject. I tried to distill in this book all the experience I gathered during my 30-year teaching and consulting career in order to present a text that systematically introduces the reader to the subject matter. It would be a personal gratification if, like its predecessor *Modern Sucker-Rod Pumping* published by PennWell in 1993, this book, too, would be chosen for graduate level courses at different petroleum engineering schools.

It is fully understood by anyone in the industry that describing multiphase flow in oil wells is the basis of solving most of the problems in gas lifting. Since this is an area where almost revolutionary achievements happened in the last 20 years, I fully describe the pressure drop calculation procedures for vertical, inclined, and horizontal wells including the latest mechanistic models. The chapter on production engineering fundamentals includes, in addition to a very detailed treatment of multiphase flow, a review of fluid properties, well inflow performance, basic hydraulics, well temperature, and systems analysis.

Further chapters systematically introduce the reader to the hardware of gas lifting, reflecting the latest developments in gas lift valve and other equipment designs. The great variety of gas lift valves, their constructional and operational details are fully discussed along with the latest achievements on describing their dynamic performance. The description of gas lift installation types helps the engineer select the right combination of well equipment. The chapter on continuous flow gas lift fully describes the different ways to optimize the wells' operation, including the latest optimization theories (lift gas allocation to wells, systems analysis, etc.). Unloading and surface control of continuous flow gas lift wells round up this chapter.

The discussion of intermittent gas lift includes conventional, chamber, and plunger-assisted installations and describes the performance, design, and optimization of such wells. A detailed treatment of the surface gas lift system follows, including the operation and design of the complete system consisting of the compressor station, and the distribution and gathering facilities. The last chapter includes practical advice on the analysis and troubleshooting of gas lift installations. All necessary calculations are fully discussed, and the many charts in the appendices are intended to help field engineers.

Nowadays, personal computers belong to every petroleum engineer's desk and this book was designed with this fact in mind. Since most design and analysis problems in gas lifting are too complex to be solved by the conventional tools of the engineer, I heavily relied on computerized solutions in the text.

While writing this book I burned a lot of midnight oil and many times had regretfully neglected those I love most—my family. Their patience and understanding is appreciated.

This is already the third project I have worked on with PennWell. Ms. Marla Patterson (PennWell) and Ms. Sue Rhodes Dodd (Amethyst Enterprises) were always most helpful and forgiving. A special "thank you" to both of them.

Gábor Takács
July 2005

1 | Introduction to Gas Lifting

1.1 Artificial Lifting

Most oil wells in the early stages of their lives flow naturally to the surface and are called flowing wells. Flowing production means that the pressure at the well bottom is sufficient to overcome the sum of pressure losses occurring along the flow path to the separator. When this criterion is not met, natural flow ends and the well dies.

Wells may die for two main reasons: either their flowing bottomhole pressure drops below the total pressure losses in the well, or the opposite happens and pressure losses in the well become greater than the bottomhole pressure needed for moving the wellstream to the surface. The first case occurs when a gradual decrease in reservoir pressure takes place because of the removal of fluids from the underground reservoir. The second case involves an increasing flow resistance in the well, generally caused by (a) an increase in the density of the flowing fluid as a result of decreased gas production; or (b) various mechanical problems like a small tubing size, downhole restrictions, etc. Surface conditions, such as separator pressure or flowline size, also directly impact total pressure losses and can prevent a well from flowing.

Artificial lifting methods are used to produce fluids from wells already dead or to increase the production rate from flowing wells; and several lifting mechanisms are available to choose from. One widely used type of artificial lift method uses a pump set below the liquid level in the well to increase the pressure so as to overcome flowing pressure losses that occur along the flow path to the surface. Other lifting methods use compressed gas, injected periodically below the liquid present in the well tubing and use the expansion energy of the gas to displace a liquid slug to the surface. The third mechanism works on a completely different principle: instead of increasing the pressure in the well, flowing pressure losses are decreased by a continuous injection of high-pressure gas into the wellstream. This enables the actual bottomhole pressure to move well fluids to the surface.

Although all artificial lift methods can be distinguished based on the previous three basic mechanisms, the usual classification is somewhat different and is discussed here.

Pumping

Pumping involves the use of some kind of a pump installed downhole to increase the pressure in the well to overcome the sum of flowing pressure losses. These methods can be classified further using several different criteria, *e.g.*, the operational principle of the pump used. However, the generally accepted classification is based on the way the downhole pump is driven and distinguishes between rod and rodless pumping.

Rod pumping methods utilize a string of rods that connects the downhole pump to the surface driving mechanism which, depending on the type of pump used, makes an oscillating or rotating movement. The first kinds of pumps to be applied in water and oil wells were of the positive-displacement type requiring an alternating vertical movement to operate. The dominant and oldest type of rod pumping is *walking-beam pumping*, or simply called *sucker-rod pumping*. It uses a positive-displacement plunger pump and its most familiar surface feature is that the rotary movement of the prime mover is converted with the help of a pivoted walking beam.

The need for producing deeper and deeper wells with increased liquid volumes necessitated the evolution of long stroke sucker-rod pumping. Several different units were developed with the common features of using the same pumps and rod strings (as in the case of beam-type units) but with substantially longer pump stroke lengths. The desired long strokes did not permit the use of a walking beam, and completely different surface driving mechanisms had to be invented. The basic types in this class are distinguished according to the type of surface drive used: pneumatic drive, hydraulic drive, or mechanical drive long-stroke pumping.

A newly emerged rod-pumping system uses a progressing cavity pump that requires the rod string to be rotated for its operation. This pump, like the plunger pumps used in other types of rod-pumping systems, also works on the principle of positive displacement but does not contain any valves.

Rodless pumping methods, as the name implies, do not have a rod string to operate the downhole pump from the surface. Accordingly, other means (besides mechanical) are used to drive the downhole pump, such as electric or hydraulic. A variety of pump types are applied with rodless pumping, including centrifugal, positive displacement, or hydraulic pumps. Electric submersible pumping (ESP) utilizes a submerged electrical motor driving a multistage centrifugal pump. Power is supplied to the motor by an electric cable run from the surface. Such units are ideally suited to produce high liquid volumes.

The other lifting systems in the rodless category all employ a high-pressure power fluid that is pumped down the hole. Hydraulic pumping was the first method developed; such units have a positive-displacement pump driven by a hydraulic engine, contained in one downhole unit. The engine or motor provides an alternating movement necessary to operate the pump section. The hydraulic turbine-driven pumping unit consists of a multistage turbine and a multistage centrifugal pump section connected in series. The turbine is supplied with power fluid from the surface and drives the centrifugal pump at high rotational speeds, which lifts well fluids to the surface.

Jet pumping, although it is a hydraulically driven method of fluid lifting, completely differs from the rodless pumping principles discussed so far. Its downhole equipment converts the energy of a high-velocity jet stream into useful work to lift well fluids. The downhole unit of a jet pump installation is the only oil well pumping equipment known today containing no moving parts.

Gas Lifting

Gas lifting uses natural gas compressed at the surface and injected in the wellstream at some downhole point. In continuous flow gas lift, a steady rate of gas is injected in the well tubing, aerating the liquid and thus reducing the pressure losses occurring along the flow path. Due to this reduction in flow resistance of the well tubing, the well's original bottomhole pressure becomes sufficient to move the gas/liquid mixture to the surface and the well starts to flow again. Therefore, continuous flow gas lifting can be considered as the continuation of flowing production.

In intermittent gas lift, gas is injected periodically into the tubing string whenever a sufficient length of liquid has accumulated at the well bottom. A relatively high volume of gas injected below the liquid column pushes that column to the surface as a slug. Gas injection is then interrupted until a new liquid slug of the proper column length builds up again. Production of well liquids, therefore, is done by cycles. The plunger-assisted version of intermittent gas lift uses a special free plunger traveling in the well tubing to separate the upward-moving liquid slug from the gas below it. These versions of gas lift physically displace the accumulated liquids from the well, a mechanism totally different from that of continuous flow gas lifting.

Comparison of Lift Methods

Although there are some other types of artificial lift known, their importance is negligible compared to those just mentioned. Thus, there are a multitude of choices available to an engineer when selecting the type of lift to be used. Some of the possible types may be ruled out by field conditions such as well depth, production rate, fluid properties, etc. Still, in general, more than one lift system turns out to be technically feasible. It is then the production engineer's responsibility to select the type of lift that provides the most profitable way of producing the desired liquid volume from the given well(s). After a decision is made concerning the lifting method to be applied, a complete design of the installation for initial and future conditions should follow.

In order to provide a rough comparison of the available artificial lifting methods, two figures demonstrate approximate maximum liquid production rates of the different installations given in the function of lifting depth [1]. Figure 1–1 shows three lifting mechanisms capable of producing exceptionally high liquid rates: gas lifting, electrical submersible pumping, and jet pumping. As seen, gas lifting (continuous flow) allows the lifting of the greatest amounts of liquid from any depth. Figure 1–2, on the other hand, includes artificial lift methods of moderate liquid capacity: hydraulic pumping, progressive cavity pumping, rod pumping, and plunger lifting. In most cases, lifting depth has a profound impact on the liquid volume lifted, and well rates rapidly decrease in deeper wells.

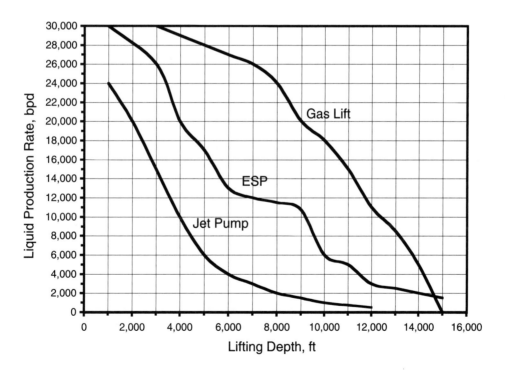

Fig. 1–1 Maximum liquid production rates vs. lifting depth for various high-rate artificial lift methods.

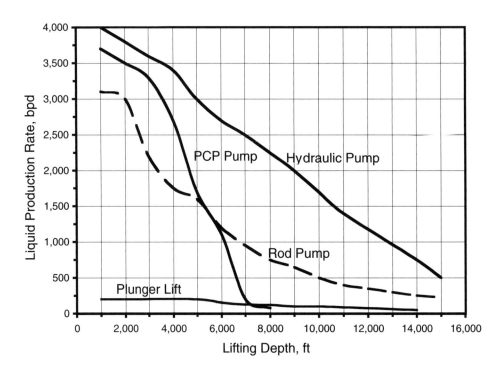

Fig. 1–2 Maximum liquid production rates vs. lifting depth for various moderate-rate artificial lift methods.

1.2 Short History of Gas Lifting

Readers interested in a detailed treatment of gas lift history are advised to consult K. E. Brown's excellent book on gas lifting. [2] In this section, a short overview of the most important developments in gas lift history is given.

The removal of water from flooded mine shafts is an ever-existing problem with miners, and the use of air to lift water to the surface was first reported in the mines of Chemnitz, Hungary around the middle of the 18th century. [3] Ever since, the lifting of great amounts of water from shafts and wells alike with the help of downhole injection of air has been a major application of airlift in the industry. In the oilfield, compressed air was first used to lift oil from wells in Pennsylvania around 1864. [4, 5] The *well blower* installation consisted of a 1-in. air conduit fixed to the tubing string, the lower end of which injected air into the wellstream. The addition of air to the produced liquids reduced the density of the mixture flowing in the tubing and thus caused a decrease of backpressure on the formation. Existing formation pressure became sufficient to move well fluids to the surface and the well started to flow again. Continuous flow air (gas) lift was born.

The first large-scale oilfield applications of the airlift technique occurred in Texas where major fields (Spindletop, etc.) were airlifted after 1900, and quickly formed companies sold compressed air in the *air-for-hire* period. The use of natural gas as the lifting medium replaced the use of air during the second half of the 1920s, and its main advantages include the following:

(a) no explosive gas mixture is formed

(b) no corrosive effects

(c) wide availability

(d) the possibility of recovering light hydrocarbon fractions previously wasted

Early gas lift systems included gas compressors driven by steam engines, steam being generated in oil-fired boilers. Single- or dual-stage compressors with suction pressures near atmospheric pressure were used with consequently low discharge pressures of a few hundred psi. Usually, no effort was spent on sizing of well piping (tubing and gas conduit) and no attention was paid to gas and power usage.

Initial installations consisted of a simple tubing string hung from the casinghead, and lift gas entered uncontrolled at the tubing shoe. Later, the use of *foot-pieces* (devices installed at the tubing shoe for regulating the injection of gas into the tubing) tried to impose some control on the amount of gas injected. These early installations were seriously limited to shallow wells because of the relatively low gas-injection pressures. Although a higher than normal *kickoff* pressure could increase the lifting depth, it was only after the introduction of *kickoff valves* and the stepwise *unloading* process that gas lifting from greater depths could be accomplished. There were hundreds of kickoff valves, working on different principles, patented from the 1910s on, only to be almost completely obsoleted by the pressure-operated gas lift valves emerging in the early 1940s. Pressure-operated valves, used as unloading or operating valves, provide proper control of injected gas volumes and are in use even today. A more detailed historical description of the evolution of gas lift valves is given in chapter 3 of this book.

Downhole equipment used in gas-lifted wells also underwent a great evolution, and the early open installations utilizing a tubing string only were gradually replaced with installations including a packer and a standing valve. The running of gas lift valves and other downhole devices became easier and more cost effective with the introduction of wireline retrievable equipment in the 1930s. New lifting methods like chamber and plunger lift were introduced to produce wells with low liquid capacities.

The surface system required in a gas-lifted field includes the compressor station, the gas distribution, and the fluid-gathering systems. It was very early (late 1920s) that the *closed rotative gas lift system* was introduced with lift gas being circulated from the compressors to the wells and back. From about this time on, natural gas associated with oil was collected and sold to gas transmission systems that required the use of compressors with higher discharge pressures. Higher injection pressures (but still less than 1,000 psi) allowed the lifting of deeper and deeper wells and extended the applicability of gas lifting. Today's deepest wells, as well as the advanced reservoir pressure maintenance programs necessitate the use of gas lift pressures of 2,000 psi and higher. Since gas lift equipment for such high-pressure ranges is increasingly offered by manufacturers, today's gas lift installations can reach depths and production rates that were prohibitive even only a few decades ago.

1.3 Basic Features of Gas Lifting

Operational Mechanisms

The two broad categories of gas lifting (continuous flow and intermittent lift) both utilize high-pressure natural gas injected from the surface to lift well fluids but work on different principles. In continuous-flow operation, lift gas is continuously injected at the proper depth into the wellstream from the casing-tubing annulus or the tubing string into the flow string, which can be the tubing string or the annulus, respectively. The injection of the proper amount of lift gas greatly reduces the density of the wellstream as well as the flowing pressure losses occurring above the injection point because the major part of vertical multiphase pressure drop is due to the change of potential energy. Accordingly, total pressure losses in the entire tubing string will also decrease, allowing the existing pressure at the well bottom to overcome them and to lift the wellstream to the surface. If the well was dead before, it starts to flow again, whereas flowing wells will produce much greater liquid volumes than before. Continuous flow gas lift, therefore, may be considered as the continuation of flowing production, and its basic operational mechanism is the reduction of flow resistance of the production string.

Intermittent gas lift, although also using compressed gas injected from the surface, works on a completely different operational mechanism. The well is produced in periodically repeated cycles, with accumulated liquid columns being physically displaced to the surface by the high-pressure lift gas injected below them. During the intermittent cycle, well fluids are first allowed to accumulate above the operating gas lift valve, then lift gas is injected through the valve and below the liquid column. If the proper injection pressure and gas volume are used, lift gas propels the liquid slug to the wellhead and into the flowline. In light of this and in contrast to continuous flow gas lifting, the operation of an intermittent gas lift installation is a transient process. As such, it is much more complicated to describe and model than the steady-state fluid lifting process taking place in a well placed on continuous flow gas lift.

Applications

During its almost 150-year history, gas lifting proved to be one of the most popular methods of lifting liquids from wells. It has been widely applied in oilfields with high gas production rates where other artificial methods are plagued with frequent failures or cannot be used at all. Gas lift is especially suited to the offshore environment where extremely high liquid volumes have to be reliably lifted. A recent application [6] describes a case where a liquid rate of 40,000 bpd is reached from 10,800 ft (true vertical depth [TVD]) deviated wells through 7-in. tubing strings.

Gas lifting can produce any well during the whole lifespan of the well from the time it dies until its abandonment. Higher liquid rates, usually associated with early production times, are achieved by continuous flow. Later, as formation pressure and liquid rates gradually decrease, easy conversion to intermittent gas lift ensures that production goals are met. Close to abandonment, chamber lift can be applied (again with a minimum of conversion costs), providing the production of the well with extremely low formation pressures.

Due to its many advantageous features, gas lifting is very popular in the oil-producing regions of the world. Statistical figures [7] from the United States in 1993 show that about 10% of the artificially produced wells are placed on gas lifting, and the vast majority of the rest are beam pumped. Since most of these wells belong to the *stripper well* category producing less than 10 bpd, the previous numbers are misleading. After excluding stripper wells, the situation completely changes because 53% of the remaining wells are gas lifted. Although these numbers refer to the number of producing wells only, the total amount of liquid produced by gas lift is also substantial. In 2003, two major producers reported [8] on the following contribution of gas lift to their total liquid production: ExxonMobil 31%, Shell 25%.

Advantages, Limitations

General advantages of using any version of gas lift can be summed up as follows [2, 7, 9]:

- In contrast to most other artificial lift methods, gas lifting offers a high degree of flexibility. This means that any gas lift installation can very easily be modified to accommodate extremely great changes in liquid production rates.

- Using only gas lift throughout their productive life, wells can successfully be depleted.

- In fields where wells produce substantial amounts of formation gas, many times gas lifting is the only choice.

- Well deviation or crooked holes do not hinder production operations.

- High well temperatures can easily be managed.

- Corrosion control in wells is easily accomplished.

- Most gas lift installations provide a full-bore tubing string through which downhole measurements and workover operations can be conducted.

- The use of wireline retrievable equipment allows for the economy of service operations.

- Surface wellhead equipment is not obtrusive and requires little space.

General disadvantages include the following:

- Usually, a sufficient amount of formation gas production throughout the life of the field is required to maintain the normal operation of the gas lift system. New developments, however, can provide other sources of lift gas like nitrogen produced from the air. [10]

- High separator pressures are very detrimental to the operation of any kind of gas lift installations.

- Energy efficiency of lifting is usually lower than for the other kinds of artificial lift methods.

Individual advantages and limitations of the different available versions of gas lifting are discussed in this book.

References

1. *Artificial Lift Systems* (brochure). Weatherford Co., Houston, TX, 1999.

2. Brown, K. E. *Gas Lift Theory and Practice.* Tulsa, OK: Petroleum Publishing Co., 1967.

3. Shaw, S. F. *Gas Lift Principles and Practices.* Houston, TX: Gulf Publishing Co., 1939.

4. *History of Petroleum Engineering.* Dallas, TX: American Petroleum Institute, 1961.

5. Cloud, W. F. *Petroleum Production.* Norman, OK: University of Oklahoma Press, 1937.

6. Duncan, G. J. and B. Beldring, "A Novel Approach to Gas Lift Design for 40,000 bpd Subsea Producers." Paper SPE 77727 presented at the Annual Technical Conference and Exhibition held in San Antonio, TX, September 29–October 2, 2002.

7. Clegg, J. D., S. M. Bucaram, and N. M. Hein, Jr., "Recommendations and Comparisons for Selecting Artificial-Lift Methods." *JPT,* December 1993: 1128–67.

8. Martinez, J., "Downhole Gas Lift and the Facility." Paper presented at the ASME/API Gas Lift Workshop, held in Houston, TX, February 4–5, 2003.

9. Neely, B., "Selection of Artificial Lift Methods." Paper SPE 10337 presented at the 56th Annual Fall Technical Conference and Exhibition held in San Antonio, TX, October 5–7, 1981.

10. Aguilar, M. A. L. and M. R.A. Monarrez, "Gas Lift with Nitrogen Injection Generated In Situ." Paper SPE 59028 presented at the International Petroleum Conference and Exhibition in Mexico, held in Villahermosa, February 1–3, 2000.

2 | Production Engineering Fundamentals

2.1 Introduction

Chapter 2 presents a concise treatment of the basic knowledge required from a practicing engineer working on production engineering problems. Production engineering principles and practices related to the design and analysis of artificial lift installations are detailed in depth. The first section discusses the most significant properties of oilfield fluids: oil, water, and natural gas. Since no artificial lift design can be made without the proper knowledge of the oil well's production capability, the section on inflow performance presents a thorough survey of the available methods for describing inflow to oil wells. A separate section deals with the single-phase flow problems encountered in production engineering design. Due to its profound importance, the description of two-phase flow phenomena occupies the largest part of the chapter. Vertical, inclined, and horizontal flow are discussed, and most of the available calculation models for finding the pressure drop in pipes and flow restrictions are presented. After the section on well temperature calculations, the chapter is concluded with a review of the latest tool of the production engineer: the systems analysis approach of evaluating well performance.

The calculation models presented in the chapter are supported by worked examples throughout the text. Later chapters rely on the concepts and procedures covered here, and a basic understanding of this chapter is advised.

2.2 Properties of Oilfield Fluids

2.2.1 Introduction

The fluids most often encountered in oil well production operations are hydrocarbons and water. These can be either in a liquid or gaseous state depending on prevailing conditions. As conditions like pressure and temperature change along the flow path (while fluids are moving from the well bottom to the surface), phase relations and physical parameters of the flowing fluids continuously change as well. Therefore, in order to determine operating conditions or to design production equipment, it is essential to describe those fluid parameters that have an effect on the process of fluid lifting. In the following, calculation methods for the determination of physical properties of oil, water, and natural gas are detailed.

Naturally, the most reliable approach to the calculation of accurate fluid properties is to use actual well data. Collection of such data usually involves pressure-volume-temperature (pVT) measurements and necessitates the use of delicate instrumentation. This is why in the majority of cases, such data are not complete or are missing entirely. In these cases, one has to rely on previously published correlations. The correlations presented in this section help the practicing engineer to solve the problems that insufficient information on fluid properties can cause. The reader is warned, however, that this is not a complete review of available methods, nor are the procedures given superior by any means to those not mentioned. Our only aim is to give some theoretical background and practical methods for calculating oilfield fluid properties. This goal also implies that only black-oil type hydrocarbon systems are investigated, as these are usually encountered in oil well production.

Before a detailed treatment of the different correlations, generally accepted definitions and relevant equations of the fluid physical properties are given.

2.2.2 Basic thermodynamic properties

Density, ρ

It gives the ratio of the mass of the fluid to the volume occupied. Usually it varies with pressure and temperature and can be calculated as

$$\rho = \frac{m}{V}$$
2.1

Specific Gravity, γ

In case of liquids, γ_l is the ratio of the liquid's density to the density of pure water, both taken at standard conditions:

$$\gamma_l = \frac{\rho_{lsc}}{\rho_{wsc}}$$
2.2

Specific gravity of gases is calculated by using air density at standard conditions:

$$\gamma_g = \frac{\rho_{gsc}}{\rho_{asc}} = \frac{\rho_{gsc}}{0.0764}$$
2.3

Specific gravity, as seen previously, is a dimensionless measure of liquid or gas density. It is widely used in correlating physical parameters and is by definition a constant for any given gas or liquid.

In the petroleum industry, gravity of liquids is often expressed in API gravity, with °API units. Specific gravity and API gravity are related by the formula:

$$\gamma_l = \frac{141.5}{131.5 + °API}$$
2.4

Viscosity, μ

Viscosity, or more properly called *dynamic viscosity*, directly relates to the fluid's resistance to flow. This resistance is caused by friction generated between moving fluid molecules. Viscosity of a given fluid usually varies with pressure and temperature and has a great effect on the pressure drop of single-phase or multiphase flows.

The customary unit of dynamic viscosity is the centipoise, *cP*.

Bubblepoint Pressure, p_b

In multicomponent hydrocarbon fluid systems, changes of pressure and temperature result in phase changes. If we start from a liquid phase and decrease the pressure at a constant temperature, the first gas bubble comes out of solution at the bubblepoint pressure of the system. At higher pressures, only liquid phase is present; at lower pressures, a two-phase mixture exists. Bubblepoint pressure, therefore, is a very important parameter and can be used to determine prevailing phase conditions.

Solution Gas-Oil Ratio, R_s

Under elevated pressures, crude oil dissolves available gas and then can release it when pressure is decreased. To quantify the gas volume a crude can dissolve at given conditions, solution gas-oil ratio (GOR) is used. This parameter gives the volume at standard conditions of the dissolved gas in a crude oil. Since oil volume is measured at atmospheric conditions in a stock tank in barrels (STB), the measurement unit of R_s is scf/STB.

R_s is a function of system composition, pressure, and temperature and is defined as:

$$R_s = \frac{V_{gdissolved}}{V_{osc}} \qquad\qquad 2.5$$

Gas solubility in water can also be described with the previous principles; but due to its lower magnitude, it is usually neglected.

Volume Factor, B

The volume factor of a given fluid (gas, oil, or water) is utilized to calculate actual fluid volumes at any condition from volumes measured at standard conditions. It includes the effects of pressure and temperature, which have an impact on actual volume. In case of liquids, the volume factor represents the effects of dissolved gases as well. The volume factor is sometimes called formation volume factor (FVF), which is a misnomer because it implies the specific conditions of the formation. The name *FVF*, therefore, should only be used to designate the value of the volume factor at reservoir conditions.

In general, volume factor is defined as follows:

$$B = \frac{V(p,T)}{V_{sc}} \qquad\qquad 2.6$$

where: $V(p,T)$ = volume at pressure p and temperature T, bbl or cu ft

V_{sc} = fluid volume at standard conditions, STB or scf

By using Equation 2.1 and substituting volume V into the previous equation, an alternate definition of volume factor can be written:

$$B = \frac{\rho_{sc}}{\rho(p,T)} \qquad\qquad 2.7$$

where: $\rho(p,T)$ = fluid density at given p,T

ρ_{sc} = fluid density at standard conditions

2.2.3 Liquid property correlations

2.2.3.1 Water. Liquids are usually compared to pure water, of which the basic properties at standard conditions are

Specific Gravity $\gamma_w = 1.00$

API gravity   10 °API

Density $\rho_{w\,sc} = 62.4$ lb/cu ft = 350 lb/bbl (1,000 kg/m^3)

Hydrostatic Gradient   0.433 psi/ft (9.8 kPa/m)

Dynamic Viscosity $\mu_w = 1$ cP (0.001 Pas)

In production engineering calculations, gas solubility in formation water is usually neglected and water viscosity is assumed to be constant. This is why in the following parts of the book, no correlations will be given for R_s and μ_w of water.

Volume Factor of Water

The volume factor of water at p and T can be approximated by the use of the correlation of Gould [1]:

$$B_w = 1.0 + 1.21 \times 10^{-4} T_x + 10^{-6} T_x^2 - 3.33 \times 10^{-6} p \qquad 2.8$$

where: $T_x = T - 60,\ °F$

- -

Example 2–1. Calculate the pressure exerted by a static column of water of height 200 ft (61 m). Average temperature and average pressure in the column are $p = 50$ psi (344.5 kPa) and $T = 120$ °F (322 K), respectively. Use a water specific gravity of 1.12.

Solution

Water density at standard conditions from Equation 2.2:

$\rho_{l\,sc} = \gamma_w\,\rho_{w\,sc} = 1.12\ 62.4 = 69.9$ lb/cu ft (1,120 kg/m^3).

The volume factor of water at the given conditions is found from Equation 2.8 with $T_x = 120 - 60 = 60$ °F:

$B_w = 1.0 + 1.2\ 10^{-4}\ 60 + 10^{-6}\ 60^2 - 3.33\ 10^{-6}\ 50 = 1.01$ bbl/STB (1.01 m^3/m^3).

The density at p and T from Equation 2.7:

$\rho(p,T) = 69.9 / 1.01 = 69.2$ lb/cu ft (1,109 kg/m^3).

The hydrostatic pressure at the bottom of the column is:

$p = \rho\,h = 69.2\ 200 = 13,840$ lb/sq ft $= 13,840 / 144 = 96$ psi (662 kPa).

- -

2.2.3.2 Crude oil.

Bubblepoint Pressure

For estimating bubblepoint pressure of crude oils, the use of the Standing correlation [2] is widely accepted:

$$p_b = 18 \left(\frac{R_b}{\gamma_g}\right)^{0.83} 10^y \qquad 2.9$$

where: y = 0.00091 T - 0.0125 °API

R_b = solution GOR at pressures above p_b, scf/STB

Example 2–2. Decide whether free gas is present in a well at the well bottom, where p = 700 psi (4.83 MPa) and T = 100° F (311 K). The produced oil is of 30 °API gravity and has a total solution GOR of R_b = 120 scf/STB (21.4 m³/m³), gas specific gravity being 0.75.

Solution

The bubblepoint pressure, using Equation 2.9:

y = 0.00091 100 - 0.01255 30 = - 0.285

p_b = 18 (120 / 0.75)0.83 10-0.284 = 631 psi (4.35 MPa).

Thus only liquid phase exists, because the given pressure is above the calculated bubblepoint pressure.

Solution GOR

If under some conditions crude oil and liberated gas are present, they must coexist in equilibrium state. This means that any small decrease in pressure results in some gas going out of solution from the oil. The oil is, therefore, at its bubblepoint pressure at the prevailing temperature and Standing's bubblepoint pressure correlation can be applied. Substituting local pressure and temperature into Equation 2.9, this can be solved for R_s:

$$R_s = \gamma_g \left(\frac{p}{18 \cdot 10^y}\right)^{1.205} \qquad 2.10$$

where: y = 0.00091 T - 0.0125 °API

R_s can be estimated also by the use of the Lasater [3] or the Vazquez-Beggs [4] correlation.

For pressures above the bubblepoint, the solution GOR is constant and the oil is undersaturated. This means that the oil can dissolve more gas, if available.

Volume Factor

Several correlations are available for the calculation of oil volume factors. [2, 4] Of these, Standing's B_o correlation [2] is discussed here, which was developed on the same database as his bubblepoint pressure correlation. It is valid for pressures less than bubblepoint pressure, *i.e.* in the two-phase region.

$$B_o = 0.972 + 1.47 \times 10^{-4} F^{1.175} \qquad 2.11$$

where: $F = R_s (\gamma_g/\gamma_o)^{0.5} + 1.25 T$

In case of undersaturated conditions, *i.e.* pressures over bubblepoint pressure, no gas phase is present. The change in oil volume is affected by thermal expansion and by compression due to actual pressure. As these effects work against each other, B_o can be estimated as a constant for any crude oil above its bubblepoint pressure.

Example 2–3. Liquid inflow to a well was calculated to be q = 600 bpd (95 m³/d) oil at downhole conditions, where p = 800 psi (5.52 MPa) and T = 120 °F (322 K). Find the crude oil volume that will be measured at the surface in the stock tank. Oil gravity is 30 °API, gas specific gravity is 0.6.

Solution

The oil volume in the stock tank can be found from the definition of oil volume factor, Equation 2.6:

$$q_{o\,sc} = q(p,T) / B_o$$

To find B_o, R_s has to be calculated first, from Equation 2.10:

$$y = 0.00091\ 120 - 0.0125\ 30 = -0.266$$

$$R_s = 0.6\ (\ 800 / 18 / 10^{-0.266}\ )^{1.205} = 121.4\ \text{scf/STB}\ (21.9\ \text{m}^3/\text{m}^3).$$

Before applying Standing's B_o correlation, oil specific gravity is calculated from API gravity, by Equation 2.4:

$$\gamma_o = 141.5 / (131.5 + 30) = 0.876$$

With the previous values, from Equation 2.11:

$$F = 121.4\ (0.6 / 0.876)^{0.5} + 1.25\ 120 = 250.5$$

$$B_o = 0.972 + 1.47\ 10^{-4}\ 250.5^{1.175} = 1.07\ \text{bbl/STB}\ (1.07\ \text{m}^3/\text{m}^3)$$

Finally, oil volume at standard conditions:

$$q_{o\,sc} = 600 / 1.07 = 561\ \text{STB/d}\ (89\ \text{m}^3/\text{d})$$

Oil volume has *shrunk* from 600 bpd (95 m³/d) at the well bottom to 561 STB/d (89 m³/d) in the stock tank. This volume change is the effect of gas going continuously out of solution in the well and in the tank, and it is also the effect of volume changes due to variations in pressure and temperature.

Viscosity

The following applies to crude oils exhibiting Newtonian behavior, as non-Newtonian oils or emulsions cannot be characterized by viscosity alone.

Factors affecting the viscosity of Newtonian oils are composition, pressure, temperature, and the amount of dissolved gas. The viscosity of dead (gasless) oils at atmospheric pressure was correlated with API gravity and temperature by Beal [5], given in Figure 2–1.

Dissolved gas volume has a very profound impact on the viscosity of live oil. A generally accepted method to account for this is the correlation developed by Chew and Connally [6]. They presented the following equation:

$$\mu_o = A\mu_{oD}^b \qquad\qquad 2.12$$

where:　　μ_o　 = live oil viscosity, cP

μ_{oD} = viscosity of dead (gasless) oil, cP

The authors gave factors A and b on a graph, in the function of solution GOR. The two curves can be closely approximated by the equations given as follows.

$$A = 0.2 + 0.810^{-0.00081R_s} \qquad\qquad 2.13$$

$$b = 0.43 + 0.5710^{-0.00072R_s} \qquad\qquad 2.14$$

2.2.4 Properties of natural gases

2.2.4.1 Behavior of gases.
Before discussing the behavior of natural gases, some considerations on ideal gases have to be detailed. An ideal gas has molecules that are negligibly small compared to the volume occupied by the gas. The kinetic theory of such gases states that the volume of a given ideal gas, its absolute pressure, its absolute temperature, and the total number of molecules are interrelated. This relationship is described by the Ideal Gas Law. From this, any parameter can be calculated, provided the other three are known.

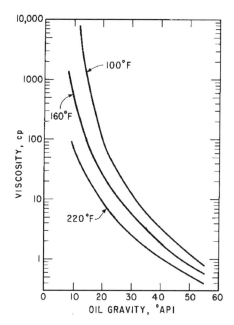

Fig. 2–1 Dead oil viscosity at atmospheric pressure.

Natural gases are gas mixtures and contain mainly hydrocarbon gases with usually lower concentrations of other components. Due to their complex nature and composition, they cannot be considered as ideal gases and cannot be described by the Ideal Gas Law. Several different methods were devised to characterize the behavior of real gases. These are called equations of state and are usually empirical equations that try to describe the relationships between the so-called state parameters of gas: absolute pressure, absolute temperature, and volume. The most frequently used and simplest equation of state for real gases is the Engineering Equation of State:

$$pV = ZnRT_a \qquad\qquad 2.15$$

where: $\quad p \quad$ = absolute pressure, psia

$\qquad\qquad V \quad$ = gas volume, cu ft

$\qquad\qquad Z \quad$ = gas deviation factor

$\qquad\qquad n \quad$ = number of moles

$\qquad\qquad R \quad$ = 10.73, gas constant

$\qquad\qquad T_a \quad$ = absolute temperature = T(°F) + 460, °R

The Engineering Equation of State differs from the Ideal Gas Law by the inclusion of the gas deviation factor only, which is sometimes referred to as compressibility or supercompressibility factor. It accounts for the deviation of the real gas volume from the volume of an ideal gas under the same conditions and is defined as

$$Z = \frac{V_{actual}}{V_{ideal}} \qquad\qquad 2.16$$

The problem of describing the behavior of natural gases was thus reduced to the proper determination of deviation factors.

The most common approach to the determination of deviation factors is based on the Theorem of Corresponding States. This principle states that real gas mixtures, like natural gases, behave similarly if their pseudoreduced parameters are identical. The definition of pseudoreduced pressure and temperature are:

$$p_{pr} = \frac{p}{p_{pc}}$$ 2.17

$$T_{pr} = \frac{T}{T_{pc}}$$ 2.18

In the previous equations, p_{pc} and T_{pc} are the pseudocritical pressure and temperature of the gas. These can be determined as weighted averages from the critical parameters of the individual gas components.

Using the previous principle, deviation factors for common natural gas mixtures can be correlated as functions of their pseudoreduced parameters. The most widely accepted Z-factor correlation was given by Standing-Katz [7] and is reproduced in Figure 2–2.

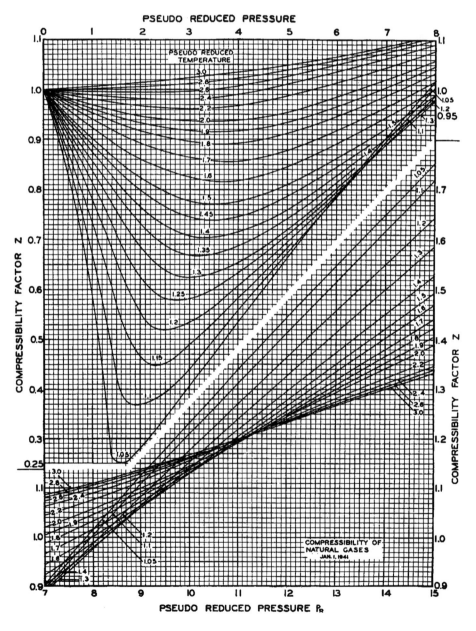

Fig. 2–2 Deviation factor for natural gases.

Example 2–4. Find the deviation factor for a natural gas at p = 1,200 psia (8.28 MPa) and T = 200 °F (366 K), if the pseudocritical parameters are p_{pc} = 630 psia (4.35 MPa) and T_{pc} = 420 °R (233 K).

Solution

The pseudoreduced parameters are

p_{pr} = 1,200 / 630 = 1.9

T_{pr} = (200 + 460) / 420 = 1.57

Using these values, from Figure 2–2:

Z = 0.85

Gas Volume Factor, B_g

The Engineering Equation of State enables the direct calculation of volume factors for gases. Equation 2.15 can be written for a given number of moles in the following form:

$$\frac{pV}{ZT_a} = \left(\frac{pV}{ZT_a} \right)_{sc}$$

2.19

This equation can be solved for B_g, which is the ratio of actual volume to the volume at standard conditions:

$$B_g = \frac{V}{V_{sc}} = \frac{p_{sc}ZT_a}{pZ_{sc}T_{sc}}$$

2.20

Substituting into this the values p_{sc} = 14.7 psia, T_{sc} = 520 °R, and Z_{sc} = 1, one arrives at

$$B_g = 0.0283\frac{ZT_a}{p}$$

2.21

Example 2–5. What is the actual volume of the gas, if its volume measured at standard conditions is 1.2 Mscf (33.9 m³)? Other data are identical to those of Example 2–4.

Solution

The volume factor of the gas, from Equation 2.21:

B_g = 0.0283 0.85 (200 + 460) / 1200 = 0.013

Actual volume is found from Equation 2.6:

$V(p,T)$ = B_g V_{sc} = 0.013 1200 = 15.6 cu ft (0.44 m³).

Gas Density

The fact that gas volume factor is an explicit function of state parameters allows a direct calculation of gas density at any conditions. Based on the definition of volume factor, gas density can be expressed from Equation 2.7, and after substituting the formula for gas volume factor, we get

$$\rho(p,T) = \frac{\rho_{sc}}{B_g} = \frac{0.0764\gamma_g p}{0.0283ZT_a} = 2.7\gamma_g\frac{p}{ZT_a}$$

2.22

The previous formula is used to find the actual density of natural gases at any pressure and temperature based on the knowledge of their specific gravities and deviation factors.

‒ ‒

Example 2–6. What is the actual density of the gas given in the previous examples, if its specific gravity is 0.75?

Solution

The deviation factor is 0.85, as found in Example 2–4.

By using Equation 2.22 we get:

$\rho(p,T)$ = 2.7 0.75 1200 / (0.85 660) = 4.3 lb/cu ft (69 kg/m³).

An alternate way of calculation is the use of the previously calculated B_g value. Gas density at standard conditions:

ρ_{sc} = 0.0764 γ_g = 0.0764 0.75 = 0.057 lb/cu ft.

Gas volume factor was found in Example 2–5 to be equal to 0.013. This allows a direct calculation of actual gas density:

$\rho(p,T) = \rho_{sc}/B_g$ = 0.057 / 0.013 = 4.4 lb/cu ft (70 kg/m³).

‒ ‒

2.2.4.2 Gas property correlations.

Pseudocritical Parameters

In most cases, gas composition is unknown and pseudocritical parameters cannot be calculated by the additive rule. The correlation of Hankinson-Thomas-Phillips [8] gives pseudocritical pressure and temperature in the function of gas specific gravity:

$$p_{pc} = 709.6 - 56.7\gamma_g \qquad \qquad 2.23$$

$$T_{pc} = 170.5 - 307.3\gamma_g \qquad \qquad 2.24$$

The previous equations are valid only for sweet natural gases, *i.e.* gases with negligible amounts of sour components.

Sour gases, like CO_2 and H_2S considerably affect critical parameters of natural gas mixtures. These effects were investigated by Wichert and Aziz [9, 10], who proposed a modified calculation method to account for the presence of sour components in the gas. They developed a correction factor, to be calculated by the following formula:

$$e = 120 \ (A^{0.9} - A^{1.6}) + 15(B^{0.5} - B^4) \qquad \qquad 2.25$$

Factors A and B are mole fractions of the components $CO_2 + H_2S$ and H_2S in the gas, respectively.

This correction factor is applied to the pseudocritical parameters (determined previously assuming the gas to be sweet) by using the equations given as follows.

$$T_m = T_{pc} - e \qquad \qquad 2.26$$

$$p_m = p_{pc} \frac{T_m}{T_{pc} + B(1-B)e} \qquad \qquad 2.27$$

where: T_m = modified pseudocritical temperature, °R

           P_m = modified pseudocritical pressure, psia

Deviation Factor

The use of the Standing-Katz chart (Fig. 2–2) is a generally accepted way to calculate deviation factors for natural gases. The inaccuracies of visual read-off and the need for computerized calculations necessitated the development of mathematical models that describe this chart. Some of these procedures involve the use of simple equations and are easy to use, *e.g.* the methods of Sarem [11] or Gopal [12]. Others use equations of state and require tedious iterative calculation schemes to find deviation factors, such as Hall-Yarborough [13] or Dranchuk et al. [14]. A complete review of the available methods is given by Takacs [15].

The following simple equation, originally proposed by Papay [16], was found to give reasonable accuracy and provides a simple calculation procedure.

$$Z = 1 - \frac{3.52 p_{pr}}{10^{0.9813 T_{pr}}} = \frac{0.274 p^2_{pr}}{10^{0.8157 T_{pr}}} \qquad\qquad 2.28$$

Example 2–7. Find the deviation factor for a 0.65 specific gravity natural gas at p = 600 psia (4.14 MPa) and T = 100 °F (311 K) with Papay's procedure. The gas contains 5% of CO_2 and 8% of H_2S.

Solution

The pseudocritical parameters from Equation 2.23 and 2.24:

p_{pc} = 709.6 - 58.7 0.65 = 671.4 psia (4.63 MPa),

T_{pc} = 170.5 + 307.3 0.65 = 370.2 °R (206 K).

The parameters A and B in the Wichert-Aziz correction for sour components:

A = (5 + 8) / 100 = 0.13

B = 8 / 100 = 0.08

The correction factor is found from Equation 2.25:

e = 120 ($0.13^{0.9}$ - $0.13^{1.6}$) + 15 ($0.08^{0.5}$ - 0.08^4) = 120 (0.159 - 0.038) + 15 (0.238 - 4.09 10^{-5}) = 14.52 + 3.57 = 18.09

The corrected pseudocritical parameters are evaluated with Equation 2.26 and 2.27:

T_m = 370.2 - 18.76 = 351.4 °R (195 K)

p_m = 570.8 351.4/ (370.2 + 0.08 (1 - 0.08) 18.76) = 235,719.1 / 371.5 = 634 psia (4.38 MPa).

Reduced parameters can be calculated using Equation 2.17 and 2.18:

p_{pr} = 600 / 634.4 = 0.946

T_{pr} = (100 + 460) / 351.4 = 1.594

Finally, Z-factor is found by Papay's equation (Equation 2.28):

Z = 1 - 3.52 0.946 / ($10^{0.9813\ 1.594}$) + 0.274 0.946^2 / ($10^{0.8157\ 1.594}$) = 1 - 0.091 + 0.012 = 0.921.

Gas Viscosity

Since dynamic viscosity is a measure of a fluid's resistance to flow, it is easy to understand that gases in general exhibit viscosities several magnitudes less than those of liquids. This is why the importance of gas viscosity in flowing pressure drop calculations is normally considered negligible. Still, both single and multiphase pressure drop calculations require a reliable estimate of gas viscosity under flowing conditions.

Based on experimental measurement on several different natural gas mixtures in a wide range of pressures and temperatures, Lee et al. [17] developed the following correlation:

$$\mu_g \, (p,T) = 0.00001 \, K \, e^{\,X \rho_g^{\,y}} \qquad\qquad 2.29$$

where:

$$K = \frac{(9.4 + 0.02M)T_a^{1.5}}{209 + 19M + T_a} \qquad\qquad 2.30$$

$$X = 3.5 + \frac{986}{T_a} + 0.01M \qquad\qquad 2.31$$

$$y = 2.4 - 0.2X \qquad\qquad 2.32$$

In the previous expressions:

μ_g = gas viscosity, cP

M = gas molecular weight, $M = 29\, \gamma_g$

T_a = absolute temperature, °R

The correlation contains the term ρ_g in g/cm^3 units. It is found from the formula given as follows:

$$\rho_g = 0.0433 \gamma_g \, \frac{p}{ZT_a} \qquad\qquad 2.33$$

- -

Example 2–8. Calculate the viscosity of a 0.7 specific gravity gas at a pressure of 2,000 psia (13.8 MPa) and a temperature of 160 °F (344 °K). The deviation factor at actual conditions is 0.75.

Solution

The molecular weight of the gas is found as:

$M = 29 \cdot 0.7 = 20.3$

The term K is calculated by Equation 2.30:

$K = [(9.4 + 0.02 \cdot 20.3) \, 620^{1.5}] / [209 + 19 \cdot 20.3 + 620] = 151{,}383 / 1{,}215 = 124.6$

X and y are found from Equation 2.31 and 2.32, respectively:

$X = 3.5 + 986 / 620 + 0.01 \cdot 20.3 = 5.29$

$y = 2.4 - 0.2 \cdot 5.29 = 1.34$

Gas density in g/cm^3 units is calculated from Equation 2.33:

$\rho_g = (0.0433 \cdot 0.7 \cdot 2{,}000) / (0.75 \cdot 620) = 60.6 / 465 = 0.13$ g/cm^3.

Finally, gas viscosity is found from Equation 2.29:

$\mu_g = 0.0001 \cdot 124.6 \, \exp[5.29 \cdot 0.13^{1.34}] = 0.012 \, \exp[0.344] = 0.017$ cP (1.7 $\cdot$ 10^{-5} Pas).

- -

2.3 Inflow Performance of Oil Wells

2.3.1 Introduction

The proper design of any artificial lift system requires an accurate knowledge of the fluid rates that can be produced from the reservoir through the given well. Present and also future production rates are needed to accomplish the following basic tasks of production engineering:

- selection of the right type of lift

- detailed design of production equipment

- estimation of future well performance

The production engineer, therefore, must have a clear understanding of the effects governing fluid inflow into a well. Lack of information may lead to over-design of production equipment or, in contrast, equipment limitations may restrict attainable liquid rates. Both of these conditions have an undesirable impact on the economy of artificial lifting and can cause improper decisions as well.

A well and a productive formation are interconnected at the sandface, the cylindrical surface where the reservoir is opened. As long as the well is kept shut in, sandface pressure equals reservoir pressure and thus no inflow occurs to the well. It is easy to see that, in analogy to flow in surface pipes, fluids in the reservoir flow only between points having different pressures. Thus, a well starts to produce when the pressure at its sandface is decreased below reservoir pressure. Fluid particles in the vicinity of the well then move in the direction of pressure decrease and, after an initial period, a stabilized rate develops. This rate is controlled mainly by the pressure prevailing at the sandface, but is also affected by a multitude of parameters such as reservoir properties (rock permeability, pay thickness, etc.), fluid properties (viscosity, density, etc.) and well completion effects (perforations, well damage). These latter parameters being constant for a given well, at least for a considerable length of time, the only means of controlling production rates is the control of bottomhole pressures. The proper description of well behavior, therefore, requires that the relationship between bottomhole pressures and the corresponding production rates be established. The resulting function is called the well's inflow performance relationship (IPR) and is usually obtained by running well tests.

This section discusses the various procedures available for the description of well inflow performance. Firstly, the most basic terms with relevant descriptions are given.

2.3.2 Basic concepts

Darcy's Law

The equation describing filtration in porous media was originally proposed by Darcy and can be written in any consistent set of units as:

$$\frac{q}{A} = -\frac{k}{\mu}\frac{dp}{dl}$$

2.34

This formula states that the rate of liquid flow, q, per cross-sectional area, A, of a given permeable media is directly proportional to permeability, k, the pressure gradient, dp/dl, and is inversely proportional to liquid viscosity. The negative sign is included because flow takes place in the direction of decreasing pressure gradients. Darcy's equation assumes a steady state, linear flow of a single-phase fluid in a homogeneous porous media saturated with the same fluid. Although these conditions are seldom met, all practical methods are based on Darcy's work.

Drainage Radius, r_e

Consider a well producing a stable fluid rate from a homogeneous formation. Fluid particles from all directions around the well flow toward the sandface. In idealized conditions, the drainage area, *i.e.* the area where fluid is moving to the well, can be considered a circle. At the outer boundary of this circle, no flow occurs and undisturbed reservoir conditions prevail. Drainage radius, r_e, is the radius of this circle and represents the greatest distance of the given well's influence on the reservoir under steady-state conditions.

Average Reservoir Pressure, p_R

The formation pressure outside the drainage area of a well equals the undisturbed reservoir pressure, p_R, which can usually be considered a steady value over longer periods of time. This is the same pressure as the bottomhole pressure measured in a shut-in well, as seen in Figure 2–3.

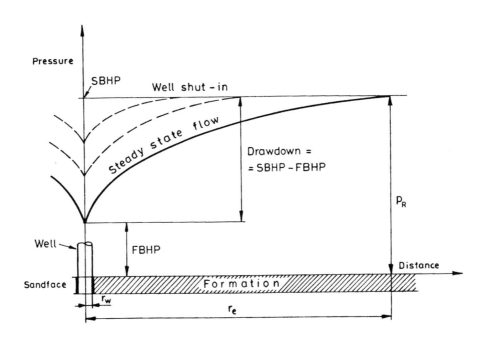

Fig. 2–3 Pressure distribution around a well in the formation.

Flowing Bottomhole Pressure (FBHP), p_{wf}

Figure 2–3 shows the pressure distribution in the reservoir around a producing well. In shut-in conditions, the average reservoir pressure, p_R prevails around the wellbore and its value can be measured in the well as SBHP. After flow has started, bottomhole pressure is decreased and pressure distribution at intermediate times is represented by the dashed lines. At steady-state conditions, the well produces a stabilized liquid rate and its bottomhole pressure attains a stable value, p_{wf}. The solid line on Figure 2–3 shows the pressure distribution under these conditions.

Pressure Drawdown

The difference between static and FBHP is called pressure drawdown. This drawdown causes the flow of formation fluids into the well and has the greatest impact on the production rate of a given well.

2.3.3 The productivity index concept

The simplest approach to describe the inflow performance of oil wells is the use of the productivity index (PI) concept. It was developed using the following simplifying assumptions:

- flow is radial around the well,

- a single-phase liquid is flowing,

- permeability distribution in the formation is homogeneous, and

- the formation is fully saturated with the given liquid.

For the previous conditions, Darcy's equation (Equation 2.34) can be solved for the production rate:

$$q = \frac{0.00708kh}{\mu B \ln\left(\frac{r_e}{r_w}\right)}(p_R - p_{wf}) \qquad\qquad 2.35$$

where: q = liquid rate, STB/d

   k = effective permeability, mD

   h = pay thickness, ft

   μ = liquid viscosity, cP

   B = liquid volume factor, bbl/STB

   r_e = drainage radius of well, ft

   r_w = radius of wellbore, ft

Most parameters on the right-hand side are constant, which permits collecting them into a single coefficient called PI:

$$q = PI(p_R - p_{wf}) \qquad\qquad 2.36$$

This equation states that liquid inflow into a well is directly proportional to pressure drawdown. It plots as a straight line on a pressure vs. rate diagram, as shown in Figure 2–4. The endpoints of the PI line are the average reservoir pressure, p_R, at a flow rate of zero and the maximum potential rate at a bottomhole flowing pressure of zero. This maximum rate is the well's absolute open flow potential (AOFP) and represents the flow rate that would occur if flowing bottomhole pressure could be reduced to zero. In practice, it is not possible to achieve this rate, and it is only used to compare the deliverability of different wells.

The use of the PI concept is quite straightforward. If the average reservoir pressure and the PI are known, use of Equation 2.36 gives the flow rate for any FBHP. The well's PI can either be calculated from reservoir parameters, or measured by taking flow rates at various FBHPs.

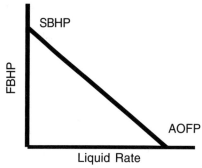

Fig. 2–4 Well performance with the PI concept.

Example 2–9. A well was tested at p_{wf} = 1,400 psi (9.7 MPa) pressure and produced q = 100 bpd (15.9 m³/d) of oil. Shut-in bottom pressure was p_{ws} = 2,000 psi (13.8 MPa). What is the well's PI and what is the oil production rate at p_{wf} = 600 psi (4.14 MPa).

Solution

Solving Equation 2.36 for PI and substituting the test data:

PI = q / (p_R - p_{wf}) = 100 / (2,000 – 1,400) = 0.17 bopd/psi (0.0039 m³/kPa/d)

The rate at 600 psi (4.14 MPa) is found from Equation 2.36:

q = PI (p_{ws} - p_{wf}) = 0.17 (2,000 - 600) = 238 bopd (37.8 m³/d).

2.3.4 Inflow performance relationships

2.3.4.1 Introduction. In most wells on artificial lift, bottomhole pressures below bubblepoint pressure are experienced. Thus, there is a gas phase present in the reservoir near the wellbore, and the assumptions that were used to develop the PI equation are no longer valid. This effect was observed by noting that the PI was not a constant as suggested by Equation 2.36. Test data from such wells indicate a downward curving line, instead of the straight line shown in Figure 2–4.

The main cause of a curved shape of inflow performance is the liberation of solution gas due to the decreased pressure in the vicinity of the wellbore. This effect creates an increasing gas saturation profile toward the well and simultaneously decreases the effective permeability to liquid. Liquid rate is accordingly decreased in comparison to single-phase conditions and the well produces less liquid than indicated by a straight-line PI. Therefore, the constant PI concept cannot be used for wells producing below the bubblepoint pressure. Such wells are characterized by their IPR curves, to be discussed in the following section.

2.3.4.2 Vogel's IPR correlation. Vogel used a numerical reservoir simulator to study the inflow performance of wells depleting solution gas drive reservoirs. He considered cases below bubblepoint pressure and varied pressure drawdowns, fluid, and rock properties. After running several combinations on the computer, Vogel found that all the calculated IPR curves exhibited the same general shape. This shape is best approximated by a dimensionless equation [18]:

$$\frac{q}{q_{max}} = 1 - 0.2 \frac{p_{wf}}{p_R} - 0.8 \left(\frac{p_{wf}}{p_R}\right)^2 \qquad 2.37$$

where: q = production rate at bottomhole pressure p_{wf}, STB/d

   q_{max} = maximum production rate, STB/d

   p_R = average reservoir pressure, psi

Equation 2.37 is graphically depicted in Figure 2–5.

Although Vogel's method was originally developed for solution gas drive reservoirs, the use of his equation is generally accepted for other drive mechanisms as well [19]. It was found to give reliable results for almost any well with a bottomhole pressure below the bubblepoint of the crude.

In order to use Vogel's method, reservoir pressure needs to be known along with a single stabilized rate and the corresponding FBHP. With these data, it is possible to construct the well's IPR curve by the procedure discussed in the following example problem.

FBHP, Fraction of Reservior Pressure

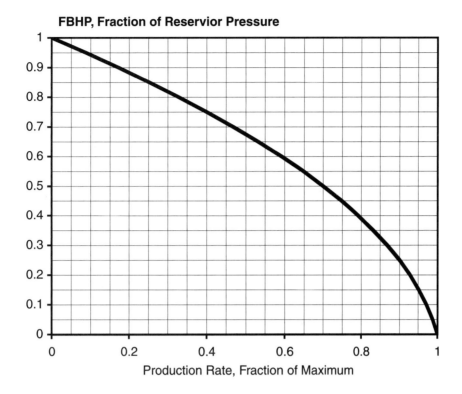

Production Rate, Fraction of Maximum

Fig. 2–5 Vogel's dimensionless inflow performance curve.

Example 2–10. Using data of the previous example find the well's AOFP and construct its IPR curve, by assuming multiphase flow in the reservoir.

Solution

Substituting the test data into Equation 2.37:

$100 / q_{max} = 1 - 0.2 \,(1,400/2,000) - 0.8 \,(1,400/2,000)^2 = 0.468.$

From the previous equation the AOFP of the well:

$q_{max} = 100 / 0.468 = 213.7$ bopd (34 m³/d)

Now find one point on the IPR curve, where $p_{wf} = 1,800$ psi (12.42 MPa) using Figure 2–5.

$p_{wf} / p_R = 1,800 / 2,000 = 0.9$

From Figure 2–5:

$q / q_{max} = 0.17$, and q = 213.7 0.17 = 36.3 bopd (5.8 m³/d).

The remaining points of the IPR curve are evaluated the same way.

Figure 2–6 shows the calculated IPR curve along with a straight line valid for PI = 0.17 (0.0039 m³/kPa/d), as found in Example 2–9. Calculated parameters for the two cases are listed as follows:

	Max. Rate	Rate at 600 psi
Vogel	213.7	185.5
Constant PI	340	238

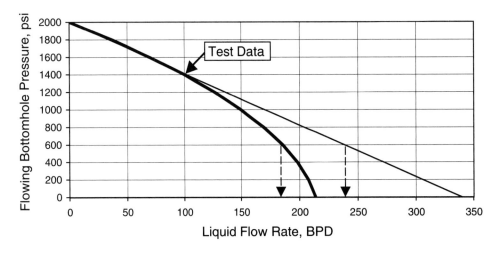

Fig. 2–6 Comparison of IPR curves for Examples 2–7 and 2–8.

Comparison of the preceding results indicates that considerable errors can occur if the constant PI method is used for conditions below the bubblepoint pressure.

2.3.4.3 Fetkovich's method. Fetkovich demonstrated that oil wells producing below the bubblepoint pressure and gas wells exhibit similar inflow performance curves [20]. The general gas well deliverability equation can thus also be applied to oil wells:

$$q = C(p_R^2 - p_{wf}^2)^n$$

2.38

Coefficients C and n in this formula are usually found by curve-fitting of multipoint well test data. Evaluation of well tests and especially isochronal tests is the main application for Fetkovich's method.

2.4 Single-phase Flow

2.4.1 Introduction

In all phases of oil production, several different kinds of fluid flow problems are encountered. These involve vertical or inclined flow of a single-phase fluid or of a multiphase mixture in well tubing, as well as horizontal or inclined flow in flowlines. A special hydraulic problem is the calculation of the pressure exerted by the static gas column present in a well's annulus. All the problems mentioned require that the engineer be able to calculate the main parameters of the particular flow, especially the pressure drop involved.

In this section, basic theories and practical methods for solving single-phase pipe flow problems are covered that relate to the design and analysis of gas lifted wells. As all topics discussed have a common background in hydraulic theory, a general treatment of the basic hydraulic concepts is given first. This includes detailed definitions of and relevant equations for commonly used parameters of pipe flow.

2.4.2 Basic principles

The General Energy Equation

Most pipe flow problems in petroleum production operations involve the calculation of the pressure along the flow path. Pressure being one form of energy, its variation over pipe length can be found from an energy balance written between two points in the pipe. The General Energy Equation describes the conservation of energy and states that the change in energy content between two points of a flowing fluid is equal to the work done on the fluid minus any energy losses. From this basic principle, the change in flowing pressure, dp, over an infinitesimal length dl can be evaluated as

$$\frac{dp}{dl} = \frac{g}{g_c} \rho \sin \alpha + \frac{\rho v dv}{g_c dl} + \left(\frac{dp}{dl}\right)_f \qquad\qquad 2.39$$

This equation gives the pressure gradient, dp/dl for the flow configuration of an inclined pipe shown in Figure 2–7. Analysis of the right-hand side terms shows that, in this general case, pressure gradient consists of three components:

- hydrostatic

- acceleration

- friction gradients

The hydrostatic gradient, also known as *elevation gradient* stands for the changes in potential energy—the acceleration gradient represents kinetic energy changes. The third component accounts for irreversible energy losses due to fluid friction. The relative magnitude of these components in the total pressure gradient depends on the number of flowing phases, pipe inclination, and other parameters [21].

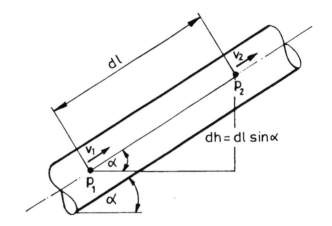

Fig. 2–7 Flow configuration for the General Energy Equation.

Flow Velocity, v

The velocity of flow is a basic hydraulic parameter and is usually calculated from the flow rate given at standard conditions. Using the definition of volume factor B to arrive at actual volumetric flow rate, velocity is calculated from:

$$v = \frac{q(p,T)}{A} + \frac{q_{sc}}{BA} \qquad\qquad 2.40$$

where: A = cross-sectional area of pipe

In multiphase flow, only a fraction of pipe cross section is occupied by one phase, because at least two phases flow simultaneously. Thus, the previous formula gives imaginary values for the individual phases, frequently denoted as superficial velocities. Although actually nonexistent, these superficial velocities are widely used in multiphase flow correlations.

Friction Factor, f

Frictional losses constitute a significant fraction of flowing pressure drops, thus their accurate determination is of utmost importance. For calculating friction gradient or frictional pressure drop, the Darcy-Weisbach equation is universally used.

$$\left(\frac{dp}{dl}\right)_f = f\,\frac{\rho v^2}{2g_c d}$$ 2.41

where: f = friction factor

g_c = 32.2, a conversion factor

d = pipe diameter

The friction factor f is a function of the Reynolds number N_{Re} and the pipe's relative roughness and is usually determined from the Moody diagram.

Reynolds Number, N_{Re}

Due to the complexity of single-phase—and especially multiphase—flow, the number of independent variables in any hydraulic problem is quite large. One commonly used way to decrease the number of variables and to facilitate easier treatment is the forming of dimensionless groups from the original variables. Such dimensionless groups or numbers are often used in correlations, the most well known one being the Reynolds number defined as follows:

$$N_{Re} = \frac{\rho v d}{\mu}$$ 2.42

This equation gives a dimensionless number in any consistent set of units. Using the customary unit of cP for viscosity, it can be transformed to

$$N_{Re} = 124\,\frac{\rho v d}{\mu}$$ 2.43

where: ρ = fluid density, lb/cu ft

v = flow velocity, ft/s

d = pipe diameter, in.

μ = fluid viscosity, cP

Moody Diagram

Friction factor, in general, was found to be a function of the Reynolds number and pipe relative roughness. Relative roughness is defined here as the ratio of the absolute roughness ε of the pipe inside wall to the pipe inside diameter:

$$k = \frac{\varepsilon}{d}$$ 2.44

Typical absolute roughness values are ε = 0.0006 in (1.5E-5 m) for new and ε = 0.009 in (2.3E-4 m) for used well tubing.

There are several formulae describing friction factors for different flow conditions. Most of them are included in the Moody diagram [22], which is a graphical presentation of Darcy-Weisbach-type f values. Use of this chart, given in Figure 2–8, is generally accepted in the petroleum industry. Figure A–1 in Appendix A contains a full-page copy of the Moody diagram.

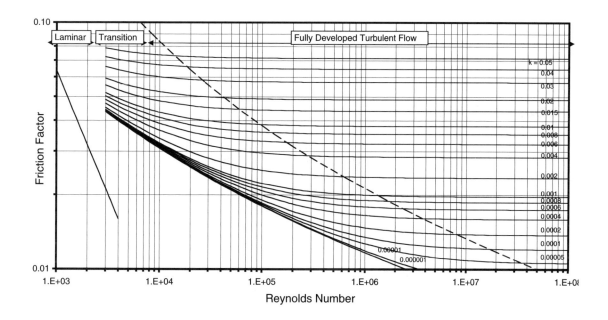

Fig. 2–8 The Moody [22] diagram: Friction factors for pipe flow.

As seen in the Moody diagram, the friction factor is a different function of the variables N_{Re} and k in different ranges. In laminar flow (*i.e.* Reynolds numbers lower than 2,000–2,300) f varies with the Reynolds number only:

$$f = \frac{64}{N_{Re}} \qquad\qquad 2.45$$

On the other hand, in fully developed turbulent flow, for Reynolds numbers higher than 2,000–2,300 and for a rough pipe, friction factor is a sole function of the pipe's relative roughness. There exists a transition region between smooth wall flow and fully developed turbulent flow where the value of friction factor is determined by both N_{Re} and k values.

For computer applications, the use of the Colebrook equation is recommended, which can be solved for friction factors in both the transition and the fully developed turbulent regions [23]. It is an implicit function and can be solved for f values using some iterative procedure, *e.g.* the Newton-Raphson scheme.

$$\frac{1}{\sqrt{f}} = 2 \log \frac{3.7}{k} - 2 \log \left(1 + \frac{9.335}{k N_{Re} \sqrt{f}} \right) \qquad\qquad 2.46$$

The inconvenience of an iteration scheme can be eliminated if an explicit formula is used to calculate friction factors. Gregory and Fogarasi [24] investigated several such models and found that the formula developed by Chen [25] gives very good results. This formula, as shown here, very accurately reproduces the Moody diagram over the entire range of conditions. It does not necessitate an iterative scheme and thus can speed up lengthy calculations (*e.g.* multiphase pressure drop calculations) involving the determination of a great number of friction factors.

$$\frac{1}{\sqrt{f}} = -2 \log \left(\frac{k}{3.7065} - \frac{5.0452}{N_{Re}} \log A \right)$$ 2.47

where:

$$A = \frac{k^{1.1098}}{2.8257} + \left(\frac{7.149}{N_{Re}} \right)^{0.8981}$$ 2.48

2.4.3 Pressure drop calculations

2.4.3.1 Single-phase liquid flow. Liquids are practically incompressible; that is why flow velocity is constant in a pipe with a constant cross section. With no change in velocity along the flow path, no kinetic energy changes occur and the General Energy Equation has to be modified accordingly. Setting the right-hand second term to zero in Equation 2.39 and substituting the expression of friction gradient from Equation 2.41 in the term $(dp/dl)_f$:

$$\frac{dp}{dl} = \frac{g}{g_c} \rho \sin\alpha + f \frac{\rho v^2}{2 g_c d}$$ 2.49

This equation gives the pressure gradient for liquid flow in an inclined pipe. It can be solved for pressure drop in a finite pipe length l, and using customary field units will take the form:

$$\Delta p = \frac{1}{144} \rho l \sin\alpha + 1.294 \times 10^{-3} f \frac{1}{d} \rho v^2$$ 2.50

where Δp = pressure drop, psi

ρ = flowing density, lb/cu ft

l = pipe length, ft

α = pipe inclination above horizontal, degrees

v = flow velocity, ft/s

d = pipe diameter, in.

In case of vertical flow $\alpha = 90°$, $\sin \alpha = 1$, and pressure drop is calculated by

$$\Delta p = \frac{1}{144} \rho l + 1.294 \times 10^{-3} f \frac{l}{d} \rho v^2$$ 2.51

For horizontal flow, there is no elevation change and inclination angle is zero, thus $\sin \alpha = 0$. Therefore, total pressure drop for horizontal flow consists of frictional losses only, and can be expressed by using the formula:

$$\Delta p = 1.294 \times 10^{-3} f \frac{l}{d} \rho v^2$$ 2.52

All the pressure drop formulae discussed previously contain the liquid flow velocity v, which is evaluated on the basis of Equation 2.40, and using field measurement units can be written as the following:

$$v = 0.0119 \frac{q_l}{Bd^2} \qquad\qquad\qquad 2.53$$

where v = flow velocity, ft/s

q_l = liquid flow rate, STB/d

d = pipe diameter, in.

B = volume factor, bbl/STB

The procedures for calculating pressure drop in single-phase liquid flow are illustrated by presenting two example problems.

- -

Example 2–11. Calculate the pressure drop occurring in a horizontal line of length 4,000 ft (1,219 m) with an inside diameter of 2 in. (51 mm) for a water flow rate of 1,000 bpd (159 m^3/d). Pipe absolute roughness is 0.00015 in. (0.0038 mm), water specific gravity is 1.05. Use B_w = 1 and μ = 1 cP (0.001 Pas).

Solution

Find the flow velocity in the pipe by using Equation 2.53:

v= 0.0119 1000 / (1 2^2) = 2.98 ft/s (0.91 m/s)

Next, the Reynolds number is found with Equation 2.43, substituting $\rho = \gamma_w \, \rho_{w\,sc} / B_w$:

N_{Re} = (124 1.05 62.4 2.98 2) / 1 =48,422.

Flow is turbulent, as N_{Re} > 2,300. Friction factor is read off from the Moody diagram at N_{Re} = 48,422 and at a relative roughness of k = 0.00015 / 2 = 7.5 10^{-5}.

From Fig 2–8:

f = 0.021

The frictional pressure drop is calculated using Equation 2.52:

Δp = 1.294 10^{-3} 0.021 (4,000 1.05 62.4 2.98^2) / 2 = 32 psi (220 kPa).

- -

Example 2–12. Calculate the wellhead pressure required to move 900 bpd (143 m^3/d) of oil into a 200 psi (1.38 MPa) pressure separator through a flowline with a length of 2,000 ft (610 m) and a diameter of 2 in (51 mm). The oil produced is gasless with a specific gravity of 0.65 and has a viscosity of 15 cP (0.015 Pas) and a B_o of 1. The flowline has a constant inclination, the separator being at a vertical distance of h = 100 ft (30.5 m) above the wellhead.

Solution

The hydrostatic pressure drop is the first term in Equation 2.50

Using $\rho_o = \gamma_o\, \rho_{w\,sc} / B_o = 0.65\ 62.4 / 1 = 40.6$ lb/cu ft (650 kg/m^3),

and $\sin \alpha = h / l = 100 / 2{,}000 = 0.05$,

we get $\Delta p_{hydr} = (\ 40.6\ 2{,}000\ 0.05\ ) / 144 = 28.2$ psi (194.3 kPa).

For calculating frictional pressure drop, flow velocity and Reynolds number are calculated by Equations 2.53 and 2.43, respectively:

$v = 0.0119\ (900/1\ 2^2) = 2.68$ ft/s (0.817 m/s),

$N_{Re} = 124\ (40.6\ 2.67\ 2) / 15 = 1{,}799.$

Flow is laminar, as N_{Re} is less than the critical value of 2,300. Friction factor for laminar flow is given by Equation 2.45 as:

$f = 64 / 1{,}799 = 0.036$

Frictional pressure drop is calculated using the second term in Equation 2.50:

$\Delta p_f = 1.294\ 10^{-3}\ 0.036\ (2{,}000\ 40.6\ 2.67^2) / 2 = 13.5$ psi (93 kPa)

Total pressure drop equals $28.2 + 13.5 = 41.7$ psi (287 kPa). The pressure required at the wellhead is the sum of separator pressure and flowing pressure drop:

$p = 200 + 41.7 = 241.7$ psi (1.67 MPa).

2.4.3.2 Single-phase gas flow. The most important feature of gas flow, when compared to the flow of liquids, is the fact that gas is highly compressible. This means that in contrast with liquid flow, flow velocity in any pipe cross section does not stay constant. At the input end of the pipe where pressure is higher, gas occupies less space, and flow velocity is less than further down the line. Consequently, gas velocity increases in the direction of flow. Therefore, friction losses will change along the pipe because they vary with the square of flow velocity. This is the reason why the calculation of flowing pressure losses cannot be done for the total pipe length at once as was the case in single-phase liquid flow.

The direction of gas flow can be vertical, inclined, or horizontal. Most often, vertical or inclined flow takes place in gas wells, a topic outside the scope of the present work. Horizontal gas flow, however, is a common occurrence in surface gas injection lines. In the following, we introduce the basics of calculating pressure drops in horizontal pipes.

Similar to liquid flow, the basic differential equation for horizontal gas flow in pipes is found from Equation 2.49 by neglecting the elevation term:

$$\frac{dp}{dl} = f\, \frac{\rho v^2}{2g_c d} \qquad\qquad 2.54$$

Gas density is found from the Engineering Equation of State, as shown previously, and given in Equation 2.22. Flow velocity is calculated based on the gas flow rate under standard conditions, q_{sc}, the pipe cross section, A, and the volume factor B_g of the gas phase:

$$v = \frac{q_{sc}B_g}{A}$$
2.55

After substitution of the previous terms in the basic equation and using field units, we get

$$\frac{dp}{dl} = 1.26 \times 10^{-5} f \gamma_g \frac{q_{sc}^2}{d^5} \frac{ZT}{p}$$
2.56

As clearly seen, pressure gradient is a function of actual pressure and thus will change along the pipe. Since the right-hand side of the equation contains pressure-dependent parameters, a direct solution is not possible. A common assumption is that friction factor, deviation factor and flow temperature are constants and can be approximated by properly selected average values. Using this approach, the previous equation becomes a separable type differential equation, the solution of which is the following:

$$p_1^2 - p_2^2 = 2.52 \times 10^{-5} \bar{f} \gamma_g \frac{q_{sc}^2}{d^5} \bar{Z}\bar{T}L$$
2.57

where: p_1 = inflow pressure, psia

p_2 = outflow pressure, psia

$\bar{f}$ = friction factor

γ_g = specific gravity of the gas

q_{sc} = gas flow rate at standard conditions, Mscf/d

d = pipe diameter, in.

$\bar{Z}$ = deviation factor

$\bar{T}$ = flowing temperature, R

L = pipe length, ft

From the previous formula, outflow pressure is easily found.

The friction factor can be estimated from the Weymouth formula:

$$f = \frac{0.032}{\sqrt[3]{d}}$$
2.58

where: d = pipe diameter, in.

It follows from the nature of the previous solution that the outflow pressure or the pressure drop in the pipe is a result of an iterative calculations scheme. Using successive approximations, assumed values of outflow pressure p_2 are reached until convergence is found. Usually, only a few iterations are needed.

Example 2–13. Find the outflow pressure in a 2 in. diameter, 6,500 ft long gas injection pipeline, if inflow pressure equals 800 psia. Gas specific gravity is 0.6, assume an isothermal flow at 80 °F, with a gas flow rate of 1,200 Mscf/d.

Solution

First some parameters being constant during the calculation process are found.

Friction factor is calculated from Equation 2.58 as

$$f = 0.032 / 2^{0.333} = 0.025$$

Pseudocritical parameters of the gas are found from Equations 2.23 and 2.24:

$$p_{pc} = 709 - 58.7 \ 0.6 = 674.8 \text{ psia, and}$$

$$T_{pc} = 170.5 + 307.3 \ 0.6 = 354.9 \text{ °R.}$$

Since flow temperature is constant, the pseudoreduced temperature will also be constant:

$$T_{pr} = (80 + 460) / 354.9 = 1.522.$$

For the first iteration, let's assume an outflow pressure equal to 800 psia. Average and pseudoreduced pressures will be

$$p_{avg} = (800 + 800) / 2 = 800 \text{ psia, and } p_{pr} = 800 / 674.8 = 1.186.$$

Now, deviation factor can be calculated. Using the Papay formula (Equation 2.28):

$$Z = 1 - 3.52 \ 1.186 / (10^{0.9815 \ 1.522}) + 0.274 \ 1.186^2 / (10^{0.8157 \ 1.522}) = 0.888.$$

Using Equation 2.57:

$$p_1^2 - p_2^2 = 2.52 \times 10^{-5} \ 0.025 \ 0.6 \ \frac{1,200^2}{2^5} \ 0.888 \ (80 + 460) \ 6,500 = 53,869.$$

From here, p_2 is found as:

$$p_2 = \sqrt{800^2 - 53.869} = 765.6 \text{ psia}$$

The calculated value differs from the assumed one, therefore another iteration is needed. Now the p_2 value just calculated is assumed as the outflow pressure. Average and pseudoreduced pressures will be

$$p_{avg} = (800 + 765.6) / 2 = 782.8 \text{ psia, and } p_{pr} = 782.5 / 674.8 = 1.16.$$

Now, deviation factor can be calculated. Using the Papay formula (Equation 2.28):

$$Z = 1 - 3.52 \ 1.16 / (10^{0.9815 \ 1.522}) + 0.274 \ 1.16^2 / (10^{0.8157 \ 1.522}) = 0.890.$$

Using Equation 2.57:

$$p_1^2 - p_2^2 = 2.52 \times 10^{-5}\ 0.025\ 0.6\ \frac{1{,}200^2}{2^5}\ 0.890\ (80 + 460)\ 6{,}500 = 53{,}987.$$

From here, p_2 is found as:

$$p_2 = \sqrt{800^2 - 53.869} = 765.6 \text{ psia}$$

Since this value is very close to the assumed one, no more iterations are needed and the outflow pressure is $p_2 = 765.5$ psia.

2.4.4 Gas gradient in annulus

2.4.4.1 Theoretical background. Usually, the gas column present in a gas-lifted or a pumping well's annulus is considered to be in a semi-static state, as flow velocities in the annular space are most often negligible. This is due to the large cross-sectional areas involved and the relatively low gas flow rates. In such cases, frictional losses can be disregarded and the pressure distribution in the column is affected only by gravitational forces. In unusual conditions (*e.g.* in wells completed with relatively small casing sizes or when extremely high gas flow rates are involved), friction losses in the well's annulus cannot be disregarded and gas pressure vs. well depth calculations should be modified accordingly.

The General Energy Equation for the case of a static gas column is written with the friction and acceleration gradient neglected, and it only contains the hydrostatic term. From Equation 2.39:

$$\frac{dp}{dl} = \frac{g}{g_c}\rho\sin\alpha \qquad\qquad 2.59$$

For an inclined pipe $dl \sin \alpha = dh$, where dh is the increment of vertical depth coordinate. Introducing this expression and using psi for the unit of pressure, we get

$$\frac{dp}{dh} = \frac{1}{144}\rho_g \qquad\qquad 2.60$$

The actual density of gas, ρ_g can be evaluated in the knowledge of its specific gravity and volume factor. Using Equation 2.3 and 2.21, gas density can be written as

$$\rho_g = \frac{\gamma_{g\,sc}}{B_g} = \frac{0.0764\gamma_g}{0.0283\ \dfrac{ZT}{p}} \qquad\qquad 2.61$$

Upon substitution of this expression into the original equation the following differential equation is reached:

$$\frac{dp}{dh} = 0.01877\gamma_g\ \frac{p}{ZT} \qquad\qquad 2.62$$

The previous formula cannot be analytically solved, since it contains the empirical functions of deviation factor Z and actual temperature T. In order to develop a numerical solution, it is assumed that Z and T are constants for a small vertical depth increment Δh. Using these assumptions, Equation 2.62 can already be solved between two points in the annulus as

$$p_1 = p_2 \exp\left(0.01877\gamma_g \frac{\Delta h}{Z_{avg}T_{avg}}\right)$$

2.63

where: p_1 = gas column pressure at depth h, psi

   p_2 = gas column pressure at depth h + Δh, psi

   Z_{avg} = average Z factor between points 1 and 2

   T_{avg} = average absolute temperature between points 1 and 2, °R

   Δh = vertical distance between points 1 and 2, ft

The previous equation provides a simple way for approximate calculations of gas column pressures. This involves a trial-and-error procedure, the main steps of which are detailed in the following. Note that the bottom pressure will be calculated using only one depth increment, i.e. Δh is set to the total height of the gas column.

1.   Assume a value for the pressure at the bottom of the column, which is to be calculated.

2.   Calculate an average pressure in the column.

3.   From temperature distribution data, find an average temperature in the column.

4.   Obtain a deviation factor for the previous conditions.

5.   Use Equation 2.63 to find a calculated value of gas column pressure.

6.   If the calculated and the assumed pressures are not close enough, use the calculated one for a new assumed pressure and repeat the calculations. This procedure is continued until the two values are sufficiently close.

—— — ——— — — — — — ——— — — — — ——— — —— — — — ——— — —— — —— — — — ——— — — — ——

Example 2–14. Calculate the gas pressure at a depth of 4,500 ft (1,372 m) if the surface pressure is 300 psia (2.07 MPa), and the gas is of 0.7 specific gravity. Wellhead temperature is 100 °F (311 K) and the temperature at the given depth is 180 °F (355 K).

Solution

Assume a p_2 of 330 psia (2.28 MPa) at the bottom of gas column.

Average pressure in the column:

p_{avg} = (300 + 330) / 2 = 315 psia (2.17 MPa).

Average temperature is calculated as

T_{avg} = (100 + 180) / 2 = 140 °F = 140 + 460 = 600 °R (333 K).

The pseudocritical parameters of the gas are calculated by Equations 2.23 and 2.24:

p_{pc} = 709.6 - 58.7 0.7 = 669 psia (4.62 MPa),

T_{pc} = 170.5 + 307.3 0.7 = 386 °R (214 K).

The pseudoreduced parameters for the average conditions are found with Equations 2.17 and 2.18:

p_{pr} = 315 / 669 = 0.47,

T_{pr} = 600 / 386 = 1.54.

Deviation factor Z is read off Figure 2–2:

Z = 0.955

Calculated p_2 is evaluated using Equation 2.63:

p_2 = 300 exp (0.01877 0.7 4500 / 0.955 / 600) = 300 exp (0.1032) = 332.6 psia (2.29 MPa)

This is not equal to the assumed value of 330 psia (2.28 MPa), thus a new trial is made. The average pressure with the new assumed bottom pressure of 332.6 psia (2.29 MPa):

p_{avg} = (300 + 332.6) / 2 = 316.3 psia (2.18 MPa).

Average and pseudoreduced temperatures did not change, pseudoreduced pressure is

p_{pr} = 316.3 / 669 = 0.473

Using Fig. 2–2:

Z = 0.952

The new value for p_2:

p_2 = 300 exp (0.01877 0.7 4500 / 0.952 / 600) = 300 exp (0.1035) = 332.7 psia (2.29 MPa).

Since this is sufficiently close to the assumed value of 332.6 psia (2.29 MPa), gas pressure at the given level is 332.7 psia (2.29 MPa).

The calculation scheme previously detailed has its inherent errors that are due to the use of only one depth step. The proper use of Equation 2.63, however, requires that the assumption of a constant Z factor and a constant temperature be met in the given depth increment Δh. This can only be assured by dividing the total gas column length into sufficiently small steps. The smaller the depth increments taken, the smaller the variation of Z and T in the incremental column height and the more accurately the actual conditions are approximated. Therefore, to ensure maximum accuracy, it is essential to use a stepwise iteration procedure with sufficiently small depth increments.

2.4.4.2 Computer solution. This calculation model easily lends itself to computer solutions. Figure 2–9 offers the flowchart of a computer program that follows the previous procedure. The basic steps of the calculations are essentially the same as detailed before, the only

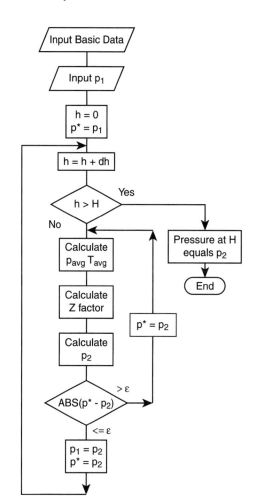

Fig. 2–9 Flowchart for calculating pressure distribution in a static gas column.

difference being that the total column length is divided into several increments. For the first increment, the pressure at the upper point equals the known surface pressure p_1. Starting from this, pressure p_2 valid at the lower point of the increment is calculated. Finding p_2 involves a trial-and-error procedure and using a first guess of $p^* = p_1$ is convenient. After the assumed value, p^* and the calculated p_2 have converged, the next depth increment is taken. Pressure p_1 at the top of the next lower increment must equal the pressure at the bottom of the previous increment. The procedure is repeated for subsequent lower depth increments until reaching the bottom of the gas column. At the last increment, the final p_2 value gives the pressure at the bottom of the gas column.

Example 2–15. Using the data of the previous example, calculate the pressure exerted by the gas column with the detailed procedure. Use a depth increment of 450 ft (137 m).

Solution

Calculation results are contained in Table 2–1. The accuracy for subsequent p_2 values was set to 0.1 psia (0.69 kPa). As seen from the table, only two trials per depth increment were necessary. The final result of 332.5 psia (2.29 MPa) compares favorably with the 332.7 psia (2.29 MPa) calculated in the previous example.

h	p1	p*	p avg	T avg	Z	p2
ft	psia	psia	psia	R		psia
450	300.00	300.00	300.00	564	0.955	303.31
		303.31	301.65	564	0.955	303.31
900	303.31	303.31	303.31	572	0.956	306.60
		306.60	304.95	572	0.956	306.60
1350	306.60	306.60	306.60	580	0.957	309.88
		309.88	308.24	580	0.957	309.88
1800	309.88	309.88	309.88	588	0.958	313.15
		313.15	311.52	588	0.958	313.15
2250	313.15	313.15	313.15	596	0.959	316.40
		316.40	314.78	596	0.958	316.40
2700	316.40	316.40	316.40	604	0.960	319.64
		319.64	318.02	604	0.959	319.64
3150	319.64	319.64	319.64	612	0.961	322.87
		322.87	321.26	612	0.960	322.87
3600	322.87	322.87	322.87	620	0.962	326.09
		326.09	324.48	620	0.962	326.09
4050	326.09	326.09	326.09	628	0.963	329.29
		329.29	327.69	628	0.963	329.29
4500	329.29	329.29	329.29	636	0.964	332.47
		332.47	330.88	636	0.964	332.48

Table 2–1 Calculation results for Example 2–12.

2.4.4.3 Universal gas gradient chart. Investigations of calculated pressure traverses in static gas columns have shown that pressure vs. depth curves valid for a given gas specific gravity and a given surface pressure can very closely be approximated by straight lines. An explanation of this observation is that the effects on the gas density of the increase in temperature down the well and of the increasing pressure tend to offset each another. The use of average pressure gradients is, therefore, a viable approach to the calculation of gas column pressures.

Figure 2–10 presents a chart constructed using the previous logic, where pressure gradients are plotted for different gas specific gravities in the function of surface pressure. A similar chart is given by Brown and Lee [26] that was developed by using a constant deviation factor, as found by the present author. Figure 2–10, on contrary, was constructed with the proper consideration of gas deviation factors and can thus provide higher accuracy.

Appendix B contains a full-page copy of the previous chart to be used for estimating annular gas column pressures.

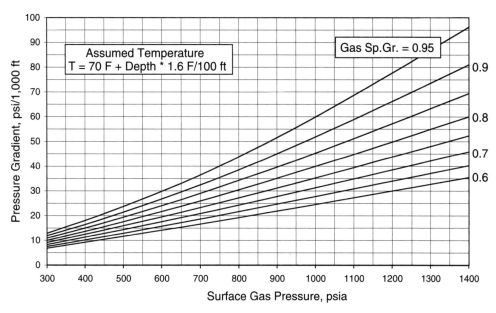

Fig. 2–10 Static gas pressure gradient chart.

Example 2–16. Use Figure 2–10 to estimate gas column pressure for the previous example.

Solution

At p_1 = 300 psia surface pressure and a specific gravity of 0.7, pressure gradient is found as 7 psia/1,000 ft (0.158 kPa/m). Gas column pressure with this value:

$$p_2 = p_1 + grad\ H / 1000 = 300 + 7\ 4.5 = 331.5\ psia\ (2.287\ MPa)$$

The discrepancy of this result from the value calculated in Example 2–15 is due to the difference of actual and chart basis temperatures.

2.4.5 Choke flow

When metering or controlling single-phase gas or liquid flow, a restriction in the flow cross section provides the necessary pressure drop that is used either to measure the flow rate or to restrict its magnitude. Metering of liquid or gas flow rates by using restrictions (such as flow nozzles or orifices) is outside the scope of this book, but many problems in gas lifting relate to controlling gas flow rates through such devices. Two main applications stand out: surface control of gas injection rates and injection gas flow through gas lift valves. In both cases, high-pressure gas is forced to flow through a restriction where a significant decrease in flow cross section causes a pressure drop across the device. The pressure drop thus created decreases the flow rate, and this phenomenon allows controlling of gas injection rates into the well.

For the reasons described previously, this chapter deals with single-phase gas flow through surface injection chokes and the ports of gas lift valves. The basic features of the two cases are the same, and they have the same theoretical background, allowing their common treatment. Surface injection gas control is achieved with the use of so-called flow beans or chokes that behave similar to the port of any valve.

2.4.5.1 Calculation of gas throughput of chokes.
Due to the relatively high decrease of cross section in a choke, gas velocity increases as compared to its value in the full cross section of the pipe upstream of the choke. According to the Equation of Continuity, flow velocity in the choke would continuously increase as the size of the choke is decreased. This tendency is, however, limited by the fact that no velocity greater than the local sound velocity can exist in any gas flow. When maximum flow velocity (the local sound velocity) occurs, the flow is called *critical* as compared to other, so-called *sub-critical* cases. Critical flow is a common occurrence and is characterized by the fact that the flow rate through the choke is independent of downstream conditions, especially the downstream pressure.

Critical conditions in a choke or a flow restriction occur, depending on the pressure conditions. The critical pressure ratio is found from the following equation [27] where an adiabatic flow across the choke is assumed:

$$\left(\frac{p_2}{p_1}\right)_{cr} = \left(\frac{2}{\kappa+1}\right)^{\frac{\kappa}{\kappa-1}}$$

2.64

where: p_2 = pressure downstream of the choke, psi

p_1 = pressure upstream of the choke, psi

κ = ratio of specific heats

The ratio of specific heats, $\kappa = c_p/c_v$, is relatively constant in the usual temperature and molecular weight ranges valid for hydrocarbon gases, its generally accepted value being $\kappa = 1.256$. Using this value, the critical pressure ratio is found to be equal to 0.554. This means that a downstream pressure of roughly half of the upstream pressure results in

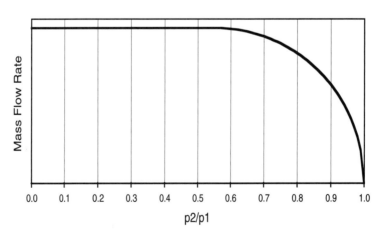

Fig. 2–11 Gas flow rate through a choke vs. pressure ratio.

a critical flow. Any further decrease in downstream pressure does not increase the gas flow rate as seen from Figure 2–11 where, for an arbitrary case, gas flow rate is plotted vs. pressure ratio.

In order to calculate gas throughput in a choke, the General Energy Equation and the Equation of Continuity must be simultaneously applied between two points in the flow path, *i.e.* one upstream of the choke and the other at the throat section. From this, gas velocity at the throat section can be found, which after being multiplied with the choke cross-sectional area gives the local volumetric gas flow rate. After correction for actual pressure and temperature conditions, the gas rate at standard conditions is found.

Omitting the lengthy derivation of the previous procedure, the final formula for calculating the gas flow rate through a choke is given as follows:

$$q_{g\,sc} = 693\ C_d\,d^2\,p_1\sqrt{\frac{1}{\gamma_g T_1}}\ \sqrt{\frac{2\kappa}{\kappa+1}}\ \sqrt{\left(\frac{p_2}{p_1}\right)^{\frac{2}{\kappa}} - \left(\frac{p_2}{p_1}\right)^{\frac{\kappa+1}{\kappa}}} \qquad\qquad 2.65$$

where: $q_{g\,sc}$ = flow rate at standard conditions (14.7 psia and 60 °F), Mscf/d

C_d = discharge coefficient

d = choke diameter, in.

p_1 = flowing pressure upstream of the choke, psia

p_2 = flowing pressure downstream of the choke, psia

γ_g = gas specific gravity

T_1 = absolute temperature upstream of the choke, °R

κ = ratio of specific heats

The value of the discharge coefficient C_d varies with the nature of the restriction (flow nozzle, orifice, etc.) and other flow parameters and must be established experimentally. For surface chokes and gas lift valve ports the value $C_d = 0.865$ suggested by the Thornhill-Craver Co. is generally used. [28] Although the formula was developed for the flow of ideal gases, it has found widespread use in gas lift practice.

The validity of the previous formula is restricted to sub-critical flow conditions, *i.e.* to the range:

$$\left(\frac{p_2}{p_1}\right)_{cr} \le \frac{p_2}{p_1} \le 1.0$$

For critical flow the term p_2/p_1 in Equation 2.65 is replaced with its critical value $(p_2/p_1)_{cr}$. By this, it is assured that gas flow rate stays constant after critical flow conditions have been reached.

– –

Example 2–17. Calculate the gas throughput if a gas of 0.65 specific gravity flows through a choke of ¼ in. (6.4 mm) diameter with an upstream pressure of 1,000 psia (6.9 MPa) and a downstream pressure of 800 psia (5.5 MPa). Flow temperature is 150 °F (339 K), use a discharge coefficient of 0.9. How does the gas rate change if the downstream pressure is reduced to 500 psia (3.5 MPa)?

Solution

First check the flow conditions for critical flow. The critical pressure ratio is found from Equation 2.64:

$$(p_2/p_1)_{cr} = (2 / 2.256)^{(1.256 / 0.256)} = 0.887^{4.906} = 0.554$$

The actual pressure ratios for the two cases:

$(p_2/p_1) = 800 / 1,000 = 0.8 > (p_2/p_1)_{cr}$, so the flow is sub-critical,

$(p_2/p_1) = 500 / 1,000 = 0.5 < (p_2/p_1)_{cr}$, so the flow is critical.

For the first, sub-critical case the actual pressure ratio is used in Equation 2.65:

$$q_{g\ sc} = 693\ 0.9\ 0.25^2\ 1,000 \sqrt{\frac{1}{0.65\ 610}} \sqrt{\frac{2.512}{0.256}} \sqrt{0.8^{1.592} - 0.8^{1.796}} = 38,981\ 0.05\ 3.13\ 0.18 =$$

$$= 1,098 \text{ Mscf/d } (31,092 \text{ m}^3/\text{d}).$$

In the second case, the critical pressure ratio is used instead of the actual one in Equation 2.65:

$$q_{g\ sc} = 693\ 0.9\ 0.25^2\ 1,000 \sqrt{\frac{1}{0.65\ 610}} \sqrt{\frac{2.512}{0.256}} \sqrt{0.5^{1.592} - 0.5^{1.796}} = 38,981\ 0.05\ 3.13\ 0.21 =$$

$$= 1,281 \text{ Mscf/d } (36,274 \text{ m}^3/\text{d}).$$

·—

2.4.5.2 Gas capacity charts. Calculation of gas flow rates through chokes is a frequently occurring problem in designing and analyzing gas lift installations. Due to its practical importance, many different kinds of graphical solutions to the problem are known. Such charts can take different forms and are based on different assumed conditions, their use being more convenient than actual hand calculations.

A universally applicable gas capacity chart is given in Figure 2–12. It was developed for the following conditions:

Gas Sp.Gr. = 0.65; C_d = 0.865 T_1 = 60 °F.

When calculating choke capacities, actual flow temperatures are always different from standard temperature. This is especially true for the chokes or ports of gas lift valves installed downhole where elevated temperatures exist. Therefore, for conditions different from the chart base, gas capacities read from the chart must be corrected. Based on the original formula, one can derive the following correction function:

$$q_{act} = q_{chart} C_{d\ act} \frac{21.25}{\sqrt{\gamma_{act} T_{act}}} \tag{2.66}$$

where: q_{act} = actual rate at standard conditions (14.7 psia and 60 °F), Mscf/d

              q_{chart} = gas flow rate read from chart, Mscf/d

              $C_{d\ act}$ = actual discharge coefficient

              γ_{act} = actual gas specific gravity

              T_{act} = actual temperature upstream of the choke, °R

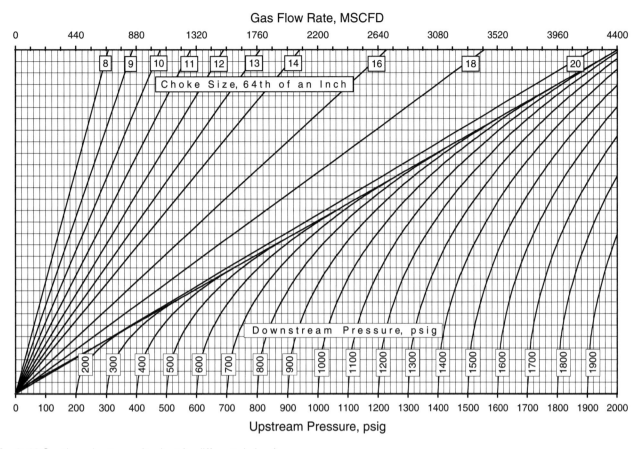

Fig. 2–12 Gas throughput capacity chart for different choke sizes.

Appendix C contains two gas capacity charts for different ranges of choke sizes, as well as a correction chart that can be used to correct gas volumes at temperatures different from chart base.

Example 2–18. Find the gas volumes for the cases given in the previous example with the use of the gas capacity chart in Figure 2–12.

Solution

For case one, start at an upstream pressure of 1,014 psia (7 MPa) and go vertically until crossing the curve valid for a downstream pressure of 814 psia (5.6 MPa). From the intersection, draw a horizontal to the left to the proper choke size (16/64 in.). Drop a vertical from here to the upper scale to find the gas flow rate of 1,140 Mscf/d (32,281 m³/d).

Since actual flow conditions differ from chart base values, a correction must be applied. According to Equation 2.65, the actual gas flow rate is

$$q_{act} = 1,140 \cdot 0.9 \frac{21.25}{\sqrt{0.65610}} = 1,140 \cdot 0.96 = 1,094 \text{ Mscf/d (30,978 m3/d).}$$

The second case involves critical flow and the vertical line from the upstream pressure should be drawn to the upper boundary line of the downstream pressures constituting critical flow. Using the same procedure as in the first case, a gas flow rate of 1,360 Mscf/d (38,511 m³/d) is read from the chart. Since flow conditions are similar to case one, again a correction factor of 0.96 is used to find the actual flow rate:

$$q_{act} = 1,360 \cdot 0.96 = 1,305 \text{ Mscf/d (36,953 m}^3\text{/d).}$$

2.5 Two-phase Flow

2.5.1 Introduction

The steady-state simultaneous flow of petroleum liquids and gases in wells and surface pipes is a common occurrence in the petroleum industry. Oil wells normally produce a mixture of fluids and gases to the surface while phase conditions usually change along the flow path. At higher pressures, especially at the well bottom, flow may be single-phase. But going higher up in the well, the continuous decrease of pressure causes dissolved gas to gradually evolve from the flowing liquid, resulting in multiphase flow. Even gas wells can produce condensed liquids and/or formation water in addition to gas. Both oil and gas wells may be vertical, may be inclined, or may contain longer horizontal sections. Surface flowlines as well as long distance pipelines are laid over hilly terrain and are used to transfer oil, gas, and water simultaneously.

Multiphase flow, however, is not restricted to the previous cases. Most artificial lifting methods involve a multiphase mixture present in well tubing, like gas lifting when the injection of high-pressure gas into the well at a specific depth makes the flow in the well a multiphase one. Other lifting methods, *e.g.* the use of rod or centrifugal pumps, also are similar. It is easy to understand then that the proper description of multiphase flow phenomena is of prime importance for petroleum engineers working in the production of oil and gas wells. It is the pressure drop or the pressure traverse curve, which the main parameters of fluid flow can be calculated with, that forms the basis of any design or optimization problem. The accurate calculation of vertical, inclined, or horizontal two-phase pressure drops not only improves the engineering work but plays a significant economic role in producing single wells and whole fields alike.

Due to the practical importance of two-phase flow problems, the world petroleum industry has made great efforts to develop accurate pressure drop computation methods. Research was started many decades ago and up to now has not yet been completed, but the number of available multiphase pressure drop correlations is significant today.

This section is devoted to the introduction and solution of two-phase flow problems encountered in the production and transport of oil, gas, and water. First, the reader is introduced to the most common principles and concepts of two-phase flow. Based on these principles, the pressure drop prediction models for oil wells are reviewed next. All available calculation models are covered including the empirical and the so-called *mechanistic* ones. The accuracy obtainable with the use of the different models is discussed, based on the latest data from the technical literature. The next section deals with horizontal and inclined flow and gives a complete review of the pressure gradient calculation methods. Finally, two-phase flow through restrictions (surface and downhole chokes, valves, etc.) is discussed in detail.

2.5.2 Basic principles

Multiphase flow is significantly more complex than single-phase flow. Single-phase flow problems are well defined and most often have analytical solutions developed over the years. Therefore, the most important task, *i.e.* the calculation of pressure drop along the pipe, can be solved with a high degree of calculation accuracy. This is far from being so if a simultaneous flow of more than one phase takes place. The introduction of a second phase makes conditions difficult to predict due to several reasons. According to Szilas [29], these can be grouped as follows:

- For any multiphase flow problem, the number of the variables affecting the pressure drop is enormous. In addition to the many thermodynamic parameters for each phase several variables describing the interaction of the two phases (interfacial tension, etc.) as well as other factors have to be taken into account. Because of this difficulty, researchers usually try to reduce the total number of variables with the introduction of nondimensional parameter groups.

- The two-phase mixture is compressible, *i.e.* its density depends on the actual flowing pressure and temperature. Since in most flow problems pressure and temperature vary in a wide range, this effect must be properly accounted for.

- Frictional pressure losses are much more difficult to describe since more than one phase is in contact with the pipe wall. The multiphase mixture's velocity distribution across the pipe is also different from single-phase laminar or turbulent flows, adding up to the difficulty of pressure drop prediction.

- In addition to frictional pressure losses, a new kind of energy loss arises: slippage loss. This is due to the great difference in densities of the liquid and gas phases and the tendency of the gas phase to overtake or slip-by the liquid phase. Under some conditions, slippage losses greatly contribute to the total pressure drop.

- Physical phenomena and pressure losses considerably vary with the spatial arrangement in the pipe of the flowing phases, conventionally called flow patterns. Since flow patterns considerably differ from each other, different procedures must be used for each of them to calculate the pressure drop.

Because of the complexity of two-phase flow problems, many investigators (especially the early ones who used empirical methods) did not try to write up the necessary differential equations but started from a simple General Energy Equation. Use of this equation allows for a simplified treatment of the two-phase flow problem and has since resulted in many calculation models ensuring engineering accuracy. A more comprehensive treatment is provided by the so-called mechanistic approach which is the prime choice of recent investigators. In the following, calculation models of both the traditional (or empirical), and the mechanistic approach is discussed.

Before the detailed treatment of vertical, inclined, and horizontal two-phase flows, the basic assumptions followed throughout the following discussions must be listed:

- Only steady-state flow is investigated. The reason for this is that in the majority of production engineering problems stabilized flow takes place. Most artificial lift applications, as well as most horizontal flow problems, involve steady-state flow; only special cases (slugging in long offshore flowlines, etc.) necessitate the more complex treatment required for transient two-phase flow.

- The temperature distribution along the flow path is known. Although it is possible to simultaneously solve the differential equation systems written for heat transfer and two-phase flow, it is seldom done. Usually, the distribution of temperature is either known or can be calculated independently. On the other hand, simultaneous calculation of the pressure and temperature distribution necessitates additional data on the thermal properties of the flowing fluids and the pipe's environment, and these are seldom available.

- The mass transfer between the two phases is considered to follow the *black-oil* type of behavior, *i.e.* the composition of the liquid phase is assumed to be constant. If the liquid phase undergoes compositional changes along the flow path, in case of volatile oils or condensate fluids, the black oil model cannot be used and equilibrium calculations must be performed. Most production engineering calculations, however, require the use of the black oil model.

2.5.2.1 Two-phase flow concepts.
This section covers some important concepts that apply to any two-phase flow problem. Understanding of these concepts will help master the solution of specific problems in vertical, horizontal, or inclined flows.

2.5.2.1.1 Flow patterns. Liquid and gas, flowing together in a pipe, can assume different geometrical configurations relative to each other. These arrangements have long been experimentally investigated through transparent pipe sections and are generally called flow patterns or flow regimes. One extreme of the observed flow patterns is the case of the continuous dense phase with a low amount of dispersed light phase contained in it. On the other end stands the flow pattern with a small quantity of the dense phase dispersed in a continuous light phase. When starting from the flow of a continuous liquid (dense) phase, introduction of a progressing amount of the gas (light) phase into the pipe will achieve all possible flow patterns.

The flow patterns observed for vertical, inclined, and horizontal pipes have many common features, but there is a basic difference. In horizontal flow, the effect of gravity is much more pronounced than in vertical flows. Due to this, gas tends to separate from the heavier liquid phase and travels at the top of the pipe; this is the stratified flow pattern. In inclined pipes, on the other hand, gravity forces usually prevent stratification.

Figure 2–13 depicts a commonly occurring flow pattern, *slug flow*, for different pipe inclinations.

To facilitate the determination of the actual flow pattern, the individual boundaries of the visually observed flow patterns are usually plotted on so-called flow pattern maps. In most cases, these diagrams have nondimensional numbers on their coordinate axes, as detailed in later sections.

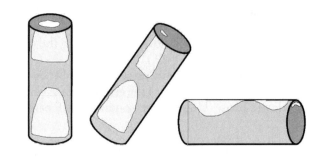

2.5.2.1.2 Superficial velocities. In two-phase flow, the velocities of the gas and liquid particles cannot be readily calculated because of their irregular motion. Even average velocities, easily found in single-phase

Fig. 2–13 Slug flow pattern in vertical, inclined, and horizontal pipes.

cases, are difficult to estimate. Instead of these, most calculation models employ a parameter called *superficial velocity* which, thanks to its simple definition, is a favorite correlating parameter.

Superficial velocity means the cross-sectional average velocity when assuming the pipe to be fully occupied with the given phase. This is calculated based on the in-situ fluid rate and the pipe's cross-sectional area as follows:

$$v_s = \frac{q_{sc}B(p,T)}{A_p}$$
2.67

If the production rates at standard conditions, the relevant thermodynamic parameters of the phases at the given pressure and temperature are known, and water production is negligible, liquid and gas superficial velocities are found as given as follows:

$$v_{sl} = 6.5 \times 10^{-5} \frac{q_{osc}B_o(p,T)}{A_p}$$
2.68

$$v_{sg} = 1.16 \times 10^{-5} \frac{q_{osc}[GOR - R_s(p,T)]B_g(p,T)}{A_p}$$
2.69

where: q_{osc} = oil flow rate, STB/d

B_o = oil volume factor at p and T, bbl/STB

B_g = gas volume factor at p and T, ft^3/scf

GOR = gas-oil ratio at standard conditions, scf/STB

R_s = solution GOR at p and T, scf/STB

A_p = pipe cross-sectional area, ft^2

Example 2–19. Find the superficial velocities in an oil well at three different points in the tubing. The vertical tubing is 8,245 ft long and 2.441 in internal diameter. Oil rate is 375 bpd with no water, production GOR = 480 scf/STB. Specific gravities of oil and gas are 0.82 and 0.916, respectively. Wellhead and bottomhole pressures and temperatures are 346.6 psi, 2,529.5 psi, and 80.3 °F, 140.3 °F, respectively. Perform the calculations at the wellhead, at the well bottom, and at a depth of 3,700 ft where the pressure is assumed to be 1,220 psi.

Solution

Let us calculate the conditions at the wellhead first, where the temperature is 80.3 °F.

Solution GOR and oil volume factor will be calculated from the Standing correlations, where the API degree of the oil is found from Equation 2.4:

$API° = 141.5 / \gamma_l - 131.5 = 41$

R_s is calculated from Equation 2.10:

$y = 0.00091\ 80.3 - 0.0125\ 41 = -0.44$

$R_s = 0.916\ (346.6 / 18 / 10^{-0.44})^{1.205} = 109.7$ scf/STB.

B_o is calculated from Equation 2.11:

$F = 109.7\ (0.916 / 0.82)^{0.5} + 1.25\ 80.3 = 216.3$

$B_o = 0.972 + 1.47\ 10^{-4}\ 216.3^{1.175} = 1.05$ bbl/STB.

Pipe cross-sectional area is:

$A_p = 2.442^2\ \pi / 4 / 144 = 0.0325$ sq ft.

Superficial liquid velocity can now be calculated from Equation 2.68 as:

$v_{sl} = 6.5\ 10^{-5}\ 375\ 1.05 / 0.0325 = 0.79$ ft/s.

In order to find gas volume, gas volume factor should be calculated first. For this, the pseudocritical parameters are determined from Equation 2.23 and 2.24:

$p_{pc} = 709.6 - 58.7\ 0.916 = 655.8$ psi, and

$T_{pc} = 170.5 + 307.3\ 0.916 = 451.9$ R.

Pseudoreduced parameters are as follows (Equations 2.17 and 2.18):

$p_{pr} = 346.6 / 655.8 = 0.528.$

$T_{pr} = (80.3 + 460) / 451.9 = 1.195.$

Gas deviation factor is determined from the Papay formula (Equation 2.28):

$Z = 1 - 3.52\ 0.528 / (10^{0.9813\ 1.195}) + 0.274\ 0.528^2 / (10^{0.8157\ 1.195}) = 1 - 0.125 + 0.008 = 0.883.$

Gas volume factor from Equation 2.21:

$B_g = 0.0283\ 0.883\ (80.3 + 460) / 346.6 = 0.0389.$

Finally, superficial gas velocity from Equation 2.69:

$v_{sg} = 1.16\ 10^{-5}\ 375\ (480 - 109.7)\ 0.0389 / 0.0325 = 1.93$ ft/s.

Calculations for the other two points are similar, main results are given as follows:

Parameter	Wellhead	At 3,770 ft	Bottom
temperature, F	80.3	107.2	140.3
R_s, scf/STB	109.7	466.9	480
B_o, bbl/STB	1.05	1.26	1.286
v_{sl}, ft/s	0.79	0.94	0.96
Z, -	0.883	0.706	-
B_g, -	0.0389	0.009	-
v_{sg}, ft/s	1.93	0.01	-

2.5.2.1.3 *Gas slippage.* When liquid and gas flow together in a vertical or inclined pipe, their velocities differ so that the gas phase always overtakes or *slips past* the liquid phase. Gas slippage, or simply called *slippage*, is a basic phenomenon in two-phase flow that must be fully understood. Slippage is caused by several factors, including the following:

- The difference in density of gas (light phase) and liquid (dense phase) results in buoyancy force acting on the gas phase that increases gas velocity.

- Energy losses occurring in the flow direction are much less in the gas phase than in liquid; thus gas moves easier.

- Since pressure decreases in the direction of flow and liquid (in contrast to gas) is practically incompressible, gas expands and its velocity increases accordingly.

All these effects result in an increase of gas velocity, relative to that of the liquid phase. This causes the mixture density to increase, in comparison with the *no-slip* case, *i.e.* when the two phases flow at identical velocities. The amount of slippage is governed by many factors including the flow pattern, phase densities, flow velocities, pipe size, etc. Slippage is significant at low liquid flow rates and diminishes with increasing liquid rates as frictional pressure drop takes over.

Gas slippage is frequently described with the *drift-flux* model which, according to Zuber and Findlay [30] gives the average cross-sectional velocity of the gas phase in a two-phase mixture as:

$$v_s = C_0 v_m + v_b \qquad\qquad 2.70$$

where: C_0 = distribution factor

v_m = mixture velocity, ft/s

v_b = bubble rise velocity, ft/s

The distribution factor C_0 reflects the velocity and concentration distribution over the pipe cross section and can attain a value between 1.0 and 1.5. Mixture velocity is the sum of the individual phases' superficial velocities. The term v_b represents the terminal rising velocity of a single gas bubble or a swarm of bubbles. It can be calculated in a variety of ways, and one commonly used solution for the bubble flow pattern was proposed by Harmathy [31]:

$$v_b = 153 \sqrt[4]{\frac{g\sigma_l(\rho_2 - \rho_g)}{\rho_l^2}} = 0.79 \sqrt[4]{\frac{\sigma_l(\rho_2 - \rho_g)}{\rho_l^2}} g \qquad 2.71$$

where: ρ_l =   liquid density, lb/cu ft

ρ_g =   gas density, lb/cu ft

σ_l =   interfacial tension, dyne/cm

The slip velocity between the phases can be calculated from the gas and liquid phase velocities:

$$v_s = v_g + v_l \qquad 2.72$$

2.5.2.1.4 Liquid holdup. Many of the early investigators did not or could not consider the slippage between the two phases and assumed that liquid and gas traveled at equal velocities. In this case, the liquid content of a pipe section is easily calculated from the actual flow rates entering the pipe section. This parameter is called the no-slip liquid holdup and is found from the simple formula:

$$\lambda_l = \frac{q_l(p,T)}{q_l(p,T) + q_g(p,T)} = \frac{v_{sl}}{v_{sl} + v_{sg}} = \frac{v_{sl}}{v_m} \qquad 2.73$$

Since the sum of the liquid and gas volumes must be equal to the pipe section's volume, the following relationship always holds for the gas void fraction, λ_g:

$$\lambda_g = 1 - \lambda_l \qquad 2.74$$

In real two-phase flows, as discussed previously, gas overtakes the liquid, and the pipe section's actual gas content will be less than the no-slip value. In order to investigate this problem, consider a relatively short pipe section containing a flowing two-phase mixture as given in Figure 2–14. Obviously, the liquid content of the section of length Δh equals the ratio of the liquid volume contained in the pipe segment to the total pipe volume. If the length of the pipe section is decreased to an infinitesimal length of dl, this ratio approaches the cross-sectional liquid content called liquid holdup and denoted by ε_l. The same is true for the gas content and the gas void fraction ε_g.

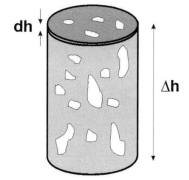

Fig. 2–14 Definition of liquid holdup.

Since the liquid and gas phases flow through a cross-sectional area of $\varepsilon_l A_p$ and $\varepsilon_g A_p$, respectively, their actual velocities are found from the next formulas:

$$v_g = \frac{v_{sg}}{\varepsilon_g} \quad \text{and} \quad v_l = \frac{v_{sl}}{\varepsilon_l}$$

Gas velocity was previously expressed by Equation 2.70 and substitution into the previous first formula gives the gas void fraction as

$$\varepsilon_g = \frac{v_{sg}}{C_o v_m + v_{sg}} \qquad 2.75$$

Liquid holdup can then be expressed as follows:

$$\varepsilon_l = 1 - \varepsilon_g = 1 - \frac{v_{sg}}{C_o v_m + v_b} \qquad 2.76$$

Another way to express liquid holdup can be derived from the definition of the slip velocity (Equation 2.72). Substituting the formulas of gas and liquid phase velocities into the slip velocity we get

$$v_s = \frac{v_{sg}}{1 - \varepsilon_l} - \frac{v_{sl}}{\varepsilon_l} \qquad\qquad 2.77$$

The previous equation can be solved for ε_l as follows:

$$\varepsilon_l = \frac{v_s - v_m + \sqrt{(v_m - v_s)^2 + 4 v_s v_{sl}}}{2 v_s} \qquad\qquad 2.78$$

Liquid holdup, calculated from either Equation 2.76 or Equation 2.78 previously, includes the effect of gas slippage and must therefore always be greater than or equal to the no-slip liquid holdup, *i.e.* $\varepsilon_l \geq \lambda_l$. The difference between the two values indicates the magnitude of the actual gas slippage. The value of ε_l gives the proportion of the pipe cross section occupied by the liquid phase in the flowing two-phase mixture. If liquid holdup is known, the two-phase mixture density can be easily calculated, which is one of the main goals of two-phase flow calculations.

2.5.2.1.5 *Mixture properties.* In order to facilitate the solution of the relevant equations describing the behavior of two-phase flows, many times the flowing mixture is substituted by an imaginary homogeneous fluid. The thermodynamic properties of this fluid must be calculated from the individual parameters of the phases. The most important such parameters are mixture density and viscosity.

Based on the previous discussions on gas slippage and liquid holdup, it is easy to understand that two-phase mixture density is found from the densities of the phases, and the liquid holdup. Depending on the liquid holdup used, one can calculate a no-slip mixture density as follows:

$$\rho_{ns} = \rho_l \lambda_l + \rho_g \lambda_g \qquad\qquad 2.79$$

or a density including the effects of gas slippage as follows:

$$\rho_m = \rho_l \varepsilon_l + \rho_g \varepsilon_g \qquad\qquad 2.80$$

The phase densities figuring in the previous equations are to be found at the actual temperature and pressure valid at the given pipe section. If water flow rate is zero, liquid density equals the in-situ density of the oil phase. This can be calculated by the following formula, which includes the effect of the gas dissolved in the oil phase at the given conditions:

$$\rho_o = \frac{350.4 \, \gamma_o + 0.0764 \gamma_g \, R_s(p,T)}{5.61 \, B_o(p,T)} \qquad\qquad 2.81$$

Gas density is determined as shown as follows (see Equation 2.22):

$$\rho_g = 2.7 \, \gamma_g \frac{p}{Z T_a} \qquad\qquad 2.82$$

where: γ_o = specific gravity of oil

γ_g = specific gravity of gas

T_a = absolute temperature, R°

Z = gas deviation factor

R_s = solution GOR at p and T, scf/STB

B_o = oil volume factor at p and T, bbl/STB

Similar to density calculations, mixture viscosity is usually found as a weighted average of the two phases' viscosities. As before, the weights applied can be based on the no-slip or the slip concept. The most common way to calculate mixture viscosity uses the no-slip holdup as follows:

$$\mu_{ns} = \mu_l \lambda_l + \mu_g \lambda_g \qquad 2.83$$

Another frequently applied expression is the following:

$$\mu_m = \mu_l^{\varepsilon_l} + \mu_g^{\varepsilon_g} \qquad 2.84$$

Finally, a formula similar to Equation 2.80 can also be used.

$$\mu_m = \mu_l \varepsilon_l + \mu_g \varepsilon_g \qquad 2.85$$

where: μ_l = liquid viscosity at p and T, cP

μ_g = gas viscosity at p and T, cP

Example 2–20. Calculate liquid holdup and mixture density at the three depths given in the previous example.

Solution

The relevant parameters calculated in Example 2–19 can be used and the wellhead case is elaborated in detail.

Oil phase in-situ density is found from Equation 2.81:

$$\rho_o = (\ 350.4 \cdot 0.82 + 0.0764 \cdot 0.916 \cdot 109.7\ ) / (\ 5.61 \cdot 1.05\ ) = 49.9 \text{ lb/cu ft.}$$

In-situ gas density is evaluated from Equation 2.82:

$$\rho_g = 2.7 \cdot 0.916 \cdot 346.6 / 0.883 / (\ 80.3 + 460\ ) = 1.79 \text{ lb/cu ft.}$$

Liquid holdup is determined with the drift-flux model, where bubble rise velocity is calculated with the Harmathy formula (Equation 2.71) as follows. Interfacial tension is approximated with a value of 8 dyne/cm.

$$v_b = 0.79\ [\ 8\ (\ 49.9 - 1.79\ ) / 49.9^2\ ]^{0.25} = 0.49 \text{ ft/s.}$$

Gas void fraction from Equation 2.75 by assuming a distribution factor of $C_0 = 1$:

$$\varepsilon_g = 1.93 / [1\ (0.79 + 1.93) + 0.49] = 0.60.$$

Liquid holdup equals:

$$\varepsilon_l = 1 - \varepsilon_g = 1 - 0.60 = 0.4.$$

Finally, mixture density from Equation 2.80:

$$\rho_m = 49.9 \cdot 0.4 + 1.79 \cdot 0.60 = 21.0 \text{ lb/cu ft.}$$

Calculations for the other two points are similar, main results are given as follows:

Parameter	Wellhead	At 3,770 ft	Bottom
oil density, lb/cu ft	49.9	45.4	44.5
gas density, lb/cu ft	1.79	7.5	-
bubble rise velocity, ft/s	0.49	0.48	-
liquid holdup, -	0.40	0.99	1.00
mixture density, lb/cu ft	21.0	44.9	44.5

2.5.2.2 Multiphase flow. Most of the pressure drop calculation models were developed for two discrete phase—liquid and gas—and are therefore properly called two-phase methods. In case the liquid phase is not pure oil or pure water but a mixture of these, the properties of this phase have to be found based on the properties of oil and water. In a lack of sufficient research results, the usual solution is to apply a mixing rule based on the actual flow rates of the components. This approach is at least questionable since it does not include the effects of slippage between oil and water, the occurrence of emulsions, etc.

If the liquid flowing is a mixture of oil and water, the superficial velocities of the liquid and gas phases are determined by the formulae as follows:

$$v_{sl} = 6.5 \times 10^{-5} \frac{q_{lsc}}{A_p} \left[B_o(p,T)\frac{1}{1+WOR} + B_w(p,T)\frac{WOR}{1+WOR} \right] \qquad 2.86$$

$$v_{sg} = 1.16 \times 10^{-5} \frac{q_{lsc}}{A_p} \left[GLR - R_s(p,T)\frac{1}{1+WOR} \right] B_g(p,T) \qquad 2.87$$

where:
- q_{lsc} = liquid flow rate, STB/d
- B_o = oil volume factor at p and T, bbl/STB
- B_w = water volume factor at p and T, bbl/STB
- B_g = gas volume factor at p and T, ft³/scf
- GLR = gas-liquid ratio at standard conditions, scf/STB
- R_s = solution GOR at p and T, scf/STB
- WOR = water-oil ratio at standard conditions
- A_p = pipe cross-sectional area, ft²

The density, viscosity, and other thermodynamic parameters of the liquid phase are found from a weighted average of the two fluids' relevant properties. This approach has its limitations since oil-water mixtures may behave very differently from this assumption. Liquid density is affected by oil slippage, a phenomenon resembling gas slippage, whereby the lighter oil particles overtake the more dense water phase. Because of this, mixture density increases beyond the value calculated from a mixing rule. On the other hand, the viscosity of oil-water mixtures may exhibit non-Newtonian behavior if an emulsion is formed.

With due regard to the previous limitations, liquid density is usually calculated from the following expression:

$$\rho_l = \rho_o \frac{WOR}{1+WOR} + \rho_w \frac{1}{1+WOR}$$

2.88

The actual density of formation water is found from its specific gravity and volume factor as follows:

$$\rho_w = \frac{350.4\gamma_w}{5.61 B_w(p,T)} = 62.5 \frac{\gamma_w}{B_w(p,T)}$$

2.89

2.5.2.3 Pressure gradient equations. In the majority of two-phase or multiphase flow problems, we are interested in calculating the pressure distribution along the pipe. Because of the complexity of the flow phenomena in any one of the flow patterns and the existence of many flow patterns, it is very difficult to write up the governing differential equations: the conservation of mass and momentum. These equations have to refer to both phases and are, therefore, difficult or even impossible to solve. To overcome this situation, one usually treats the multiphase mixture as a homogeneous fluid and writes up the General Energy (Bernoulli) Equation for this hypothetical phase. This equation describes the conservation of energy between two points (lying close to each other) in an inclined pipe and can be solved for the pressure gradient (see also Section 2.4.2):

$$\frac{dp}{dh} = \frac{g}{g_c}\rho_m\sin\alpha + \left(\frac{dp}{dl}\right)_f + \frac{\rho_m v dv}{g_c dl}$$

2.90

where: α = pipe inclination angle, measured from the horizontal, radians

In a vertical pipe, $\alpha = 90°$ and $\sin \alpha = 1.0$, hence $dl = dh$, and the previous equation can be rewritten in function of the vertical distance h as

$$\frac{dp}{dh} = \frac{g}{g_c}\rho_m + \left(\frac{dp}{dh}\right)_f + \frac{\rho_m v dv}{g_c dh}$$

2.91

As seen, pressure gradient dp/dl or dp/dh in multiphase flow is composed of three terms representing the different kinds of energy changes occurring in any pipe. Of these, the second one represents the irreversible pressure losses due to fluid friction, which is conveniently expressed with the Darcy-Weisbach equation (see Equation 2.41). In the order of their relative importance, the three components of the total pressure gradient are listed as follows for an inclined and a vertical pipe, respectively, using field measurement units:

$$\left(\frac{dp}{dl}\right)_{el} = \frac{1}{144}\rho_m\sin\alpha \qquad\qquad \left(\frac{dp}{dh}\right)_{el} = \frac{1}{144}\rho_m \qquad elevation\ term,$$

$$\left(\frac{dp}{dl}\right)_f = 1.294 \times 10^{-3} f \frac{\rho_m v^2}{d} \qquad\qquad \left(\frac{dp}{dh}\right)_f = 1.294 \times 10^{-3} f \frac{\rho_m v^2}{d} \qquad frictional\ term,\ and$$

$$\left(\frac{dp}{dl}\right)_{kin} = 2.16 \times 10^{-4} \frac{\rho_m v dv}{dl} \qquad\qquad \left(\frac{dp}{dh}\right)_{kin} = 2.16 \times 10^{-4} \frac{\rho_m v dv}{dh} \qquad kinetic\ term.$$

The hydrostatic component represents the change in potential energy in the multiphase flow due to the gravitational force acting on the mixture. It is usually found from the density of the mixture including the effect of gas slippage. This means the use of the mixture density formula based on the actual value of the liquid holdup (see Equation 2.80).

The frictional component stands for the irreversible pressure losses occurring in the pipe due to fluid friction on the pipe inner wall. As discussed earlier, these losses are difficult to describe since more than one phase contacts the pipe wall. Velocity distribution across the pipe is also different from those in single-phase flow, adding up to the difficulty of pressure drop prediction. These are the reasons why different investigators use widely different formulas for the calculation of frictional losses and for the determination of two-phase friction factors.

The kinetic or acceleration component of the pressure drop represents the kinetic energy changes of the flowing mixture and is proportional to the changes in flow velocity. It can, therefore, be significant only in cases when flow velocity undergoes great changes over a short section of the pipe. Such conditions can only exist at very high flow velocities in certain flow patterns. This is why the kinetic term is completely ignored by many investigators and is only considered in definite flow patterns by others.

The relative magnitude of the pressure gradient equation's three terms in the total pressure gradient may be different in various cases. In vertical or slightly inclined oil wells, acceleration is negligible and friction makes up only about 10%, the major part being the elevation term. In high-rate gas wells, however, friction losses constitute a greater part of the total pressure drop; and the kinetic term cannot be ignored, especially near the wellhead where small changes in flowing pressure can incur great changes in flow velocity.

In a completely horizontal pipe, elevation change is nil and most of the total pressure drop comes from frictional losses. In pipes or pipelines transporting a multiphase mixture with a low gas content, acceleration losses may be completely ignored. On the other hand, if the medium transported is mostly gas, kinetic energy losses have to be properly considered. In pipelines laid over hilly terrain or in offshore flowlines transporting a multiphase mixture, elevation changes cannot be neglected and the total pressure drop may include the effects of all three components.

2.5.3 Two-phase flow in oil wells

2.5.3.1 Background theories. This section covers the most important kind of multiphase flow problems encountered in production practice, *i.e.* flow in oil wells. The importance of a proper description of flow behavior in oil wells cannot be overemphasized because the pressure drop in the well is the greatest and most decisive part of the total pressure drop in the production system. Starting from the boundary of the drainage area, formation fluids first flow in the porous media surrounding the well and then enter the wellbore. From here, vertical or inclined flow occurs in the well until the wellstream reaches the surface. Finally, the surface-laid flowline transports well fluids to a separator holding a constant pressure. Along this flow path, flowing pressure of the wellstream continuously decreases because flow through every system component entails a different measure of pressure loss. If pressure drops in the reservoir, the well, and the flowline are compared, the greatest amount is always found to occur in the well. The reason for this is the great elevation difference from the well bottom to the wellhead and the consequently high hydrostatic term in the pressure gradient equation.

As discussed previously, two-phase flow in oil wells is a very important topic in production engineering in general. Gas lifting, in particular, heavily relies on the proper calculation of pressure drops in oil well tubing, making this topic of prime importance for anyone involved in gas lifting. Without a proper knowledge of multiphase behavior and without a complete command of the engineering tools predicting pressure drops in oil wells, no gas lift design or analysis is possible.

For the previous reasons, the present section gives a comprehensive treatment of the topic. First, background theories and concepts specific to oil well flow are detailed. Then, all available models for pressure drop calculation are described, covering the early empirical and the up-to-date mechanistic models in historical order. The once popular gradient curves, graphical presentations of the pressure traverse in oil wells, are also described and discussed. Finally, the accuracies obtainable from the different calculation models are analyzed and considerations on the production engineer's proper attitude toward those are described.

2.5.3.1.1 Vertical flow patterns. The flow patterns (flow regimes) in vertical pipes have long been visually observed in transparent pipe sections. Depending on the phase velocities and in-situ fluid parameters, the basic patterns early investigators defined are: bubble, slug, transition, and mist. They occur in the given sequence if gas flow rate is continuously increased for a constant liquid flow rate. This is very close to the conditions in an oil well because solution gas gradually escapes from the oil on its way to the surface. At or near the well bottom, therefore, single-phase liquid flow exists if gas is still in solution at that pressure. Higher up, solution gas gradually escapes, resulting in an increased gas flow velocity. Gas already liberated from the solution expands due to a reduction of flowing pressure in the upward direction, further increasing gas velocity. The continuous increase in gas velocity makes it happen that all possible flow patterns may be observed in a single well.

The following classification reflects the results of the latest research and follows the universally accepted method of flow pattern prediction proposed by Barnea [32]. Flow patterns are discussed in the natural order they occur in a vertical oil well, starting from the well bottom.

Bubble Flow

At low to medium gas flow velocities, the gas phase takes the form of uniformly distributed discrete bubbles rising in the continuous liquid phase (see Fig. 2–15). The bubbles (due to their low density) tend to overtake the liquid particles, and gas slippage occurs. Flowing density calculations for this pattern, therefore, must include the effects of slippage. Liquid is in contact with the pipe wall, making the calculation of frictional losses relatively simple.

Fig. 2–15 Bubble flow pattern.

The previous discussion is only valid for low gas and liquid velocities usually found in relatively large diameter tubing. If gas velocity is increased by an increased gas rate for the same low- to medium-liquid velocity, the individual small bubbles start to coalesce into larger *Taylor* bubbles and the flow pattern progressively changes into slug flow. This transition is found to occur at gas void fractions of $\varepsilon_g \geq 0.25$. If, on the other hand, liquid velocity is increased, the high turbulent forces break up bigger gas bubbles and, at the same time, prevent the coalescence of smaller ones. Thus, even above a gas void fraction of $\varepsilon_g = 0.25$, slug flow cannot develop, but transition from bubble into dispersed bubble flow will occur.

Dispersed Bubble Flow

If liquid velocity is relatively high at low gas velocities, the gas phase exists in very small bubbles evenly distributed in the continuous liquid phase moving at a high velocity. Since the gas bubbles are taken by the continuous liquid phase, both phases travel at the same velocity and no slippage occurs. The multiphase mixture behaves like a homogeneous phase. Therefore, mixture density for the dispersed bubble flow pattern is easily found from the no-slip liquid holdup, since $\varepsilon_l = \lambda_l$. Note that this case requires a very simplified treatment and it is close in nature to single-phase liquid flow (the liquid phase being in contact with the pipe wall) as before.

For higher gas velocities, dispersed bubble flow cannot be sustained because the increased number of gas bubbles becomes so tightly packed that coalescence occurs even at the existing high turbulence levels. The flow pattern then changes into churn flow at gas void fractions of about $\varepsilon_g \geq 0.52$,

Slug Flow

In slug flow (see Fig. 2–16), the continuous liquid phase present in bubble and dispersed bubble flows starts to diminish, and liquid slugs and large gas bubbles follow each other in succession. The large bullet-shaped elongated Taylor bubbles contain most of the gas phase and occupy almost the total pipe cross-sectional area. They are surrounded by a thin falling liquid film at the pipe wall. The Taylor bubbles are separated by liquid slugs occupying the full pipe cross section and containing small amounts of distributed gas bubbles. Even this basic description shows that mixture density and friction loss calculations necessitate a different approach as compared to bubble flows. As will be detailed in later sections, a *slug unit* consisting of a Taylor bubble and a liquid slug must be investigated for pressure drop calculations.

Fig. 2–16 Slug flow pattern.

If gas rate is increased in slug flow, the gas rate is increased, the Taylor bubbles become larger and, at the same time, the amount of dispersed gas bubbles in the liquid slug increases. When the gas void fraction reaches a critical value, the liquid slugs start to disintegrate and a new flow pattern, transition, or churn flow develops. This happens if the gas void fraction in the liquid slug is about $\varepsilon_g = 0.52$.

Transition (Churn) Flow

At gas rates greater than those in slug flow, both the Taylor bubbles and the liquid slugs are continuously destroyed by the high shear forces, and neither phase is continuous (see Fig. 2–17). As liquid slugs collapse, they are lifted by smaller Taylor bubbles of distorted shape. The liquid phase, therefore, makes an oscillatory motion in constantly alternating directions.

When gas rates further increase, a transition to annular flow takes place. This is governed by two mechanisms: liquid film stability and bridging. At lower liquid rates, increased gas velocities can stabilize the liquid film and annular flow can develop. At higher liquid rates, higher gas velocities are needed to overcome the bridging of the liquid phase and to start the onset of annular flow.

Fig. 2–17 Transition (churn) flow pattern.

Annular (Mist) Flow

At extremely high gas flow velocities, mist or annular flow occurs where the gas phase is continuous in the pipe core. Liquid is present at the pipe wall as a wavy liquid film and as small entrained droplets in the gas core (see Fig. 2–18). Liquid droplets entrained in the gas core are transported at the same high velocity as that of the gas particles eliminating slippage between the phases. Core density, therefore, can be calculated from the no-slip holdup, λ_l. Frictional losses occur on the interface of the gas core and the wavy liquid film covering the pipe wall, this makes the determination of film thickness of prime importance.

Fig. 2–18 Annular (mist) flow pattern.

Flow Pattern Maps

Flow patterns are visually observed and cataloged in laboratory conditions. In order to classify the conditions under which the different flow patterns occur, one usually employs flow pattern maps, *i.e.* graphical representations of the ranges of occurrence for each pattern. The first such map, describing vertical flow in oil wells, was proposed by Duns-Ros [33] who used two nondimensional parameter groups given as follows to describe the flow patterns:

$$N_{lv} = v_{sl} \sqrt[4]{\frac{\rho_l}{g \sigma_l}} = 1.938 v_{sl} \sqrt[4]{\frac{\rho_l}{\sigma_l}} \qquad\qquad 2.92$$

$$N_{gv} = v_{sg} \sqrt[4]{\frac{\rho_l}{g \sigma_l}} = 1.938 v_{sg} \sqrt[4]{\frac{\rho_l}{\sigma_l}} \qquad\qquad 2.93$$

The previous parameters are called the liquid and gas velocity number, respectively. Depending on their values and using the Duns-Ros flow pattern map (Fig. 2–19), one can find the actual flow pattern at any conditions. Use of this flow pattern map was widespread until recently but up-to-date mechanistic models use the most comprehensive prediction proposed by Barnea [32]. Figure 2–20 shows the map used by Kaya et al. [34], which is also based on Barnea's work and predicts the five vertical flow patterns described previously. It is important to note that in this flow pattern map, the boundaries of the different patterns are given in the function of the two phases' superficial velocities.

— · — · — · — · — · — · — · — · — · — · — · — · — · — · — · — · —

Example 2–21. Determine the flow patterns by the Duns-Ros method at the three points in the well given in Example 2–19.

Solution

The parameters calculated in the previous examples can be utilized to find the flow patterns on the Duns-Ros flow pattern map.

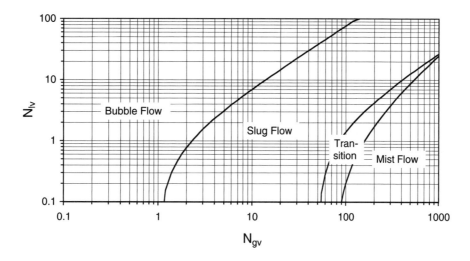

Fig. 2–19 Vertical flow pattern map according to Duns-Ros. [33]

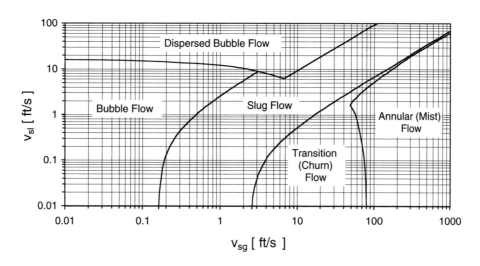

Fig. 2–20 Vertical flow pattern map as proposed by Kaya et al. [34]

For the wellhead case, dimensionless liquid velocity number is calculated from Equation 2.92:

$N_{lv} = 0.79 [49.9 / 8]^{0.25} = 2.42$.

Gas velocity number from Equation 2.93:

$N_{gv} = 1.93 [49.9 / 8]^{0.25} = 5.92$.

After plotting the previous coordinates on the flow pattern map, one finds the flow pattern to be slug.

The results of similar calculations for the rest of the cases are

Parameter	Wellhead	At 3,770 ft	Bottom
N_{lv} , -	2.42	2.82	2.97
N_{gv} , -	5.92	0.05	0
Flow pattern	Slug	Bubble	Single

2.5.3.1.2 Liquid holdup calculations. In vertical or inclined multiphase flow in oil wells, the greatest part of the pressure drop occurs due to the weight of the multiphase mixture present in the pipe that represents the hydrostatic term in the basic pressure drop equation (see Equation 2.90). In the majority of cases, the hydrostatic or elevation term makes up about 90% of the total pressure drop in the well. At extremely high liquid flow rates, of course, the contribution of fluid friction increases, but the total pressure drop is usually determined by this term.

As seen from the basic pressure drop equation (Equation 2.90), the elevation term includes the in-situ mixture density that, in turn, can be found from the actual liquid holdup. Therefore, a proper calculation of the hydrostatic pressure losses heavily relies on the determination of the proper value of the liquid holdup in the given pipe section. Due to its great importance, the proper calculation of liquid holdup is the strong foundation of any pressure drop prediction model.

As detailed in Two-Phase Flow Concepts (Section 2.5.2.1), there are two basic approaches to liquid holdup calculations: use of the bubble rise velocity (Equation 2.70) or the slip velocity formula (Equation 2.72). The ways different investigators find liquid holdup will be detailed later in the discussions of the individual pressure drop calculation models. Table 2–2 gives a summary of the different formulas and includes some of the most important correlations. Note that the table includes the parameters C_o, v_b, and v_s required for finding liquid holdup, ε_l.

Model	Correlation	C_0	v_b
Bubble Flow Pattern			
Duns-Ros			$v_s = S_I \sqrt[4]{\dfrac{g\sigma}{\rho_l}}$
Orkiszewski	Griffith		$v_s = 0.8$ ft/s
Aziz et.al.	Zuber et al.	1.2	$1.41 \sqrt[4]{\dfrac{\sigma g(\rho_l - \rho_g)}{\rho_l^2}}$
Chierici et al.	Griffith		$v_s = 0.8$ ft/s
Ansari et al.	Modified Harmathy	1.2	$1.53 \sqrt[4]{\dfrac{\sigma g(\rho_l - \rho_g)}{\rho_l^2}}\, \varepsilon_l^{0.5}$
Chokshi et al.	Taitel	1.18	$1.41 \sqrt[4]{\dfrac{\sigma g(\rho_l - \rho_g)}{\rho_l^2}}$
Hasan-Kabir	Harmathy	1.2	$1.5 \sqrt[4]{\dfrac{\sigma g(\rho_l - \rho_g)}{\rho_l^2}}$
Slug Flow Pattern			
Duns-Ros			$v_s = S_{II} \sqrt[4]{\dfrac{g\sigma}{\rho_l}}$
Orkiszewski	Griffith-Wallis	1.0	$C_1 C_2 \sqrt{gd}$
Aziz et al.	Wallis	1.2	$C \sqrt{\dfrac{gd(\rho_l - \rho_g)}{\rho_l}}$
Chierici et al.	Griffith-Wallis Nicklin et al.	1.0	$C_1 C_2 \sqrt{gd}$
Hasan-Kabir	Wallis	1.2	$C \sqrt{\dfrac{gd(\rho_l - \rho_g)}{\rho_l}}$

Table 2–2 Gas phase or slippage velocity correlations used in some of the two-phase vertical pressure drop calculation models.

2.5.3.1.3 Frictional and kinetic losses. Frictional pressure losses in vertical or inclined oil wells are usually rather low when compared to the hydrostatic or elevation pressure drop. In wells producing medium to high liquid rates, frictional drops amount to a maximum of 10% of the total pressure drop. This number, of course, increases for extremely high liquid production rates.

Due to its limited impact, the calculation of frictional pressure losses does not play a central role in most of the pressure drop prediction models. This treatment is justified by the fact that even great calculation errors in the frictional drop do not bring about comparable errors in the total pressure drop. In contrast, small inaccuracies in the hydrostatic term can cause relatively large errors in the calculated pressure drop. For this reason, most investigators concentrate on the accurate prediction of the elevation term and treat the frictional term less thoroughly.

Investigation of the frictional term in the pressure gradient equation (Equation 2.90) reveals that the three basic parameters determining frictional pressure drop are: friction factor f, mixture density ρ_m, and mixture velocity v_m. Friction factor determination involves the finding of the Reynolds number, N_{Re}. The parameters listed here are developed by some of the investigators as given in Table 2–3. As seen in the bubble and slug flow patterns, there is a great variety of opinions on the proper calculation of the necessary parameters. Mist flow is not included because of its low importance in oil well flow.

Model	N_{Re}	f	ρ	v^2
Bubble Flow Pattern				
Duns-Ros	$\dfrac{\rho_l v_{sl} d}{\mu_l}$	Moody + correction	ρ_l	$v_{sl}^2\left(1+\dfrac{v_{sg}}{v_{sl}}\right)$
Orkiszewski	$\dfrac{\rho_l v_{sl} d}{\varepsilon_l \mu_l}$	Moody	ρ_l	$\dfrac{v_{sl}^2}{\varepsilon_l^2}$
Aziz et al.	$\dfrac{\rho_l v_m d}{\mu_l}$	Moody	$\varepsilon_g \rho_g + (1-\varepsilon_g)\rho_l$	v_m^2
Chierici et al.	$\dfrac{\rho_l v_l d}{\mu_l}$	Moody	ρ_l	v_{sl}^2
Ansari et al.	$\dfrac{\rho_m v_m d}{\mu_m}$	Moody	$\varepsilon_g \rho_g + (1-\varepsilon_g)\rho_l$	v_m^2
Chokshi et al.	$\dfrac{\rho_m v_m d}{\mu_m}$	Moody	$\varepsilon_g \rho_g + (1-\varepsilon_g)\rho_l$	v_m^2
Hasan-Kabir	$\dfrac{\rho_m v_m d}{\mu_m}$	Moody	$\varepsilon_g \rho_g + (1-\varepsilon_g)\rho_l$	v_m^2
Slug Flow Pattern				
Duns-Ros	$\dfrac{\rho_l v_{sl} d}{\mu_l}$	Moody + correction	ρ_l	$v_{sl}^2\left(1+\dfrac{v_{sg}}{v_{sl}}\right)$
Orkiszewski	$\dfrac{\rho_l v_m d}{\mu_l}$	Moody	ρ_l	$v_m^2\left[\dfrac{v_l+v_b}{v_m+v_b}+\Gamma\right]$
Aziz et al.	$\dfrac{\rho_l v_m d}{\mu_l}$	Moody	$\varepsilon_g \rho_g + (1-\varepsilon_g)\rho_l$	$v_m^2\left[\dfrac{l_s}{l_b+l_s}\right]$
Chierici et al.	$\dfrac{\rho_l v_m d}{\mu_l}$	Moody	ρ_l	$v_m^2(1-\varepsilon_g)$
Ansari et al.	$\dfrac{\rho_m v_m d}{\mu_{LS}}$	Moody	$\rho_{LS}(1-\beta)$	v_m^2
Chokshi et al.	$\dfrac{\rho_{LS} v_m d}{\mu_{LS}}$	Moody	$\rho_{LS}(1-\beta)$	v_m^2
Hasan-Kabir	$\dfrac{\rho_m v_m d}{\mu_m}$	Moody	$\varepsilon_l \rho_l$	v_m^2

Table 2–3 Details of frictional pressure drop calculations used in some of the two-phase vertical pressure drop calculation models.

Kinetic or acceleration losses represent changes in the kinetic energy of the flowing mixture and are even less pronounced than friction losses. Since kinetic energy is proportional to the square of flow velocity, the acceleration term of the basic equation (Equation 2.90) can only be significant if rapid flow velocity changes occur over short sections of the flow path. This does not normally happen in bubble or slug flow where acceleration effects are usually neglected. Most investigators completely ignore the kinetic term; others consider it in mist flow only. In the mist flow pattern, usually present in gas wells producing a wet gas phase, kinetic effects cannot be neglected. This is especially true near the wellhead where flowing pressure rapidly decreases, causing the rapidly expanding mixture's velocity to undergo large changes.

2.5.3.2 Inclined and annulus flow. The majority of calculation models developed for two-phase flow in oil wells assumes a vertical well trajectory. This approach usually is justified in wells having only small deviations from the vertical. Many wells, however, are either intentionally or unintentionally drilled as deviated wells. In such cases, the following two basic problems arise: the flow path significantly increases (as compared to the vertical case) and flow patterns change with changes in pipe inclination).

In case of slightly deviated wells, a simple approximate solution is applied that accounts for the change in flow path length only. Here, the use of the proper pressure gradient equation (Equation 2.90) automatically accounts for pipe inclination in the hydrostatic and frictional pressure drop terms. Note that this solution presupposes that well inclination is very close to the vertical and that flow patterns valid at vertical flow are present.

For larger inclinations associated with directionally drilled wells, the previous solution is ineffective and can result in large calculation errors because the predicted flow pattern is a function of pipe angle. As the pipe is inclined from the vertical, the flow pattern map's boundaries gradually change and new flow patterns can also arise. The most significant change is the emergence of the stratified flow patterns where the two phases flow more or less independently of each other. These changes, of course, must be properly followed and described in the pressure drop calculation model.

The effect of pipe inclination on the prevailing flow pattern was first investigated by Gould et al. [35], who made a qualitative study of the problem. Present-day pressure drop prediction models developed for deviated wells try to properly include the effects of well inclination on the occurrence of flow patterns. Most of them utilize the unified flow pattern model of Barnea [32] valid for any pipe angle from the vertical to the horizontal. The most important models will be detailed in later sections.

The flow of a two-phase mixture in a well's casing-tubing annulus poses another special case. For many years, flow in annuli was treated like pipe flow by utilizing the hydraulic diameter concept. This parameter is frequently used in solving single-phase flow problems and is defined as four times the cross-sectional area divided by the wetted perimeter of the annulus.

$$d_h = 4 \frac{\dfrac{D^2\pi}{4} - \dfrac{d^2\pi}{4}}{D\pi + d\pi} = D - d \qquad\qquad 2.94$$

where: D = casing inside diameter (ID), in.

d = tubing outside diameter (OD), in.

The casing-tubing annulus, however, cannot completely be characterized by the hydraulic diameter alone because it can have a concentric or eccentric configuration, depending on the position of the tubing inside the casing string. Flow patterns in well annuli differ from those present in pipes especially in eccentric configurations. At the present, only a few pressure drop prediction models are available for annulus flow, and the reader is directed to the excellent book of Brill and Mukherjee [36] where a complete treatment of the topic can be found.

2.5.3.3 Empirical correlations.

2.5.3.3.1 Introduction. Early empirical models treated the multiphase flow problem as the flow of a homogeneous mixture of liquid and gas. This approach completely disregarded the slippage between the phases. Because of the poor physical model adopted, calculation accuracy was low. Later, the evolution of the calculation models brought about the appearance of empirical liquid holdup correlations to account for the slippage between the phases. More advanced models tried to include also the effects of the different spatial arrangements of the two phases, *i.e.* the flow patterns. All these improvements, however, could not significantly improve the inadequacies of the homogeneous model.

Empirical two-phase pressure drop correlations were first classified by Orkiszewski [37] according to the next criteria, later adopted by others. [38]

Group I

- Slip between phases is not considered, both phases are assumed to travel at the same velocity.

- Flow patterns are not distinguished, general formulas are given for mixture density and friction factor determination.

- Pressure losses are described by a single energy loss factor.

Group II

- Flowing density calculations include the effects of slippage.

- No flow patterns are distinguished.

Group III

- Flow patterns are considered and calculation of mixture density and friction factor varies with the flow pattern.

- Flowing mixture density calculations include the effects of slippage.

Some of the better-known correlations can be classified into the previous groups as follows.

Group	Correlation	Reference
I	Poettmann-Carpenter	39
	Baxendell	40
	Baxendell-Thomas	41
	Fancher-Brown	42
	Gaither et al.	43
	Hagedorn-Brown I	44
	Cornish	45
II	Hagedorn-Brown II	46
III	Duns-Ros	33
	Orkiszewski	37
	Beggs-Brill	47
	Mukherjee-Brill	48

In the following part of this section, the most important empirical correlations used for prediction of the pressure drop in oil wells are described in the order of their publication.

2.5.3.3.2 Poettmann-Carpenter and related correlations.

Summary

This was the first practical calculation model for vertical two-phase flow. It disregards slippage losses and the effects of viscosity and does not consider flow patterns. It was very popular in the 1960s and 1970s and many pressure traverse curve sets (gradient curves) were based on its use.

Poettmann and Carpenter [39] developed their correlation for vertical two-phase flow based on 49 pressure drop measurements. Their conceptual model ignores the existence of flow patterns and assumes that all pressure losses can be correlated by a friction-type term. Since slippage and acceleration losses are disregarded, the no-slip mixture density is used in the basic equation. Starting from Equation 2.91, valid for a vertical pipe section, after leaving out the kinetic term, and expressing pressure gradient in psi/ft units we get:

$$\frac{dp}{dh} = \frac{1}{144}\,\rho_{ns} + 1.294 \times 10^{-3}\,f\,\frac{\rho_{ns}v_m^2}{d} \qquad\qquad 2.95$$

where: ρ_{ns} = no-slip mixture density, lb/cu ft

f = energy loss factor -

v_m = mixture velocity, ft/s

d = pipe diameter, in.

Based on their field measurements, the authors back-calculated the friction—or more properly called *energy loss factor f*—figuring in their basic equation and tried to correlate it with the flow parameters. The final energy loss factor correlation is shown in Figure 2–21 where the abscissa is a viscosity-less Reynolds number.

For a long time until the early 1960s, this was the only available correlation for calculating the pressure drop in oil wells. At those times, many gas lift handbooks contained gradient curve sheets developed with the use of this correlation. Today, the use of the Poettmann-Carpenter correlation is outdated because, due to its theoretical shortcomings, calculation accuracy is low. Satisfactory results can be expected for high-rate wells only where a no-slip condition is present.

Several attempts were made to increase the accuracy of the Poettmann-Carpenter correlation while preserving its basic principles. Baxendell and Thomas [41] used data from high-rate oil wells to modify the original *f*-curve. They plotted the energy loss factor as given in Figure 2–22. Fancher and Brown [42] made extensive measurements in an 8,000 ft test well to develop a new energy loss factor correlation. They found that *f* values greatly depended on the well's gas-liquid ratio (GLR) and presented three different curves for three ranges of the well's producing GLR, as seen in Figure 2–23.

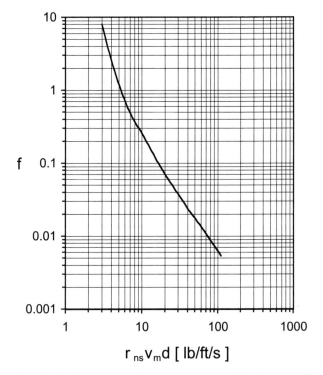

Fig. 2–21 Energy loss factor according to Poettmann and Carpenter. [39]

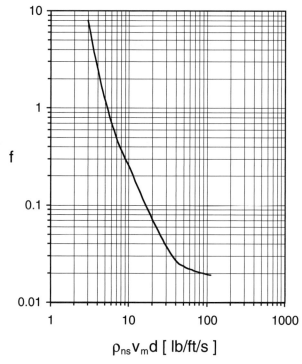

Fig. 2–22 Energy loss factor according to Baxendell and Thomas. [41]

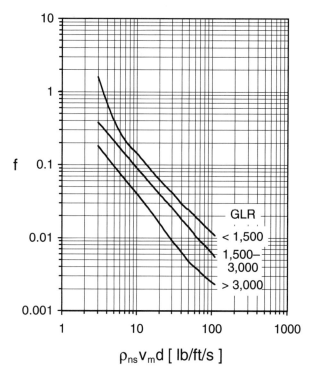

Fig. 2–23 Energy loss factor according to Fancher and Brown. [42]

Example 2–22. Calculate the pressure gradients according to Poettmann and Carpenter, Baxendell and Thomas, and Fancher and Brown for the three depths in the well described in Example 2–19.

Solution

Most of the parameters needed for the solution were calculated in previous examples and will be used in the following. One point for the Poettmann-Carpenter method is described in detail.

To find the no-slip mixture density ρ_{ns}, first the no-slip liquid holdup is calculated from Equation 2.73:

$\lambda_l = 0.79 / (0.79 + 1.93) = 0.29$.

No-slip mixture density is determined according to Equation 2.79:

$\rho_{ns} = 49.9 \ 0.29 + 1.79 \ (1 - 0.29) = 15.77$ lb/cu ft.

The abscissa of the energy loss factor correlation (Fig. 2–21) is found as

$\rho_{ns} \ v_m \ d = 15.77 \ (0.79 + 1.93) \ 2.441 / 12 = 8.73$ lb/ft/s.

The f value read off Figure 2–21 is

$f = 0.32$.

The pressure gradient at this point is found from Equation 2.95:

$dp/dh = 15.77 / 144 + 1.294 \times 10^{-3} \ 0.32 \ 15.77 \ (0.79 + 1.93)^2 / 2.442 = 0.1095 + 0.0019 = 0.129$ psi/ft.

After performing similar calculations for the rest of the cases, the following results were received:

Parameter	Wellhead	At 3,770 ft	Bottom
Poettmann-Carpenter			
f, -	0.32	0.32	0.32
gradient, psi/ft	0.129	0.318	0.316
Baxendell-Thomas			
f, -	0.31	0.31	0.31
gradient, psi/ft	0.129	0.317	0.316
Fancher-Brown			
f, -	0.17	0.17	0.17
gradient, psi/ft	0.120	0.314	0.313

2.5.3.3.3 Duns-Ros correlation.

Summary

Based on an extensive laboratory project in vertical flow where liquid holdup was actually measured, the authors used dimensional analysis to derive their correlation. Duns and Ros were the first to distinguish flow patterns and to develop different correlations for each pattern. Kinetic effects are considered in mist flow only.

Although originally published in 1961 by Ros [49], this correlation is conventionally called the Duns-Ros model, after their paper. [33]

The Duns-Ros calculation model is based on a very extensive laboratory project conducted and funded by the Shell Company. The experimental pipe was a vertical flow string of about 180 ft length with diameters of 3.2 cm, 8.02 cm, or 14.23 cm. The authors noted that above the point of injection of liquid and gas into the vertical pipe, pressure gradients much greater than those higher above can be observed due to entrance effects. Such effects are also present in real wells but are negligible because of the great depths involved. To eliminate entrance effects, a pipe section of at least three times the length of the measuring section (33 ft) was installed upstream of the measuring pipe. Measurements were carried out at atmospheric pressure using air, water, and various oils.

At closely controlled conditions, liquid and gas rates, flowing temperature, and pressure at several points were measured in a total of 4,000 tests. The actual flow pattern was observed and recorded in the transparent measuring section. Duns-Ros were the first to actually measure in their experiments the liquid holdup using a radioactive tracer material mixed into the liquid.

Processing of their extensive experimental data base (a total of more than 20,000 pressure gradient data) required a comprehensive theoretical treatment of the flow phenomena. Duns-Ros were the first to perform a dimensional analysis of multiphase flow and showed that a total of nine dimensionless groups could be established, based on the 12 most important independent variables. Out of these, 4 groups were identified as significant. As will be seen later, the authors correlated their experimental data almost solely using these 4 dimensionless groups:

$$N_{lv} = v_{sl} \sqrt[4]{\frac{\rho_l}{g \sigma_l}} = 1.938 \, v_{sl} \sqrt[4]{\frac{\rho_l}{\sigma_l}} \quad \text{liquid velocity number} \qquad 2.96$$

$$N_{gv} = v_{sg} \sqrt[4]{\frac{\rho_l}{g\sigma_l}} = 1.938 v_{sg} \sqrt[4]{\frac{\rho_l}{\sigma_l}} \qquad \text{gas velocity number}$$

2.97

$$N_d = d \sqrt{\frac{\rho_l g}{\sigma_l}} = 10.1 d \sqrt{\frac{\rho_l}{\sigma_l}} \qquad \text{pipe diameter number}$$

2.98

$$N_l = \mu_l \sqrt[4]{\frac{g}{\rho_l \sigma_l^3}} = 0.157 \mu_l \sqrt[4]{\frac{1}{\rho_l \sigma_l^3}} \qquad \text{liquid viscosity number}$$

2.99

where: ρ_l = liquid density, lb/cu ft

σ_l = interfacial tension, dyne/cm

v_{sl} = liquid superficial velocity, ft/s

v_{sg} = gas superficial velocity, ft/s

μ_l = liquid viscosity, cP

d = pipe diameter, in.

Duns and Ros were the first to consider the occurrence of flow patterns they observed in their experimental equipment. They also showed that flow phenomena in the different flow patterns are completely different from each other. In order to properly describe multiphase flow, therefore, one must use different calculation models in the different flow patterns. This is the approach still adopted by all present-day investigators of multiphase flow.

Basic Equation

Originally, Duns-Ros presented all their relevant equations in nondimensional forms and defined their basic equation for pressure drop prediction in a dimensionless form as well. When deriving their basic equation, the kinetic term of the pressure gradient equation (Equation 2.91) can be expressed as follows:

$$\left(\frac{dp}{dh} \right)_{kin} = (\rho_l v_{sl} + \rho_g v_{sg}) \frac{v_{sg}}{144 \, p \, g_c} \frac{dp}{dh}$$

2.100

Substituting into the general pressure drop equation and solving for the pressure gradient we get

$$\frac{dp}{dh} = \frac{\left(\dfrac{dp}{dh} \right)_{el} + \left(\dfrac{dp}{dh} \right)_f}{1 - (\rho_l v_{sl} + \rho_g v_{sg}) \dfrac{v_{sg}}{144 \, p \, g_c}} = \frac{\left(\dfrac{dp}{dh} \right)_{el} + \left(\dfrac{dp}{dh} \right)_f}{1 - E_k}$$

2.101

where: E_k = dimensionless kinetic term

According to Duns-Ros, the acceleration or kinetic term is significant in mist flow only. This means that in bubble and slug flow the value of the dimensionless kinetic term E_k equals zero.

Flow Pattern Map

The flow pattern map used by the authors is given in Figure 2–19. The coordinate axes are the gas and liquid velocity numbers defined before. As seen, four flow patterns are distinguished: bubble, slug, transition, and mist. The boundaries of the flow patterns (shown in the figure for air-oil flow in an 8-cm pipe) are found from the following equations:

$$N_{gv} = L_1 + L_2 N_{lv} \qquad \textit{bubble-slug boundary} \qquad\qquad 2.102$$

$$N_{gv} = 50 + 36 N_{lv} \qquad \textit{slug-transition boundary} \qquad\qquad 2.103$$

$$N_{gv} = 75 + 84 N_{lv}^{0.75} \qquad \textit{transition-mist boundary} \qquad\qquad 2.104$$

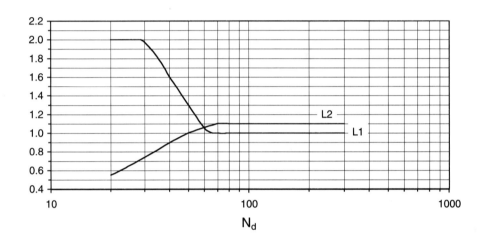

Fig. 2–24 Bubble-slug flow pattern transition parameters.

Functions L_1 and L_2 previously are evaluated from Figure 2–24 in the function of the pipe diameter number N_d.

Liquid Holdup Calculation Basics

According to Duns-Ros' general approach, measured liquid holdup data were correlated by using the dimensionless groups defined previously as independent, and the dimensionless slip velocity number as the dependent parameter. The latter is defined as follows:

$$S = v_s \sqrt[4]{\frac{\rho_l}{g\sigma_l}} = 1.938 v_s \sqrt[4]{\frac{\rho_l}{\sigma_l}} \qquad\qquad 2.105$$

After calculating the dimensionless slip velocity number, the real slip velocity v_s can be expressed from the previous equation:

$$v_s = 0.52 S \sqrt[4]{\frac{\rho_l}{\sigma_l}} \qquad\qquad 2.106$$

As derived in Two-Phase Flow Concepts (Section 2.5.2.1), knowledge of the slip velocity enables one to find liquid holdup (see Equation 2.78) from the formula reproduced here:

$$\varepsilon_l = \frac{v_s - v_m + \sqrt{(v_m - v_s)^2 + 4 v_s v_{sl}}}{2 v_s} \qquad\qquad 2.107$$

In the knowledge of the liquid holdup, mixture density ρ_m is easily calculated using the basic Equation 2.80, *i.e.* $\rho_m = \rho_l \, \varepsilon_l + \rho_g \, (1 - \varepsilon_l)$.

The previous general calculation procedure is used in the bubble and slug flow patterns. Mist flow requires a simplified treatment as will be detailed later.

Bubble Flow

The dimensionless slip velocity number for bubble flow is found from the formula

$$S = F_1 + F_2 N_{lv} + F_3' \left(\frac{N_{gv}}{1 + N_{lv}} \right)^2 \qquad\qquad 2.108$$

Functions F_1, F_2, F_3, and F_4 are found from the correlations given in Figure 2–25 and Figure 2–26, and F_3' is evaluated from

$$F_3' = F_3 - \frac{F_3}{N_d} \qquad\qquad 2.109$$

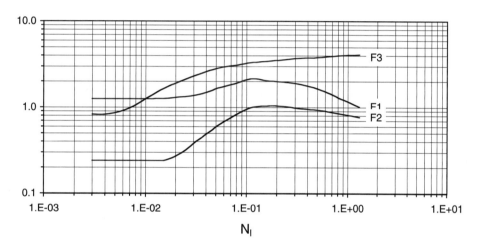

Fig. 2–25 Duns-Ros parameters F_1, F_2, F_3.

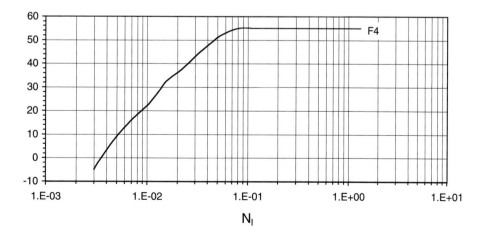

Fig. 2–26 Duns-Ros parameter F_4.

The frictional pressure gradient is found from the Darcy-Weisbach equation (Equation 2.41) and expressed in customary field units:

$$\left(\frac{dp}{db}\right)_f = 1.294 \times 10^{-3} f \frac{\rho_l v^2_{sl}}{d}\left(1 + \frac{v_{sg}}{v_{sl}}\right)$$

2.110

where:

ρ_l = liquid density, lb/cu ft

f = friction factor

v_{sl} = liquid superficial velocity, ft/s

v_{sg} = gas superficial velocity, ft/s

d = pipe diameter, in.

The friction factor f in the previous equation is a combination of the Moody friction factor f_1, calculated for a Reynolds number of

$$N_{Re} = 124 \frac{\rho_l v_{sl} d}{\mu_l}$$

2.111

where: μ_l = liquid viscosity, cP.

and factor f_2, which corrects for the GLR and is found from Figure 2–27; and another factor f_3, important for greater liquid viscosities and calculated from the formula as follows:

$$F_3 = 1 + \frac{f_1}{4}\sqrt{\frac{v_{sg}}{50 v_{sl}}}$$

2.112

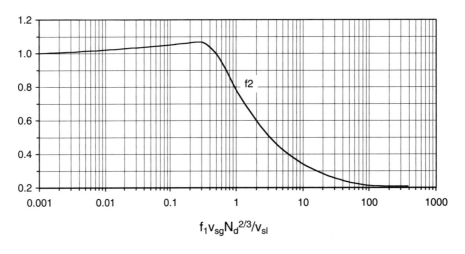

Fig. 2–27 Duns-Ros friction factor parameter f_2.

Finally, the friction factor to be used in Equation 2.109 is expressed as:

$$f = f_1 - \frac{f_2}{f_3}$$

2.113

Slug Flow

In slug flow, the authors used the same calculation model for the determination of friction losses as in bubble flow. The calculation of liquid holdup follows the same general procedure, but the nondimensional slip velocity is evaluated from

$$S = (1+F_5) \; \frac{N_{gv}^{0.982} + F_6'}{(1+F_7 \, N_{lv})^2}$$

2.114

In this expression, parameters F_5 and F_7 are given in Figure 2–28, F_6 in Figure 2–29, and F_6' is found from the formula

$$F_6' = 0.029 \, N_d + F_6$$

2.115

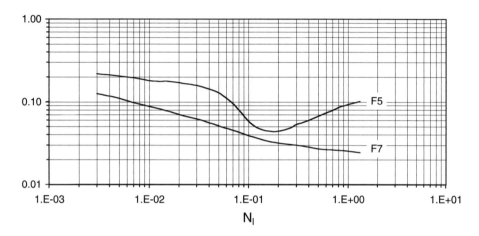

Fig. 2–28 Duns-Ros parameters F_5, F_7.

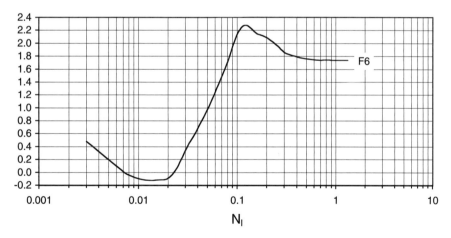

Fig. 2–29 Duns-Ros parameter F_6.

Mist Flow

In mist flow, according to the concept of Duns-Ros, the high velocity gas stream contains and transports the liquid phase in the form of evenly dispersed small droplets. Phase velocities thus being practically equal, no gas slippage occurs and the mixture density equals the no-slip value given as follows. This is the density to be used in the elevation term of the pressure gradient equation:

$$\rho_m = \rho_l \lambda_l + \rho_g (1 - \lambda_l)$$

2.116

where: $\lambda_l = v_{sl}/v_m$ no-slip liquid holdup

Calculation of the frictional pressure gradient in mist flow is complicated by the fact that the gas stream flowing in the core of the pipe contacts a liquid film covering the pipe's inside wall. The liquid film can develop waves due to the drag of the high-velocity gas. These waves define the roughness to be used in the calculation of the friction factor. The frictional term is written for the gas phase as follows:

$$\left(\frac{dp}{dh}\right)_f = 1.294 \times 10^{-3} \; f \frac{\rho_g v^2_{sg}}{d}$$

2.117

where: ρ_g = gas density, lb/cu ft

f = friction factor

v_{sg} = gas superficial velocity, ft/s

d = pipe diameter, in.

The Reynolds number is also written for the gas phase:

$$N_{Re} = 124 \frac{\rho_g v_{sg} d}{\mu_g}$$

2.118

where: μ_g = gas viscosity, cP

Duns and Ros found that the roughness of the liquid film is governed by two dimensionless numbers, the Weber, and a viscosity number defined as follows:

$$N_{We} = \frac{\rho_g v^2_{sg} \varepsilon}{\sigma_l}$$

2.119

$$N_\mu = \frac{\mu_l^2}{\rho_l \sigma_l \varepsilon}$$

2.120

where: ρ_g = gas density, lb/cu ft

v_{sg} = gas superficial velocity, ft/s

σ_l = interfacial tension, dyne/cm

μ_l = liquid viscosity, cP

ε = liquid wave absolute roughness, in.

The relative roughness of the liquid film covering the pipe wall is never smaller than the value valid for the pipe. Its maximum value can reach 0.5, when two opposing waves touch each other and gas flow is choked. Between these two extremes, Duns and Ros proposed the use of the following formulas to find the relative roughness of the liquid waves:

$$k = \frac{0.0749 \sigma_l}{\rho_g v^2_{sg} d}$$
for the range $\quad N_{We} N_\mu \leq 0.005 \quad$ 2.121

$$k = \frac{0.3731 \sigma_l}{\rho_g v^2_{sg} d} (N_{We} N_\mu)^{0.302}$$
for the range $\quad N_{We} N_\mu \leq 0.005 \quad$ 2.122

where: $\quad k$ = relative roughness of the liquid wave

If the equations result in a relative roughness above the maximum value of $k = 0.05$ found in the Moody diagram (Fig. 2–8), the following extrapolation can be used:

$$f = 4 \left\{ \frac{1}{[4\log(0.27\,k)]^2} + 0.067\,k^{1.73} \right\}$$
2.123

In mist flow, acceleration or kinetic effects cannot be ignored and the total pressure gradient should be calculated with the use of Equation 2.101.

Transition Flow

As seen in Figure 2–19, there is a transition between slug and mist flows often called the *transition region*. This is where liquid slugs are continuously destroyed by the high velocity gas flow, finally leading to an annular or mist flow pattern. For the calculation of pressure gradient, Duns and Ros proposed a weighted average of the gradients found for slug and mist flows. If the gradients for slug and mist flow are known, the gradient for the transition region is calculated from:

$$\frac{dp}{dh} = B \left(\frac{dp}{dh} \right)_{slug} + (1 + B) \left(\frac{dp}{dh} \right)_{mist}$$
2.124

The weighting factor B represents a linear interpolation between the boundaries mist-transition and slug-transition and can be calculated from the formula:

$$B = \frac{(75 + 84\,N_{lv}^{0.75}) - N_{gv}}{(75 + 84\,N_{lv}^{0.75}) - (50 + 36\,N_{lv})}$$
2.125

— - — - — - — - — - — - — - — - — - — - — - — - — - — - — - — - — - — - —

Example 2–23. Calculate the pressure gradients with the Duns-Ros procedure for the three depths in the well described in Example 2–19.

Solution

The most important flow parameters at the three points were calculated in previous examples and the wellhead case is described in detail.

First, the flow pattern must be determined. Liquid and gas velocity numbers from Example 2–21 are

$N_{lv} = 2.42$, and $N_{gv} = 5.92$.

For the determination of the bubble-slug boundary, variables L_1 and L_2 must be found in the function of the pipe diameter number defined by Equation 2.98:

$N_d = 10.1\ 2.441\ (49.9\,/\,8)^{0.5} = 61.6.$

From Figure 2–24: L_1 = 1.08 and L_2 = 1.02.

The boundary equation from Equation 2.102:

N_{gv} = 1.08 + 1.02 2.42 = 3.55, flow is slug because the actual value of N_{gv} = 5.92 is greater than this.

Liquid holdup calculations follow and liquid viscosity number is calculated from Equation 2.99 by assuming a liquid viscosity of 3 cP:

N_l = 0.157 3 / (49.9 8^3)$^{0.25}$ = 0.037

Factors F_5 ... F_7 are read off from Figures 2–28 and 2–29:

F_5 = 0.13; F_6 = 0.67; F_7 = 0.054.

Factor F_6' is evaluated from Equation 2.115:

F_6' = 0.029 61.6 + 0.67 = 2.456.

Now, Equation 2.114 is used to find the dimensionless slip velocity:

S = (1 + 0.13) (5.920.982 + 2.456) / (1 + 0.054 2.42) = 7.23.

Slip velocity and from Equation 2.106:

v_s = 0.52 7.23 (8 / 49.9)$^{0.25}$ = 2.38 ft/s

Equation 2.107 is used to calculated liquid holdup:

$$\varepsilon_l = \frac{2.38 - 0.79 - 1.93 + \sqrt{(0.79 + 1.93 - 2.38)^2 + 4\ 2.38\ 0.79}}{2\ 2.38} = 0.509$$

Flowing mixture density is expressed with the liquid holdup from Equation 2.80:

ρ_m = 0.509 49.9 + (1- 0.509) 1.79 = 26.29 lb/cu ft.

Elevation pressure gradient is easily found:

$(dp/db)_{el}$ = 26.29 / 144 = 0.183 psi/ft.

Calculation of the frictional pressure gradient follows, and the Reynolds number defined by Equation 2.111 is determined first:

N_{Re} = 124 49.9 0.79 2.441 / 3 = 3979.

Friction factor f_1 is read off from Figure 2–8 as f_1 = 0.04.

To find f_2, first the abscissa of Figure 2–27 is calculated as 1.52, than f_2 is read off as f_2 = 0.63.

Another correction factor is found from Equation 2.112:

f_3 = 1 + 0.04 (1.93 / 50 / 0.79)$^{0.5}$ / 4 = 1.0.

Finally, friction factor is evaluated by Equation 2.113:

f = 0.04 0.63 / 1 = 0.025.

Frictional pressure gradient is defined by Equation 2.110:

$(dp/dh)_f$ = 0.001294 0.025 49.9 0.79^2 / 2.441 (1 + 1.93 / 0/79) = 0.001 psi/ft.

Since Duns and Ros neglect the kinetic term in slug flow, total pressure gradient is the sum of the elevation and frictional components:

dp/dh = 0.183 + 0.001 = 0.184 psi/ft.

For the other points in the example well, only the final results are given as follows.

Parameter	Wellhead	At 3,770 ft	Bottom
N_{lv}	2.42	2.82	2.87
N_{gv}	5.92	1.62	0.0
N_{gv} boundary	3.55	4.05	-
flow pattern	Slug	Bubble	Single
F_1, F_2, F_3, F_3', F_4	-	1.4; 0.5; 2.5; 1.68; 48	-
F_5, F_6, F_6', F_7	0.13; 0.67; 246; 0.054	-	-
S	7.23	2.81	-
slip velocity, ft/s	2.38	0.95	-
liquid holdup, -	0.51	0.99	1.0
mixture density, lb/cu ft	26.3	45.0	44.5
elevation gradient, psi/ft	0.183	0.313	0.309
Reynolds number, -	3979	4316	4329
f_1, f_2, f_3	0.04; 0.63; 1.0	0.04; 1.01; 1.0	0.036; -; -
Friction factor, -	0.025	0.037	0.036
Frictional gradient, psi/ft	0.001	0.0008	0.0008
pressure gradient, psi/ft	0.184	0.314	0.310

_._ _._

2.5.3.3.4 _Hagedorn-Brown correlation._

Summary

The correlation is based on an extensive number of measurements in a 1,500 ft deep test well. Flow patterns are not distinguished, and liquid holdup is a correlating parameter only. Due to the weaknesses of the original model, only the modified correlation is used today.

The first paper of Hagedorn-Brown [44] examined the effect of viscosity on pressure drop and developed an energy loss factor correlation similar to that of Poettmann-Carpenter. In their later paper [46], they reported on a large-scale experimental project to investigate vertical multiphase flow. The authors conducted their measurements in a 1,500 ft deep vertical well and used three different industrial tubing sizes: 1 in., $1\frac{1}{4}$ in., and $1\frac{1}{2}$ in. Water and crude oils with different viscosities were used along with air as the gas phase. Flowing pressure was measured along the tubing length at four depths. An extensive number of experiments (475 test runs) was performed and the authors also used the 106 test data of Fancher-Brown [42]. The resulting pressure drop calculation model became one of the more successful theories of vertical flow for many decades.

Basic Equation

The authors considered the kinetic component of the pressure drop to be significant and used the following basic equation to express pressure gradient in psi/ft units:

$$\frac{dp}{dh} = \frac{1}{144} \rho_m + 1.294 \times 10^{-3} f \frac{\rho_{ns}^2 v_m^2}{\rho_m d} + 2.16 \times 10^{-4} \rho_m \frac{v_m dv_m}{dh} \qquad 2.126$$

where: ρ_{ns} = no-slip mixture density, lb/cu ft

ρ_m = mixture density, lb/cu ft

f = friction factor -

v_m = mixture velocity, ft/s

d = pipe diameter, in.

Hagedorn and Brown did not consider the effects of flow patterns, hence they proposed a simplified calculation scheme independent of the prevailing flow pattern. Since liquid holdup was not a measured parameter, the authors had to devise a procedure to back-calculate its value. Starting from the measured pressure drop, frictional losses were calculated first using single-phase procedures. The difference of the measured and calculated frictional losses was attributed to the elevation gradient including slippage losses. Using this approach, the authors developed their liquid holdup correlation as detailed later.

Liquid Holdup

As discussed previously, the liquid holdup correlation of Hagedorn-Brown does not reflect actual holdup measurements; the holdup values they use are considered correlating parameters only. To describe their data, the authors used the same four dimensionless parameters as Duns and Ros (see Equations 2.96–2.99).

The function ε_l/ψ is found from Figure 2–30, where p and p_{sc} are the actual and the standard pressures, respectively. The abscissa contains the parameter CN_l to be read from Figure 2–31. Finally, parameter ψ is read from Figure 2–32. Liquid holdup then is calculated as

$$\varepsilon_l = \frac{\varepsilon_l}{\psi} \psi \qquad 2.127$$

Based on the liquid holdup, mixture density is easily found again from Equation 2.80.

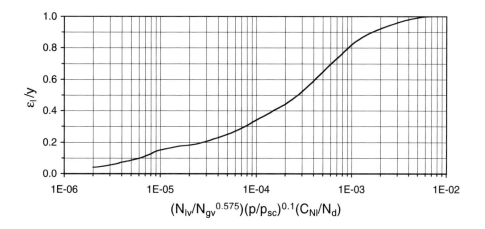

Fig. 2–30 Hagedorn-Brown [46] correlation for ε_l/ψ.

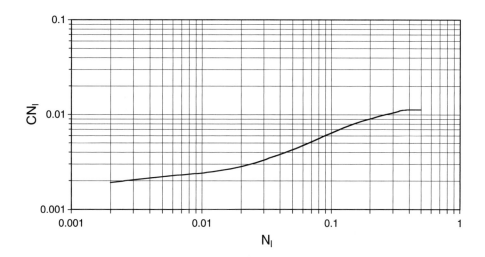

Fig. 2–31 Hagedorn-Brown [46] correlation for CN_l.

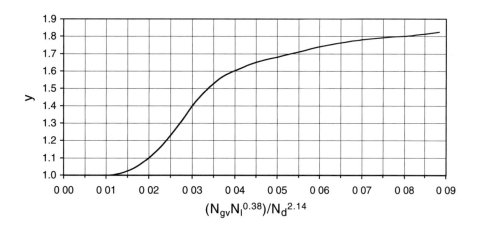

Fig. 2–32 Hagedorn-Brown [46] correlation for Ψ.

Frictional Term

To find the frictional pressure gradient, friction factor f is found from the Moody-diagram (Fig. 2–8) based on the Reynolds number defined as follows:

$$N_{Re} = 124 \; \frac{\rho_{ns} \, v_m d}{\mu_m} \qquad\qquad 2.128$$

where mixture viscosity is calculated from

$$\mu_m = \mu_l^{\varepsilon_l} + \mu_g^{(1-\varepsilon_l)} \qquad\qquad 2.129$$

Improvements

The imperfections of the original Hagedorn-Brown correlation must have been surfaced to the authors shortly after the publication of their paper. This is proved by the fact that the extensive collection of vertical multiphase pressure traverse curves included in Brown's book on gas lifting [50] was developed after several modifications to the original theory. The nature of the modifications was revealed by Brown [51] in 1977 only, and is discussed as follows.

There was a general observation that the correlation predicted too low gradients for low liquid rates and low GLRs. The discrepancy was pronounced for larger pipe sizes. As it turned out, this was caused by calculated liquid holdup values significantly lower than the no-slip holdup. This situation, however, as discussed in Section 2.5.2, is physically impossible because in upward multiphase flow, liquid holdup is always greater than the no-slip value. The possible cause of the problem is the basic approach of the authors who in their experiments did not measure holdup but back-calculated it. The proposed correction is to compare calculated and no-slip holdups and to use the no-slip holdup every time it is greater than that found from the original correlation.

To further improve calculation accuracy, Brown [51] suggests an additional modification of the original correlation. This involves the introduction of flow pattern predictions according to Griffith-Wallis [52]. If the actual flow pattern is bubble, the use of the Griffith correlation [53] valid for this flow pattern is suggested. According to Griffith, a constant slip velocity of $v_s = 0.8$ ft/s can be used for bubble flow. Using this value and Equation 2.78, liquid holdup can readily be calculated. The same approach is used by Orkiszewski [37] as detailed in a later section.

Since it is recommended by Brown [51] that the previous modifications always be used for an improved accuracy, the Hagedorn-Brown correlation is exclusively used in its modified form today.

_ . _

Example 2–24. Calculate the pressure gradients according to the original Hagedorn-Brown procedure for the cases in the previous examples.

Solution

Only the wellhead case is detailed in the following, for which the next parameters calculated in previous examples are used.

$v_{sl} = 0.79$; $v_{sg} = 1.93$; $\rho_l = 49.9$ lb/cu ft; $\rho_g = 1.79$ lb/cu ft; $\rho_{ns} = 15.8$ lb/cu ft;

$N_{lv} = 2.42$; $N_{gv} = 5.92$; $N_d = 61.6$; $N_l = 0.37$.

First calculate the liquid holdup. For this, CN_l is found from Figure 2–31 for $N_l = 0.37$ as $CN_l = 0.011$.

Now the abscissa of Figure 2–30 can be calculated for the wellhead pressure of p = 346.6 psi:

$(2.42 / 5.92^{0.575}) (346.6 / 14.7)^{0.1} (0.011 / 61.6) = 0.00021$

From the previous figure, $\varepsilon_l/\psi = 0.44$ is read off.

To find the correction ψ, the abscissa of Figure 2–32 is evaluated first:

$(5.92 \; 0.37^{0.38}) / 61.6^{2.14} = 0.00025.$

At this value, $\psi = 1.0$ is read from Figure 2–32.

Liquid holdup is found from Equation 2.127:

$\varepsilon_l = 0.44 \; 1.0 = 0.44.$

Now mixture density can be calculated from the universally applicable Equation 2.8:

$\rho_m = 49.9 \; 0.44 + 1.79 \; (1 - 0.44) = 22.97 \text{ lb/cu ft.}$

Elevation pressure gradient is easily found:

$(dp/dh)_{el} = 22.97 / 144 = 0.159 \text{ psi/ft.}$

To determine the frictional gradient, assume a gas viscosity of $\mu_g = 0.02$ cP, and calculate the mixture viscosity from Equation 2.129:

$\mu_m = 3^{0.44} \; 0.02^{(1 - 0.44)} = 0.18 \text{ cP.}$

Reynolds number is found from Equation 2.128:

$N_{Re} = 124 \; 15.8 \; (0.79 + 1.93) \; 2.441 / 0.18 = 71,620.$

Using the Moody diagram (Fig. 2–8) and a relative roughness of 0.0002, friction factor is read as $f = 0.021$.

Frictional pressure gradient is the second term of Equation 2.126:

$(dp/dh)_f = 0.001294 \; 0.021 \; 15.8^2 \; (0.79 + 1.93)^2 / (22.97 \; 2.441) = 0.00089 \text{ psi/ft.}$

Neglecting the kinetic term, total pressure gradient is the sum of the elevation and frictional components:

$dp/dh = 0.160 \text{ psi/ft.}$

For the remaining points in the example well only final results are given as follows.

Parameter	Wellhead	At 3,770 ft	Bottom
CN_l, -	0.011	0.011	0.011
ε_l/ψ, -	0.44	0.98	-
ψ, -	1.0	1.0	-
liquid holdup, -	0.44	0.98	1.0
mixture density, lb/cu ft	22.97	44.63	44.47
elevation gradient, psi/ft	0.159	0.309	0.309
mixture viscosity, cP	0.18	2.71	3
Reynolds number, -	71,620	4,785	4,328
friction factor, -	0.21	0.04	0.04
frictional gradient, psi/ft	0.0008	0.0008	0.0008
pressure gradient, psi/ft	0.160	0.311	0.310

2.5.3.3.5 Orkiszewski correlation.

Summary

This correlation is a composite of several previous models with modifications based on 148 field measurements. The Duns-Ros flow pattern map is used with a redefined boundary between bubble and slug flows. An improved calculation model for slug flow is presented. It is a very popular correlation, mainly due to its improved accuracy over previous models.

The approach of Orkiszewski [37] was completely different from that of previous investigators of vertical multiphase flow. He started from a thorough study of the available correlations and tried to obtain their accuracies. For this reason, he compiled a data bank consisting of a total of 148 pressure drop measurements collected from 22 heavy-oil wells, all data of Poettmann-Carpenter [39], and selected data from other sources (Fancher-Brown [42], Baxendell-Thomas [41], Hagedorn-Brown [44]). Based on his calculations he selected for each flow pattern, the correlation giving the most accurate calculation results. After some modifications to some of the correlations, he proposed a composite calculation model that came to be known by his name. In summary, his correlation is set up from the following:

Flow pattern	Correlation used
bubble flow	Griffith [53]
slug flow	modified Griffith-Wallis [52]
Transition flow	Duns-Ros [33]
mist flow	Duns-Ros [33]

Basic Equation

The basic formula for the calculation of the pressure gradient is the familiar equation containing the three general components of the total pressure drop:

$$\frac{dp}{dh} = \frac{\left(\dfrac{dp}{dh}\right)_{el} + \left(\dfrac{dp}{dh}\right)_{f}}{1 - E_k} \qquad 2.130$$

According to Orkiszewski, the kinetic term is negligible in all but the mist flow pattern.

Flow Pattern Map

Orkiszewski modified the Duns-Ros flow pattern map [33] in the bubble-slug transition as shown in Figure 2–33. The boundary between bubble and slug flow is adopted from the proposal of Griffith-Wallis [52] as given as follows:

$$\lambda_g = 1.071 - 0.2662 \, \frac{v_m^2}{d} \qquad 2.131$$

where: $\lambda_g = v_{sg}/v_m$ = no-slip gas void fraction, -

$v_m = v_{sg} + v_{sl}$ = mixture superficial velocity, ft/s

d = pipe diameter, in.

The formula on the right hand side is constrained to be ≥ 0.13.

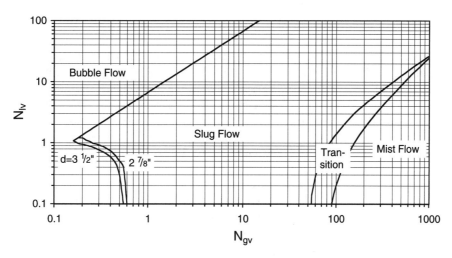

Fig. 2–33 Flow pattern map of Orkiszewski. [37]

Bubble Flow

In bubble flow Orkiszewski assumed a constant slip velocity of v_s = 0.8 ft/s between the flowing phases, as proposed by Griffith [53]. Based on this, liquid holdup is found from the formula reproduced here (see the derivation of Equation 2.78):

$$\varepsilon_l = \frac{v_s - v_m + \sqrt{(v_m - v_s)^2 + 4v_s v_{sl}}}{2v_s}$$

2.132

The elevation component of the pressure gradient is easy to calculate after the mixture density is determined from the liquid holdup:

$$\left(\frac{dp}{dh}\right)_{el} = \frac{1}{144}\left[\rho_l \varepsilon_l + \rho_g (1 - \varepsilon_l)\right]$$

2.133

The frictional component of the pressure gradient in bubble flow is calculated by Orkiszewski as follows:

$$\left(\frac{dp}{dh}\right)_f = 1.294 \times 10^{-3} f \frac{\rho_l}{d}\left(\frac{v_{sl}}{\varepsilon_l}\right)^2$$

2.134

where: ρ_l =   liquid density, lb/cu ft

f =   friction factor, -

v_{sl} =   liquid superficial velocity, ft/s

ε_l =   liquid holdup, -

d =   pipe diameter, in.

Friction factor f is determined from the Moody-diagram (Fig. 2–8) based on the Reynolds number defined as

$$N_{Re} = 124 \frac{\rho_l d}{\mu_l} \frac{v_{sl}}{\varepsilon_l}$$

2.135

where: μ_l = liquid viscosity, cP

Slug Flow

Orkiszewski considered the slug flow pattern to be the most important one because in most of the wells he investigated, a large portion of the tubing length was covered by this flow pattern. Griffith and Wallis [52] made a comprehensive investigation of slug flow and developed a method to calculate the mixture density of a slug unit consisting of a gas bubble and a liquid slug. When evaluating this model using the measured data of 148 wells, Orkiszewski found a significant discrepancy in mixture densities. This behavior was suspected to come from some basic flaw in the model, and a thorough investigation revealed that Griffith and Wallis disregarded the effects of the liquid film around the Taylor bubbles, and the droplets entrained in the Taylor bubbles.

After the necessary modification to the original formula of Griffith-Wallis [52], pressure gradient in slug flow is found from the following formula, where the second term in the mixture density represents the change of flowing density due to the effect of the liquid film and the entrained liquid:

$$\left(\frac{dp}{db}\right)_{el} = \frac{1}{144}\,\rho_m = \frac{1}{144}\left(\frac{\rho_l\,(v_{sl} + v_b) + \rho_g v_{sg}}{v_m + v_b} + \Gamma\,\rho_l\right) \tag{2.136}$$

In the previous formula, two parameters need clarification. The term v_b is the rise velocity of the gas bubbles and is calculated from the correlations proposed by the original authors. The basic formula given by Griffith and Wallis [52] has the form:

$$v_b = C_1 C_2 \sqrt{gd} \tag{2.137}$$

where: d = pipe diameter, ft

Factors C_1 and C_2 are determined from Figure 2–34 and Figure 2–35, respectively, in the function of the following Reynolds numbers:

$$N_{Reb} = 124\,\frac{\rho_l v_b d}{\mu_l} \tag{2.138}$$

$$N_{Rel} = 124\,\frac{\rho_l v_m d}{\mu_l} \tag{2.139}$$

where: ρ_l = liquid density, lb/cu ft

v_m = mixture superficial velocity, ft/s

v_b = bubble rise velocity, ft/s

d = pipe diameter, in.

μ_l = liquid viscosity, cP

In many cases, the Reynolds number range in Figure 2–35 is exceeded, and Orkiszewski proposed the use of the following extrapolation equations (note that pipe diameter d is in feet) to find the bubble rise velocity.

For $N_{Reb} \le 3{,}000$:

$$v_b = \left(0.546 + 8.74 \times 10^{-6}\,N_{Rel}\right)\sqrt{gd} \tag{2.140}$$

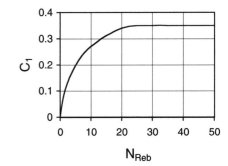

Fig. 2–34 Griffith-Wallis [52] C_1 parameter.

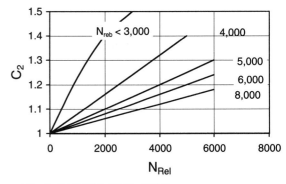

Fig. 2–35 Griffith-Wallis [52] C_2 parameter.

For $N_{Reb} \geq 8,000$:

$$v_b = \left(0.35 + 8.74 \times 10^{-6}\, N_{Rel}\right) \sqrt{gd}$$
2.141

For $3,000 < N_{Reb} < 8,000$:

$$v_b = \frac{1}{2} \left(v_{bs} + \sqrt{v_{bs}^2 + \frac{13.59 \mu_l}{\rho_l \sqrt{d}}} \right)$$
2.142

where:

$$v_{bs} = \left(0.251 + 8.74 \times 10^{-6}\, N_{Rel}\right) \sqrt{gd}$$
2.143

Since the Reynolds number N_{Reb} depends on the bubble rise velocity v_b, which is to be determined, the calculation of bubble rise velocity requires a simple iterative procedure. First, a good initial guess of v_b is made and N_{Reb} is found, then v_b is calculated from Equation 2.137 or the extrapolation functions. If the assumed and calculated values are close enough, the final value of v_b is found. Otherwise, a new iteration step is performed until the v_b values converge.

The other parameter needing explanation in Equation 2.136 is called the liquid distribution coefficient, Γ. As discussed before, it represents the change in mixture density due to the effects of the liquid film and the entrained liquid droplets in the gas bubble. This correction to the original Griffith-Wallis model was introduced by Orkiszewski who developed the equations given as follows. The proposed correlations were developed using measured data taken from Hagedorn-Brown [44]. There are altogether four formulas presented and their use depends on the value of the mixture velocity and the continuous liquid phase, as given in the next table.

Continuous liquid phase	v_m	Use
water	< 10 ft/s	Eq. 2.144
water	> 10 ft/s	Eq. 2.145
oil	< 10 ft/s	Eq. 2.146
oil	> 10 ft/s	Eq. 2.147

$$\Gamma = \frac{0.013 \log \mu_l}{d^{1.38}} - 0.681 + 0.232 \log v_m - 0.428 \log d$$
2.144

$$\Gamma = \frac{0.045 \log \mu_l}{d^{0.799}} - 0.709 + 0.162 \log v_m - 0.888 \log d$$
2.145

$$\Gamma = \frac{0.0127 \log(\mu_l + 1)}{d^{1.415}} - 0.284 + 0.167 \log v_m + 0.113 \log d$$
2.146

$$\Gamma = \frac{0.0274 \log(\mu_l + 1)}{d^{1.371}} + 0.161 + 0.569 \log d + A$$
2.147

where:

$$A = -\log v_m \left[\frac{0.01 \log(\mu_l + 1)}{d^{1.571}} + 0.397 + 0.63 \log d \right]$$

In the previous formulas, pipe diameter is in feet. The calculated value of the liquid distribution coefficient Γ is limited by the following functions, depending on the value of the mixture velocity:

If mixture velocity is   **Constraint**

$\leq$ 10 ft/s $\Gamma = \geq - 0.065 v_m$

$>$ 10 ft/s $\Gamma = \geq - \dfrac{v_b}{v_m + v_b} \left(1 - \dfrac{\rho_m}{\rho_l} \right)$

The frictional component of the pressure gradient was also modified by Orkiszewski, as compared to the original Griffith-Wallis model. The liquid distribution coefficient, as before, represents the effects of the liquid film on the pipe wall and the entrained liquid in the gas bubble:

$$\left(\frac{dp}{dh} \right)_f = 1.294 \times 10^{-3} f \frac{\rho_l v^2_m}{d} \left[\left(\frac{v_{sg} + v_b}{v_m + v_b} \right) + \Gamma \right] \qquad \text{2.148}$$

where: ρ_l = liquid density, lb/cu ft

           f = friction factor, -

           v_m = mixture superficial velocity, ft/s

           v_{sl} = liquid superficial velocity, ft/s

           v_b = bubble rise velocity, ft/s

           d = pipe diameter, in.

Friction factor f is determined from the Moody-diagram (Fig. 2–8) based on the Reynolds number defined as

$$N_{Re} = 124 \, \frac{\rho_l v_m d}{\mu_l} \qquad \text{2.149}$$

where: μ_l = liquid viscosity, cP

Other Flow Patterns

For the mist and the transition flow patterns, Orkiszewski recommended the use of the Duns-Ros [33] calculation models.

Improvements

There are two modifications in use with the Orkiszewski correlation, and both relate to the calculation of the liquid distribution coefficient, Γ.

It was found that the correlations given for the determination of Γ lead to a discontinuity at a superficial mixture velocity of v_m = 10 ft/s. This holds for both pairs of equations given for the continuous water and oil case. The effect of these discontinuities on the prediction of pressure drop is that calculations may not converge because the pressure gradient, too, is discontinuous at around a mixture velocity of 10 ft/s. At the suggestion of Triggia, Brill [54] presented a modified set of equations that eliminate the discontinuity of the original set. So instead of Equation 2.145, use Equation 2.150; and instead of Equation 2.147, use Equation 2.151. By doing so, pressure gradient discontinuities are eliminated and no convergence problems will be met. Note that pipe diameter is in feet units again.

$$\Gamma = \frac{0.013 \log \mu_l}{d^{1.38}} - 0.287 - 0.428 \log d - 0.162 \log v_m \qquad \text{2.150}$$

$$\Gamma = \frac{0.0127 \log(\mu_l + 1)}{d^{1.38}} - 0.117 + 0.113 \log d + C(1 - \log v_m) \qquad 2.151$$

where:

$$C = \frac{0.01 \log(\mu_l + 1)}{d^{1.571}} + 0.397 + 0.63 \log d$$

The other improvement is applied for cases with high flow rates when high superficial mixture velocities are encountered. In such cases, liquid distribution coefficients may become too-large negative numbers, and Equation 2.136 gives a mixture density lower than the no-slip value. Therefore, it should be continuously checked whether this condition occurs and actual density must not be allowed to drop below the no-slip density.

--- --- --- --- --- --- --- --- --- --- --- --- --- --- --- --- --- --- --- ---

Example 2–25. Calculate the pressure gradients in the well described in previous examples according to Orkiszewski.

Solution

As before, only the wellhead case is detailed for which the next parameters calculated in previous examples are used:

$v_{sl} = 0.79$ ft/s; $v_{sg} = 1.93$ ft/s; $\rho_l = 49.9$ lb/cu ft; $\rho_g = 1.79$ lb/cu ft.

For the determination of the prevailing flow pattern, no-slip void fraction, λ_g is to be compared to the bubble-slug boundary. Actual void fraction is:

$\lambda_g = 1.93 / (0.79 + 1.93) = 0.71$

The boundary condition is evaluated from Equation 2.131:

$\lambda_g = 1.071 - 0.2662 (0.79 + 1.93)^2 / 2.441 = 0.26.$

Since the actual void fraction is greater than the boundary value, the flow pattern is slug.

Liquid holdup calculations in slug flow require the determination of bubble rise velocity v_b. For the iterative solution assume an initial value of $v_b = 1$ ft/s. Based on this value, the required Reynolds numbers are found from Equation 2.138 and 2.139:

$N_{Reb} = 124 \ 49.9 \ 1 \ 2.441 / 3 = 5,036$, and

$N_{Rel} = 124 \ 49.9 \ (\ 0.79 + 1.93 \ ) \ 2.441 / 3 = 13,705.$

Since the value of N_{Reb} exceeds the ranges of Figure 2–35, v_b must be evaluated with the extrapolation equations proposed by Orkiszewski. Note that $3,000 < N_{Reb} < 8,000$ and Eqs. 2.143 and 2.142 have to be used:

$v_{bs} = (\ 0.251 + 8.74 \times 10^{-6} \ 13,705 \ ) \ (\ 32.17 \ 2.441 / 12 \ )^{0.5} = 0.949$ and

$$v_b = \frac{1}{2} \left(0.949 + \sqrt{0.949^2 + \frac{13.59 \ 3}{49.9 \sqrt{\frac{2.441}{12}}}} \right) = 1.297 \ \textbf{ft/s}$$

Checking of the new v_b is done through the calculation of a new N_{Reb} value:

$N_{Reb} = 124 \ 49.9 \ 1.29 \ 2.441 / 3 = 6,286$, which also falls in the original ranges, therefore the calculated value is right.

The next problem is the calculation of the liquid distribution factor, Γ. Since only oil is flowing with a velocity of less than 10 ft/s, Equation 2.146 has to be used:

$$\Gamma = \frac{0.0127 \log(3 + 1)}{\left(\frac{2.441}{12}\right)^{1.415}} - 0.284 + 0.167 \log(0.79 + 1.93) + 0.113 \log\left(\frac{2.441}{12}\right) = -0.217$$

The constraint for this case is:

$\Gamma \geq - 0.065 \ (\ 0.79 + 1.93\ ) = -0.177$, so the final value is $\Gamma = -0.177$.

The elevation pressure gradient is found from Equation 2.136 as

$$\left(\frac{dp}{db}\right)_{el} = \frac{1}{144}\left(\frac{49.9\ (0.79 + 1.297) + 1.79 \quad 1.93}{0.79 + 1.93 + 1.293} - 0.177 \ 49.9\right) = 0.125 \ \textbf{ft/s}$$

Calculation of the frictional term follows with the determination of the Reynolds number according to Equation 2.149:

$N_{Re} = 124 \ 49.9 \ (\ 0.79 + 1.93\ ) \ 2.441 \ / \ 3 = 13,706.$

Using the Moody diagram (Fig. 2–8) and a relative roughness of 0.0002, friction factor is read as $f = 0.03$.

Frictional pressure gradient from Equation 2.148:

$$\left(\frac{dp}{db}\right)_{f} = 1.294 \times 10^{-3} \ 0.03 \ \frac{49.9\ (0.79 + 1.93)^2}{2.441}\left[\left(\frac{0.79 + 1.297}{(0.79 + 1.93) + 1.297}\right) - 0.177\right] = 0.002 \ \text{psi/ft}$$

Finally, total pressure gradient is the sum of the elevation and frictional terms:

$dp/db = 0.125 + 0.002 = 0.127$ psi/ft.

For the remaining points in the example well only final results are given as follows.

Parameter	Wellhead	At 3,770 ft	Bottom
$\lambda_{g'}$ -	0.71	0.02	-
λ_g limit, -	0.26	0.97	-
flow pattern	slug	bubble	single
bubble rise velocity, ft/s	1.297	-	-
liquid distr. factor, -	-0.177	-	-
slip velocity, ft/s	-	0.8	-
mixture density, lb/cu ft	18.0	45.1	44.47
elevation gradient, psi/ft	0.125	0.313	0.309
Reynolds number, -	13,705	4,357	4,329
friction factor, -	0.03	0.04	0.04
frictional gradient, psi/ft	0.002	0.0009	0.0009
pressure gradient, psi/ft	0.127	0.314	0.310

2.5.3.3.6 Chierici et al. correlation.

Summary

Practically identical to the Orkiszewski model. The only difference is in slug flow, where the authors eliminated the discontinuities in liquid holdup occurring when the original model is applied.

Chierici et al. [55] adopted the Orkiszewski model for the prediction of flow patterns and the calculation of pressure gradients in all but the slug flow pattern. Therefore, this correlation can be regarded as a slightly modified version of the original. For this reason, only the modifications are detailed as follows.

In slug flow, Chierici et al. used the drift-flux approach for the determination of the liquid holdup. According to this theory, liquid holdup is found from the following basic equation (see Equation 2.76):

$$\varepsilon_l = 1 - \frac{v_{sg}}{C_0 v_m + v_b} \qquad 2.152$$

where: C_0 = distribution factor, -

v_m = mixture superficial velocity, ft/s

v_{sg} = gas superficial velocity, ft/s

v_b = bubble rise velocity, ft/s

The authors assumed the distribution factor to be $C_0 = 1$, and calculated the bubble rise velocity v_b as suggested by Griffith and Wallis [52]:

$$v_b = C_1 C_2 \sqrt{gd} \qquad 2.153$$

where: d = pipe diameter, ft

As with the Orkiszewski model, C_1 is found from Figure 2–34. When finding the other parameter C_2, the authors showed that the method proposed by Orkiszewski is defective. There are no problems in the $N_{Rel} \leq 6,000$ range, but at higher Reynolds numbers the extrapolation functions given by Orkiszewski (see Equations 2.140 –2.143) result in discontinuities.

To prevent the discontinuities caused by the Orkiszewski extrapolations of Figure 2–35, Chierici et al. proposed the use of the following formula:

$$C_2 = \frac{1}{1 - 0.2 \frac{v_m}{v_b}} \qquad 2.154$$

The authors proved that the use of the previous equation eliminates the discontinuities of the calculated C_2 values and ensures the convergence of pressure gradient calculations in slug flow.

2.5.3.3.7 Beggs-Brill correlation.

Summary

The first correlation developed for all pipe inclination angles was based on a great number of data gathered from a large-scale flow loop. Flow patterns are determined for a horizontal direction only and are solely used as correlating parameters. Its main strength is that it enables a simple treatment of inclined wells and the description of the well-flowline system.

In order to describe inclined multiphase flow in directionally drilled wells, the authors conducted a wide-scale experimental study. [47] Their setup was a complete fluid loop in which two 90-ft-long pipes (including 45-ft-long transparent acrylic pipes for flow pattern recognition) were joined by a short flexible pipe. The two measuring sections were 45 ft long each and could be inclined at any angle, providing an uphill and a downhill section at the same time. Pipe diameters were 1 in. and 1 ½ in.; fluids used were air and water. During test runs, liquid holdup and pressure drop were measured with a great accuracy. Each set of flow measurements was started with horizontal flow, then pipe inclination was increased in increments up to 90° from the horizontal, i.e. vertical flow. This approach allowed for the determination of the effects of inclination on the liquid holdup and the pressure gradient. In all, 584 flow tests were run.

Basic Equation

The basic equation is written up with the pipe length l instead of the vertical depth coordinate because the pipe is considered to be inclined. Beggs and Brill took into account all three components of the multiphase pressure drop: elevation, friction, and acceleration:

$$\frac{dp}{dl} = \frac{\left(\frac{dp}{d}\right)_{el} + \left(\frac{dp}{d}\right)_{f}}{1 - E_k} \qquad\qquad 2.155$$

The components of the pressure gradient are expressed as follows:

$$\left(\frac{dp}{dl}\right)_{el} = \frac{1}{144}\rho_m \sin\alpha = \frac{1}{144}\left[\rho_l \varepsilon_l(\alpha) + \rho_g(1 - \varepsilon_l(\alpha))\right]\sin\alpha \qquad\qquad 2.156$$

$$\left(\frac{dp}{dl}\right)_{f} = 1.294 \times 10^{-3} f \frac{\rho_{ns} v_m^2}{d} \qquad\qquad 2.157$$

$$E_k = 2.16 \times 10^{-4} f \frac{v_m v_{sg} \rho_{ns}}{p} \qquad\qquad 2.158$$

where: ρ_m = mixture density, lb/cu ft

ρ_{ns} = no-slip mixture density, lb/cu ft

$\varepsilon_l(\alpha)$ = liquid holdup at an inclination angle of α, -

α = pipe inclination angle, measured from the horizontal, rad

f = friction factor, -

v_m = mixture superficial velocity, ft/s

v_{sg} = gas superficial velocity, ft/s

p = pressure, psi

d = pipe diameter, in.

Flow Patterns

In the Beggs-Brill correlation, flow patterns are determined for horizontal flow only, using an empirically developed flow pattern map. Since this map is valid for horizontal flow only, actual flow patterns in inclined or vertical cases cannot be predicted. In this approach, flow pattern is a correlating parameter only and does not indicate an actual flow pattern.

The authors used two parameters to describe the horizontal flow pattern transitions: the no-slip liquid holdup λ_l and the mixture Froude number N_{Fr}, defined as follows:

$$\lambda_l = \frac{v_{sl}}{v_m} \qquad\qquad 2.159$$

$$N_{Fr} = \frac{v_m^2}{g\,d} = 0.373\,\frac{v_m^2}{d} \qquad\qquad 2.160$$

where: v_{sl} = liquid superficial velocity, ft/s

v_m = mixture superficial velocity, ft/s

d = pipe diameter, in.

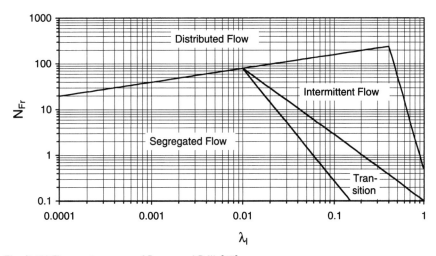

Fig. 2–36 Flow pattern map of Beggs and Brill. [47]

The flow patterns distinguished by Beggs and Brill are designated according to the possible horizontal flow patterns as: segregated, intermittent, and distributed patterns. The original flow pattern map was later modified [56] by the authors, who included a transition flow pattern between the segregated and intermittent patterns. The modified map is shown in Figure 2–36.

The boundaries of the flow patterns, as shown in the figure, are described by the following functions:

$$N_{Fr} = 316\,\lambda_l^{0.302} \qquad\qquad\qquad \textit{segregated-distributed} \qquad\qquad 2.161$$

$$N_{Fr} = 9.25 \times 10^{-4}\,\lambda_l^{-2.468} \qquad\qquad\qquad \textit{segregated-transition} \qquad\qquad 2.162$$

$$N_{Fr} = 0.1\,\lambda_l^{-1.452} \qquad\qquad\qquad \textit{transition-intermittent} \qquad\qquad 2.163$$

$$N_{Fr} = 0.5\,\lambda_l^{-6.738} \qquad\qquad\qquad \textit{intermittent-distributed} \qquad\qquad 2.164$$

Liquid Holdup Calculation

The authors developed a single liquid holdup formula for the three basic horizontal flow patterns. After the holdup for the horizontal case is found, a correction is applied to calculate its value at the given inclination angle. Experimental data showed that liquid holdup varies with the inclination angle in such a way that a maximum and a minimum is found at the inclination angles of approximately +50° and -50°, respectively. In summary, liquid holdup for an inclined pipe is determined in two steps:

(1) the holdup that would exist if the pipe were horizontal is found

(2) this value is corrected for the actual inclination angle

The horizontal liquid holdup is found for all three basic flow patterns from the following formula, where the coefficients vary according to the flow pattern:

$$\varepsilon_l(0) = \frac{a\,\lambda_l^b}{N_{Fr}^c} \qquad\qquad 2.165$$

The coefficients a, b, and c are given in Table 2–4, and the calculated holdup is restricted to $\varepsilon_l(0) \geq \lambda_l$.

Based on its value for the horizontal case, holdup for the actual inclination angle α is calculated from the following equation:

$$\varepsilon_l(\alpha) = \varepsilon_l(0)\psi \qquad\qquad 2.166$$

Flow Pattern	a	b	c
Segregated	0.980	0.4846	0.0868
Intermittent	0.845	0.5351	0.0173
Distributed	1.065	0.5824	0.0609

Table 2–4 Horizontal liquid holdup coefficients. [47]

In the previous equation, inclination correction factor Ψ is defined by the following expressions:

$$\psi = 1 + C\left[\sin(1.8\alpha) - 0.333\sin^3(1.8\alpha)\right] \qquad\qquad 2.167$$

$$C = (1 + \lambda_l)\ln\left(e\,\lambda_l^f N_{lv}^g N_{Fr}^h\right) \qquad\qquad 2.168$$

The coefficients e, f, g, and h for the different flow patterns are given in Table 2–5, and the calculated value of factor C is restricted to $C \geq 0$.

For vertical flow, i.e. $\alpha = 90°$ the correction function ψ is simplified to:

$$\psi = 1 + 0.3\,C \qquad\qquad 2.169$$

Flow Pattern	e	f	g	h
Segregated Uphill	0.011	-3.7680	3.5390	-1.6140
Intermittent Uphill	2.960	0.3050	-0.4473	0.0978
Distributed Uphill	No correction required, ψ = 1.0			
All Patterns Downhill	4.700	-0.3692	0.1244	-0.5056

Table 2–5 Inclination correction coefficients. [47]

If in the transition flow pattern, liquid holdup is calculated from a weighted average of its values valid in the segregated and intermittent flow patterns:

$$\varepsilon_l(\alpha)_{\text{transition}} = B\,\varepsilon_l(\alpha)_{\text{segregated}} + (1 - B)\varepsilon_l(\alpha)_{\text{intermittent}} \qquad\qquad 2.170$$

where:

$$B = \frac{0.1\lambda_l^{-1.452} - N_{Fr}}{0.1\lambda_l^{-1.452} - 9.25 \times 10^{-4}\,\lambda_l^{-2.468}}$$

Determination of the Friction Factor

Similar to liquid holdup calculations, the Beggs and Brill correlation uses a single formula to calculate the friction factor in each flow pattern. In the following expression, normalizing friction factor f_n is determined by the assumption that the pipe wall is smooth. It can be found either from the Moody diagram (Fig. 2–8) or with the use of an appropriate equation.

$$f = f_n\frac{f}{f_n} \qquad\qquad 2.171$$

The Reynolds number to be used has the form:

$$N_{Re} = 124 \frac{\rho_{ns} v_m d}{\mu_{ns}}$$ 2.172

where no-slip mixture viscosity (see Equation 2.83) is calculated from:

$$\mu_{ns} = \mu_l \lambda_l + \mu_g (1 - \lambda_l)$$ 2.173

The ratio of friction factors figuring in Equation 2.171 was approximated by the authors with the following formula, based on their experimental data:

$$\frac{f}{f_n} = e^s$$ 2.174

where:

$$s = \frac{\ln y}{-0.0523 + 3.182 \ln y - 0.8725(\ln y)^2 + 0.01853(\ln y)^4}$$ 2.175

$$y = \frac{\lambda_l}{[\varepsilon_l(\alpha)]^2}$$ 2.176

Beggs and Brill found that the function defined by Equation 2.175 is discontinuous in the range $1 < y < 1.2$, and proposed the use of the following formula in that range:

$$s = \ln(2.2y - 1.2)$$ 2.177

Improvements

As mentioned before, the first modification of the original correlation was the inclusion of a transition flow pattern as proposed by Payne et al. [56]. The same authors, based on the results of their extensive experiments in a more than 500-ft-long horizontal flow loop with a diameter of 2 in., found that the Beggs and Brill correlation under-predicted frictional losses because the original measurements were made in almost smooth pipes. They recommended [57] the use of a normalizing friction factor (defined in Equation 2.171) to be determined for the pipe's actual roughness.

Liquid holdups calculated with the original correlation were also found to deviate from measured values. Payne et al. [57] showed that calculated holdups always over-predict actual ones, and they introduced appropriate correction factors for uphill and downhill flow cases:

$$\varepsilon_l(\alpha) = 0.924\varepsilon_l(\alpha)_{original} \qquad \qquad \text{for uphill flow}$$ 2.178

$$\varepsilon_l(\alpha) = 0.685\varepsilon_l(\alpha)_{original} \qquad \qquad \text{for downhill flow}$$ 2.179

It must be noted that the Beggs-Brill correlation is always used with the previous modifications.

– $\cdot$ –– $\cdot$ –– $\cdot$ –– $\cdot$ –– $\cdot$ –– $\cdot$ –– $\cdot$ –– $\cdot$ –– $\cdot$ –– $\cdot$ –– $\cdot$ –– $\cdot$ –– $\cdot$ –– $\cdot$ –– $\cdot$ –– $\cdot$ –– $\cdot$ –

Example 2–26. Calculate the pressure gradients in the well of the previous examples according to the original Beggs-Brill procedure.

Solution

As before, only the wellhead case is detailed for which the next parameters calculated in previous examples are used:

$$v_{sl} = 0.79 \text{ ft/s}; \ v_{sg} = 1.93 \text{ ft/s}; \ v_m = 2.72 \text{ ft/s}; \ \rho_l = 49.9 \text{ lb/cu ft}; \ \rho_g = 1.79 \text{ lb/cu ft}, \ N_{lv} = 2.42.$$

To establish the flow pattern, no-slip liquid holdup and Froude number have to be calculated from Equation 2.159 and Equation 2.160:

$\lambda_l = 0.79 / 2.72 = 0.290$, and

$N_{Fr} = 0.373\ 2.72^2 / 2.441 = 1.13$.

As read from Figure 2–36, the flow pattern is intermittent.

Liquid holdup is calculated from Equation 2.165 with the proper coefficients chosen from Table 2–4:

$\varepsilon_l(0) = (0.845\ 0.290^{0.5351}) / 1.13^{0.0173} = 0.435$

Compared to the no-slip holdup, it is seen that $\varepsilon_l(0) > \lambda_l$, so the value is verified.

Since the well is vertical, correction for pipe inclination is calculated from Equation 2.169 with the C factor found from Equation 2.168:

$C = (1 - 0.29)\ \ln(2.96\ 0.29^{0.305}\ 2.42^{-0.4473}\ 1.13^{0.0978}) = 0.23$, and

$\psi = 1 + 0.3\ 0.23 = 1.069$.

Liquid holdup at the given inclination angle is found from Equation 2.166 as follows:

$\varepsilon_l(\alpha) = 0.435\ 1.069 = 0.465$.

Mixture density is found as:

$\rho_m = 49.9\ 0.465 + 1.79\ (1 - 0.465) = 24.18$ lb/cu ft.

The elevation component of the pressure gradient is calculated from Equation 2.156:

$(dp/dl)_{el} = 24.18 / 144 = 0.168$ psi/ft.

The calculations for the friction factor involve a Reynolds number including the no-slip parameters of the multiphase mixture. No-slip density and viscosity are calculated with the no-slip liquid holdup:

$\rho_{ns} = 49.9\ 0.29 + 1.79\ (1 - 0.29) = 15.77$ lb/cu ft, and

$\mu_{ns} = 3\ 0.29 + 0.02\ (1 - 0.29) = 0.89$ cP.

Reynolds number according to Equation 2.172:

$N_{Re} = 124\ 15.77\ 2.72\ 2.441 / 0.89 = 14,672$.

From the Moody diagram (Fig. 2–8) and assuming a smooth pipe wall, a normalizing friction factor is read as $f_n = 0.028$.

To calculate the ratio of friction factors, the factor y must be first found from Equation 2.176:

$y = 0.290 / 0.435^2 = 1.53$

From this, $\ln(y) = 0.4279$ and the parameter s in Equation 2.175 is

$s = 0.4279 / (-0.0523 + 3.182\ 0.4279 - 0.8725\ 0.4279^2 + 0.01853\ 0.4279^4) = 0.372$.

Friction factor ratio is found from Equation 2.174:

$f / f_n = \exp (0.372) = 1.45.$

The friction factor to be used in the frictional pressure gradient term is thus, according to Equation 2.171:

$f = 0.028 \ 1.45 = 0.04.$

The frictional component of the pressure gradient is calculated from Equation 2.157:

$(dp/dl)_f = 0.001294 \ 0.04 \ 15.77 \ 2.72^2 / 2.441 = 0.0025$ psi/ft.

To include the kinetic effects, the dimensionless kinetic term is evaluated from Equation 2.158:

$E_k = 2.16 \ 10^{-4} \ 2.72 \ 1.93 \ 15.77 / 346.6 = 5.2 \ 10^{-5}.$

Finally, the total pressure gradient is calculated from Equation 2.155:

$dp/dl = (0.168 + 0.0025) / (1 - 5.2 \ 10^{-5}) = 0.170$ psi/ft.

For the remaining points in the example well only final results are given as follows.

Parameter	Wellhead	At 3,770 ft	Bottom
no-slip holdup, -	0.290	0.983	1.0
Froude number, -	1.13	0.14	-
flow pattern	Intermittent	intermittent	single
calculated $\varepsilon_l(0)$, -	0.435	0.866	-
horizontal liquid holdup, -	0.435	0.983	1
inclination corr. factor, -	1.069	1.002	1
$\varepsilon_l(\alpha)$, -	0.465	0.985	1
mixture density, lb/cu ft	24.2	44.8	44.5
elevation gradient, psi/ft	0.167	0.311	0.309
no-slip mixt. density, lb/cu ft	15.8	44.7	44.5
no-slip mixt. viscosity, cP	0.88	2.95	3
Reynolds number, -	14,672	4,403	4,329
Moody friction factor, -	0.028	0.038	0.04
y, -	1.53	1.02	-
s, -	0.372	0.037	-
f/f_n, -	1.45	1.04	-
friction factor, -	0.040	0.039	0.04
frictional gradient, psi/ft	0.0025	0.0009	0.0009
dim.-less kin. term, -	5.2 10-5	1.2 10-7	0.0
pressure gradient, psi/ft	0.170	0.312	0.310

2.5.3.3.8 Cornish correlation.

Summary

This correlation was developed for high liquid rates (more than 5,000 bpd) and large tubulars. Flow patterns and slippage are disregarded and the calculation model requires pVT measurements. It found little use in practice.

The author investigated high-rate flowing wells and proposed a simplified calculation model to calculate pressure drop in such wells. [45] He assumed the effect of flow patterns to be negligible because in these conditions, flow is always dispersed bubble. Additional assumptions are

- slippage can be disregarded in high-rate wells,

- acceleration effects must be included,

- the friction factor should be found from the Moody diagram.

Based on these assumptions, the fundamental equation of Cornish can be written as:

$$\frac{dp}{dh} = \frac{\left(\dfrac{dp}{dh}\right)_{el} + \left(\dfrac{dp}{dh}\right)_{f}}{1 - E_k}$$

2.180

The elevation and frictional components of the pressure gradient are

$$\left(\frac{dp}{dh}\right)_{el} = \frac{1}{144}\,\rho_m = \frac{1}{144}\left[\rho_l\lambda_l + \rho_g\left(1 - \lambda_l\right)\right]$$

2.181

$$\left(\frac{dp}{dh}\right)_{f} = 1.294 \times 10^{-3}\,f\,\frac{\rho_{ns}v_m^2}{d}$$

2.182

where: ρ_{ns} = no-slip mixture density, lb/cu ft

λ_l = no-slip liquid holdup, -

f = friction factor, -

v_m = mixture superficial velocity, ft/s

d = pipe diameter, in.

The author presented a new formula for the dimensionless kinetic term E_k, which is a variation of the expression given in Equation 2.158.

Flowing no-slip mixture density ρ_{ns} is found from pVT measurements. The author proposed the use of mixture density vs. flowing pressure diagrams from which the densities read at the required pressures should be substituted into the basic equation. The other approach is to use the definition of mixture density and the no-slip mixture holdup:

$$\lambda_l = \frac{v_{sl}}{v_m}$$

2.183

Frictional losses are calculated on the basis of a friction factor found from the Moody diagram (Fig. 2–8) where the Reynolds number is defined as

$$N_{Re} = 124\,\frac{\rho_{ns}v_m d}{\mu_{ns}}$$

2.184

No-slip mixture viscosity is calculated from

$$\mu_{ns} = \mu_l^{\lambda_l} + \mu_g^{(1-\lambda_l)}$$ 2.185

Since the correlation model was developed for large liquid rates often produced through annuli, the author presents an average absolute roughness formula for annular flow.

2.5.3.3.9 Mukherjee-Brill correlation.

Summary

Developed for all pipe inclination angles, this correlation was the first to consider the change of flow patterns with pipe inclination. Using a large experimental data base, the proposed correlation covers all uphill and downhill flow patterns.

The objective of the authors [48] was to improve the Beggs-Brill model while using almost the same test facility. They used a U-shaped steel pipe of 1.5-in. internal diameter that could be lowered or raised to ensure any pipe inclination between 0° and 90° from the horizontal. This setup allowed for the investigation of uphill and downhill flow at the same time so the effects of pipe inclination on the pressure drop could be investigated. The dual pipes had 32-ft-long measuring sections in addition to the 22-ft-long entrance sections. Liquid holdup was measured with capacitance-type sensors and flow patterns were observed through transparent pipe sections. Fluids used were air and different oils, and approximately 1,000 pressure drop measurements were carried out. [58]

Basic Equation

The basic equation is written up for an inclined pipe and is practically identical to that of Beggs and Brill [47]. The formulas presented as follows are valid for the conventional uphill flow patterns only, because the stratified flow pattern (present in downhill and horizontal flow) requires different treatment.

$$\frac{dp}{dl} = \frac{\left(\dfrac{dp}{dl}\right)_{el} + \left(\dfrac{dp}{dl}\right)_f}{1 - E_k}$$ 2.186

Pressure gradient components are expressed as follows:

$$\left(\frac{dp}{dl}\right)_{el} = \frac{1}{144}\rho_m \sin\alpha = \frac{1}{144}\left[\rho_l \varepsilon_l(\alpha) + \rho_g\left(1 - \varepsilon_l(\alpha)\right)\right]\sin\alpha$$ 2.187

$$\left(\frac{dp}{dl}\right)_f = 1.294 \times 10^{-3} f \frac{\rho_m v_m^2}{d}$$ 2.188

$$E_k = 2.16 \times 10^{-4} \frac{v_m v_{sg} \rho_m}{p}$$ 2.189

where: ρ_m = mixture density, lb/cu ft

$\varepsilon_l(\alpha)$ = liquid holdup at an inclination angle of α, -

α = pipe inclination angle, measured from the horizontal, rad

f = friction factor, -

v_m = mixture superficial velocity, ft/s

v_{sg} = gas superficial velocity, ft/s

p = pressure, psi

d = pipe diameter, in.

Flow Patterns

Mukherjee and Brill were the first to consider the effect of pipe inclination angle on flow pattern transitions in oil wells. In addition to the conventional flow patterns (bubble, slug, and mist) observed in vertical or inclined pipes, they included the stratified flow, present in horizontal or downhill ($-90° \leq \alpha \leq 0°$) flow configurations. Their flow pattern boundary equations vary with pipe inclination, and different flow pattern maps can be constructed for different pipe inclinations. Figure 2–37 presents the flow pattern map valid for uphill flow where the effect of inclination is clearly seen on the bubble-slug flow transition, whereas the slug-mist transition is independent of the inclination angle.

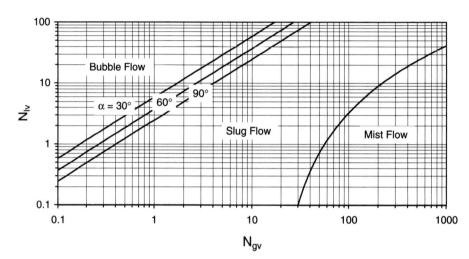

Fig. 2–37 Flow pattern map of Mukherjee and Brill [48] for uphill flow.

As seen in Figure 2–37, the flow pattern map utilizes the familiar dimensionless liquid N_{lv} and gas velocity N_{gv} numbers. The boundary between the slug and mist (annular) flow patterns was found to vary with liquid viscosity but to be independent of pipe inclination angle, as described by the expression:

$$N_{gv} = 10^{\left[1.401 - 2.694N_l + 0.521N_{lv}^{0.329}\right]}$$

2.190

The boundary of the bubble and slug flow patterns in uphill flow ($0° < \alpha \leq 90°$) varies with the pipe inclination angle α as follows:

$$N_{lv} = 10^{\left[\log N_{gv} + 0.940 + 0.074\sin\alpha - 0.855\sin^2\alpha + 3.695N_l\right]}$$

2.191

If the flow direction is horizontal or downhill ($-90° \leq \alpha \leq 0°$), the bubble-slug transition was found to occur along the next curve:

$$N_{gv} = 10^{\left[0.431 - 3.003N_l - 1.138\log N_{lv}\sin\alpha - 0.429(\log N_{lv})^2\sin\alpha + 1.132\sin\alpha\right]}$$

2.192

For the same flow configuration (horizontal or downhill) the boundary of the stratified flow pattern is given by the equation:

$$N_{lv} = 10^{\left[0.321 - 0.017N_{gv} - 4.267\sin\alpha - 2.972N_l - 0.033(\log N_{gv})^2 - 3.925\sin^2\alpha\right]}$$

2.193

In the previous formulas, the dimensionless parameters N_{lv}, N_{gv}, and N_l are defined by Equations 2.96, 2.97, and 2.99, respectively.

Calculation of Liquid Holdup

Mukherjee and Brill made more than 1,500 liquid holdup measurements in a broad range of inclination angles ($-90° \leq \alpha \leq 90°$) and, after a regression analysis of their data, proposed the following general equation for the calculation of liquid holdup at any inclination angle:

$$\varepsilon_l(\alpha) = \exp\left[(C_1 C_2\sin\alpha + C_3\sin^2\alpha + C_4 N_l^2)\frac{N_{gv}^{C_5}}{N_{lv}^{C_6}}\right]$$

2.194

The authors presented three sets of data for the coefficients C_1 to C_6, one for all flow patterns in uphill and horizontal flow, one for downhill stratified flow, and one for the other downhill flow patterns. The relevant coefficients are given in Table 2–6.

Flow Direction	Flow Pattern	Coefficients in Eq. 2.194					
		C1	C2	C3	C4	C5	C6
Uphill and Horizontal	All	-0.380113	0.129875	-0.119788	2.343227	0.475685	0.288657
Downhill	Stratified	-1.330282	4.808139	4.171584	56.262268	0.079951	0.504887
	Other	-0.516644	0.789805	0.551627	15.519214	0.371771	0.393952

Table 2–6 Liquid holdup coefficients according to Mukherjee-Brill. [48]

It should be noted that, identically to the Beggs-Brill correlation [47], liquid holdup varies with pipe inclination according to the group $(C_1 + C_2 \sin \alpha + C_3 \sin^2 \alpha)$. As discussed before, this shape ensures that liquid holdup has a maximum and a minimum at the inclination angles of approximately +50° and -50°, as observed by Beggs-Brill and later by Mukherjee-Brill.

Calculation of Friction Losses

Frictional pressure drops are calculated differently for the different flow patterns. For bubble and slug flow, Equation 2.188 is used with the friction factor f determined from the Moody diagram (Fig. 2–8) for the following Reynolds number:

$$N_{Re} = 124 \frac{\rho_{ns} v_m d}{\mu_{ns}}$$

2.195

In the previous formula, no-slip mixture density and viscosity, respectively, are calculated from

$$\rho_{ns} = \rho_l \lambda_l + \rho_g (1 - \lambda_l)$$

2.196

$$\mu_{ns} = \mu_l \lambda_l + \mu_g (1 - \lambda_l)$$

2.197

In annular (mist) flow, the original equation (Equation 2.188) is slightly modified because no slippage occurs between the small liquid droplets and the high velocity gas stream present in the pipe core:

$$\left(\frac{dp}{dl} \right)_f = 1.294 \times 10^{-3} f \frac{\rho_{ns} v_m^2}{d}$$

2.198

Friction factor for mist flow is calculated from an empirical correlation proposed by Mukherjee and Brill, based on the description of their measurement data:

$$f = f_{ns} + f_R$$

2.199

In this formula f_{ns} is a no-slip friction factor found from the Moody diagram (Fig. 2–8), determined for the no-slip Reynolds number defined in Equation 2.195. The friction factor ratio f_R is found from Table 2–7, as a function of the holdup ratio defined as:

$$\varepsilon_R = \frac{\lambda_l}{\varepsilon_l(\alpha)}$$

2.200

ε_R	0.01	0.20	0.30	0.40	0.50	0.70	1.00	10.0
f_R	1.00	0.98	1.20	1.25	1.30	1.25	1.00	1.00

Table 2–7 Friction factor ratios according to Mukherjee-Brill. [48]

Stratified flow, occurring in horizontal or downhill flow configurations only, required a different treatment. The authors assume that liquid and gas flow separately with liquid occupying the bottom part of the pipe and apply momentum balance equations for each phase. The resulting equations are not reproduced here.

—··—··—··—··—··—··—··—··—··—··—··—··—··—··—··—··—··—··—··—

Example 2–27. Calculate the pressure gradient according to the Mukherjee-Brill correlation for the three points in the well of the previous examples.

Solution

Again, only the wellhead case is detailed with the next input parameters calculated in previous examples:

$v_{sl} = 0.79$ ft/s; $v_{sg} = 1.93$ ft/s; $v_m = 2.72$ ft/s; $\rho_l = 49.9$ lb/cu ft; $\rho_g = 1.79$ lb/cu ft.

$\rho_{ns} = 15.77$ lb/cu ft; $\mu_{ns} = 0.89$ cP.

$N_{lv} = 2.42$; $N_{gv} = 5.92$; $N_l = 0.37$.

To find the prevailing flow pattern, the bubble-slug boundary condition can be found from Equation 2.191 with $\sin\alpha = 1$, because the well is vertical:

$$N_{lv} = 10^{[\log 5.92 + 0.940 + 0.074 - 0.855 + 3.695\ 0.37]} = 20.85.$$

The flow pattern is slug, because this value is greater than the actual $N_{lv} = 2.41$.

Liquid holdup is calculated by substituting $\sin\alpha = 1$ into Equation 2.194:

$$\varepsilon_l = \exp\left[\left(-.380113 + 0.129875 - 0.119788 + 2.343227\ 0.37^2\right)\frac{5.92^{0.475685}}{2.42^{0.288657}}\right] = 0.537$$

From liquid holdup, mixture density is found as:

$$\rho_m = 49.9\ 0.537 + 1.79\ (1 - 0.537) = 27.65 \text{ lb/cu ft.}$$

The elevation term of the pressure gradient is calculated from Equation 2.187:

$$(dp/dl)_{el} = 27.65\ /\ 144 = 0.192 \text{ psi/ft.}$$

Calculation of the friction component starts with the calculation of the Reynolds number from Equation 2.195:

$$N_{Re} = 124\ 15.77\ 2.72\ 2.441\ /\ 0.89 = 14{,}672.$$

Figure 2–8 is used to read off a friction factor of $f = 0.028$.

Frictional pressure gradient is evaluated according to Equation 2.188:

$$(dp/dl)_f = 0.001294\ 0.028\ 27.65\ 2.72^2\ /\ 2.441 = 0.0011 \text{ psi/ft.}$$

Kinetic effects are included by the dimensionless kinetic term from Equation 2.189:

$$E_k = 2.16\ 10^{-4}\ 2.72\ 1.93\ 27.65\ /\ 346.6 = 9.1\ 10^{-5}.$$

Total pressure gradient is calculated from Equation 2.186:

$$dp/dl = (0.192 + 0.0011)\ /\ (1 - 9.1\ 10^{-5}) = 0.193 \text{ psi/ft.}$$

For the remaining points only final results are given as follows.

Parameter	Wellhead	At 3,770 ft	Bottom
N_{lv}, -	2.42	2.82	2.87
N_{gv}, -	5.91	0.05	-
N_l, -	0.195	0.100	0.099
N_{lv} boundary, -	20.85	0.16	-
flow pattern	slug	bubble	single
liquid holdup, -	0.537	0.941	1
mixture density, lb/cu ft	27.7	43.2	44.5
elevation gradient, psi/ft	0.192	0.299	0.309
no-slip mixt. density, lb/cu ft	15.8	44.7	44.5
no-slip mixt. viscosity, cP	0.88	2.95	3
Reynolds number, -	14,672	4,403	4,329
Moody friction factor, -	0.028	0.038	0.04
friction factor, -	0.028	0.038	0.04
frictional gradient, psi/ft	0.0011	0.0008	0.0009
dim.less kin. term, -	9.05 10-5	1.19 10-7	0.0
pressure gradient, psi/ft	0.193	0.300	0.310

2.5.3.4 Mechanistic models.

2.5.3.4.1 Introduction. The continuous efforts of researchers and practicing engineers to improve the accuracy of pressure drop predictions have indicated that empirical calculation methods, by their nature, can never cover all parameter ranges that may exist in field operations. Fundamental hydraulic research, as well as adaptation of the achievements of the abundant literature sources in chemical engineering and nuclear industries, has gradually shifted the emphasis from empirical experimentation to a more comprehensive analysis of the multiphase flow problem. This is the reason why the modeling approach is exclusively utilized in the current research of multiphase flow behavior. Investigators adopting this approach model the basic physics of the multiphase mixture's flow and develop appropriate fundamental relationships between the basic parameters. At the same time, they try to eliminate empirical correlations in order to widen the ranges of applicability.

The *mechanistic* approach detailed previously was first used in the pioneering work of Aziz et al. [59] in 1972. Their calculation model involves the prediction of the actual flow pattern based on a simplified flow pattern map. Multiphase mixture density, friction factor, and pressure gradient are then evaluated from comprehensive equations valid for the flow pattern just determined. Although the authors tried to properly describe the fundamental physical phenomena, some of their solutions (flow pattern map, etc.) lacked the necessary theoretical basis, as compared to the models of today.

It was found that the failure of the empirical methods could partly be attributed to the features of the flow pattern maps they employ. These maps use arbitrarily chosen dimensionless groups as coordinates and are limited in their applicability by the experimental data base used for their construction. This is why all the up-to-date mechanistic models employ theoretically sound solutions for flow pattern recognition. The most popular recent mechanistic model for flow pattern determination is the one proposed by Barnea [32] who identifies four distinct flow patterns (bubble, slug, churn, and annular) and formulates the appropriate transition boundaries based on a thorough mechanistic modeling.

In summary, mechanistic models for vertical multiphase pressure drop calculations are characterized by a comprehensive determination of the prevailing flow pattern at several depths in the tubing. The calculation of the basic flow parameters (mixture density, friction factor, etc.) is then executed by using appropriate formulas developed from mechanical modeling of the particular flow pattern.

As mentioned previously, the earliest mechanistic model was published by Aziz et al. [59] Their approach was adopted by others after a considerable time delay, since the next such model of Hasan and Kabir [60] came out in 1988 only. The recent procedures of Ansari et al. [61] and Chokshi et al. [62] represent the state-of-the-art in vertical pressure drop calculations. However, further research and the emergence of more universally applicable pressure drop prediction methods can be expected in the future. [63, 34]

2.5.3.4.2 Aziz-Govier-Fogarasi model.

Summary

The predecessor of present-day mechanistic models was the first not based on experimental data. It focuses on the bubble and slug flow patterns, although flow pattern prediction is a weakness of the model. The approach of the authors to the determination of pressure gradient components in different flow patterns is still followed today.

The authors' objective was to develop a sound mechanistically based multiphase pressure drop prediction model for vertical oil wells. Instead of an experimental approach, they tried to set up a simple calculation scheme for the most important flow patterns: bubble and slug flow. [59] All calculations are based on the identification of the flow patterns, and different methods selected from the abundant literature sources are utilized in the different patterns. Since no experimental data were used to develop the model, the proposed calculation scheme does not contain any inherent errors prone to in previous experimental correlations.

Basic Equation

The basic formula of the pressure gradient is the familiar equation containing the three general components of the total pressure drop:

$$\frac{dp}{dh} = \frac{\left(\dfrac{dp}{dh}\right)_{el} + \left(\dfrac{dp}{dh}\right)_{f}}{1 - E_k} \qquad\qquad 2.201$$

Aziz et al. disregard the kinetic term in all but the mist flow pattern. In those cases, the dimensionless kinetic term E_k equals zero.

Flow Pattern Map

The authors use their own flow pattern map shown in Figure 2–38. The coordinate axes of the map, in contrast to most other maps, are dimensioned variables as given as follows:

$$N_x = v_{sg} \left(\frac{\rho_g}{0.0764}\right)^{1/3} \left[\left(\frac{72}{\sigma_l}\right)\left(\frac{\rho_l}{62.4}\right)\right]^{1/4} \qquad\qquad 2.202$$

$$N_y = v_{sl} \left[\left(\frac{72}{\sigma_l} \right) \left(\frac{\rho_l}{62.4} \right) \right]^{1/4}$$ 2.203

where: ρ_l = liquid density, lb/cu ft

ρ_g = gas density, lb/cu ft

v_{sl} = liquid superficial velocity, ft/s

v_{sg} = gas superficial velocity, ft/s

σ_l = interfacial tension, dyne/cm

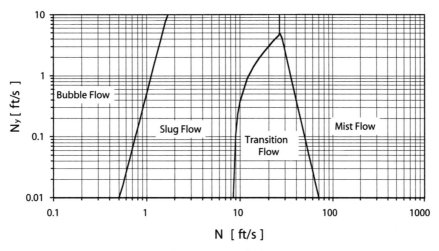

Fig. 2–38 Flow pattern map of Aziz-Govier-Fogarasi. [59]

Four flow patterns are distinguished: bubble, slug, mist, and transition. The boundary equations for the different patterns are given as follows:

$N_x = 0.51(100N_y)^{0.172}$ *bubble-slug boundary* 2.204

$N_x = 8.6 + 3.8N_y$ for $N_y \leq 4$ *slug-transition boundary* 2.205

$N_x = 70(100N_y)^{-0.152}$ for $N_y \leq 4$ *transition-mist boundary* 2.206

$N_x = 26$ for $N_y > 4$ *slug-mist boundary* 2.207

Bubble Flow

Gas slippage in bubble flow is described by the drift-flux model and the Zuber-Findlay [30] formula is used to calculate the average cross-sectional gas velocity in the two-phase mixture:

$v_g = 1.2v_m + v_b$ 2.208

Bubble rise velocity v_b is predicted from the following formula:

$$v_b = 1.41 \sqrt[4]{\frac{g\sigma_l(\rho_l - \rho_g)}{\rho_l^2}} = 0.728 \sqrt[4]{\frac{\sigma_l(\rho_l - \rho_g)}{\rho_l^2}}$$ 2.209

where: ρ_l = liquid density, lb/cu ft

ρ_g = gas density, lb/cu ft

σ_l = interfacial tension, dyne/cm

As described in Section 2.5.2.1, knowledge of the average gas velocity allows a direct determination of the liquid holdup (see derivation of Equation 2.76) from

$$\varepsilon_l = 1 - \frac{v_{sg}}{1.2\, v_m + v_b}$$ 2.210

The hydrostatic or elevation component of the pressure gradient equation is easily found, based on the liquid holdup just calculated:

$$\left(\frac{dp}{db}\right)_{el} = \frac{1}{144}\left[\rho_l\varepsilon_l + \rho_g\,(1 - \varepsilon_l)\right]$$ 2.211

Calculation of the frictional term in bubble flow, according to Aziz et al., is done by using the formula:

$$\left(\frac{dp}{db}\right)_f = 1.294 \times 10^{-3} f\,\frac{\rho_m v_m^2}{d}$$ 2.212

Friction factor f is determined from the Moody diagram (Fig. 2–8), using the Reynolds number:

$$N_{Re} = 124\,\frac{\rho_l v_m d}{\mu_l}$$ 2.213

Since the acceleration term in bubble flow is considered negligible, the total pressure gradient is the sum of the elevation and frictional terms.

Slug Flow

The determination of the liquid holdup follows the same approach as in bubble flow, only the bubble rise velocity v_b is substituted by the terminal rise velocity of a Taylor bubble. Aziz et al. use the formula proposed by Wallis [64]:

$$v_b = C\,\sqrt{\frac{gd(\rho_l - \rho_g)}{\rho_l}} = 1.637\,C\,\sqrt{\frac{d(\rho_l - \rho_g)}{\rho_l}}$$ 2.214

Proportionality factor, C, is calculated from the expression:

$$C = 0.345[1 - \exp\,(-0.029N)]\left[1 - \exp\!\left(\frac{3.37 - N_E}{m}\right)\right]$$ 2.215

The definition of the two new dimensionless numbers: N_E Eotvos number, and N viscosity number are given as

$$N_E = \frac{gd^2(\rho_l - \rho_g)}{\sigma_l} = 101.4\,\frac{d^2(\rho_l - \rho_g)}{\sigma_l}$$ 2.216

$$N = \frac{\sqrt{d^3 g \rho_l (\rho_l - \rho_g)}}{\mu_l} = 203 \frac{\sqrt{d^3 \rho_l (\rho_l - \rho_g)}}{\mu_l}$$

2.217

where:

ρ_l = liquid density, lb/cu ft

ρ_g = gas density, lb/cu ft

σ_l = interfacial tension, dyne/cm

μ_l = liquid viscosity, cP

d = pipe diameter, in.

The factor m is found from the table as follows, as a function of N:

N	≤ 18	$18 > N > 250$	≥ 250
m	25	$69\,N^{-0.35}$	10

After C is known, Taylor bubble velocity v_b can be found, and liquid holdup is calculated from Equation 2.210. Elevation pressure gradient is than determined from Equation 2.211.

Since the acceleration term is again disregarded, only frictional pressure gradient needs to be calculated from

$$\left(\frac{dp}{dh}\right)_f = 1.294 \times 10^{-3} f \frac{\rho_l \varepsilon_l v_m^2}{d}$$

2.218

Friction factor f is determined from the Moody diagram (Fig. 2–8), using the same Reynolds number as in bubble flow. (Equation 2.213)

Other Flow Patterns

Aziz et al. considered bubble and slug flow as the most important flow patterns in vertical oil well flow and did not provide calculation models for other flow patterns. In mist and transition flow, they recommended the use of the Duns-Ros methods. [33] Because flow pattern boundaries are different in the two models, the interpolation formula proposed for transition flow by Duns-Ros had to be modified. Total pressure gradient in transition flow, therefore, is determined from the equations given as follows:

$$\frac{dp}{dh} = B \left(\frac{dp}{dh}\right)_{slug} + (1 + B)\left(\frac{dp}{dh}\right)_{mist}$$

2.219

$$B = \frac{70(100N_y)^{-0.152} - N_x}{70(100N_y)^{-0.152} - (8.6 + 3.8N_y)}$$

2.220

Improvements

The only modification of the original Aziz et al. procedure was proposed by Al-Najjar and Al-Soof [65], who indicated that calculation accuracy can be improved if the Duns-Ros [33] flow pattern map is used. They based this conclusion on data from production through the casing-tubing annulus.

Example 2–28. Calculate the pressure gradient according to the Aziz et al. model for the three points in the well of the previous examples.

Solution

The wellhead case is detailed with the next input parameters calculated in previous examples:

v_{sl} = 0.79 ft/s; v_{sg} = 1.93 ft/s; v_m = 2.72 ft/s; ρ_l = 49.9 lb/cu ft; ρ_g = 1.79 lb/cu ft.

To find the flow pattern, the coordinates of the Aziz et al. flow pattern map have to be determined according to Equations 2.202 and 2.203:

N_x = 1.93 (1.79 / 0.0764)$^{1/3}$ (72 49.9 / 62.4 /8)$^{0.25}$ = 9.06.

N_y = 0.79 (72 49.9 / 62.4 /8)$^{0.25}$ = 1.29.

The bubble-slug and the slug-transition boundaries are calculated from Equation 2.204 and 2.205:

N_x = 0.51 (100 1.29)$^{0.172}$ = 1.18, and

N_x = 8.6 + 3.8 1.29 = 13.52.

The flow pattern is slug, because the actual N_x lies between the two boundary values.

In order to calculate the bubble rise velocity, two dimensionless parameters, N_E and N are found from Equation 2.216 and 2.217:

N_E = 101.4 2.441^2 (49.9 – 1.79) / 8 = 3634, and

N = 203 (2.4413 49.9 (49.9 – 1.79))$^{0.5}$ / 3 = 12648.

Factor m is found in the function of N as m = 10.

The proportionality factor of Equation 2.215:

$$C = 0.345[1 - \exp(-0.029\ 12648)]\left[1 - \exp\left(\frac{3.37 - 3634}{10}\right)\right] = 0.345$$

The bubble rise velocity can now be calculated (Equation 1.214):

v_b = 1.637 0.345 (2.441 (49.9 – 1.79) / 49.9)$^{0.5}$ = 0.866 ft/s.

Liquid holdup from Equation 2.210:

ε_l = 1 – 1.93 / (1.2 2.72 + 0.866) = 0.533.

From this, mixture density is found as:

ρ_m = 49.9 0.533 + 1.79 (1 – 0.533) = 27.43 lb/cu ft.

The elevation term of the pressure gradient is calculated from Equation 2.211:

$(dp/db)_{el}$ = 27.43 / 144 = 0.190 psi/ft.

Calculation of the friction component starts with finding of the Reynolds number from Equation 2. 213:

N_{Re} = 124 49.9 2.72 2.441 / 3 = 13,706.

From Figure 2–8 a friction factor of f = 0.029 is read off.

Frictional pressure gradient is evaluated according to Equation 2.218:

$(dp/dh)_f$ = 0.001294 0.029 49.9 0.533 2.72^2 / 2.441 = 0.003 psi/ft.

Kinetic effects are disregarded in slug flow and the total pressure gradient is found from Equation 2.201 with E_k = 0:

dp/dh = 0.190 + 0.003 = 0.193 psi/ft.

For the remaining points only final results are given as follows.

Parameter	Wellhead	At 3,770 ft	Bottom
$N_{x'}$ -	9.06	0.12	0
$N_{y'}$ -	1.29	1.51	1.53
flow pattern	Slug	bubble	single
bubble rise velocity, ft/s	0.866	0.456	-
liquid holdup, -	0.533	0.989	1
mixture density, lb/cu ft	27.4	45.0	44.5
elevation gradient, psi/ft	0.190	0.313	0.309
Reynolds number, -	13,706	4,390	4,329
friction factor, -	0.029	0.039	0.04
frictional gradient, psi/ft	0.003	0.0008	0.0008
pressure gradient, psi/ft	0.193	0.313	0.310

2.5.3.4.3 Hasan-Kabir model.

Summary

The first model based fully on mechanistic considerations utilizes a comprehensive flow pattern recognition scheme and accounts for the effects of pipe inclination. Bubble, slug, and annular mist flows are treated, and pressure gradient calculations cover multiphase flow in annuli as well.

The authors published several papers detailing their calculation model for vertical wells [60, 66] and inclined wells [67, 68] including annulus flow configurations. They used the latest flow pattern determination models and the achievements of current hydraulic research.

Basic Equation

Pressure gradient is found from the familiar equation valid for an inclined pipe and contains the three general components of the total pressure drop:

$$\frac{dp}{dl} = \frac{\left(\dfrac{dp}{dl}\right)_{el} + \left(\dfrac{dp}{dl}\right)_f}{1 - E_k}$$

2.221

As usual, the kinetic term is disregarded in all but the mist flow pattern. In those cases the dimensionless kinetic term E_k equals zero. The hydrostatic or elevation term of the pressure gradient is detailed as follows:

$$\left(\frac{dp}{dh}\right)_{el} = \frac{1}{144} \rho_m \sin \alpha = \frac{1}{144} \left[\rho_l \varepsilon_l + \rho_g (1 - \varepsilon_l) \right] \sin \alpha$$

2.222

In the frictional gradient term, different densities and velocities are used in the different flow patterns, but the formula has the following general form:

$$\left(\frac{dp}{dl}\right)_f = 1.294 \times 10\text{--}3 \, f \, \frac{\rho v^2}{d}$$

2.223

Flow Pattern Map

The authors distinguish the usual five flow patterns: bubble, dispersed bubble, slug, churn, and mist (annular). Basically, the determination of flow patterns follows the universally accepted formulas of Barnea [32] with some exceptions. The flow pattern map used by Hasan and Kabir is displayed in Figure 2–39. As seen, the map has fluid velocities plotted at its axes, but investigation of the transition formulas detailed later will show those boundaries to vary with the fluid properties. The map shown in the figure is valid for a vertical pipe and the flow taking place at atmospheric conditions.

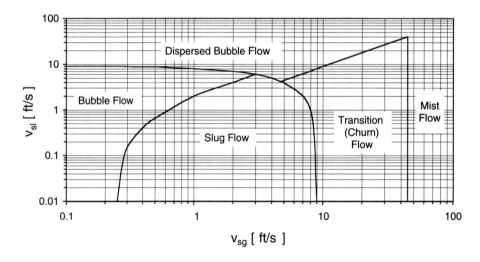

Fig. 2–39 Flow pattern map of Hasan and Kabir [67] valid for vertical flow.

When compiling the formulas for the flow pattern transitions, the authors concluded that only the bubble-slug transition is affected by the pipe inclination. The other transitions can accurately be predicted by the equations valid for vertical flow. In the following, the formulas for the relevant boundaries are given.

$$v_m^{1.12} = 7.78 d^{0.48} \sqrt{\frac{(\rho_l - \rho_g)}{\sigma_l}} = \left[\frac{\sigma_l}{\rho_l} \right]^{0.6} \left[\frac{\rho_l}{\mu_l} \right]^{0.08}$$ *bubble-dispersed bubble*   2.224

$$v_{gs} = (0.429v_{sl} + 0.537v_b) \sin\alpha \qquad \textit{bubble-slug} \qquad 2.225$$

In the previous formula, bubble rise velocity, v_b, is calculated from the Harmathy [31] formula (Eq. 2.71) since it was found to be independent of pipe inclination.

$$\rho_g v_{sg}^2 = 0.0089(\rho_l v_{sl}^2)^{1.7} \quad \text{if} \quad \rho_l v_{sl}^2 < 34 \frac{\text{lb}}{\text{ft s}^2} \qquad \textit{slug-churn} \qquad 2.226$$

$$\rho_g v_{sg}^2 = 11.5 \log(\rho_l v_{sl}^2) - 13.6 \quad \text{if} \quad \rho_l v_{sl}^2 > 34 \frac{\text{lb}}{\text{ft s}^2} \qquad \textit{slug-churn} \qquad 2.227$$

$$v_{sg} = 1.6 \sqrt[4]{\frac{\sigma_l(\rho_l - \rho_g)}{\rho_g^2}} \qquad \textit{churn-mist} \qquad 2.228$$

$$v_{sg} = 1.083v_{sl} \qquad \textit{dispersed bubble-mist} \qquad 2.229$$

where:
ρ_l = liquid density, lb/cu ft

ρ_g = gas density, lb/cu ft

σ_l = interfacial tension, dyne/cm

ρ_m = mixture density, lb/cu ft

α = pipe inclination angle, measured from the horizontal, rad

v_m = mixture superficial velocity, ft/s

v_{sg} = gas superficial velocity, ft/s

v_{sl} = liquid superficial velocity, ft/s

μ_l = liquid viscosity, cP

d = pipe diameter, in.

Bubble, Dispersed Bubble Flows

Hasan and Kabir proposed the same calculation procedure for the bubble and dispersed bubble flow patterns. Their approach to describe these flow patterns is similar to that of Aziz et al. [59]

Liquid holdup is determined with the use of the drift-flux model (see Section 2.5.2.1) from:

$$\varepsilon_l = 1 - \frac{v_{sg}}{C_0 v_m + v_b} \qquad 2.230$$

where:
$C_0 = 1.2$ for tubing flow

$C_0 = 1.2 + 0.371 \dfrac{d_t}{d_c}$ for flow in a casing-tubing annulus

d_t = tubing outside diameter, in.

d_c = casing inside diameter, in.

Bubble rise velocity v_b was found to be insensitive to pipe inclination angle, and the use of the Harmathy [31] formula is proposed:

$$v_b = 0.79 \sqrt[4]{\frac{\sigma_l (\rho_l - \rho_g)}{\rho_l^2}}$$

2.231

where: ρ_l = liquid density, lb/cu ft

ρ_g = gas density, lb/cu ft

σ_l = interfacial tension, dyne/cm

For calculating the frictional pressure gradient, the authors use the properties of the multiphase mixture in Equation 2.223:

$$\left(\frac{dp}{dl}\right)_f = 1.294 \times 10^{-3} f \frac{\rho_m v_m^2}{d}$$

2.232

The friction factor f is determined from the Moody diagram (Fig. 2–8), using the Reynolds number:

$$N_{Re} = 124 \frac{\rho_l v_m d}{\mu_l}$$

2.233

Slug Flow

The same basic approach (the drift-flux model) is applied to slug flow as done in bubble flow but the properties of large Taylor bubbles occupying almost the total pipe cross-sectional area have to be used. The liquid holdup formula, therefore, must be modified to

$$\varepsilon_l = 1 \frac{v_{sg}}{C_1 v_m + v_T}$$

2.234

where: $C_1 = 1.2$ for tubing flow

$C_1 = 1.2 + 0.7 \dfrac{d_t}{d_c}$ for flow in a casing-tubing annulus

The terminal rise velocity v_T of the Taylor bubbles present in slug flow was found to be significantly influenced by the inclination angle of the pipe. For a vertical pipe, terminal rise velocity of a Taylor bubble is calculated, similar to Aziz et al. [59], from Equation 2.214. To find the rise velocity in an inclined pipe, a balance of the buoyancy and drag forces acting on the Taylor bubble submerged in the two-phase mixture yields the following formula:

$$v_T = C_2 \sqrt{\frac{d(\rho_l - \rho_g)}{\rho_l}} = \sqrt{\sin\alpha} \, (1 + \cos\alpha)^{1.2}$$

2.235

where: $C_2 = 0.573$ for tubing flow

$C_2 = 0.573 + 0.16 \dfrac{d_t}{d_c}$ for flow in a casing-tubing annulus

d_t = tubing outside diameter, in.

d_c = casing inside diameter, in.

Frictional pressure gradient is calculated according to Aziz et al. [59]:

$$\left(\frac{dp}{dl}\right)_f = 1.294 \times 10^{-3} f \frac{\rho_l \varepsilon_l v_m^2}{d}$$

2.236

Friction factor f is determined from the Moody diagram (Fig. 2–8), using the same Reynolds number as in bubble flow. (Equation 2.233)

Churn Flow

The authors did not examine the churn flow pattern in detail and proposed that the calculation model developed for slug flow is used. The only modification was changing of the values of C_1 in Equation 2.234 to reflect the chaotic nature of the flow by assuming a flatter gas concentration profile across the pipe. The new values are

$C_1 = 1.15$ for tubing flow

$C_1 = 1.15 + 0.7 \dfrac{d_t}{d_c}$ for flow in a casing-tubing annulus

Mist (Annular) Flow

In mist flow, the gas and liquid phases flow in two well-defined parts of the pipe. In the central core, a high-velocity gas stream flows, which includes small dispersed liquid droplets of the same velocity. At the pipe wall, on the other hand, a slower-moving wavy liquid film is situated. Because of this flow configuration, the elevation component of the pressure gradient is almost entirely determined by the properties of the central core, while the frictional part occurs on the liquid film. It is also easy to see that pipe inclination does not affect the flow conditions because of the high flow velocities.

In order to calculate the elevation component, the liquid holdup in the gas core has to be found. Since the liquid droplets taken by the gas stream exhibit the same velocity, no slippage between the gas and liquid phases occurs here. Therefore, a no-slip liquid holdup must be calculated, which from simple volumetric considerations takes the form:

$$\lambda_{lc} = 1 - \frac{E v_{sl}}{v_{sg} + E v_{sl}}$$

2.237

In the previous formula liquid entrainment E is defined as the fraction of the total liquid content of the wellstream entrained in the gas core. Assuming fully developed turbulent flow in the liquid film, Hasan and Kabir propose the following formulas to calculate liquid entrainment in the gas core:

$$E = 184 \times 10^{-4} v_{crit}^{2.86} \qquad\qquad\qquad \text{if} \quad v_{crit} < 13 \frac{ft}{s}$$

2.238

$$E = 0.857 \log v_{crit} - 0.642 \qquad\qquad\qquad \text{if} \quad v_{crit} > 13 \frac{ft}{s}$$

2.239

Gas core liquid entrainment is limited to values $E \leq 1.0$ since part of the total liquid content is situated in the liquid film covering the pipe wall. The definition of the critical vapor velocity v_{crit} figuring in the previous formulas is

$$v_{crit} = 10^4 \times \frac{v_{sg}\mu_g}{\sigma_l}\sqrt{\frac{\rho_g}{\rho_l}}$$

2.240

where: ρ_l = liquid density, lb/cu ft

ρ_g = gas density, lb/cu ft

σ_l = interfacial tension, dyne/cm

v_{sg} = gas superficial velocity, ft/s

μ_g = gas viscosity, cP

The elevation term can now be calculated from Equation 2.222, by substituting into the mixture density the gas core density defined as follows:

$$\rho_m = \rho_c = \rho_l\lambda_{lc} + \rho_g(1 - \lambda_{lc})$$

2.241

The frictional pressure drop occurs due to the high-velocity flow of the gas core on the wavy liquid film covering the pipe inside wall. Since the thickness of this film is not very significant (typically less than 5% of the pipe diameter), the basic formula to be used in this special case neglects the changes in diameter:

$$\left(\frac{dp}{dl}\right)_f = 1.294 \times 10^{-3} f_c \frac{\rho_c}{d}\left(\frac{v_{sg}}{1 - \lambda_{lc}}\right)^2$$

2.242

where: ρ_c = gas core no-slip density, lb/cu ft

v_{sg} = gas superficial velocity, ft/s

λ_{lc} = liquid holdup in gas core, -

f_c = liquid film friction factor, -

d = pipe diameter, in.

In the previous formula, friction factor in the liquid film is found, according to Hasan and Kabir from the formula:

$$f_c = 0.024\ (1 + 75\lambda_{lc})\sqrt[4]{\frac{\mu_g}{\rho_g v_{sg} d}}$$

2.243

where: ρ_g = gas density, lb/cu ft

μ_g = gas viscosity, cP

Gas core density being defined by Equation 2.240, frictional pressure gradient is easily found from Equation 2.241.

In mist flow, acceleration effects cannot be neglected and the dimensionless kinetic term E_k must be included in the basic equation. (Equation 2.221) Hasan and Kabir proposed the following formula:

$$E_k = 2.16 \times 10^{-4} \frac{v_{sg}^2 \rho_c}{p} \qquad\qquad 2.244$$

where: p = pressure, psi

Example 2–29. Calculate the pressure gradient according to the Hasan-Kabir model for the three points in the well of the previous examples.

Solution

Only the wellhead case is detailed with the next input parameters taken from previous examples:

v_{sl} = 0.79 ft/s; v_{sg} = 1.93 ft/s; v_m = 2.72 ft/s; ρ_l = 49.9 lb/cu ft; ρ_g = 1.79 lb/cu ft.

First check if the conditions for dispersed bubble flow are met by calculating the required mixture velocity from Equation 2.224:

$$v_m^{1.12} = 7.78 \; 2.441^{0.48} \sqrt{\frac{(49.9 - 1.79)}{8}} = \left[\frac{8}{49.9}\right]^{0.6} \left[\frac{49.9}{3}\right]^{0.08} = 14.54$$

From this, v_m = 10.9 ft/s which is less than the actual v_m = 2.72 ft/s, therefore the flow pattern cannot be dispersed bubble.

Now the bubble-slug transition will be checked but first the bubble rise velocity must be calculated from Equation 2.71:

v_b = 0.79 [8 (49.9 – 1.79) / 49.9^2]$^{0.25}$ = 0.495 ft/s.

The gas superficial velocity required for slug flow from Equation 2.225, where sinα = 1:

v_{gs} = 0.429 0.79 + 0.357 0.495 = 0.516 ft/s.

Since the previous value is less than the actual v_{gs} = 1.93 ft/s, the flow pattern is slug.

In order to find the liquid holdup, first the rising velocity of the Taylor bubbles is calculated from Equation 2.235, again with the substitution of sinα = 1:

v_T = 0.573 [2.441 (49.9 – 1.79) / 49.9]$^{0.5}$ = 0.88 ft/s.

Liquid holdup from Equation 2.234:

ε_l = 1 – 1.93 / (1.2 2.72 + 0.88) = 0.534.

Mixture density can now be calculated:

ρ_m = 49.9 0.534 + 1.79 (1 – 0.534) = 27.5 lb/cu ft.

Elevation gradient from Equation 2.222:

$(dp/dl)_{el}$ = 27.5 /144 = 0.191 psi/ft.

For frictional pressure gradient calculation the Reynolds number is found from Equation 2.232:

N_{Re} = 124 49.9 2.72 2.441 / 3 = 13706.

From Figure 2–8, a friction factor of f = 0.029 is read off.

Frictional pressure gradient is evaluated according to Equation 2.236:

$(dp/dl)_f$ = 0.001294 0.029 49.9 0.543 2.72^2 / 2.441 = 0.003 psi/ft.

Total pressure gradient is the sum of the two components just calculated:

dp/dl = 0.19 + 0.003 = 0.193 psi/ft.

For the remaining points only final results are given as follows.

Parameter	Wellhead	At 3,770 ft	Bottom
bubble-dispersed v_m limit, ft/s	10.9	10.2	-
bubble rise velocity, ft/s	0.495	0.489	-
bubble-slug v_{gs} limit, ft/s	0.516	0.679	-
flow pattern	slug	bubble	single
Taylor bubble rise velocity, ft/s	0.879	-	-
liquid holdup, -	0.534	0.990	1
mixture density, lb/cu ft	27.5	45.0	44.5
elevation gradient, psi/ft	0.190	0.313	0.309
Reynolds number, -	13,706	4,391	4,329
friction factor, -	0.029	0.039	0.04
frictional gradient, psi/ft	0.003	0.0008	0.0008
pressure gradient, psi/ft	0.194	0.313	0.310

2.5.3.4.4 *Further models.* In recent years, there is a growing amount of publications dealing with mechanistic modeling of multiphase well flow. These are getting more and more complicated and are founded on the latest research results published in the professional literature. It remains to be seen if the increased calculation demand usually associated with these models is justified by an increase in prediction accuracy. The reader is advised to consult Section 2.5.3.6, where published accuracies of empirical and mechanistic models are compared.

In the following paragraphs, short descriptions of the most recent models are given.

Ansari et al. [61] developed their calculation model for vertical uphill flow. The flow patterns recognized are bubble, dispersed bubble, slug, churn, and annular mist. In slug flow, two different calculation procedures are given for the two kinds of flow: fully developed and developing slug flow. For fully developed slug flow—very important in vertical flow—mass balance equations result in a set of eight equations to be solved for eight unknowns. Churn (transition) flow is treated as part of slug flow. In mist flow, a Newton-Raphson iterative procedure is used to find the thickness of the falling liquid film present on the inside pipe wall.

Chokshi, Schmidt, and Doty [62] report on an extensive experimental project in a vertical, 1,348-ft-long flow pipe of 3½ in. diameter. In addition to downhole pressure transducers, a gamma-ray densitometer provided accurate data on liquid holdup. Measurement data were used for comparison purposes only, and the pressure gradient formulas developed by the authors reflect a fully mechanistic approach. Only three flow patterns—bubble, slug, and annular mist—are considered, and the flow is assumed to be vertical. In bubble flow, the drift-flux model is followed; and a set of seven equations is to be solved for slug flow. For mist flow, the separate flow of a gas core and a liquid film is described, with liquid film thickness as the main variable.

Kaya, Sarica, and Brill [34] presented a calculation model for deviated wells and have included the effect of pipe inclination on flow pattern transitions using the Barnea [32] equations. Bubble, dispersed bubble, slug, churn, and annular mist flow patterns are distinguished. For bubble flow, the drift-flux model is used; for dispersed bubble flow, the no-slip model is used. They use, with significant modifications, the Chokshi, Schmidt, and Doty [62] model in slug flow. Churn flow is treated similar to slug flow, and the Ansari et al. [61] model is adopted for annular mist flow.

The pressure drop calculation model of Gomez et al. [63] represents a new approach to solving multiphase flow problems because it treats horizontal, inclined, and vertical flows with one single unified mechanistic model. This approach has definite advantages when a simultaneous solution of wellbore and surface flow problems is sought, like in Nodal Analyses.

2.5.3.5 Calculation of pressure traverses.

Previous sections described the numerous ways pressure gradients in multiphase oil well flow can be determined. Although this is the most difficult part of multiphase flow calculations, it is the pressure drop in the well or the pressure distribution along the tubing that production engineers are mostly interested in. The present section details how the basic pressure gradient equations are solved to reach that goal.

The general pressure gradient equations for vertical or inclined multiphase flow (see Section 2.5.2.3) are ordinary, non-algebraic differential equations that can be written in the following symbolic form:

$$\frac{dp}{dh} = f(p, T) = f(p, h) \qquad\qquad 2.245$$

As seen, pressure gradient is basically a function of the prevailing pressure and temperature. However, flowing temperature depends on well depth (this function is assumed to be known in our treatment) and this is the reason why the right-hand side of the equation can be expressed as a function of pressure and flow length. The analytical solution of the previous differential equation is impossible because the right-hand side contains many empirical functions (deviation factor, friction factor, etc.) and further it cannot normally be expressed in an analytical form.

Thus, some numerical method must be employed for the solution, *i.e.* for the calculation of pressure distribution along the flow pipe. The method most often used involves a trial-and-error (or iterative) procedure ideally suited to computer calculations. It is based on the conversion of the basic differential equation into a difference equation using finite differences. After conversion, the right-hand side of Equation 2.245 is evaluated at the average conditions of the given increment as shown here. Provided these increments are small enough, then the change of the flowing gradient is negligible and the new equation closely approximates the original one:

$$\frac{\Delta p}{\Delta h} = f(p_{avg}, h_{avg}) \qquad\qquad 2.246$$

This equation, in the knowledge of a previously set pressure or depth increment, allows the calculation of the unknown increment. The only complication is that the pressure gradient at average conditions (the right-hand side) depends on average temperature, which in turn is a function of well depth. Thus an iterative scheme must be used to find one increment from the other. The details of a complete pressure distribution calculation are given as follows.

The calculation of a flowing pressure traverse in a multiphase oil well starts with dividing the total well into pressure or depth increments, depending on the approach desired. Figure 2–40 shows a flowchart for calculating the bottomhole pressure based on set pressure increments. Calculations start from the known wellhead conditions $p_1 = p_{wh}$ and $h_1 = 0$. First, the actual pressure increment Δp is set, whose proper selection will be discussed later. Then, in order to calculate an average temperature, which in turn is a function of the depth increment to be sought, an initial value of Δh needs to be estimated. After average pressure in the actual calculation step is found, average temperature is calculated from the temperature-depth function that is assumed to be known. Since average conditions in the actual calculation step are known, the pVT properties (densities, viscosities, etc.) of the flowing fluids can be calculated. The next step is to evaluate the pressure gradient using the appropriate procedure prescribed in the utilized pressure drop calculation model.

The right-hand side of Equation 2.246 being known the depth increment Δh_{calc}, corresponding to the previously set pressure increment is easily calculated from that formula. This increment is compared to its assumed value and if they differ by more than the previously set accuracy ε a new iteration step is performed with an assumed depth increment equal to the value just calculated. This iterative process is continued until the last two values of the depth increment match with the required accuracy. At this point, the depth increment belonging to the actual pressure increment is known, specifying the pressure and depth at a lower point in the well. If the well bottom is not yet reached, a new step is processed and the calculation of a new depth increment follows by repeating the previous procedure. When the well bottom is reached, bottomhole pressure is easily calculated by interpolation.

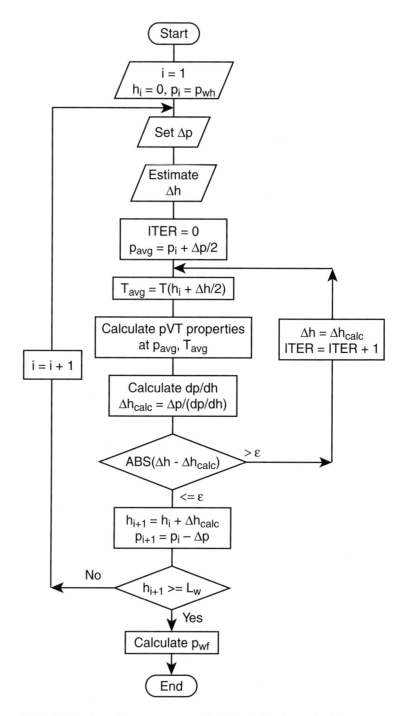

Fig. 2–40 Flowchart of pressure traverse calculation for iterating on depth increment.

In case the wellhead pressure is to be calculated from a known bottomhole pressure, the previously detailed process is slightly modified. In this case, calculations are started from the well bottom with the average pressures and temperatures calculated accordingly.

The other approach to pressure traverse calculations is to divide the flow pipe into a series of depth increments and calculate the corresponding pressure increments. Again, Equation 2.246 is used but now it is solved for the pressure increment Δp_{calc}. The flowchart of this calculation model is given in Figure 2–41 for a case where wellhead pressure is sought. Calculations follow the one described previously and are started from the conditions at the well bottom. Again, calculation of bottomhole pressures requires only minor changes in the procedure given.

The two basic solutions (iterating on depth or on pressure increment) detailed previously offer different advantages and drawbacks. Setting the pressure increment and then finding the corresponding depth increment may not converge when the pressure gradient approaches zero, a condition often occurring in downward flow. In such cases, the solution of Equation 2.246 has pressure gradient in the denominator and may thus result in an infinite depth increment. On the other hand, the advantage of this approach is that no iterations at all are required if flowing temperature is constant along the pipe. The other approach (iterating on pressure) allows one to exactly follow the well's inclination profile since depth increments can be selected at will. In addition, no interpolation is required at the end of the calculation process. Further, the convergence problem mentioned previously does not exist because the solution of Equation 2.246 for Δp involves multiplication only.

As with every numerical solution, the accuracy of the previous procedures relies on the proper selection of step sizes and iteration tolerances. Naturally, the smaller the pressure or depth increment, the more accurate the solution. Small increments, on the other hand, require more steps to be taken and increase calculation time. Consequently, the proper selection of increment sizes is a matter of finding a balance between the required accuracy and computation time. Some guidelines are given as follows:

- For obtaining sufficient accuracy, the basic rule is that increments should be selected so that pressure gradient is essentially constant inside the increment.

- Pressure increments should vary with well depth so that they decrease at smaller depths. This is recommended because flowing gradients decrease in the same direction. Usually, increments of about 10% of the actual pressure are used.

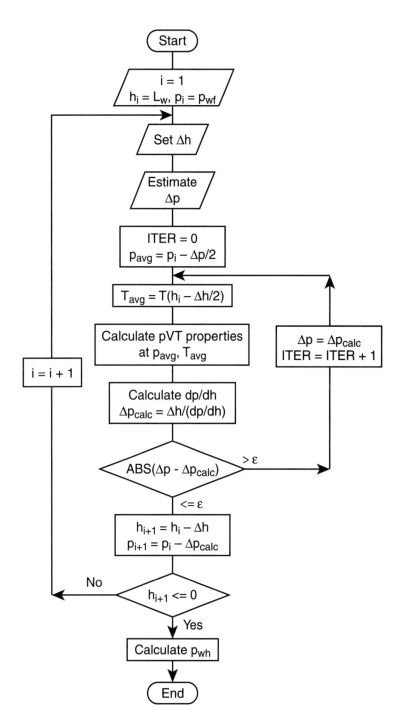

Fig. 2–41 Flowchart of pressure traverse calculation for iterating on pressure increment.

- Selection of depth increments should consider the well's deviation profile by varying the increment sizes.

- The well should be divided into segments with a new segment wherever a significant change in inclination angle or fluid rates occurs. The latter happens at gas lift injection points or in case of commingled production.

2.5.3.6 Accuracy and selection of pressure drop calculation models.

2.5.3.6.1 Introduction. Previous parts of Section 2.5.3 described in detail the many calculation models available to present-day petroleum engineers dealing with multiphase flow problems. As seen from the long list of procedures, there is a big choice available today. This section investigates the accuracies obtainable from the use of the different models and presents the right philosophy to be adopted by the production engineer.

Before the detailed treatment of the objectives set forth previously, let us examine the behavior of empirical models. As discussed before, empirical correlations were originally developed under widely different physical circumstances. This fact could readily be seen in Table 2–8 where the main parameters of experimental data used by the different authors are shown. Grouping of the models is done according to the criteria described in Section 2.5.3.3.1. Experimental data may originate from laboratory measurements, special field tests, or routine field measurements. The pipe lengths, sizes, and liquids investigated are listed along with the total number of measured pressure traverse curves. The table proves that data of the different authors represent very different conditions.

AUTHOR	Publ. Year	Group	Data	Pipe Length	Nominal Pipe Diameters, in							Water Cut, %	# of Data
					1	1 1/4	1 1/2	2 3/8	2 7/8	3 1/2	Ann.		
P. - CARPENTER	1952	I	field	various								0 - 98	49
GILBERT	1954		field									0	
BAXENDELL	1958	I	field										50
BAX. - THOMAS	1961	I	field	6000'								0	25
DUNS - ROS	1963	III	lab	33'	ID = 3.2 cm, 8.02 cm, and 14.23 cm pipes							0 & 100	4000
FANCH. - BROWN	1963	I	test	8000'								95	106
GAITHER ET AL.	1963	I	test	1000'								100	139
HAG. - BROWN I	1964	I	test	1500'								0 & 100	175
HAG. - BROWN II	1965	II	test	1500'								0 & 100	581
ORKISZEWSKI	1967	III	field	various								0 & 100	148
BEGGS - BRILL	1973	III	test	45'								100	584
MUKH. - BRILL	1985	III	test	32'								0	1000

Table 2–8 Summary of experimental data sets used by the authors of empirical multiphase pressure drop correlations.

The difference in original conditions is shown for two further aspects in Figure 2–42 and Figure 2–43, where the liquid flow rate and GLR ranges of the basic data are displayed for each correlation.

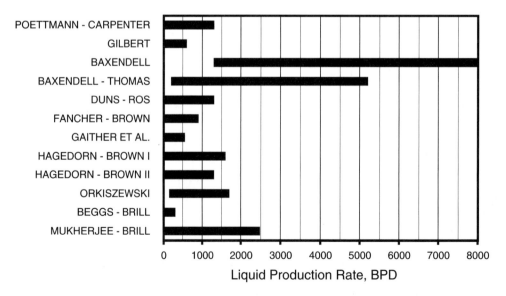

Fig. 2–42 Liquid flow rate ranges of different empirical multiphase pressure drop correlations.

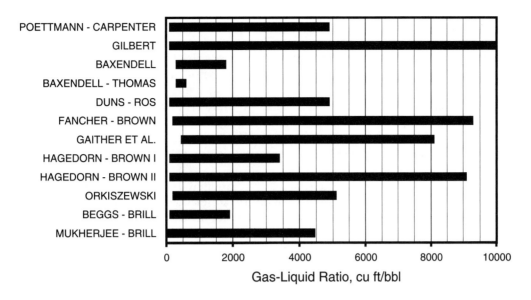

Fig. 2–43 GLR ranges of different empirical multiphase pressure drop correlations.

The table and the two figures given previously indicate that authors used data from widely different physical conditions (pipe sizes, flow rate, GLR-ranges, etc.) to develop their respective pressure drop prediction methods. Consequently, if one correlation is used outside its original conditions, predictions with heavy errors are likely to occur. Although this is quite obvious, in practice all correlations are being used irrespective of the flow conditions at hand.

2.5.3.6.2 Possible sources of prediction errors. As it is generally known, calculated pressure drops usually deviate from those actually measured in a well. Prediction errors can be attributed to a host of known and unknown issues. The classification presented as follows is an approach to sort the possible sources of calculation errors in some larger categories. A detailed analysis of the described sources of error would allow one to draw important conclusions for the further improvement of accuracy in vertical two-phase pressure drop calculations.

Characteristics of the Physical Model

When developing a calculation model for multiphase pressure drop determination, it is never possible to include all conceivable features actually affecting the pressure drop. Therefore, the behavior of the real and thus extremely complex multiphase flow must be approximated by a properly chosen simplified physical model. The physical model adapted has a decisive effect on the accuracy and behavior of the calculation model based on it. The more comprehensive the physical model applied, the more likely it is that flow conditions are properly described.

It was shown before that available calculation methods utilize different physical models for the description of multiphase flow phenomena. Out of the many possible simplifications and approximations used to build the physical model, two criteria are found to play an important role in the determination of calculation accuracy.

- Physical phenomena are essentially different in the various flow patterns occurring in multiphase vertical flow, considerations for this effect must be included in every computation method. Mechanistic models, by their nature, take care of this problem, but some of the earlier empirical methods lack this feature resulting in lower calculation accuracy.

- Under certain conditions, the effect of gas slippage in liquid can be very important. In such cases correlations considering no slip between phases (e.g. Poettmann-Carpenter and its improvements) may perform poorly.

Errors in Fluid Properties

When performing the calculation of a multiphase pressure traverse several thermodynamic parameters of the flowing fluids (oil, water, natural gas) must be known in the wide ranges of pressure and temperature occurring in a well. But experimental pVT data on these are usually available at reservoir temperature and at projected reservoir pressures only. A good example is oil volume factor B_o, commonly called FVF by reservoir engineers, usually measured at reservoir conditions only. For production engineering calculations, however, B_o must be known in the wide ranges of pressure and temperature starting from wellhead to bottomhole conditions. This situation also holds for other thermodynamic parameters like solution GOR, oil viscosity, bubblepoint pressure, etc. for which measured data are seldom available.

In a lack of experimental data, which is the general case, the required thermodynamic properties of oil, water, and gas have to be computed using standard petroleum industry correlations. Use of such correlations instead of measurement data inevitably introduces errors and may heavily affect the accuracy of calculated pressure drops.

Of the many fluid properties required for pressure drop calculations, the value of the bubblepoint pressure plays the most decisive role as shown by several authors. [55, 69] This is primarily due to the fact that the value of the bubblepoint pressure determines the point in the well below which single-phase flow prevails and above which multiphase flow starts. Since calculation accuracy in single-phase flow is excellent but is much worse in multiphase flow, a small error in the value of bubblepoint pressure can bring about great errors in the calculated pressure drop. Standard bubblepoint pressure correlations, however, may give 50% or higher prediction errors [36], thus further increasing the deviation of calculated and measured pressure drops. Because of this, any multiphase pressure drop model may perform differently if, in lack of experimental data, different correlations are used for the calculation of bubblepoint pressure. This effect is greatly affected by the direction of calculation (from wellhead to bottom or vice versa) but that will be discussed later. Consequently, use of measured bubblepoint pressure data may greatly improve calculation accuracy, but this holds for the rest of the fluid properties as well.

Most of the pressure drop calculation models were developed for two discrete phases, a liquid and a gas one and are therefore properly called two-phase methods. In case the liquid phase is not pure oil or water but a mixture of these, the properties of this phase have to be found based on the properties of oil and water. In lack of sufficient research results, the usual solution is to apply a mixing rule based on the actual flow rates of the components. This approach is at least questionable since it does not include the effects of slippage between oil and water, the occurrence of emulsions, etc. Therefore, water-cut oil production usually increases the errors in calculated pressure drops and is a subject of future research.

Empirical Correlations

All empirical pressure drop correlations include several empirical functions applicable only in certain specified ranges, outside which their performance is at least questionable. There is no use to list all such limited correlations used in all pressure drop models because even a superficial study of the calculation processes can reveal them. These are responsible for the general observation that any correlation, if used outside its original ranges, will produce increased calculation errors. Mechanistic models, on the contrary, may exhibit much wider ranges of applicability, partly because they almost completely eliminate the use of empirical correlations.

Calculation Direction

It is a widely observed fact that the accuracy of any multiphase flow model depends also on the direction the calculations are made as compared to the flow direction. If pressure drop is calculated along the flow direction (wellhead pressure is calculated from a known bottomhole pressure), the error between measured and calculated pressure drops is always lower as compared to calculations done in the other direction. Therefore, any multiphase flow model performs better if the solution of the gradient equation is done starting from the bottom of the well, *i.e.* along the flow direction. The explanation of this behavior lies in the great difference of calculation accuracies valid in single-phase and multiphase flows. As discussed before, single-phase liquid flow is relatively easy to describe, and pressure drop calculations are quite accurate. In comparison, multiphase flow is much more complex and greater calculation errors can be expected.

Since single-phase flow usually occurs at or above the well bottom, calculations started from there involve the determination of single-phase pressure drops first. Moving higher up the hole, at a depth corresponding to the local bubblepoint, pressure flow type changes to a multiphase one and calculation accuracy rapidly deteriorates. Obviously, the cumulative calculation error will depend on the reliability of the bubblepoint pressure value and the relative depth occupied by single-phase flow, as first shown by Gregory et al. [70]. If, as is generally the case, a sufficient tubing length is occupied by single-phase flow, the accuracy of the calculated total pressure drop will clearly be determined by the excellent accuracy of single-phase flow description. Thus, pressure drop calculations done in the flow direction are relatively more accurate.

If, on the contrary, calculations are started at the wellhead, initial calculation steps involve much greater calculation errors due to the existence of multiphase flow at and below that point. Since these errors are cumulative and add up in successive calculation steps, the greater accuracy achieved in the single-phase region below the depth corresponding to the bubblepoint pressure does not prevent an excessive final total error.

Although the error in pressure drop calculations varies with the direction of calculations, the relative ranking of the different calculation models is not affected by this phenomenon. This fact can be observed from the results of those investigators who compared calculation accuracies for both cases, like Gregory et al. [70], Chokshi et al. [62], and Kaya et al. [34]

Special Conditions

Most of the vertical pressure drop calculation models were developed for average oilfield fluids. This is why special conditions like emulsions, non-Newtonian flow behavior, excessive scale or wax deposition on the tubing wall, etc. can pose severe problems. Predictions in such cases could be doubtful.

Special treatment is required for two commonly occurring cases: flow in inclined wells and production through casing-tubing annuli. Although there exist some approximate solutions to deal with such wells, calculation errors can be high unless a specifically developed flow model is used.

Errors in Measured Data

Empirical and mechanistic pressure drop calculation models alike were developed based on accurately and reliably measured flow parameters. This is why errors in the required input data can seriously affect the prediction accuracy of any pressure drop calculation model. Evaluation of any flow model, therefore, must be made by using reliable measured values of the actual flow parameters. Special care has to be taken to ensure that measurements in the well are carried out under steady-state conditions. Also, the accuracy of gas flow rates has to be checked because much higher inaccuracies can be expected in gas volume measurements than in liquid rates. In summary, pressure drop calculation errors stemming from poor input data will add to those of the calculation model itself and the model's true accuracy may be masked.

2.5.3.6.3 *Results of published evaluations.* Over the years, a great number of authors have investigated the accuracy and applicability of the various pressure drop correlations for different conditions. Table 2–9 contains a compilation of the results of such studies published so far. Due to insufficient data in some of the original papers, the table contains the following statistical parameters only:

$$d = \frac{\sum\limits_{i=1}^{N} d_i}{N} \qquad\qquad \textit{average error} \qquad\qquad 2.247$$

$$\sigma = \sum\limits_{i=1}^{N} \sqrt{\frac{(d_i - d)^2}{N-1}} \qquad\qquad \textit{standard deviation of errors} \qquad\qquad 2.248$$

$$d_a = \frac{\sum\limits_{i=1}^{N} |d_i|}{N} \qquad\qquad \textit{average absolute error} \qquad\qquad 2.249$$

where N is the total number of cases considered.

Calculation error is defined as the relative error in calculated pressure drops, by

$$d_i = \frac{\Delta p_{calc} - \Delta p_{meas}}{\Delta p_{meas}} \times 100 \qquad\qquad 2.250$$

where Δp_{calc} = calculated pressure drop in well

Δp_{meas} = measured pressure drop for same conditions

Table 2–9 shows the previous statistical parameters as given by the various investigators of the different correlations. Negative average errors indicate calculated pressure drops lower than measured ones, positive ones show

Model		37	69	59	73	71,74	75	45	70	48	76	77	78	61	61	62	72	63	34
Poettmann-Carpenter	d				-6.3	107.3						24.8							
	d_a											28.9							
	σ				9.6	195.7						28.2							
	N				77	726						323							
Baxendell-Thomas	d				-1.7	108.3		-5.1				19.0							
	d_a							5.1				27.2							
	σ				6.4	195.1		2.9				31.7							
	N				77	726		10				323							
Fancher-Brown	d				5.5							25.3							
	d_a											30.9							
	σ					36.1						32.5							
	N					726						323							
Hagedorn-Brown II	d	0.7	-17.8	-16.2	-3.6	1.3		-8.9	-4.3	3.6	8.8	14.4	-4.0	3.3	3.8	1.0		-11.7	-0.1
	d_a							8.8				25.4	7.6	10.9	12.3	14.7	7.5	14.5	11.4
	σ	24.2	25.9	26.6	17.6	26.1		5.8	25.9	9.7	19.1	30.0	8.9	15.2	17.1	20.5		12.1	15.7
	N	148	44	38	77	726		10	104	130	90	323	212	1026	728	1710	414	21	1380
Duns-Ros	d	2.4	0.6	-2.1		15.4		0.2	11.2		18.0	-25.7	2.1	6.6	8.0	12.9			
	d_a							4.4				34.6	8.6	14.7	15.0	21.6	7.6		
	σ	27.0	21.7	19.9		50.2		6.9	22.9		36.4	49.8	14.2	21.9	22.8	31.3			
	N	148	44	38		427		10	104		90	323	212	1050	734	1712	414		
Orkiszewski	d	-0.8	2.6	2.1	0.0	8.6	2.4	1.9	-0.4	-5.2	27.0	7.8	0.1	11.2	16.5	11.9			
	d_a							3.7				23.0	11.9	21.1	27.3	27.2	8.2		
	σ	10.8	21.1	19.8	9.7	35.7	16.2	4.4	18.3	30.5	56.4	29.1	19.4	39.5	46.7	42.6			
	N	148	44	47	65	726	35	10	104	130	90	323	212	892	596	1478	414		
Aziz et al.	d			4.4		-8.2	9.9		1.5		-11.8	3.0	2.1			-1.1			-2.5
	d_a											21.6	8.1			23.8	15.5		14.6
	σ			19.6		34.7	13.9		19.4		32.7	26.7	13.3			32.7			20.0
	N			48		726	35		104		90	323	212			1710	414		1401
Chierici et al.	d					42.8													
	d_a																		
	σ					43.9													
	N					726													
Beggs-Brill	d					17.8	3.5		14.4	-10.5		-5.4	11.7	8.2	13.6	13.9			
	d_a											21.9	13.5	16.7	18.1	23.4	6.7		
	σ					27.6	7.4		27.4	18.2		29.4	18.2	23.0	23.3	32.4			
	N					726	35		104	130		323	212	1008	685	1711	414		
Cornish	d							1.3											
	d_a							1.4											
	σ							0.9											
	N							10											
Mukherjee-Brill	d									-3.3		-19.2		17.1	16.3	28.7			
	d_a											30.5		20.9	20.5	32.9			
	σ									15.3		38.1		22.0	22.6	36.4			
	N									130		323		837	637	1710			
Hasan-Kabir	d												-3.7	-0.1	-3.9	29.2			0.5
	d_a												13.6	15.0	13.6	38.0	9.6		15.5
	σ												20.4	21.0	18.1	53.8			21.6
	N												212	920	686	1703	414		1249
Ansari et al.	d												2.5	-7.3	-1.2	-2.9		-16.1	-8.5
	d_a												7.6	14.3	10.1	19.8		17.5	14.7
	σ												11.8	18.6	14.5	27.5		14.0	18.2
	N												212	1079	750	1697		21	1423
Chokshi et al.	d															1.5		-10.5	-3.9
	d_a															16.8		12.3	11.4
	σ															25.6		12.2	15.3
	N															1711		21	1413
Gomez et al.	d																	-5.2	
	d_a																	13.1	
	σ																	14.7	
	N																	21	
Kaya et al.	d																		-2.3
	d_a																		10.3
	σ																		13.9
	N																		1407

Table 2–9 Accuracies of vertical multiphase pressure drop calculation models according to various literature sources.

that the particular correlation tends to overestimate actual pressure drops. The rows in the table refer to the different vertical pressure drop calculation models and the columns correspond to the different published comparisons, headed by their reference numbers.

In each column, *i.e.* for each investigation, the model found to give the least amount of error is indicated by a solid border around the corresponding cell. As seen, the Orkiszewski [37] correlation collected the most number of first places, five, followed by Hagedorn-Brown's [46] four. A more thorough study, however, reveals that Orkiszewski's wins all come from investigations involving relatively low numbers of cases whereby the model shows higher errors in the investigations involving greater numbers of cases. Therefore, the Orkiszewki correlation clearly cannot be declared as an overall best achiever.

The Hagedorn-Brown [46] correlation, on the other hand, suffers from a basic bias that comes up with all investigations using the Tulsa University Fluid Flow Projects (TUFFP) multiphase flow data bank [36]. The reason is that the TUFFP data bank includes a great number of data originally taken from Hagedorn's doctorate thesis with accurately measured pressure drop values that are definite favorites with other researchers as well. Consequently, investigators using this data bank, perhaps unknowingly, give an unfair advantage to the Hagedorn-Brown model, making their findings severely biased. This is the case with the results of Lawson-Brill [71] whose 726 sets of well data contained 346 sets taken from Hagedorn's thesis. Ansari et al. [61], realizing this fact, have even published two sets of statistical parameters in their evaluation (see the two columns labeled 61 in Table 2–9: one for all data, the other for data excluding those of Hagedorn's. Based on these considerations, the Hagedorn-Brown model achieves only two best places and thus cannot be named a winner either.

Let us now find the extreme values of calculation error and investigate the absolute errors, where available. The worst errors in the table (more than 100%) are given for the Poettmann-Carpenter and the Baxendell-Thomas correlations, found by Lawson-Brill [71]. It is interesting to see that the next highest error in the whole table is around 38% only. The smallest errors, on the other hand, were all found by Cornish (Cornish 1.4%, Orkiszewski 3.7%) but the low number of experimental data (only 10 sets) makes his results at least unreliable. The next best error was achieved by the Beggs-Brill correlation, found as 6.7% for 414 data sets by Aggour [72].

If the ranges of observed errors are investigated for the two basic groups of pressure drop calculation models—the empirical and the mechanistic ones—after exclusion of the unrealistic or unreliable values detailed previously, we get:

	min. abs. error	max. abs. error
Empirical models	6.7%	34.6%
Mechanistic models	7.6%	38%

The previous numbers indicate that the accuracy of mechanistic models still does not substantially exceed that of the empirical ones. This observation is in great contradiction to the claims of several authors who vindicate substantially better accuracies to mechanistic models. The older empirical correlations, therefore, cannot be ruled out when seeking the most accurate vertical pressure drop prediction model.

Since the previous discussions have not resulted in a clear and reliable ranking of the available vertical multiphase pressure drop calculation models, let us try a different approach. It is a well-known fact that statistical evaluations become more reliable when the number of investigated cases is increased. Because of this, we should focus on the findings of those investigators who dealt with the greatest number of measured pressure drop data.

All the following three investigations utilized the very comprehensive TUFFP multiphase flow data bank: Ansari et al. [61] used 755 sets of vertical well data (see the second column, 61) Chokshi et al. [62] had 1712 vertical and inclined wells, and Kaya et al. [34] used 1429 vertical wells. The numbers N in the table only show the number of successful calculations; the total number can be more, like in this case. Their findings on the accuracy of the Hagedorn-Brown, Duns-Ros, Orkiszewski, Hasan-Kabir, Ansari et al., Chokshi et al., and Kaya et al. pressure drop calculation models are shown in Table 2–10. From this table, it is clearly seen that the average absolute errors found for the different pressure drop prediction models by each investigator are quite similar. Based on the individual errors and the numbers of cases involved, composite errors valid for the total number of investigated cases could be calculated for each model.

The results are shown in the bottom row of the table. The average errors thus received integrate the results of reliable investigations and, therefore, can be considered as the most comprehensive indicators of the accuracy of vertical multiphase pressure drop calculation models.

Investigation		Hagedorn -Brown	Duns- Ros	Orki- szewski	Hasan- Kabir	Ansari et al.	Chokshi et al.	Kaya et al.
Ansari et al.	d_a	12.3%	15.0%	27.3%	13.6%	10.1%		
Ref. 61	N	728	734	596	686	750		
Chokshi et al.	d_a	14.7%	21.6%	27.2%	38.0%	19.8%	16.8%	
Ref. 62	N	1710	1712	1478	1703	1697	1711	
Kaya et al.	d_a	11.4%			15.5%	14.7%	11.4%	10.3%
Ref. 34	N	1380			1249	1423	1413	1407
Composite	d_a	13.1%	19.6%	27.2%	25.7%	16.0%	14.4%	10.3%
Results	N	3818	2446	2074	3638	3870	3124	1407

Table 2–10 Accuracies of selected vertical pressure drop calculation models.

Although the total combined number of cases varies from 3870 to 1407, the calculated overall errors given in Table 2–10 present a reliable comparison of the selected models. Basically, two conclusions can be drawn from these data:

(a) none of the models can be declared a clear winner

(b) mechanistic models do not seem to substantially improve calculation accuracy (as observed earlier)

2.5.3.6.4 Selection of an optimum model. In preceding sections, based on the findings of previously published evaluations, establishment of the calculation accuracies of the different pressure drop models was attempted. For this reason, all available data on calculation accuracies were collected in Table 2–9, which contains the most important statistical parameters of the different investigations. As discussed previously, none of the calculation models was found to be the most accurate one. Another, perhaps more important, conclusion is that the accuracy of the same model substantially varies from one investigation to the other, *i.e.* with changes in the parameter ranges of measurement data used for verification. Therefore, no multiphase pressure drop calculation model can achieve the same accuracy for all the possible conditions encountered in practice.

It follows from the previous that it does not make sense to ask which pressure drop model has the highest accuracy or which is the best. There is no overall best method, and all efforts to find it are sure to fail. Nevertheless, production engineers, in order to increase the accuracy and reliability of their designs and analyses, need to use the model giving the least calculation error for the conditions of the problem at hand. This model can and must be found and it may then be considered as the optimal one for the given case. The field engineer's philosophy on vertical pressure drop calculations must always be based on this consideration.

Now that the right approach toward multiphase vertical pressure drop prediction models has been defined, the main requirements for finding the proper one are discussed. First, the complex nature of pressure drop calculation methods practically presupposes the use of high-speed digital computers. For conducting any analysis on the application of the various models, one then needs appropriate computer programs. These can be found in several software packages abundantly available on the market.

In addition to the computer programs, a sufficiently large number of experimental pressure drop data has to be established. These have to be taken from conditions similar to those of the future application, *i.e.* from the same field and possibly the same wells. Care should be taken to ensure the reliability and accuracy of the measured data, because the use of poor quality data can considerably distort the accuracy of a pressure drop calculation model. If a sufficient number of reasonably accurate data is used, the selection of the optimum pressure drop prediction method is straightforward. First, pressure drops for all data sets using all calculation models investigated are calculated. Then, statistical parameters of the calculation errors obtained for the various models are determined. Finally, the calculation model with the best statistical parameters is chosen. This model is then considered the optimum one for the conditions at hand and must be used in all production engineering calculations involving the determination of vertical pressure drops.

2.5.3.6.5 Conclusions. By presenting an analysis of the findings of all previously published evaluations, Section 2.5.3.6 provided the required insight and proper attitude to petroleum engineers who face the problem of predicting multiphase pressure drops in oil wells. Because of the great importance of the topic, it is useful to list the basic conclusions:

1. None of the available vertical multiphase pressure drop calculation models is generally applicable because prediction errors considerably vary in different ranges of flow parameters.

2. There is no overall best calculation method, and all efforts to find one are bound to fail.

3. In spite of the claims found in the literature, the introduction of mechanistic models did not deliver a breakthrough yet, because their accuracy does not substantially exceed that of the empirical models.

2.5.3.7 Gradient curves. Previous sections have shown that calculating pressure traverses in oil wells is a cumbersome and tedious task, and one usually has to rely on the use of computer programs. Before the advent of computers, this was rarely feasible in the oilfield, and the practicing engineer had to find some simple solution to the problem. This is the reason why pre-calculated pressure traverse sheets were prepared and used in engineering calculations, eliminating the difficulty and time demand of multiphase flow calculations. Such sheets contain several pressure traverses for selected combinations of flow parameters like pipe size, liquid rate, etc. They are conventionally called *gradient curves,* but a more precise definition would be multiphase pressure traverse curve sheets.

2.5.3.7.1 Gilbert's gradient curves. The first gradient curves were introduced by Gilbert [79] who in 1954 proposed an alternative to the calculation of multiphase flow pressure drops. His objective was to provide the field engineer with a simple tool that could be used to find wellhead or bottomhole pressures in flowing oil wells with a reasonable accuracy. The model he proposed did not require detailed calculations, only the use of ready-made gradient curve sheets.

Based on field measurements and some theoretical background, Gilbert prepared several sheets with families of curves depicting the pressure distribution in vertical oil wells. The coordinate system he employed was flowing pressure vs. well depth, and each family of curves was valid for a given liquid rate, tubing diameter, and fluid parameters. The individual curves on any sheet have GLR as a parameter.

One gradient curve sheet from Gilbert's collection is shown in Figure 2–44, valid for a tubing ID of 1.9 in. and an oil rate of 600 bpd. The two axes correspond to flowing pressure and vertical depth, their use allows for an easily understandable treatment of the flowing pressure traverse. The family of curves shown in the sheet has surface or production GLR as a parameter. At a GLR of zero, the pressure traverse is a straight line because single-phase hydrostatic and frictional losses are present only. Those being linear with pipe length, their sum must also be a linear function of well depth. The other curves with increasing GLRs are no longer straight lines but show increased curvatures.

The sheet contains two sets of curves and set *B* should be shifted vertically on the depth axis to the zero depth. The reason is that curves in set *B,* if plotted among set *A* curves, would cross the original ones. The two sets of curves are divided by a curve with an arrow placed to its GLR. This curve is a very important one because it represents the minimum flowing pressures for the given set of conditions. For the given liquid rate and tubing size, the GLR corresponding to this curve is considered the optimum GLR

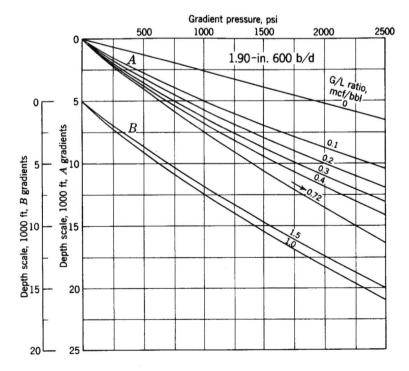

Fig. 2–44 Sample gradient curve sheet from Gilbert's collection. [79]

because it ensures the minimum of multiphase flow pressure losses in the given well. It should be noted that the pressure traverse curves with higher-than-optimum GLR values represent increased pressure drops and correspondingly higher flowing pressures.

Since Gilbert's gradient curves were empirically developed from actual pressure measurements, they precisely model actual flow conditions. The effect of increasing GLRs on the prevailing pressure drop, therefore, can theoretically be explained. To do this, the components of multiphase pressure drop have to be investigated. The hydrostatic or elevation term, as known, decreases as more and more gas is present in the two-phase mixture, *i.e.* with increased GLRs. At the same time, friction losses increase with the square of mixture velocity. The sum of these two effects must therefore exhibit a minimum at a given GLR value, the optimum GLR defined previously.

Gilbert's gradient curves were developed for the conditions of a given oilfield but have since been used for other fields as well. This practice inevitably introduces errors in pressure drop calculations, but the ease and the sound foundation of the solution usually offsets those. When using the sheets, one must keep in mind that the vertical scale means relative depth only and does not correspond to actual well depths. Therefore, all calculations must be started from the pressure axis that represents actual flowing pressures.

The steps of finding wellhead or bottomhole pressures with the help of gradient curves are summarized as follows:

1. The right sheet needs to be selected, based on the actual liquid rate and tubing size.

2. Using the selected sheet, the pressure traverse curve with the given GLR is to be found.

3. On the curve selected in Step 2, the point belonging to the known pressure (wellhead or bottomhole) is marked.

4. Tubing length is measured upward (if wellhead pressure is required) or downward (if bottomhole pressure is required) from the depth belonging to the point determined in Step 3.

5. The unknown pressure (wellhead or bottomhole) is read from the curve at the new depth calculated in Step 4.

Example 2–30. Find a well's wellhead and bottomhole pressure, respectively, with the help of Gilbert's gradient curves. The liquid rate is 600 bpd, the tubing size is 1.9 in. ID, and the well is 7,250 ft deep. Other data are:

(A)   flowing bottomhole pressure equals 2,000 psi, GLR = 400 scf/bbl

(B)   wellhead pressure is 725 psi, GLR = 1,000 scf/bbl.

Solution

Since the liquid rate and tubing size data coincide with those of Figure 2–44, calculations can be performed on that sheet.

(A)   On the gradient curve sheet, the curve with the parameter GLR = 0.4 Mscf/bbl is selected and all subsequent calculations are performed on this curve. Starting from the flowing bottomhole pressure (2,000 psi) on the pressure scale, a vertical is dropped to intersect the selected gradient curve. From the intersection, a horizontal is drawn to the depth scale to find a relative depth of 11,760 ft. Well depth is measured upwards from here to get to 4,760 ft. A horizontal line at this depth intersects the gradient curve at a pressure of 560 psi, which is the well's wellhead pressure. The part of the selected gradient curve between the two points belonging to the wellhead and bottomhole pressures constitutes the pressure distribution in the given well.

(B)   In this case, the gradient curve with the GLR value of 1 Mscf/bbl is to be used. The graphical solution is started from the known wellhead pressure and the well's depth is measured downward this time. The flowing bottomhole pressure read from the chart is 2,060 psi.

2.5.3.7.2 Other collections of gradient curves. The relative ease of using Gilbert's gradient curves was quickly realized by production engineers who started to apply the same approach to the slowly emerging pressure drop calculation models in the late 1950s and 1960s. Mainframe computers programmed to the latest pressure drop calculation models were used to prepare gradient curve sheets similar to Gilbert's. This approach enabled field engineers to use the latest achievements in multiphase flow research without the inherent burden of performing complex calculations. This is the reason why several companies came up with their version of gradient curve sheet collections. Gas lift design and analysis depended on the use of such curves till the late 1980s when the wide availability and great computing power of personal computers mostly eliminated their use.

Table 2–11 contains the most important features of gradient curve collections developed by different individuals and companies in the past. As seen, early sets used the Poettmann-Carpenter correlation, then the Hagedorn-Brown correlation (in its modified form) was frequented, while the latest collection is based on a mechanistic model. Usually, different sheets are given for some predetermined values of the water cut. As with the Gilbert curves, each sheet represents a given combination of pipe diameter and liquid rate. Example sheets from some of the collections are displayed in the following figures:

Source	Figure
CAMCO Gas Lift Manual	Fig. 2–45
Gas Lift Theory and Practice	Fig. 2–46
MERLA Gas Lift Manual	Fig. 2–47

Source	Year	Ref. No.	Model Used	Reversal	Water Cut %	Tubing IDs in	No. of Sheets
API Drill. and Prod. Practice	1954	79	Gilbert	-	0	1.38; 1.5; 1.99; 2.44; 2.99	24
US Industries Handbook of Gas Lift	1959	80	P-C.	removed	0; 100	1; 1.38; 1.5 1.99; 2.44; 2.99	94
CAMCO Gas Lift Manual	1962	81	P-C	removed	0; 50; 100	1.5; 1.99; 2.44; 2.99	185
Gas Lift Theory and Practice	1967	50	mod. H-B	removed	0; 50; 100	1; 1.25; 1.5 1.75; 1.99; 2.44; 2.99; 4	369
MERLA Gas Lift Manual	1970	82	Duns-Ros	included			n/a
Two-Phase Flow in Pipes	1978	83	mod. H-B	removed	0; 100	1.99; 4.5	30
CIM Special Vol. 20	1979	84	Aziz et al.	included	0; 50	1.99; 2.44; 2.99; 4	64
Technology of Artificial Lift Methods Vol. 3a	1980	85	mod. H-B	removed	0; 50; 90	1; 1.25; 1.5 1.75; 1.99; 2.44; 2.99; 3.5; 4; 4.5; 4.9; 5.9; 6.3 8.9; 12	729
Production Optimization	1991	86	mod. H-B	removed	0; 50; 90	1.99; 2.44; 2.99; 4	239
CEALC	1994	87	Chokshi et al.	included	0; 50; 90	1.75; 1.99; 2.44; 3.5	78

Table 2–11 Features of available vertical multiphase gradient curve collections.

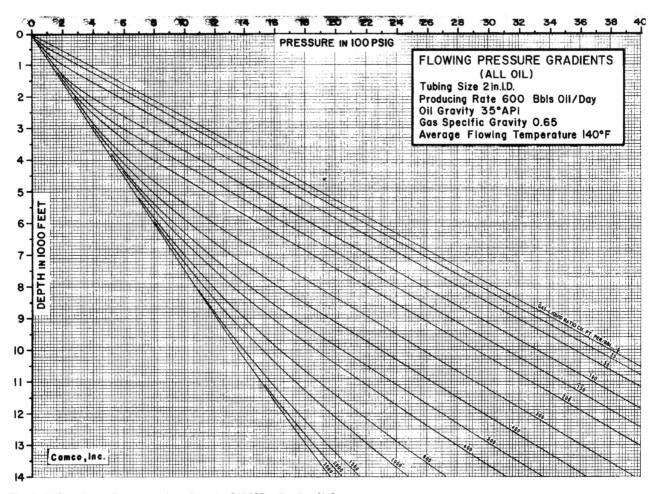

Fig. 2–45 Sample gradient curve sheet from the CAMCO collection. [81]

Each available gradient curve collection was developed under specific conditions: fluid densities, average flowing temperatures, and other fluid and flow parameters. Their use under different conditions, therefore, may introduce considerable errors in pressure drop predictions. For example, curves prepared for a low density oil cannot be expected to match the flow conditions of heavy oils. In spite of this well-recognized fact, gradient curves are often used irrespective of their original conditions. The proper approach, of course, would be to develop different sets for different fields. However, the widespread use of personal computers today in the oilfield almost completely eliminated the need to rely on gradient curves and, at the same time, considerably increased the accuracy of field calculations.

Comparison of gradient curves from different sources can reveal two interesting facts that can be seen in the example sheets as well. These are the reversal of curvature and the existence of an optimum GLR value, both discussed as follows.

Reversal of curvature is exhibited, for example, on the MERLA gradient curves (Fig. 2–47), which were based on the Duns-Ros correlation. At low flowing pressures and high GLRs, the curvature of the gradient curves changes from concave upward at higher pressures to concave downward. This behavior is caused by the increase in the acceleration component of the pressure gradient, which increases the total pressure drop at low pressures and high gas contents. Although experimentally observed and verified, the reversal was removed from many of the available gradient curve sets (see Table 2–11) because developers felt the correlations (especially the Poettmann-Carpenter) over-predicted this effect. It should be noted, however, that the more accurate calculation models (Duns-Ros, etc.) correctly predict this behavior and there is no need to remove the reversal from the gradient curves.

If gradient curve sheets based on the correlations of Poettmann-Carpenter or Hagedorn-Brown are compared to those developed with the use of more advanced pressure drop calculation models, one finds that the former do not exhibit an optimum GLR. As seen in Figures 2–45 and 2–46, increasing the GLR entails a continuous decrease in the pressure gradient. This is contrary to the behavior of the Gilbert or other curves developed by the utilization of the more advanced multiphase flow models that exhibit an optimum GLR with a minimum flowing gradient.

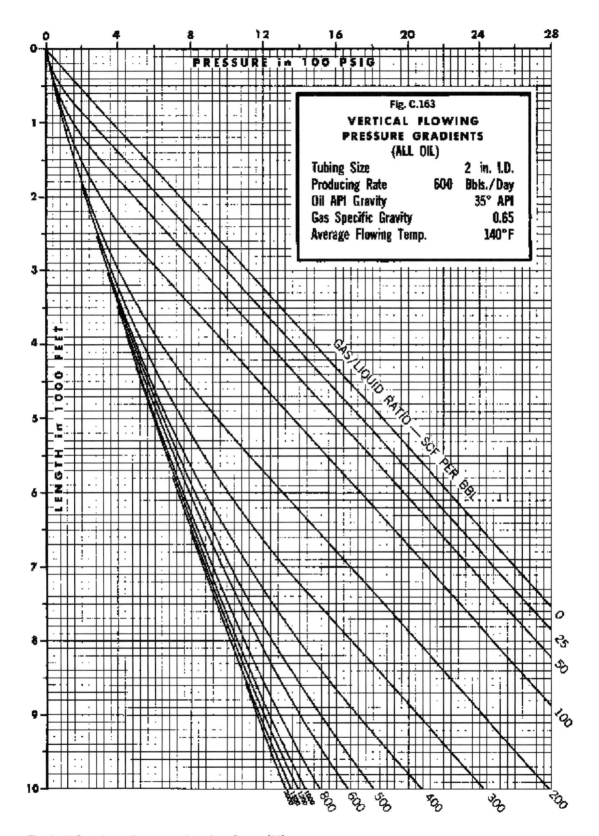

Fig. 2–46 Sample gradient curve sheet from Brown. [50]

126

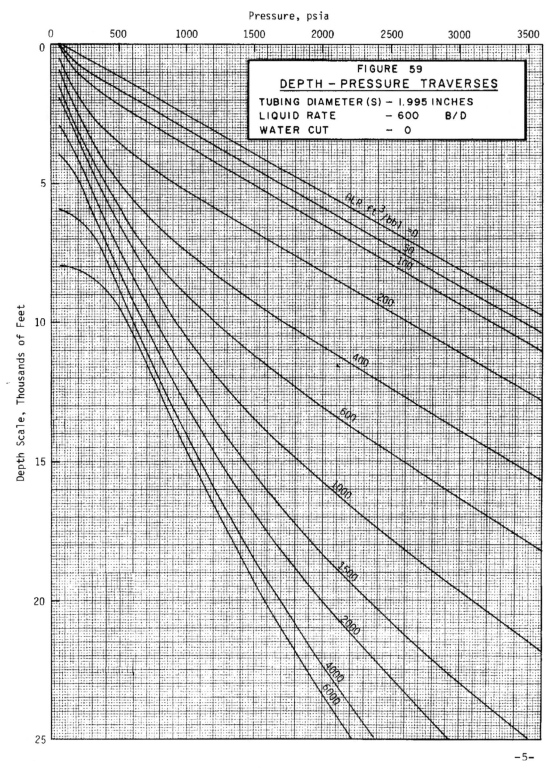

Fig. 2–47 Sample gradient curve sheet from MERLA. [82]

The occurrence of an optimum GLR value is shown in Figure 2–48, where calculated flowing bottomhole pressures in a 7,000-ft-deep well completed with a 1.99 in. ID tubing and having a wellhead pressure of 200 psi are compared. As clearly seen, gradient curves based on the Poettmann-Carpenter (CAMCO) or the Hagedorn-Brown (Brown) correlations predict continuously decreasing bottomhole pressures as GLR values are increased. On the other hand, the Gilbert or Duns-Ros (MERLA) curves correctly predict the occurrence of an optimum GLR where the bottomhole pressure and the total pressure drop is at a minimum.

Since the gradient curve with the optimum GLR represents the minimum amount of multiphase pressure drop for the given liquid rate and tubing size, it automatically ensures the most efficient use of produced or injected gas. Therefore, to achieve optimum fluid lifting conditions in flowing or gas lifted wells, one always tries to maintain this optimum GLR. This task, however, cannot be solved if a gradient curve sheet not exhibiting an optimum GLR value is used.

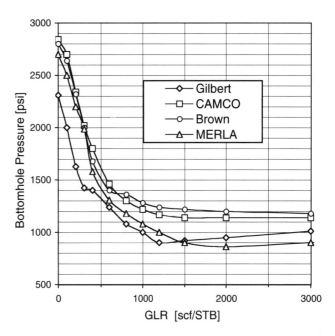

Fig. 2–48 Bottomhole pressure vs. GLR in an example well calculated from different gradient curves.

2.5.4 Horizontal and inclined flow

2.5.4.1 Introduction. In production operations, horizontal and inclined two-phase flow usually occurs in surface flowlines connecting the wellhead to a field separator. The pressure drop in the flowline, if compared to that arising in the well tubing, is of a lesser magnitude. This situation is because flowlines are usually horizontal or near horizontal, therefore no hydrostatic losses are present in the pipe. In summary, horizontal two-phase pressure drop in flowlines is generally of minor importance in the total production system consisting of the formation, the well, and the surface equipment. Exceptions to this rule are offshore wells where the length and especially the profile of the flowline do not meet the previous criteria and considerable pressure drops can be experienced because of inclined two-phase flow. The same way, substantial pressure losses can arise in long-distance pipelines transporting a two-phase mixture over a hilly terrain.

For the purposes of this chapter, i.e. the discussion of production engineering basics, only conventional wells situated on land locations are considered. In such cases, the importance of two-phase pressure drops in surface lines is much less pronounced than those occurring in wells. Accordingly, much less space is devoted to the presentation of pressure drop calculation models.

In order to develop the basic formula for calculating two-phase pressure drops in horizontal or inclined pipes, one solves the General Energy (Bernoulli) Equation for the pressure gradient (see also Section 2.4.2). The resulting formula is, according to Equation 2.90:

$$\frac{dp}{dl} = \frac{g}{g_c}\rho_m \sin\alpha + \left(\frac{dp}{dl}\right)_f + \frac{\rho_m v\, dv}{g_c dl} \qquad\qquad 2.251$$

where: α = pipe inclination angle, measured from the horizontal, radians

As seen, pressure gradient dp/dl is composed of the three familiar terms representing the different kinds of energy changes occurring in the pipe, which are, using field measurement units:

$$\left(\frac{dp}{dl}\right)_{el} = \frac{1}{144}\rho_m \sin\alpha \qquad\qquad\qquad \textit{elevation term,}$$

$$\left(\frac{dp}{dl}\right)_{f} = 1.294\times10^{-3}f\,\frac{\rho_m v^2}{d} \qquad\qquad \textit{frictional term, and}$$

$$\left(\frac{dp}{dl}\right)_{kin} = 2.16\times10^{-4}\,\frac{\rho_m v\,dv}{dl} \qquad\qquad \textit{kinetic term.}$$

In case of truly horizontal flows, the elevation term is zero, and only friction and acceleration losses have to be considered. Most of the times, changes in kinetic energy are also negligible making the total pressure drop equal to frictional losses. Therefore, for conventional onshore wells the pressure distribution along surface-laid flowlines is calculated from the frictional pressure drop alone.

According to the considerations given previously, the following sections deal with true horizontal two-phase flow and discuss the most important calculation models. This treatment is justified by the fact that calculation models for inclined two-phase flow were already discussed in Section 2.5.3. The pressure drop calculation models covered include those of Beggs and Brill [47], Mukherjee and Brill [48], Hasan and Kabir [67, 68]. All these models enable one to calculate two-phase pressure gradients at any angle, *i.e.* from horizontal to vertical.

2.5.4.2 Horizontal flow patterns. In horizontal flow, the great difference in liquid and gas densities may result in gravitational segregation of the phases, and flow patterns change accordingly. Because of this gravitational effect, a new flow pattern group—stratified (or segregated) flow—arises in addition to the ones observed in vertical flow.

According to latest research, seven flow patterns are distinguished that can be grouped into four classes as shown in Figure 2–49. This schematic diagram depicts the occurrence of the different flow patterns in relation to gas and liquid flow rates. At low liquid flow rates, increasing the gas flow rate results in smooth then wavy stratified flows, then annular wave flow. The stratified flow patterns involve a simultaneous segregated flow of liquid at the bottom and gas at the top of the pipe. If gas flow rates are increased for medium liquid flow rates, intermittent flow patterns arise starting with elongated bubble then slug flows and finally reaching annular mist flow. At very high liquid flow rates, liquid flow velocity is so high that the gas phase is dispersed in form of small bubbles constituting the dispersed bubble flow pattern.

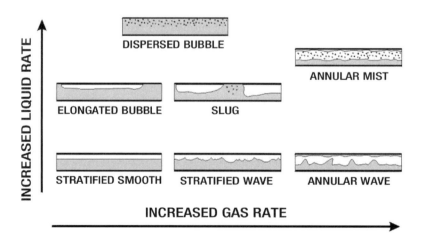

Fig. 2–49 Horizontal two-phase flow patterns.

In summary, horizontal flow patterns can be classified as follows:

Stratified group	Stratified smooth
	Stratified wavy
Intermittent group	Elongated bubble
	Slug
Annular group	Annular wavy
	Annular mist
Dispersed group	Dispersed bubble

Several flow pattern maps are available to determine the actual flow pattern in a given case. Of these, the one shown in Figure 2–50 and developed by Mandhane, Gregory, and Aziz [88] is based on a great number of experimental data. The axes on the map are the superficial velocities of the gas and liquid phases.

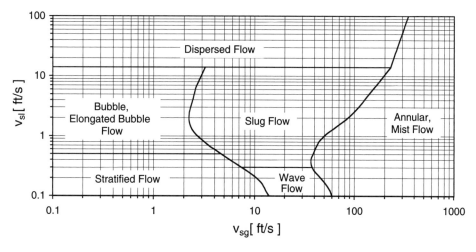

Fig. 2–50 Horizontal two-phase flow pattern map of Mandhane, Gregory, and Aziz. [88]

The latest and most widely used horizontal flow pattern map was proposed by Barnea [32] who presented a unified model for all pipe inclinations. Her map as given in Figure 2–51 is generally accepted and followed in mechanistic models of two-phase flow. Again, flow pattern boundaries are plotted vs. the superficial velocities of the flowing phases.

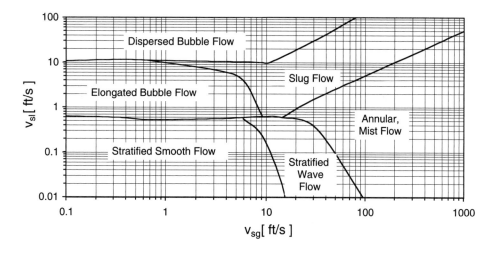

Fig. 2–51 Horizontal two-phase flow pattern map of Barnea. [32]

2.5.4.3 Empirical correlations. Several empirical correlations for horizontal pressure drop calculations are available including those of Beggs–Brill, and Mukherjee–Brill, discussed in detail in Sections 2.5.3.3.7 and 2.5.3.3.9, respectively. The present section describes only two other models, selected on their historical (Lockhart-Martinelli) and comprehensive (Dukler) features.

2.5.4.3.1 Lockhart-Martinelli correlation.

Summary

The first calculation model for horizontal two-phase flow, it does not consider flow patterns or acceleration. It offers a very simplified calculation scheme backed by limited measurement data from 1-in. pipes.

Lockhart and Martinelli [89] used data from other authors along with their own experiments, all conducted in 1-in. and smaller diameter horizontal pipes. Their theoretical derivations resulted in the following formulas to find the two-phase pressure gradient, $(dp/dl)_{TP}$ where any of the two equations may be used:

$$\left(\frac{dp}{dl}\right)_{TP} = \Phi_g^2 \left(\frac{dp}{dl}\right)_g \quad or \qquad \text{2.252}$$

$$\left(\frac{dp}{dl}\right)_{TP} = \Phi_l^2 \left(\frac{dp}{dl}\right)_l \qquad \text{2.253}$$

In the previous formulas, pressure gradients valid for the gas and liquid phases are calculated as if the given phase flowed alone through the full cross-sectional area of the pipe. The relevant equations are as follows:

$$\left(\frac{dp}{dl}\right)_g = 1.294 \times 10^{-3} f_g \frac{\rho_g v_{sg}^2}{d} \qquad \text{2.254}$$

$$\left(\frac{dp}{dl}\right)_l = 1.294 \times 10^{-3} f_l \frac{\rho_l v_{sl}^2}{d} \qquad \text{2.255}$$

where: ρ_g = gas density, lb/cu ft

ρ_l = liquid density, lb/cu ft

f_g = friction factor for gas flow, -

f_l = friction factor for liquid flow, -

v_{sg} = gas superficial velocity, ft/s

v_{sl} = liquid superficial velocity, ft/s

d = pipe diameter, in.

The friction factors for the gas and liquid cases are evaluated from the Moody diagram (Fig. 2–8) using the Reynolds numbers defined as follows:

$$N_{Re\,g} = 124 \frac{\rho_g v_{sg} d}{\mu_g} \qquad \text{2.256}$$

$$N_{\text{Re } l} = 124 \frac{\rho_l v_{sl} d}{\mu_l}$$

2.257

where: μ_l = liquid viscosity, cP

μ_g = gas viscosity, cP

Correction factors Φ_g and Φ_l were correlated by Lockhart and Martinelli with the following parameter as the independent variable:

$$X = \sqrt{\frac{\left(\dfrac{dp}{dl}\right)_l}{\left(\dfrac{dp}{dl}\right)_g}}$$

2.258

The two pressure gradients figuring in the previous formula refer to the separate flow of liquid and gas and are evaluated from Equations 2.253 and 2.254, respectively.

The authors' correlation for the correction factors Φ_g and Φ_l is given in Figure 2–52. As seen, several curves are given for each factor. Selection of the proper curve is based on the flow regimes valid for the separate flow of liquid and gas. Those can be either turbulent or laminar, depending on the relevant Reynolds numbers (see Equations 2.256 and 2.257). Lockhart and Martinelli assumed a Reynolds number of 1,000 as the lower limit of turbulent flow conditions. In the figure, the first designation refers to the liquid, the second one to the gas phase. Thus, *turb.–lam.* means turbulent liquid and laminar gas flows.

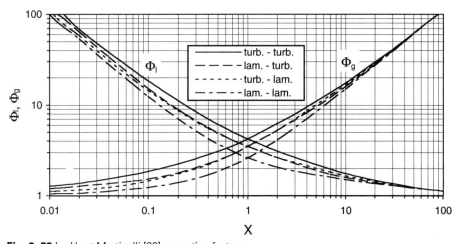

Fig. 2–52 Lockhart-Martinelli [89] correction factors.

Improvement

In 1974, Love [90] proposed a modification of the original correction factor diagram. His arguments for the modifications were

(a) the original diagram was developed for small pipe sizes (1 in. maximum) only

(b) Lockhart-Martinelli assumed a smooth pipe in their derivations.

These conditions, according to Love, introduced errors because, on the one hand, kinetic losses in larger diameter pipes are smaller, and the other hand, friction losses in real pipes are larger.

The author applied two modifications to the original diagram: he used the Moody diagram for friction factor calculations as well as corrected the kinetic component of the pressure drop. Using an extensive data bank of experimental results, Love presented Figure 2–53 with the corrected curves. As before, the first designation refers to the liquid, the second one to the gas phase's flow regime (laminar or turbulent). The author claimed that calculation accuracy for pipe sizes more than 1 in. has substantially increased.

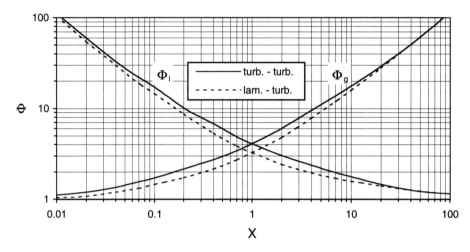

Fig. 2–53 Love's correction [90] of the Lockhart-Martinelli correlation.

Example 2–31. Find the pressure gradient according to the original Lockhart-Martinelli procedure in a flowline at the wellhead side. The flowline is of 2-in. internal diameter, oil rate is 375 bpd with no water, production GOR is 480 scf/STB. Specific gravities of oil and gas are 0.82 and 0.916, respectively. Wellhead pressure is 300 psi, and assume a constant average temperature of 80 °F, where oil and gas viscosities are 2 cP, and 0.02 cP, respectively.

Solution

In order to calculate superficial velocities, solution GOR and oil volume factor will be calculated from the Standing correlations, where the API degree of the oil is found from Equation 2.4:

$API° = 141.5 / \gamma_l - 131.5 = 41$

R_s is calculated from Equation 2.10:

$y = 0.00091 \ 80 - 0.0125 \ 41 = - 0.44$

$R_s = 0.916 \ (\ 300 / 18 / 10^{-0.44} \ )^{1.205} = 92.3$ scf/STB.

B_o is calculated from Equation 2.11:

$F = 92.3 \ (\ 0.916 / 0.82 \ )^{0.5} + 1.25 \ 80 = 197.5$

$B_o = 0.972 + 1.47 \ 10^{-4} \ 197.5^{1.175} = 1.05$ bbl/STB.

Pipe cross-sectional area is:

$A_p = 2^2 \ \pi / 4 / 144 = 0.022$ sq ft.

Superficial liquid velocity can now be calculated from Equation 2.68 as:

v_{sl} = 6.5 10^{-5} 375 1.05 / 0.022 = 1.17 ft/s.

In order to find gas volume, gas volume factor should be calculated first. For this, the pseudocritical parameters are determined from Equations 2.23 and 2.24:

p_{pc} = 709.6 – 58.7 0.916 = 655.8 psi, and

T_{pc} = 170.5 + 307.3 0.916 = 451.9 R.

Pseudoreduced parameters are as follows (Equations 2.17 and 2.18):

p_{pr} = 300 / 655.8 = 0.457.

T_{pr} = (80 + 460) / 451.9 = 1.195.

Gas deviation factor is determined from the Papay formula (Equation 2.28):

Z = 1 - 3.52 0.457 / ($10^{0.9813\ 1.195}$) + 0.274 0.457^2 / ($10^{0.8157\ 1.195}$) = 0.898.

Gas volume factor from Equation 2.21:

B_g = 0.0283 0.898 (80 + 460) / 300 = 0.0457.

Now, superficial gas velocity from Equation 2.69:

v_{sg} = 1.16 10^{-5} 375 (480 – 92.3) 0.0457 / 0.022 = 3.54 ft/s.

Superficial mixture velocity is easily found:

v_m = 1.17 + 3.54 = 4.71 ft/s.

Liquid and gas phase in-situ densities are found from Equation 2.81 and Equation 2.82, respectively:

ρ_l = (350.4 0.82 + 0.0764 0.916 92.3) / (5.61 1.05) = 50.1 lb/cu ft.

ρ_g = 2.7 0.916 300 / 0.898 / (80 +460) = 1.53 lb/cu ft.

Gas and liquid Reynolds numbers are calculated from Equations 2.256 and 2.257:

N_{Reg} = 124 1.53 3.54 2 / 0.02 = 67,161 (turbulent), and

N_{Rel} = 124 50.1 1.17 2 / 2 = 7,269 (turbulent).

Friction factors from the Moody diagram (Fig. 2–8) for an assumed relative roughness of 0.0009 are f_g = 0.02 and f_l = 0.035. Frictional pressure drops for the phases are found from Equations 2.254 and 2.255:

$(dp/dl)_g$ = 1.294 10^{-3} 0.02 1.53 3.54^2 / 2 = 0.00025 psi/ft, and

$(dp/dl)_l$ = 1.294 10^{-3} 0.035 50.1 1.17^2 / 2 = 0.00155 psi/ft.

The abscissa of Figure 2–52, according to Equation 2.258:

X = (0.00155 / 0.00025$)^{0.5}$ = 2.5.

The correction factors from Figure 2–52 are read for the *turb.–turb.* case as:

Φ_g = 7, and Φ_l = 2.8.

Finally, two-phase pressure gradient is found from any of Equation 2.252 or 2.253:

$(dp/dl)_{TP} = \Phi_g^2 \, 0.0025 = 0.012$ psi/ft, and

$(dp/dl)_{TP} = \Phi_l^2 \, 0.00155 = 0.012$ psi/ft.

The two formulas gave identical results, as discussed before.

— —

2.5.4.3.2 Dukler correlation.

Summary

The correlation is based on a wide range of measured data. Flow patterns are not considered; the kinetic term is included. The author proposed an iterative procedure for liquid holdup calculations. It can be used for pipe diameters of up to 5 1/2 in.

The author, working for an AGA-API joint project, developed several correlations for horizontal two-phase flow. The universally accepted one [91] is the *constant slip* model, which is discussed in the following paragraphs. Dukler compiled a data bank consisting of more than 20,000 measurement data taken from different sources. After careful culling, 2,620 pressure drop measurements constituted the basis for the correlation developed. Since data came from a pipe diameter range of 1 in. to 5 1/2 in., and liquid viscosities of 1–20 cP, a wide range of practical applications is covered.

Dukler did not consider the effect of flow patterns on the pressure gradient. His basic equation disregards the elevation term but includes the kinetic term, as seen here:

$$\frac{dp}{dl} = \frac{\left(\dfrac{dp}{dl}\right)_f}{1 - E_k} \qquad\qquad 2.259$$

Pressure gradient components are expressed as follows:

$$\left(\frac{dp}{dl}\right)_f = 1.294 \times 10^{-3} f \, \frac{\rho_{ns} v_m^2}{d} \qquad\qquad 2.260$$

$$E_k = 2.16 \times 10^{-4} \, \frac{1}{p} \left(\left[\frac{\rho_g v_{sg}^2}{\varepsilon_g} + \frac{\rho_l v_{sl}^2}{\varepsilon_l} \right]_2 - \left[\frac{\rho_g v_{sg}^2}{\varepsilon_g} + \frac{\rho_l v_{sl}^2}{\varepsilon_l} \right]_1 \right) \qquad\qquad 2.261$$

where:
E_k = dimensionless kinetic term, -

ρ_{ns} = no-slip mixture density, lb/cu ft

v_m = mixture superficial velocity, ft/s

ε_l = liquid holdup, -

ε_g = gas void fraction, -

f = friction factor, -

v_{sg} = gas superficial velocity, ft/s

v_{sl} = liquid superficial velocity, ft/s

ρ_g = gas density, lb/cu ft

ρ_l = liquid density, lb/cu ft

p = pressure, psi

d = pipe diameter, in.

The no-slip mixture density ρ_{ns} is defined by Dukler as

$$\rho_{ns} = \frac{\rho_l \lambda_l^2}{\varepsilon_l} + \frac{\rho_g \lambda_g^2}{\varepsilon_g}$$

2.262

In this formula, no-slip liquid holdup and gas void fraction are evaluated from

$$\lambda_l = \frac{v_{sl}}{v_{sl} + v_{sg}}$$

2.263

$$\lambda_l = \frac{v_{sg}}{v_{sl} + v_{sg}}$$

2.264

Dukler developed his own correlation for the determination of friction factors and presented a normalized friction factor f/f_n in the function of the no-slip liquid holdup λ_l shown in Figure 2–54. The parameter f_n is obtained from the equation as follows:

$$f_n = 0.0056 + 0.5 \, No_{Re n}^{-0.32}$$

2.265

No-slip Reynolds number in the above formula is defined as:

$$N_{Re\,n} = 144 \frac{\rho_{ns} v_m d}{\mu_{ns}}$$

2.266

No-slip mixture viscosity is evaluated from:

$$\mu_{ns} = \mu_l \lambda_l + \mu_g \lambda_g$$

2.267

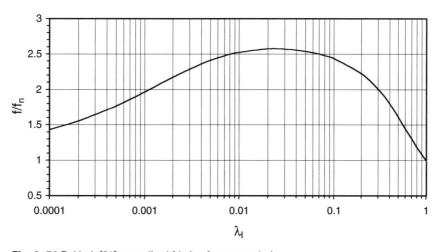

Fig. 2–54 Dukler's [91] normalized friction factor correlation.

As seen from the formulas proposed by Dukler, liquid holdup is a basic parameter for the determination of the Reynolds number, the friction factor, and the no-slip mixture density. Using a wide experimental database, the author developed a correlation for the calculation of the liquid holdup. The resulting chart is displayed in Figure 2–55.

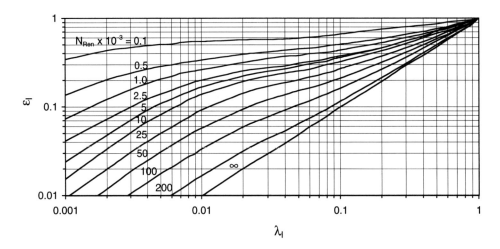

Fig. 2–55 Dukler's [91] liquid holdup correlation.

The calculation of liquid holdup from Figure 2–55 requires an iterative solution since holdup is plotted vs. no-slip Reynolds number which, in turn, depends on liquid holdup. The solution is a simple trial-and-error process consisting of:

(1) assuming a value for ε_l

(2) calculating no-slip Reynolds number from Equation 2.266

(3) determining ε_l from Figure 2–55

(4) comparing the assumed and calculated ε_l values

If not sufficiently close, assume the just calculated value and repeat the procedure until they agree with the proper accuracy.

Example 2–32. Calculate the pressure gradient for the previous example (Example 2–31) using the Dukler correlation and neglecting acceleration losses.

Solution

Calculation data taken from the previous example are:

v_{sl} = 1.17 ft/s; v_{sg} = 3.54 ft/s; v_m = 4.71 ft/s; ρ_l = 50.1 lb/cu ft; ρ_g = 1.53 lb/cu ft.

No-slip liquid holdup and gas void fraction are evaluated from Equations 2.263 and 2.264, respectively:

λ_l = 1.17 / 4.71 = 0.248,

λ_g = 3.54 / 4.71 = 0.752.

From Equation 2.267, no-slip mixture viscosity is

μ_{ns} = 2 0.48 + 0.02 0.752 = 0.51 cP.

After these preliminary calculations are done, the iterative solution for the liquid holdup can be started. For the first trial, assume $\varepsilon_l = 0.6$ and calculate the no-slip mixture density from Equation 2.262:

$$\rho_{ns} = 50.1\ (0.248)^2\ /\ 0.6 + 1.53\ (0.752)^2\ /\ 0.6 = 7.31\ \text{lb/cu ft.}$$

Reynolds number can now be calculated from Equation 2.266:

$$N_{Ren} = 144\ 7.31\ 2\ /\ 0.51 = 19{,}354.$$

Liquid holdup can be read from Figure 2–55 as $\varepsilon_l = 0.46$, which is far from the assumed value. A new iteration step is needed and $\varepsilon_l = 0.46$ is assumed. No-slip mixture density from Equation 2.262:

$$\rho_{ns} = 50.1\ (0.248)^2\ /\ 0.46 + 1.53\ (0.752)^2\ /\ 0.46 = 8.32\ \text{lb/cu ft.}$$

Reynolds number can now be calculated from Equation 2.266:

$$N_{Ren} = 144\ 8.32\ 2\ /\ 0.51 = 22{,}018.$$

Liquid holdup is read from Figure 2–55 as $\varepsilon_l = 0.45$. This sufficiently close to the assumed value and will be used in subsequent calculations.

To find the friction factor, first the ordinate of Figure 2–54 is read as $f/f_n = 2.15$. Normalizing friction factor is evaluated from Equation 2.265:

$$f_n = 0.0056 + 0.5\ 22{,}018^{-0.32} = 0.026.$$

Friction factor is the product of the previous two values:

$$f = 2.15\ 0.026 = 0.056.$$

Finally, pressure gradient equals the frictional gradient from Equation 2.260:

$$dp/dl = 1.294\ 10^{-3}\ 0.056\ 8.32\ 4.71^2\ /2 = 0.007\ \text{psi/ft.}$$

- -

2.5.4.4 Calculation of pressure traverses. Similar to vertical pressure drop calculations, determination of the pressure distribution along a horizontal pipe involves the solution of the basic pressure gradient equation. As done with vertical pipes, a trial-and-error, iterative procedure must be used. For illustrating the solution, a flowchart is shown in Figure 2–56 depicting the calculation of the outflow pressure from the inflow pressure in a horizontal pipe and using the Dukler correlation.

As seen, the total pipe length is divided in a number of segments of length Δl each, and calculations are made for each segment. First, a Δp pressure drop is assumed in the first segment and average conditions are determined. After calculating the pVT properties of the phases, iterative solution for the actual liquid holdup follows. For this, an initial value ε_l^* is assumed, which allows the calculation of the no-slip Reynolds number N_{Ren}. Reading Figure 2–55, a better approximation for ε_l is found. If this value closely matches the assumed one, no more iteration step is needed. If not, a new iteration step is performed.

After the liquid holdup has converged, the pressure gradient is evaluated using the procedure of Dukler. Based on the gradient, pressure increment Δp_{calc} is calculated and compared to the value originally assumed. If the difference is less than the previously set tolerance ε, end pressure of the first calculation segment p_{i+1} is found. Now the next segment is calculated by repeating the procedure just described. After a sufficient number of segments have been processed the length of the pipe L is reached and the outflow pressure is found.

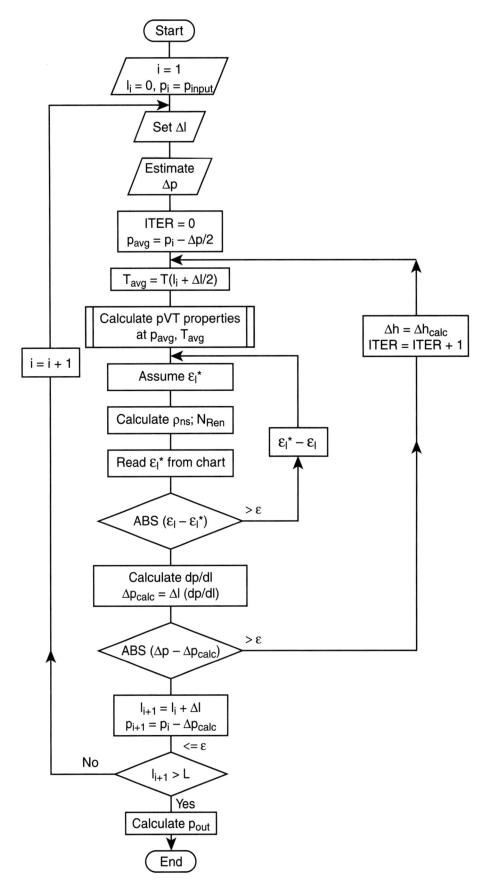

Fig. 2–56 Flowchart of horizontal two-phase pressure traverse calculations according to the Dukler correlation.

2.5.5 Flow through restrictions

Similar to single-phase cases (see Section 2.4.5), flow restrictions are used to control well production rates at the wellhead where multiphase flow usually takes place. Additional functions of wellhead or production chokes include:

- maintaining a stable wellhead pressure by preventing downstream pressure fluctuations to act on the well

- protecting surface equipment from severe slugging

- limiting pressure drawdown on the well's sandface

Wellhead chokes are divided in two broad categories: fixed or positive and variable or adjustable chokes. Positive chokes have a replaceable, bean-type orifice as shown in Figure 2–57. The housing commonly used in the field has an elbow upstream of the choke bean that facilitates removing and replacing the bean. Flow beans are usually 6-in. long and have an inner bore in fractional increments of $1/64^{th}$ of an inch with a maximum size of $1/2$ in. Where sand or corrosive fluids are produced, stainless steel, tungsten carbide or ceramic beans are used to extend operational life.

Just like in single-phase flow through a restriction, multiphase flow also can occur in a critical or a sub-critical state. In critical flow, velocity in the bean's throat attains the sonic velocity valid for the multiphase mixture at the prevailing conditions. Flow rate through the choke reaches its maximum value and downstream pressure fluctuations cannot propagate upstream. Under these conditions, changes in downstream pressure do not involve changes in the choke's throughput rate.

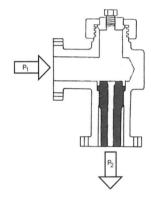

Fig. 2–57 Typical arrangement of a wellhead choke.

In sub-critical flow conditions, *i.e.* when flow velocity in the throat of the choke is below the velocity of sound, flow rate of the multiphase mixture depends on the pressure differential across the choke. At the same time, changes in the downstream pressure are reflected in changes of the upstream pressure. A wellhead choke under these conditions, therefore, cannot prevent pressure fluctuations in the surface flowline to act on the well bottom and the well's production rate will not stabilize.

The occurrence of critical flow in a flow restriction depends on the flow conditions and fluid properties. To distinguish between critical and sub-critical flows, critical pressure ratio is introduced, which is defined as the ratio of downstream and upstream pressures. If critical pressure ratio is known, the type of flow is determined from the relations given as follows:

$$\left(\frac{p_2}{p_1}\right)_{act} > \left(\frac{p_2}{p_1}\right)_{cr} \qquad \qquad \textit{the flow is sub-critical, and} \qquad \qquad 2.268$$

$$\left(\frac{p_2}{p_1}\right)_{act} \leq \left(\frac{p_2}{p_1}\right)_{cr} \qquad \qquad \textit{the flow is critical.} \qquad \qquad 2.269$$

2.5.5.1 Critical flow correlations. There are several correlations available for the calculation of multiphase flow rates through fixed chokes under critical flow conditions. Most follow the formula originally proposed by Gilbert [79] who specified the critical pressure ratio as $(p_2/p_1)_{cr} = 0.59$. His empirically developed equation is reproduced as follows:

$$q_{l\,sc} = \frac{p_1\,d^{1.89}}{3.86 \times 10^{-3}\,\mathbf{GLR}^{\,0.546}} \qquad \qquad 2.270$$

Ros [92] developed a similar formula:

$$q_{l\,sc\,n} = \frac{p_1 d^2}{4.25 \times 10^{-3}\,GLR^{0.5}}$$

2.271

where: $q_{l\,sc}$ = liquid flow rate, STB/d

p_1 = upstream pressure, psi

GLR = gas-liquid ratio, scf/STB

d = choke diameter, in.

Example 2–33. Determine the required choke size for a well producing 1,200 bpd liquid with a GLR of 800 scf/STB at a wellhead pressure of 900 psi. Assume critical flow conditions and use both the Gilbert and the Ros formulas.

Solution

For the Gilbert case, solution of Equation 2.270 for the choke diameter results:

$$d = \left(\frac{3.86 \times 10^{-3}\,GLR^{0.546}\,q_{l\,sc}}{p_1} \right)^{\frac{1}{1.89}}$$

Substituting the data in the previous formula:

$$d = (\,3.86\ 10^{-3}\ 800^{0.546}\ 1200\,/\,900\,)^{0.529} = 0.424\ \text{in} \approx 27/64\ \text{in.}$$

Using the Ros formula, Equation 2.271 can be expressed as

$$d = \left(\frac{4.25 \times 10^{-3}\,GLR^{0.5}\,q_{l\,sc}}{p_1} \right)^{0.5}$$

After substitution we get:

$$d = (\,4.25\ 10^{-3}\ 800^{0.5}\ 1200\,/\,900\,)^{0.5} = 0.4\ \text{in} \approx 26/64\ \text{in.}$$

Omana et al. [93] conducted an extensive experimental study of water and gas flow in chokes of maximum 14/64 in. size and water rates of up to 800 bpd. The authors first defined several dimensionless groups that facilitated the description of the flow behavior of chokes. Measurement data were then fitted by multiple regression analysis methods utilizing these dimensionless groups. The resultant expression is given as follows:

$$N_{ql} = 0.263\,N_\rho^{-3.49}\,N_{pl}^{3.19} \left(\frac{1}{1+R} \right)^{0.657} N_d^{1.8}$$

2.272

The definitions of the dimensionless groups are as follows:

$$N_{ql} = 1.84\,q_{l\,sc} \left(\frac{\rho_l}{\sigma_l} \right)^{1.25}$$

2.273

$$N_\rho = \frac{\rho_g}{\rho_l}$$

2.274

$$N_{pl} = 1.74 \times 10^{-2} \frac{p_1}{\sqrt{\rho_l \sigma_l}}$$

2.275

$$N_d = 10.07 \, d \sqrt{\frac{\rho_l}{\sigma_l}}$$

2.276

where: $q_{l \, sc}$ = liquid flow rate, STB/d

ρ_g = gas density at upstream conditions, lb/cu ft

ρ_l = liquid density at upstream conditions, lb/cu ft

p_1 = upstream pressure, psi

σ_l = interfacial tension, dyne/cm

d = choke diameter, in.

The calculation of the volumetric GLR at upstream conditions R is based on the production GLR:

$$R = \frac{GLR \, B_g}{561}$$

2.277

where: GLR = gas-liquid ratio, scf/STB

B_g = gas volume factor, -

The authors specified the critical pressure ratio as $(p_2/p_1)_{cr} = 0.546$.

Example 2–34. Calculate the water flow rate across a choke with a diameter of 10/64 in. if upstream and downstream pressures are p_1 = 600 psi, and p_2 = 300 psi, respectively. Flowing temperature is 80 °F, the well produces with a GLR of 500 scf/STB, liquid and gas densities upstream of the choke are ρ_l = 60 lb/cu ft, and ρ_g = 2 lb/cu ft, respectively, the interfacial tension is σ_l = 60 dynes/cm. Assume a gas deviation factor of Z = 0.91.

Solution

First check for critical flow conditions.

p_2/p_1 = 300 / 600 = 0.5 < 0.546, the flow is critical.

Calculate the dimensionless input parameters from Equations 2.274–2.276:

N_ρ = 2 / 60 = 0.033,

N_{p1} = 1.74 10^{-2} 600 / (60 60)$^{0.5}$ = 0.174,

N_d = 10.07 10 / 64 (60 / 60)$^{0.5}$ = 1.57.

In order to find the volumetric GLR, B_g is found from Equation 2.21:

B_g = 0.0283 0.91 (80 + 460) / 600 = 0.0232.

R can now be calculated from Equation 2.277:

$R = 500 \; 0.0232 / 5.61 = 2.07$ scf/scf.

The dimensionless rate is calculated from the Omana et al. correlation (Equation 2.272):

$N_{ql} = 0.263 \; 0.033^{-3.49} \; 0.174^{3.19} \; [\; 1 / (1 + 2.07) \;]^{0.657} \; 1.57^{1.8} = 158.8$.

From this, using the definition of dimensionless rate (Equation 2.273), water flow rate can be expressed:

$q_{l\,sc} = 158.8 / 1.84 \; (\; 60 / 60 \;)^{1.25} = 86$ bpd.

- -

2.5.5.2 Critical flow criteria. The correlations described in the previous section assumed that critical multiphase flow occurs at a constant critical pressure ratio. Comprehensive research has shown, however, that the sonic or wave propagation velocity in a multiphase mixture depends on both the gas and liquid properties and can be less than any of the phases' sonic velocities. This was first demonstrated by Guzhov and Medvediev [94] who investigated throat velocities in wellhead chokes while varying the gas content of the mixture. They clearly proved that the critical pressure ratio decreases as in-situ GLR decreases. As seen in Figure 2–58, at a high enough GLR, critical pressure of the mixture is very close to the value valid for single-phase gas flow. If gas content of the flowing mixture decreases, critical pressure ratio decreases as well.

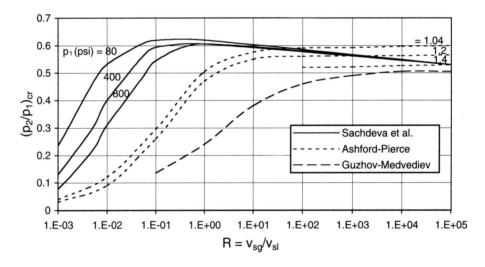

Fig. 2–58 Variation of multiphase critical pressure ratio with the gas content of the mixture.

Figure 2–58 also contains critical pressure ratios calculated from the model of Ashford and Pierce [95]. The same general trend is observed for different values of κ (ratio of specific heats) representing different kinds of natural gases. The inclusion of data from Sachdeva et al. [96] demonstrates the effect of upstream pressure on the critical pressure ratio. At higher pressures, the more dense gas increases the sonic velocity of the mixture and critical pressure ratio decreases accordingly. The effect of temperature is the opposite: higher flowing temperatures cause the critical ratio to increase.

In the following, the calculation model for finding the critical pressure ratio, as proposed by Sachdeva et al. [96] is detailed. Their basic formula is given as follows:

$$\left(\frac{p_2}{p_1} \right)_{cr} = \left(\frac{A}{B} \right)^{\frac{\kappa}{\kappa - 1}}$$

2.278

The terms A and B are evaluated from:

$$A = \frac{\kappa}{\kappa - 1} + \frac{(1 - x_1)\, \rho_{g1}\left[1 - \left(\dfrac{p_2}{p_1}\right)_{cr}\right]}{x_1\, \rho_{l1}} \qquad 2.279$$

$$B = \frac{\kappa}{\kappa - 1} + \frac{n}{2} + \frac{n(1 - x_1)\rho_{g2}}{x_1\, \rho_{l1}} + \frac{n}{2}\left[\frac{(1 - x_1)\rho_{g2}}{x_1\, \rho_{l1}}\right]^2 \qquad 2.280$$

where: ρ_{g1} = gas density at upstream conditions, lb/cu ft

ρ_{g2} = gas density at downstream conditions, lb/cu ft

ρ_{l1} = liquid density at upstream conditions, lb/cu ft

κ = ratio of specific heats for the gas, -.

The mass flow fraction of the gas phase in the total stream at upstream conditions x_1 can be evaluated from the following formula:

$$x_1 = \frac{0.0764\gamma_g\left(GLR - R_{s1}\dfrac{1}{1 + WOR}\right)}{0.0764\gamma_g\left(GLR - R_{s1}\dfrac{1}{1 + WOR}\right) + 5.615\left(B_{o1}\rho_{o1}\dfrac{1}{1 + WOR} + B_{w1}\rho_{w1}\dfrac{1}{1 + WOR}\right)} \qquad 2.281$$

where: γ_g = gas specific gravity, -

GLR = producing gas-liquid ratio, scf/STB

WOR = production water-oil ratio, -

ρ_{o1} = oil density at upstream conditions, lb/cu ft

ρ_{w1} = water density at upstream conditions, lb/cu ft

R_{s1} = solution GOR at upstream conditions, scf/STB

B_{o1} = oil volume factor at upstream conditions, -

B_{w1} = water volume factor at upstream conditions, -

Since flow in the choke throat is usually considered to occur under polytropic conditions, the polytropic exponent needs to be calculated from:

$$n = 1 + \frac{x_1(c_p - c_v)}{x_1 c_v + (1 - x_1)c_l} \qquad 2.282$$

where: c_p = specific heat of the gas at constant pressure, Btu/lb/°F

c_v = specific heat of the gas at constant volume, Btu/lb/°F

c_l = specific heat of the liquid, Btu/lb/°F

Solution of Equation 2.278 is only possible by iteration since the formula is implicit for the critical pressure ratio, $(p_2/p_1)_{cr}$. A direct substitution method can be applied after an initial guess (a starting value of $[p_2/p_1]_{cr} = 0.5$ is recommended) is made. The use of the formula will be shown in a later example.

If only single-phase gas flows through the choke, mass flow fraction of the gas phase is $x_1 = 1$, and the polytropic exponent from Equation 2.282 becomes $n = \kappa$. Substituting these values in the basic equation (Equation 2.278), one receives the well-known equation for the critical pressure ratio valid for gas flow (see Equation 2.64):

$$\left(\frac{p_2}{p_1}\right)_{cr} = \left(\frac{2}{\kappa-1}\right)^{\frac{\kappa}{\kappa-1}} \qquad 2.283$$

2.5.5.3 General calculation models. In order to calculate the fluid rate through a choke for either critical or sub-critical flow, one has to investigate the behavior of multiphase flow in a restriction. Just like in single-phase flow, the sudden decrease of cross-sectional area in the throat of the choke involves a great increase in flow velocity and, at the same time, a reduction in pressure. Generally adopted assumptions on flow conditions are

- both phases flow with the same velocity in the throat

- friction is negligible, the flow behavior is dominated by changes in kinetic energy

- expansion of the gas phase in the throat is polytropic

Based on the previous main assumptions, several authors developed general calculation models for the determination of fluid rates through chokes. [95, 96, 97] Their basic concept is that mass flow rate w is calculated from the knowledge of the velocity and density of the mixture, all valid at throat conditions:

$$w = C_d A v_2 \rho_2 \qquad 2.284$$

In the previous formula, A is the cross-sectional area of the choke bore, and C_d is the discharge coefficient. The purpose of introducing the discharge coefficient is to cover all errors and inaccuracies of the particular flow model. Thermodynamic principles imply that this value always be $C_d < 1.0$. As showed by Sachdeva et al., for a comprehensive model, the discharge coefficient is a sole function of the way the choke is installed in relation to flow. For regular wellhead chokes (see Fig. 2–57) with an elbow upstream of the choke bean, $C_d = 0.75$ is recommended, while for those without upstream perturbations $C_d = 0.85$ should be used.

In the following, the flow rate calculation model of Sachdeva et al. [96] is described and an example problem is given. The authors developed the formula reproduced as follows for the determination of liquid flow rates in either critical or sub-critical flow conditions:

$$q_{lsc} = C_d \frac{0.525d^2}{m_2} \rho_{m2} \sqrt{p_1} \left[\frac{(1-x_1)\left(1-\dfrac{p_2}{p_1}\right)}{\rho_{l1}} + \frac{x_1 \kappa \left(1-\left(\dfrac{p_2}{p_1}\right)^{\frac{\kappa}{\kappa-1}}\right)}{\rho_{gl}(\kappa-1)} \right]^{0.5} \qquad 2.285$$

The terms ρ_{m2} and m_2 are defined as follows, gas mass flow ratio x_1 is defined by Equation 2.281, p_1 is upstream pressure in psi, and d is choke diameter in inches. All other parameters were defined earlier.

$$\rho_{m2} = \left[\frac{x_1}{\rho_{g1}\left(\dfrac{p_2}{p_1}\right)^{\frac{1}{\kappa}}} + \frac{1-x_1}{\rho_{l1}} \right]^{-1} \qquad 2.286$$

$$m_2 = 8.84 \times 10^{-7}\, \gamma_g \left(GLR - R_{s2}\, \frac{1}{1 + WOR} \right) + 6.5 \times 10^{-5} \left(\rho_{o2} B_{o2}\, \frac{1}{1 + WOR} + \rho_{w2} B_{w2}\, \frac{WOR}{1 + WOR} \right) \qquad 2.287$$

The formula for liquid flow rate at standard conditions (Equation 2.285) is valid for both critical and sub-critical cases. After the actual case is found by comparing the actual pressure ratio with the critical value, the critical ratio is used instead of the actual one for critical flow. In sub-critical flow, actual pressure ratio is used.

Example 2–35. Calculate the liquid production rate of a well if wellhead pressure is p_1 = 800 psi, the wellhead choke size is 20/64 in. (0.313 in), and the pressure downstream of the choke is p_2 = 400 psi. The well produces no water with GLR = 1,000 scf/STB, oil and gas specific gravities being 0.83 and 0.7, respectively. Heat capacities are c_p = 0.54 Btu/lb/F, c_v = 0.42 Btu/lb/F, and c_l = 0.5 Btu/lb/F, the ratio of specific heats is κ = 1.286. Other parameters at upstream and downstream conditions:

parameter	upstream	downstream
ρ_o, lb/cu ft	47	49
ρ_g, lb/cu ft	4.0	2.5
R_s, scf/STB	220	120
B_o, -	1.13	1.06

Solution

The first task is to find the critical pressure ratio. Gas mass fraction is found from Equation 2.281 by substituting WOR = 0:

$$x_1 = [\, 0.0764\ 0.7\ (\, 1{,}000 - 220\,)\,] / [\, 0.0764\ 0.7\ (\, 1{,}000 - 220\,) + 5.615\ (\, 1.13\ 47\,)\,] =$$

$$= 0.1227.$$

The polytropic exponent is evaluated from Equation 2.282:

$$n = 1 + 1.227\ (0.54 - 0.42) / [1.227\ 0.42 + (1 - 0.1227)\ 0.5] = 1.03.$$

As discussed before, critical pressure ratio is found by an iterative process. First assume a value of $(p_2/p_1)_{cr}$ = 0.5. Using this value, parameters A and B are calculated from Equations 2.279 and 2.280:

$$A = 1.286 / (1.286 - 1) + [\, (\, 1 - 0.1227\,)\ 4\ (\, 1 - 0.5\,)\,] / (\, 0.1227\ 47\,) = 4.804.$$

$$B = 1.286 / (1.286 - 1) + 1.03 / 2 + 1.03\ (\, 1 - 0.1227\,)\ 2.5 / (\, 0.1227\ 47\,) +$$

$$+ 1.03 / 2\ [\, (\, 1 - 0.1227\,)\ 2.5 / (0.1227\ 47\,)\,]^2 = 5.481.$$

The calculated value of the critical pressure ratio is found from Equation 2.278:

$$\kappa / (\kappa - 1) = 1.286 / 0.286 = 4.497.$$

$$(p_2/p_1)_{cr} = (\, 4.804 / 5.481\,)^{4.497} = 0.553.$$

As seen, the calculated and assumed values are different, and new iteration steps are needed. Omitting detailed results, the converged value is $(p_2/p_1)_{cr}$ = 0.54.

Since the actual pressure ratio (400/800 = 0.5) is less than the critical one, the flow through the choke is critical and the critical ratio is to be used in the flow rate calculation. The parameters ρ_{m2} and m_2 are evaluated from Equations 2.286 and 2.287:

$$\rho_{m2} = [\ 0.1227\ /\ (\ 4\ 0.54^{1/286}\ ) + (\ 1 - 0.1227\ )\ /\ 47\ ]^{-1} = 14.66\ \text{lb/cu ft.}$$

$$m_2 = 8.84\ 10^{-7}\ 0.7\ (\ 1{,}000 - 120\ ) + 6.5\ 10^{-5}\ (\ 49\ 1.06\ ) = 0.004.$$

Now, liquid flow rate is calculated from Equation 2.285 by using $C_d = 0.75$:

$$q_{l\,sc} = 0.75\ 0.525\ 0.313^2\ 14.66\ 800^{0.5}\ /\ 0.004\ [\ (\ 1 - 0.1227\ )\ (\ 1 - 0.54\ )\ /\ 47\ +$$

$$+\ 0.1227\ 1.286\ (\ 1 - 0.54^{(1.286\,-1)/1.286}\ )\ /\ 2.5\ /\ (\ 1.286\,-1\ )\ ]^{0.5} = 659\ \text{STB/d.}$$

2.6 Well Temperature

2.6.1 Introduction

The temperature in the earth's undisturbed crust continuously increases with depth and is characterized by the geothermal gradient. This also holds for a well shut-in for a sufficiently long period. However, as soon as the well starts to produce fluids from an underground reservoir, well temperature continuously changes until a steady-state condition is reached. This is because hot fluids flowing to the well bottom and rising in the well progressively heat up the well's surroundings. Temperature distribution in producing wells, therefore, will always be affected by the factors governing the heat transfer process.

Temperature distribution in and around a producing well is schematically depicted in Figure 2–59, which shows flowing temperatures in the function of the radial distance from the well's centerline at an arbitrary depth. During shut-in, fluid in the tubing string cools down to the geothermal temperature T_r and a uniform temperature profile is observed. When the well starts to produce, the rock at the wall of the wellbore is still at the geothermal temperature and well fluids lose a considerable amount of their heat content. Therefore, at startup conditions, fluid temperature is much less than the inflow temperature of the given pipe section. As seen from the figure, most of the temperature drop occurs across the annulus (usually filled up with water) and the cement sheath. Due to the steel's high thermal conductivity, temperature losses across the tubing and the casing are usually negligible.

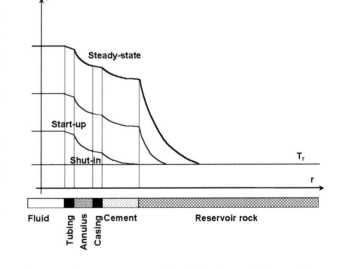

Fig. 2–59 Radial temperature distribution in a producing well.

As production time increases, the reservoir rocks surrounding the well start to heat up and the temperature outside of the wellbore increases, but it quickly converges to the geothermal temperature at greater radial distances from the well. While the well's surroundings continuously heat up, heat flow from the flowing fluids decreases accordingly, and the temperature of the flowing fluids in the tubing continuously rises. At one point, depending on the well's flow rate and the time elapsed, steady-state conditions are reached. In this case, heat lost from the well fluids equals the heat transferred into the well's surroundings and temperature conditions stabilize. The temperature of the fluids flowing in the well tubing reaches its maximum and maintains this value as long as the flow rate is constant.

Accurate prediction of well temperatures is an important task in solving production engineering problems. Without proper temperature data, accuracy of multiphase vertical pressure drop calculations can be low because pVT parameters of the flowing fluids heavily depend on temperature. The errors thus introduced impact on the design of flowing as well as gas lifted wells. In gas lifting, in addition, proper setting of pressure operated gas lift valves requires knowledge of the flowing temperature at each valve's running depth. Furthermore, injection of hot fluids in wells for any purpose also necessitates the calculation of well temperatures.

Due to the many applications over the years, many different models were developed for well temperature calculations. The present section describes the most important procedures and details their main assumptions and limitations.

2.6.2 Ramey's model

The first practical calculation model was developed by Ramey [98] who studied the temperature conditions of water injection wells. In his derivation of the energy conservation equation he assumed that

(a) the change in kinetic energy is zero for liquid injection or production

(b) changes in enthalpy are approximately equal to changes in potential energy and they cancel out

The result is that energy conditions are governed by the balance between the heat lost by the liquid and the heat transferred to the well's surroundings.

The amount of heat transferred in steady-state conditions, in general, is proportional to the temperature difference, the cross-sectional area, and the overall heat transfer coefficient. In case of a producing or injection well, temperature difference equals that existing between the injected or produced fluid and the formation, cross-sectional area is represented by the different pipe elements of the well completion. The overall heat transfer coefficient, as will be seen, includes the effects of different forms of heat transfer occurring across the well completion.

In order to study the different forms of radial heat transfer through a well, consider the cross section of a usual well completion displayed in Figure 2–60 and discuss the different mechanisms of heat transfer. [99]

- Inside the tubing string, heat in the flowing liquid is transferred to the tubing inside wall by natural convection. This is a very effective process without much heat loss, and its contribution to the overall heat transfer coefficient is negligible.

- In the material of the tubing, heat conduction between the inside and outside wall of the pipe takes place. Thermal conductivity of steel (about 600 Btu/d/ft/F) is much greater than that of any material in the well completion, making heat loss negligible in this part of the system.

- There are three different mechanisms possible in the annulus depending on the material present:

    - If the annulus contains gas only, heat can be transferred to the casing inside wall by radiation. In usual cases, this effect is disregarded.

    - If the annulus contains an insulating material, heat conduction through this insulation must properly be taken into account.

    - Natural convection of heat occurs if the annulus contains a liquid or gas. Though this can be calculated, its effect is usually disregarded.

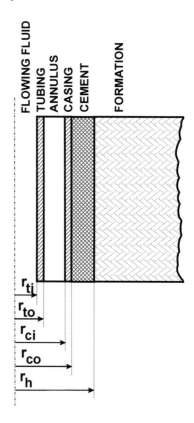

Fig. 2–60 Cross section of a typical well completion.

- Across the material of the casing, heat conduction takes place but, like in the case of the tubing, the great thermal conductivity of steel makes its contribution to the overall heat transfer coefficient negligible.

- From the casing outside wall to the radius of the wellbore, heat is transferred by conduction in the cement sheath. Usually, its effect plays the biggest role in the overall heat transfer coefficient.

In summary, the overall heat transfer coefficient's components of practical importance are the effects of heat conduction in the cement sheath and in the annulus material. Most of the calculation models consider these factors only and use the following general formula:

$$U = \left[\frac{r_{ti}}{12} \frac{\ln\left(\frac{r_{ci}}{r_{to}}\right)}{k_{an}} + \frac{r_{ti}}{12} \frac{\ln\left(\frac{r_b}{r_{co}}\right)}{k_{cem}} \right]^{-1}$$

2.288

where: U = overall heat transfer coefficient, Btu/d/ft²/F

r_{ti} = tubing inside radius, in.

r_{to} = tubing outside radius, in.

r_{ci} = casing inside radius, in.

r_{co} = casing outside radius, in.

r_b = wellbore inside radius, in.

k_{an} = thermal conductivity of annulus material, Btu/d/ft/F

k_{cem} = thermal conductivity of cement, Btu/d/ft/F

The heat transfer coefficient just described is utilized to calculate the steady-state heat loss through the well completion. Additionally, heat is lost to the surroundings of the well, *i.e.* the reservoir rock. Dissipation of heat in the rock media, however, cannot be described by the model used so far because it is of a transient nature. At the startup of fluid flow, due to the large temperature gradient across the formation, heat flow is large. But as the rocks heat up, it decreases and stabilizes. Description of this phenomenon was done by Carslaw–Jaeger [100] who tabulated their results. Based on their work, Ramey proposed the use of the following approximating function for longer production times (more than a week):

$$f(t) = \ln \frac{24 \sqrt{a\,t}}{r_b} - 0.29$$

2.289

where: a = thermal diffusivity of earth, ft²/h

t = time, hours

r_b = wellbore inside radius, in.

The final formula of Ramey [98] for calculating flowing temperature in an inclined well, at a vertical distance of z, measured from the well bottom is the following:

$$T_f(z) = T_{bb} - g_g\, z\sin\alpha + A\, g_g\, \sin\alpha \left[1 - \exp\left(-\frac{z}{A}\right)\right]$$

2.290

The factor A is called the relaxation distance and is calculated as:

$$A = \frac{6\, w\, c_l}{\pi\, r_{ti}\, k_r\, U}\left[k_r + \frac{r_{ti}\, U\, f(t)}{12}\right]$$

2.291

where: T_{bb} = bottomhole temperature, F

g_g = geothermal gradient, F/ft

z = vertical distance, measured from the well bottom, ft

α = well inclination from the horizontal, deg

w = mass flow rate of produced liquid, lb/d

c_l = specific heat of produced liquid, Btu/lb/F

r_{ti} = tubing inside radius, in.

k_r = thermal conductivity of rock material, Btu/d/ft/F

Mass flow rate w and specific heat c_l of the flowing fluid is calculated from the formulas as follows:

$$w = 350\,(q_o\, \gamma_o + q_w\, \gamma_w)$$

2.292

$$c_l = c_o\frac{1}{1 + WOR} + c_w\frac{WOR}{1 + WOR}$$

2.293

where: q_o = oil flow rate, STB/d

q_w = water flow rate, STB/d

γ_o = specific gravity of oil, -

γ_w = specific gravity of water, -

c_o = specific heat of oil, Btu/lb/F

c_w = specific heat of water, Btu/lb/F

WOR = producing water-oil ratio, -.

Example 2–36. Calculate the flowing wellhead temperature in a 5,985 ft deep vertical oil well with an annulus filled up with water. Bottomhole temperature is T_{bh} = 173 F, and the geothermal gradient equals 0.0106 F/ft. The well had been previously produced for 30 days. Other pertinent data are as follows:

r_b = 3.75 in	q_o = 2,219 STB/d
r_{co} = 2.75 in	q_w = 11 STB/d
r_{ci} = 2.475 in	q_g = 1.76 MMscf/d
r_{to} = 1.75	oil Sp.Gr. = 0.91
r_{ti} = 1.496 in	gas Sp.Gr. = 0.91
k_r = 33.6 Btu/d/ft/F	water Sp. Gr. = 1
k_{cem} = 12 Btu/d/ft/F	c_o = 0.485 Btu/lb/F
k_{wat} = 9.19 Btu/d/ft/F	c_w = 1 Btu/lb/F

Solution

First calculate the overall heat transfer coefficient U. Since the annulus is filled up with water, Equation 2.288 is solved with $k_{an} = k_{wat}$:

U = [1.496/12 ln(2.475/1.75) / 9.19 + 1.496/12 ln(3.75/2.75) / 12]$^{-1}$ = 126 Btu/d/ft^2/F.

The transient heat conduction function is found from Equation 2.289 with an average heat diffusivity value of a = 0.04 ft^2/h and a production time of t = 30 24 = 720 hours:

$f(t)$ = ln [24 (0.04 720)$^{0.5}$ / 3.75] – 0.29 = 3.25.

Liquid mass flow rate of the well is calculated from Equation 2.292 as:

w = 350 (2,219 0.91 + 11 1) = 710,601 lb/d.

Specific heat of the water-oil mixture, using Equation 2.293 and the well's WOR:

WOR = 11 / 2,219 = 0.005,

c_l = 0.485 / (1 + 0.005) + 1 0.005 / (1 + 0.005) = 0.487 Btu/lb/F.

The relaxation distance can now be calculated (Equation 2.291):

A = 6 710,601 0.487 / (π 1.496 33.6 126) [33.6 + 1.496 126 3.25 / 12) = 8,832 ft.

Wellhead temperature is found at a vertical distance of z = 5,985 ft from Equation 2.290:

T_f = 173 – 0.0106 5,985 sin(90) + 8,832 0.0106 sin(90) [1 – exp(– 5,985/8,832)] = 156 F.

Figure 2–61 shows the temperature distribution in the example well, as calculated from the Ramey model for several production times.

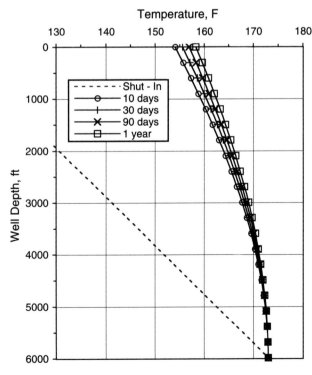

Fig. 2–61 Calculated well temperatures from the Ramey model for Example 2–36.

2.6.2.1 Modification by Hasan-Kabir. Hasan-Kabir [101] investigated the transient heat conduction time function $f(t)$ (Equation 2.289) proposed by Ramey. They pointed out that the original approximation was developed for small well radii and is accurate for larger times only. The authors conducted a study of the rigorous and Ramey's solution and introduced new formulas ensuring a better approximation:

$$f(t) = 1.1281 \sqrt{t_d} \left(1 - 0.3 \sqrt{t_d}\right) \text{ for } 10^{-10} \le t_d \le 1.5 \qquad 2.294$$

$$f(t) = (0.4063 + 0.5 \ln t_d) \left(1 + \frac{0.6}{t_d}\right) \text{ for } t_d > 1.5 \qquad 2.295$$

Definition of dimensionless time, t_d, is as follows:

$$t_d = \frac{144 \, a \, t}{r_b^2} \qquad 2.296$$

where: a = thermal diffusivity of earth, ft^2/h

 t = time, hours

 r_b = wellbore inside radius, in.

An analysis of Hasan-Kabir's data by the present author shows that the previous formulas improve calculated $f(t)$ values for times less than one day ($t_d \approx 1.5$). At greater times, the formula proposed by Ramey (Equation 2.289) gives practically identical results to those proposed by Hasan-Kabir.

2.6.3 The Shiu-Beggs correlation

Shiu and Beggs [102] investigated the temperature distribution in flowing oil wells producing a multiphase mixture. They adapted the calculation model of Ramey and proposed an empirical correlation for the determination of the relaxation distance A. Based on a total of 270 well data from three regions, they calculated the relaxation distance for each case. The authors assumed that all wells were produced for a sufficiently long time so that the transient time function $f(t)$ did not change. After a multiple regression analysis, they developed the following formula:

$$A = C_0 w^{c_1} \rho_l^{c_2} d^{c_3} p_{wb}^{c_4} API^{c_5} \gamma_g^{c_6} \qquad 2.297$$

where: ρ_l = liquid density, lb/cu ft

 d = tubing inside diameter, in,

 p_{wb} = wellhead pressure, psig

 API = API degree of oil, -

 γ_g = specific gravity of gas, -

In the previous correlation, in contrast to Ramey's formula, the well's mass flow rate is calculated for the multiphase mixture in lb/sec units:

$$w = \frac{1}{86400} \left[350 \left(q_o \gamma_o + q_w \gamma_w\right) + 0.0764 \, q_g \gamma_g\right] \qquad 2.298$$

Liquid density is calculated from the weighted average of the oil's and water's densities:

$$\rho_l = \frac{62.4}{q_o + q_w} \left(q_o \gamma_o + q_w \gamma_w\right) \qquad 2.299$$

The authors gave two sets of coefficients C_i, one for a known (Set 1), and one for an unknown (Set 2) wellhead pressure and claimed that both gave acceptable results. The coefficients are as follows:

Coefficient	Set 1	Set 2
C_0	0.0063	0.0149
C_1	0.4882	0.5253
C_2	2.9150	2.9303
C_3	-0.3476	-0.2904
C_4	0.2219	0
C_5	0.2519	0.2608
C_6	4.7240	4.4146

Example 2–37. Calculate the wellhead flowing temperature for the preceding example using the Shiu-Beggs correlation if the wellhead pressure is p_{wh} = 150 psig.

Solution

The well's mass flow rate includes the mass of the produced gas, according to Equation 2.298:

$$w = [\ 350\ (\ 2{,}219\ 0.91 + 11\ 1\ ) + 0.0764\ 1{,}762{,}000\ 0.7\ ]\ /\ 86{,}400 = 9.3\ \text{lb/sec.}$$

Liquid density is found from Equation 2.299:

$$\rho_l = [\ 62.4\ (\ 2{,}219\ 0.91 + 11\ 1\ )\ ]\ /\ (\ 2{,}219 + 11\ ) = 56.8\ \text{lb/cu ft.}$$

The oil's API gravity is found from Equation 2.4:

$$API = 141.5\ /\ 0.91 - 131.5 = 24\ \text{API.}$$

The relaxation distance predicted from Shiu-Beggs' correlation of Equation 2.297, using the Set 1 coefficients:

$$A = 0.0063\ 9.3^{0.4882}\ 56.8^{2.915}\ 2.992^{-0.3476}$$
$$150^{0.2219}\ 24^{0.2519}\ 0.7^{4.7240} = 2{,}089\ \text{ft.}$$

Wellhead temperature is found at a vertical distance of z = 5,985 ft from Equation 2.290:

$$T_f = 173 - 0.0106\ 5{,}985\ \sin(90) + 2{,}089$$
$$0.0106\ \sin(90)\ [\ 1 - \exp(\ -\ 5{,}985/2{,}089)\ ] =$$

$$= 130\ \text{F.}$$

The temperature distribution in the example well is shown in Figure 2–62 for the two sets of coefficients of the Shiu-Beggs relaxation distance correlation.

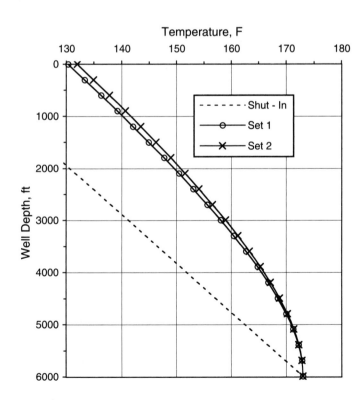

Fig. 2–62 Calculated well temperatures from the Shiu-Beggs model for Example 2–37.

2.6.4 The model of Sagar-Doty-Schmidt

Sagar, Doty, and Schmidt [103] built a comprehensive mathematical model for the description of temperature conditions in wells producing multiphase mixtures. They assumed steady-sate conditions and included the effects of all the possible sources having an impact on well temperature, such as: heat transfer in the well completion and in the formation, the changes in kinetic and potential energy, and the Joule-Thomson effect. Their final differential equation, therefore, is a strong foundation for accurate temperature calculations. However, solution of this equation requires knowledge of the wellstream's composition, and the pressure distribution in the well, which are seldom accurately known.

In order to develop a calculation model suitable for practical applications, the authors proposed a simplified solution and presented a correlation based on 392 measured temperature profiles. The theoretical background for the simplification lies in the fact that the Joule-Thomson and the kinetic effects are usually smaller than the other terms figuring in the energy balance. The authors combined these two effects into a single parameter and, using their experimental database developed a correlation for it.

Without the lengthy derivation, we reproduce the authors' final formula for the calculation of flowing temperature of the multiphase mixture at a vertical distance of z, measured from the well bottom:

$$T_f(z) = T_r(z)\frac{A\sin\alpha}{Jc_l} + AF_c + Ag_g\sin\alpha +$$

$$+ \exp\left(-\frac{z-z_0}{A}\right)\left[T_f(z_0) - T_r(z_0) + \frac{A\sin\alpha}{Jc_l} - AF_c - Ag_g\sin\alpha\right] \qquad 2.300$$

where: $T_r(z)$ = rock temperature at z, F

$T_f(z_0)$ = flowing fluid temperature at z_0, F

z = vertical distance, measured from the well bottom, ft

z_0 = vertical distance, where temperatures are known, ft

J = mechanical equivalent of heat, 778 ft-lb/Btu

g_g = geothermal gradient, F/ft

α = well inclination from the horizontal, deg.,

c_l = specific heat of produced liquid, Btu/lb/F,

F_c = correction factor, -

A = relaxation distance, ft

Relaxation distance A is calculated identically to Ramey's formula, Equation 2.291 with the overall heat transfer coefficient U defined by Equation 2.288. Specific heat of the liquid is calculated from Equation 2.293, and the well's mass flow rate includes the mass of the produced gas, as found from Equation 2.298.

Rock temperatures at the known and the required vertical distances z_0 and z, respectively, are found from the bottomhole temperature and the geothermal gradient:

$$T_r(z_0) = T_{bb} - g_g z_0 \sin\alpha \qquad 2.301$$

$$T_r(z) = T_{bb} - g_g z \sin\alpha \qquad 2.302$$

As discussed previously, correction factor F_c represents the sum of kinetic and Joule-Thomson effects. Based on the authors' extensive measured data, this factor was found to be $F_c = 0$, when $w \geq 5$ lb/sec, and for lower mass flow rates the following correlation is proposed:

$$F_c = -2.987 \times 10^{-3} + 1.006 \times 10^{-6} \, p_{wh} + 1.906 \times 10^{-4} \, w - 1.047 \times 10^{-6} \, GLR + 3.229 \times 10^{-5}$$

$$API + 4.009 \times 10^{-3} \, \gamma_g - 0.3551 \, g_g \qquad\qquad 2.303$$

where: p_{wh} = wellhead pressure, psig

w = total mass flow rate, lb/sec

GLR = producing gas-liquid ratio, scf/STB

API = API degree of oil, -

γ_g = specific gravity of gas, -

g_g = geothermal gradient, F/ft

Sagar, Doty, and Schmidt modified the transient time function $f(t)$ and derived the following formula valid for the usual wellbore sizes and for times exceeding one week:

$$f(t) = 3.53 - 0.272 \, r_h \qquad\qquad 2.304$$

where: r_h = wellbore inside radius, in.

Example 2–38. Find the wellhead temperature for the previous example well by using the Sagar-Doty-Schmidt calculation model.

Solution

The authors use the same definition of the overall heat transfer coefficient as Ramey, this is why the value from Example 2–36 can be used as $U = 126$ Btu/d/ft²/F. Similarly, mass flow rate is found from Example 2–37: $w = 9.315$ lb/sec $= 804,816$ lb/d.

Transient time function is defined by Sagar-Doty-Schmidt by Equation 3.304:

$f(t) = 3.53 - 0.272 \; 3.75 = 2.51$.

Using the $c_l = 0.487$ Btu/lb/F value from the Ramey example, relaxation distance is calculated from Equation 2.291:

$A = 6 \; 804,816 \; 0.487 \, / \, (\pi \; 1.496 \; 33.6 \; 126) \, [\; 33.6 + 1.496 \; 126 \; 2.51 \, / \, 12 \;] = 8,634$ ft.

Temperature calculation is started at the well bottom, where $z_0 = 0$ and the rock temperature at this point is found from Equation 2.301:

$T_r(0) = 173$ F.

The wellhead corresponds to an elevation of $z = 5,985$ ft, and the rock's geothermal temperature at this point is, using Equation 2.302:

$T_r = 173 - 0.0106 \; 5,985 \; \sin(90) = 109$ F.

The correction factor's value is $F_c = 0$, since the mass flow rate is greater than 5 lb/sec. Finally, the wellhead temperature is found from Equation 2.300, where the flowing temperature at $z_0 = 0$ equals the bottomhole temperature, *i.e.* $T_f(z_0) = 173$ F:

T_f = 109 - 8,634 sin(90) / 778 / 0.487 + 8,634 0.0106 sin(90) + exp(-5,985 / 8,634) [173 − 173 + 8,634 sin(90) / 778 / 0.487 - 8,634 0.0106 sin(90)] = 144 F.

Figure 2–63 shows the temperature distribution in the example well.

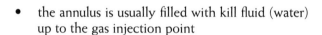

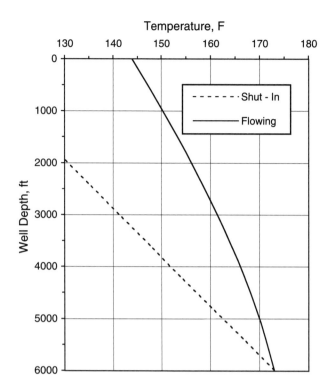

Fig. 2–63 Calculated well temperatures for Example 2–38 from the Sagar-Doty-Schmidt model.

2.6.5 Gas lifted wells

The proper information on flowing temperature distribution is of high importance in continuous flow gas lift wells because unloading valve string designs require exact temperatures at each valve depth. All the well temperature calculation models described before can be used to establish the temperature profile with the following basic considerations:

- the annulus is usually filled with kill fluid (water) up to the gas injection point

- above the gas injection point, injection gas with a low heat capacity occupies the casing-tubing annulus

- gas rates are different above and below the injection point, and the mass flow rates change accordingly

Taking into account these peculiarities of gas lifted wells, one can develop a calculation scheme for the determination of the temperature distribution. The first step of such a procedure should be to divide the well in two sections at the depth of the gas injection. The bottom section is taken first, and starting from the bottomhole temperature, the temperature at the injection point is found. In this well section, the annulus is assumed to be filled with water. For the next well section, the starting fluid temperature equals that just calculated at the injection depth. Heat transfer in this section is characterized by an increased heat insulation of the gas present in the annulus, and flowing fluid temperature is determined with the use of a higher mass flow rate, caused by the injection gas volume.

The use of the previous procedure is recommended for continuous flow gas lifted wells. Its only fault is that the Joule-Thomson effect occurring across the operating gas lift valve is disregarded. This effect causes a relatively high negative peak in the local temperature at the valve setting depth. Since this cooling is very much localized and does not considerably affect the fluid temperature even a short distance above the injection point, its elimination can be justified in most cases.

Example 2–39. Calculate the wellhead temperature in an 8,000 ft deep vertical well placed on continuous flow gas lift, using the Sagar-Doty-Schmidt model. A lift gas volume of 400,000 scf/d is injected at a depth of 4,000 ft, and the annulus is filled up with water to this level. Bottomhole temperature is T_{bh} = 190 F, and the geothermal gradient equals 0.015 F/ft. Other pertinent data are as follows:

r_h = 4.5 in	p_{wh} = 100 psi
r_{co} = 3.5 in	q_o = 500 STB/d
r_{ci} = 3.138 in	q_w = 500 STB/d
r_{to} = 1.1875 in	q_g = 200 Mscf/d
r_{ti} = 0.9975 in	oil Sp.Gr. = 0.85
k_r = 33.6 Btu/d/lb/F	gas Sp.Gr. = 0.7
k_{cem} = 12 Btu/d/ft/F	water Sp. Gr. = 1
k_{wat} = 9.192 Btu/d/ft/F	c_o = 0.485 Btu/lb/F
k_{gas} = 0.504 Btu/d/ft/F	c_w = 1 Btu/lb/F

Solution

The well is divided at the gas injection point in two depth sections and temperature calculations are performed starting from the well bottom.

For the section below the gas injection point, the total mass flow rate is found from Equation 2.298, using the gas rate produced from the formation only:

w = [350 (500 0.85 + 500 1) + 0.0764 200,000 0.7] / 86,400 = 3.87 lb/sec.

Correction factor, F_c, is found from Equation 2.303 (because w ≤ 5 lb/sec) with a GLR = 200 scf/STB and an API gravity of 35 API:

F_c = -0.002987 + 1.006 10^{-6} 100 + 1.906 10^{-4} 3.8 – 1.047 10^{-6} 200 + 3.229 10^{-5} 35 +

+ 4.009 10^{-3} 0.7 - 0.3551 0.015 = - 0.0037.

The overall heat transfer coefficient U is calculated from Equation 2.288, and, since the annulus is filled up with water, k_{an} = 9.192 Btu/d/lb/F:

U = [0.9975/12 ln(3.138/1.1875) / 9.192 + 0.9975/12 ln(4.5/3.5) / 12]$^{-1}$ =

= 95 Btu/d/ft^2/F.

Transient time function is defined by Equation 2.304:

$f(t)$ = 3.53 – 0.272 4.5 = 2.306.

Specific heat of the liquid, using Equation 2.293 and the well's WOR = 1, is:

c_l = 0.485 / (1 + 1) + 1 1 / (1 + 1) = 0.743 Btu/lb/F.

Relaxation distance is calculated from Equation 2.291:

$A = 6\ 3.87\ 86,400\ 0.743 / (\pi\ 0.9975\ 33.6\ 95)[33.6 + 0.9975\ 95\ 2.306 / 12] =$

$= 7,718$ ft.

Calculation for this section starts at the well bottom, where $z_0 = 0$ and the rock temperature is found from Equation 2.301:

$T_r(o) = 190$ F.

The liquid's flowing temperature here is $T_f(o) = 190$ F.

The gas injection point corresponds to an elevation of $z = 8,000 - 4,000 = 4.000$ ft, and the rock's geothermal temperature at this point is, using Equation 2.302:

$T_r = 190 - 0.015\ 4,000\ \sin(90) = 130$ F.

Flowing temperature at the depth of gas injection is found from Equation 2.300:

$T_f = 130 - 7,718\ \sin(90) / 778 / 0.743 - 7,718\ 0.0037 + 7,718\ 0.015\ \sin(90) +$

$+ \exp(-4,000 / 7,718)\ [190 - 190 + 7,718\ \sin(90) / 778 / 0.743 + 7,718\ 0.0037 -$

$- 7,718\ 0.015\ \sin(90)\ ] = \ 160$ F.

For the well section above the gas injection point, total mass flow rate is found from Equation 2.298, using the total gas rate (produced plus injected):

$w = [350\ (500\ 0.85 + 500\ 1) + 0.0764\ 600,000\ 0.7] / 86,400 = 4.12$ lb/sec.

Since $w \le 5$ lb/sec, correction factor, F_c, is found from Equation 2.303 with a total $GLR = 600$ scf/STB:

$F_c = -0.002987 + 1.006\ 10^{-6}\ 100 + 1.906\ 10^{-4}\ 4.12 - 1.047\ 10^{-6}\ 600 + 3.229\ 10^{-5}\ 35 +$

$+ 4.009\ 10^{-3}\ 0.7 - 0.3551\ 0.015 = - 0.0041.$

The overall heat transfer coefficient U is calculated from Equation 2.288. The annulus above the injection point contains gas, hence $k_{an} = 0.504$ Btu/d/lb/F:

$U = [\ 0.9975/12\ \ln(\ 3.138/1.1875) / 0.504 + 0.9975/12\ \ln(\ 4.5/3.5) / 12\ ]^{-1} =$

$= 6.17$ Btu/d/ft²/F.

Transient time function and specific heat of the liquid are the same as in the previous well section. Relaxation distance is calculated from Equation 2.291:

$A = 6\ 4.12\ 86,400\ 0.743 / (\pi\ 0.9975\ 33.6\ 6.17)[33.6 + 0.9975\ 6.17\ 2.306 / 12] =$

$= 84,841$ ft.

Calculation for this section starts at the injection point, where $z_0 = 4,000$ and the rock temperature at this point is found from Equation 2.301:

$T_r(o) = 190 - 0.015\ 4,000\ \sin(90) = 130$ F.

The liquid's flowing temperature at this depth equals the outflow temperature of the previous well section, i.e. $T_f(0) = 160$ F.

The wellhead corresponds to an elevation of $z = 8,000$ ft, and the rock's geothermal temperature at that point is, using Equation 2.302:

$$T_r = 190 - 0.015 \ 8,000 \ \sin(90) = 70 \ \text{F}.$$

Finally, wellhead temperature is found from Equation 2.300:

$$T_f = 70 - 84,841 \ \sin(90) \ / \ 778 \ / \ 0.743 - 84,841 \ 0.0042 + 84,841 \ 0.015 \ \sin(90) +$$

$$+ \exp(-4,000 \ / \ 84,841) \ [159 - 130 + 84,841 \ \sin(90) \ / \ 778 \ / \ 0.743 + 84,841 \ 0.0042 -$$

$$- 84,841 \ 0.015 \ \sin(90) \ ] = 134 \ \text{F}.$$

The temperature distribution in the example gas lifted well is given in Figure 2–64. It is clearly seen that fluid temperature follows different shapes below and above the point of gas injection. Cooling is less above the injection point, because the injection gas present in the annulus is a better heat insulator than the water below the gas entry point.

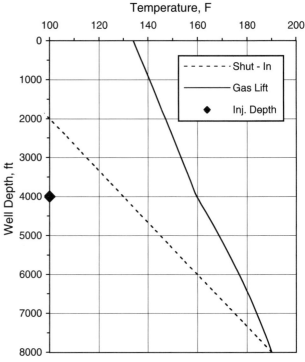

Fig. 2–64 Temperature distribution in a gas lifted well for Example 2–39.

2.7 Systems Analysis Basics

Systems analysis of producing oil and gas wells (often called Nodal Analysis) is the latest addition to the petroleum engineer's arsenal of design and analysis tools. The methodology and calculation procedures developed in the last two decades are based on the recognition that the underground reservoir, the producing well, and the surface liquid and gas handling equipment constitute a complex, interrelated system. Accordingly, any process in any element of the system entails changes that occur not only in the given part but also in the system as a whole. This section introduces the basic principles of systems analysis as adapted to the description of producing well behavior.

2.7.1 Introduction

Petroleum fluids found in an underground reservoir move through a complex system to reach their destinations on the surface. This system is called the production system and comprises the following main components: the reservoir, the producing well, the surface flowline, and the separator. Some of these can further be divided into smaller elements (for example, the well, besides the tubing string, may contain safety and/or gas lift valves, as well as other components). The production system is thus a system of interconnected and interacting elements that all have their own specific performance relationships, but each, in turn, also depends upon and influences the other elements. In order to produce fluids from the well, all components of the system must work together. Thus, the solution of any fluid production problem requires that the production system be treated as a complete entity.

The outlines of this principle were first given by Gilbert [79], the father of production engineering, in the 1950s. He described the interaction of the reservoir, the well, and the wellhead choke and proposed a system-oriented solution for determining the production rate of a flowing well. The practical use of Gilbert's ideas was limited, mainly due to the limitations of the methods available in his time for modeling the performance of the system's elements. During the last decades, however, research into the behavior of the individual hydraulic elements of oil and gas wells has been very intensive. Because of this progress, there exist today several different theories, calculation, and design procedures that reliably model the performance of each element of a production system. Good examples for this are the numerous correlations available for calculating pressure traverses in vertical and horizontal pipes.

The wide selection of available calculation models and the advent of computers, which eased the burden of the necessary calculations, led to the reappearance of Gilbert's ideas in the early 1980s. [104, 105] The new contributions aim at the numerical simulation of the production system's hydraulic behavior but also enable the optimization of the system to produce the desired flow rate most economically. Although most of the investigators study the production system only, the basic concepts have already been identified by Szilas [106] for the integration of these achievements into the description of a whole field's behavior.

The systems analysis methods and procedures mentioned previously were named "Nodal Analysis" by K.E. Brown, and the term has generally been accepted. A full treatment of Nodal Analysis principles has been given by Beggs. [86] The application of this theory to flowing and gas-lifted wells can have immediate practical and economical advantages.

2.7.2 The production system

The production system of any oil or gas well comprises part of the reservoir, the system transporting well fluids to the surface, and the surface separation equipment. A more detailed classification of the system's components would include:

- the part of the formation between the boundary of the drainage area and the wellbore

- the well's completion, *i.e.* the perforations, gravel pack, etc.

- the vertical or inclined tubing string

- any downhole restrictions placed in the tubing, like downhole safety valves, etc.

- artificial lift equipment, if required for lifting well fluids to the surface

- the production choke usually placed at the wellhead

- the flowline with its additional equipment

- the separator where primary separation of well fluids is made

All the previous components have their own performance relationships that describe their behavior under different flow conditions. The formation is characterized by the laws of flow in porous media; whereas in most of the other components, single-phase or multiphase flow in pipes takes place. Accordingly, a proper description of the total system's behavior cannot be achieved without a thorough investigation of each component's performance. The different calculation models developed for the characterization of the system's components, therefore, provide a firm foundation for systems analysis. This is the reason why, in previous sections, most of the practical methods describing the performance of the productions system's individual components were discussed in depth.

Consider a simple flowing oil well with only some of the components described previously. The schematic drawing of this simplified case is shown in Figure 2–65. Fluids in the formation flow to the well from as far as the boundary of the drainage area, 1. After entering the well through the sandface, vertical flow in the tubing starts at the well bottom, 2. In the tubing string vertical or inclined tubing flow takes place up to the wellhead, 3. A surface choke bean is installed at 4, which is also the intake point to the flowline. Horizontal or inclined flow in the flowline leads into the separator, 5.

The points in the flowing well's production system designated by the numbers in Figure 2–65 constitute the *nodes* of the systems analysis theory. They separate the different components of the system: the formation, the tubing string, the flowline, etc. Between any two node points, flowing pressure decreases in the direction of the flow and the pressure drop can be calculated based on the characteristics of the given component. By this way, the performance of any component is established as the variation of the flow rate with the pressure drop across the given component. This task is accomplished through the application of the different pressure drop calculation models valid for the different kinds of hydraulic components.

Two node points deserve special considerations: the boundary of the drainage area (point 1) and the separator (point 5). These points constitute the two endpoints of the production system and their pressures are considered constant for longer periods of time. Formation pressure at the outer boundary of the drainage area of the well, of course, changes with the depletion of the reservoir but for production engineering purposes involving short-, and medium-time periods, this change can be disregarded. Separator pressure, at the same time, is usually set for the whole life of the field and is held constant. Thus, the two endpoint pressures—the average reservoir pressure and the separator pressure—can duly be considered constant values for the purposes of systems analysis.

Since any oil or gas well's production system can be divided into its components by using appropriately placed nodes, systems analysis methods can widely be used in practice. This approach has found its use in solving the everyday design and analysis problems of production engineers.

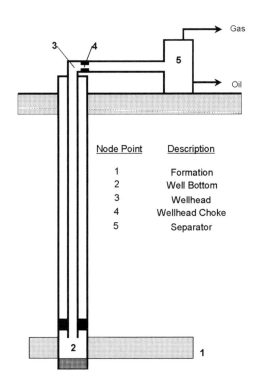

Node Point	Description
1	Formation
2	Well Bottom
3	Wellhead
4	Wellhead Choke
5	Separator

Fig. 2–65 The production system of a flowing well with node points.

2.7.3 Basic principles

One of the many objectives of systems analysis is the determination of the flow rate of a given production system. The solution of this problem is illustrated here through the example of a flowing well.

As discussed in connection with Figure 2–65, an oil well can be considered a series-connected hydraulic system made up of its components, which are bracketed by the appropriately placed nodes. Evaluation of the total system's performance permits the following conclusions to be made:

- Mass flow rate throughout the system is constant, although phase conditions change with changes in pressure and temperature.

- Pressure decreases in the direction of flow because of the energy losses occurring in the various system components.

- At node points, input pressure to the next component must equal the output pressure of the previous component.

- System parameters being constant for considerable periods of time are:

  - the endpoint pressures at the separator and in the reservoir

  - the wellbore and surface geometry data (pipe diameters, lengths, etc.)

  - the composition of the fluid entering the well bottom

Taking into account these specific features of the production system, a procedure can be devised to find the flow rate at which the system will produce. This starts with dividing the system into two subsystems at an appropriately selected node called the *solution node*. The next step is to find pressure vs. rate curves for each subsystem. Construction of these functions is started from the node points with known pressures at the separator and at the well bottom. The intersection of the two curves gives the cooperation of the subsystems and thus the desired rate.

A simple example is shown in Figure 2–66. [107] The well is a low producer and has no surface or downhole chokes installed. The well's production system is divided at Node 2 (see Fig. 2–65) with one subsystem consisting of the flowline and the tubing string and the other being the formation. The pressure vs. rate diagram of the formation is the familiar IPR curve. The other curve is constructed by summing the separator pressure and the pressure drops in the flowline (wellhead pressure curve) and by further adding to these values the pressure drops occurring in the tubing string. The resulting curve is the tubing intake pressure vs. production rate. The system's final rate is found at the intersection of this curve with the IPR curve and is 320 BOPD in the present example.

The same procedure can be followed starting from a different node position but will give the same result. This shows that systems analysis allows a flexibility to study different situations.

2.7.4 Application to gas lifting

The systems analysis principles outlined previously can readily be applied to the analysis of continuous flow gas lift wells. This is facilitated by the fact that a continuous multiphase flow takes place in the tubing string. A schematic drawing of a well placed on continuous flow gas lift is given in Figure 2–67. Obviously, the well can be treated similar to a flowing well with the only difference that the tubing string is divided in two sections with the dividing point placed at the depth of the gas injection. Consequently, the tubing string will have two sections where the flowing multiphase mixture's gas content is different. The section below the gas injection point contains the gas produced from the formation only, whereas the one above the injection point, in addition to formation gas, contains the injected gas volume as well.

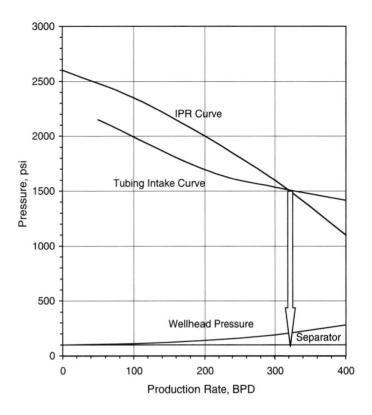

Fig. 2–66 Example systems analysis for a flowing oil well. [after 107]

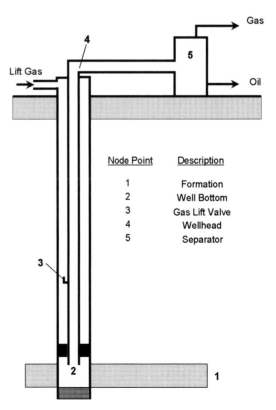

Fig. 2–67 The production system of a continuous flow gas lift well.

The methods of analysis are similar to those applied to flowing wells. The first important task is the proper selection of the solution node where the system is divided into two subsystems. The selection depends on the type of the problem to be solved. For example, if an analysis of the parameters governing the well's inflow performance is desired than the proper choice is the sandface. If, on the other hand, the capacity of the surface system (flowline, separator, etc.) is selected as the controlling parameter, placing the solution node at the wellhead is advised.

The solution node being selected, description of the two subsystem's behavior follows. Starting from the points with the known constant pressures (at the separator and at the well's drainage boundary), flowing pressures in the different hydraulic elements of the production system are calculated. The resulting two performance curves describe the behavior of the two subsystems and their common solution will give the well's production rate under the conditions studied.

References

1. Gould, T. L., "Vertical Two-Phase Steam-Water Flow in Geothermal Wells." *JPT*, Aug. 1974: 833–42.

2. Standing, M. B., "A Pressure-Volume-Temperature Correlation for Mixtures of California Oils and Gases." *API Drilling and Production Practice*, 1947: 275–86.

3. Lasater, J. A., "Bubblepoint Pressure Correlation." *Trans. AIME*, Vol. 213, 1958: 379–81.

4. Vazquez, M. and H. D. Beggs, "Correlations for Fluid Physical Property Prediction." Paper SPE 6719 presented at the 52nd Annual Fall Technical Conference and Exhibition of SPE, Denver, CO, Oct. 9–12, 1977.

5. Beal, C., "The Viscosity of Air, Natural Gas, Crude Oil and Its Associated Gases at Oilfield Temperatures and Pressures." *Trans. AIME*, Vol. 165, 1946: 94–112.

6. Chew, J. and C. H. Connally, Jr., "A Viscosity Correlation for Gas-Saturated Crude Oils." *Trans. AIME*, Vol. 216, 1959: 23–5.

7. Standing, M. B. and D. L. Katz, "Density of Natural Gases." *Trans. AIME*, Vol. 146, 1942: 140–9.

8. Hankinson, R. W., L. K. Thomas, and K. A. Phillips, "Predict Natural Gas Properties." *Hydrocarbon Processing*, April 1969: 106–8.

9. Wichert, E. and K. Aziz, "Compressibility Factor for Sour Natural Gases." *Can. J. Chem. Eng.*, April 1971: 267–73.

10. Wichert, E. and K. Aziz, "Calculate Z's for Sour Gases." *Hydrocarbon Processing*. May 1972: 119–22.

11. Sarem, A. M., "Z-factor Equation Developed for Use in Digital Computers." *OGJ*, July 20, 1959: 64–6.

12. Gopal, V. N., "Gas Z-factor Equations Developed for Computer." *OGJ*, Aug. 8, 1977: 58–60.

13. Yarborough, L. and K. R. Hall, "How to Solve Equation of State for Z-factors." *OGJ*, Feb. 18, 1974: 86–88.

14. Dranchuk, P. M., R. A. Purvis, and D. B. Robinson, "Computer Calculations of Natural Gas Compressibility Factors Using the Standing and Katz Correlation." Inst. of Petroleum Technical Series No. IP 74-008, 1974.

15. Takács, G., "Comparing Methods for Calculating Z-factor." *OGJ*, May 15, 1989: 43–6.

16. Papay, J., "Change of Technological Parameters in Producing Gas Fields." (in Hungarian) Proc. OGIL Budapest 1968: 267–73.

17. Lee, A. L., M. H. Gonzalez, and B. E. Eaking, "The Viscosity of Natural Gases." *JPT*, Aug. 1966: 997–1000.

18. Vogel, J. V., "Inflow Performance Relationships for Solution-Gas Drive Wells." *JPT*, Jan. 1968: 83–92.

19. Patton, L. D. and M. Goland, "Generalized IPR Curves for Predicting Well Behavior." *PEI*, Sept. 1980: 92–102.

20. Fetkovich, M. J., "The Isochronal Testing of Oil Wells." Paper SPE 4529 presented at the 48th Annual Fall Meeting of SPE, Las Vegas, NV, Sept. 30–Oct. 3, 1973.

21. Bradley, H. B. (Ed.) *Petroleum Engineering Handbook*. Chapter 34. Society of Petroleum Engineers, 1987.

22. Moody, L. F., "Friction Factors for Pipe Flow." *Trans. ASME*, Vol. 66, 1944: 671.

23. Fogarasi, M., "Further on the Calculation of Friction Factors for Use in Flowing Gas Wells." *J.Can.Petr.Techn.* April–June 1975: 53–4.

24. Gregory, G. A. and M. Fogarasi, "Alternate to Standard Friction Factor Equation." *OGJ*, April 1, 1985: 120–7.

25. Chen, N. H., "An Explicit Equation for Friction Factor in Pipe." Ind. Eng. Chem. Fund., 18, 296, 1979.

26. Brown, K. E. and R. L. Lee, "Easy-to-Use Charts Simplify Intermittent Gas Lift Design." *World Oil,* Feb. 1, 1968: 44–50.

27. Bobok, E., *"Fluid Mechanics for Petroleum Engineers."* Amsterdam-Oxford-New York-Tokyo: Elsevier, 1993.

28. Cook, H. L. and F. H. Dotterweich, "Report in Calibration of Positive Choke Beans Manufactured by Thornhill-Craver Company." College of Arts and Industries, Houston, TX, August 1946.

29. Szilas, A. P. *Production and Transport Of Oil And Gas.* 2nd Ed. Part A. Elsevier Publishing Co., 1985.

30. Zuber, N. and J. A. Findlay, "Average Volumetric Concentration in Two-Phase Flow Systems." *J. Heat Transfer,* Nov. 1965: 453–68.

31. Harmathy, T. Z., "Velocity of Large Drops and Bubbles in Media of Infinite or Restricted Extent." *AIChE Journal,* June 1960: 281–88.

32. Barnea, D., "A Unified Model for Predicting Flow-Pattern Transitions for the Whole Range of Pipe Inclinations." *Int. J. of Multiphase Flow. No. 1,* 1987: 1–12.

33. Duns, H. Jr. and J. C. J. Ros, "Vertical Flow of Gas and Liquid Mixtures in Wells." Proc. 6th World Petroleum Congress, Frankfurt, 1963: 451–65

34. Kaya, A. S., C. Sarica, and J. P. Brill, "Comprehensive Mechanistic Modeling of Two-Phase Flow in Deviated Wells." SPE 56522 presented at the SPE Annual Technical Conference and Exhibition held in Houston, TX, October 3–6, 1999.

35. Gould, Th. L., M. R. Tek, and D. L. Katz, "Two-Phase Flow through Vertical, Inclined, or Curved Pipe." *JPT,* Aug. 1974: 915–26.

36. Brill, J. P. and H. Mukherjee, *Multiphase Flow in Wells.* Monograph No. 17, Society of Petroleum Engineers, Dallas, TX, 1999.

37. Orkiszewski, J., "Predicting Two-Phase Pressure Drops in Vertical Pipe." *JPT,* June 1967: 829–38.

38. Brill, J. P. and H. D. Beggs, *Two-Phase Flow in Pipes.* 6th Ed. University of Tulsa, 1994.

39. Poettmann, F. H. and P. G. Carpenter, "The Multiphase Flow of Gas, Oil and Water through Vertical Flow Strings with Application to the Design of Gas-Lift Installations." *API Drilling and Production Practice,* 1952: 257–317.

40. Baxendell, P. B., "Producing Wells on Casing Flow-An Analysis of Flowing Pressure Gradients." *JPT,* August 1958: 59.

41. Baxendell, P. B. and R. Thomas, "The Calculation of Pressure Gradients in High-Rate Flowing Wells." *JPT,* October 1961: 1023–8.

42. Fancher, G. R. Jr. and K. E. Brown, "Prediction of Pressure Gradients for Multiphase Flow in Tubing." *SPEJ,* March 1963: 59–69.

43. Gaither, O. D., H. W. Winkler, and C. V. Kirkpatrick, "Single and Two-Phase Fluid Flow in Small Vertical Conduits Including Annular Configurations." *JPT*, March 1963: 309–20.

44. Hagedorn, A. R. and K. E. Brown, "The Effect of Liquid Viscosity in Vertical Two-Phase Flow." *JPT*, February 1964: 203–10.

45. Cornish, R. E., "The Vertical Multiphase Flow of Oil and Gas at High Rates." *JPT*, July 1976: 825–31.

46. Hagedorn, A. R. and K. E. Brown, "Experimental Study of Pressure Gradients Occurring During Continuous Two-Phase Flow in Small Diameter Vertical Conduits." *JPT*, April 1965: 475–84.

47. Beggs, H. D. and J. P. Brill, "A Study of Two-Phase Flow in Inclined Pipes." *JPT*, May 1973: 607–17.

48. Mukherjee, H. and J. P. Brill, "Pressure Drop Correlations for Inclined Two-Phase Flow." *J. Energy Res. Tech.*, December 1985: 549–54.

49. Ros, N. C. J., "Simultaneous Flow of Gas and Liquid as Encountered in Well Tubing." *JPT*, October 1961: 1037–49.

50. Brown, K. E. *Gas Lift Theory and Practice*. Tulsa, OK: Petroleum Publishing Co., 1967.

51. Brown, K. E. *The Technology of Artificial Lift Methods*. Vol.1. Tulsa, OK: Petroleum Publishing Co., 1977.

52. Griffith, P. and G. B. Wallis, "Two-Phase Slug Flow." *J. of Heat Transfer*. August 1961: 307–20.

53. Griffith, P., "Two-Phase Flow in Pipes." Special Summer Program, MIT, Cambridge, Massachusetts, 1962.

54. Brill, J. P., "Discontinuities in the Orkiszewski Correlation for Predicting Pressure Gradients in Wells." *J. Energy Resources Techn.* March 1989: 34–6.

55. Chierici, G. L., G. M. Ciucci, and G. Sclocchi, "Two-Phase Vertical Flow in Oil Wells-Prediction of Pressure Drop." *JPT*, August 1974: 927–38.

56. Payne, G. A., C. M. Palmer, J. P. Brill, and H. D. Beggs, "Evaluation of Inclined-Pipe, Two-Phase Liquid Holdup and Pressure-Loss Correlations using Experimental Data." Paper SPE 6874, presented at the 52nd Annual Fall Technical Conference and Exhibition, Denver, CO, Oct. 9–12, 1977.

57. Payne, G. A., C. M. Palmer, J. P. Brill, and H. D. Beggs, "Evaluation of Inclined-Pipe, Two-Phase Liquid Holdup and Pressure-Loss Correlations using Experimental Data." *JPT*, September 1979: 1198–1208.

58. Mukherjee, H. and J. P. Brill, "Liquid Holdup Correlations for Inclined Two-Phase Flow." *JPT*, May 1983: 1003–8.

59. Aziz, K., G. W. Govier, and M. Fogarasi, "Pressure Drop in Wells Producing Oil and Gas." *JCPT*, July–September 1972: 38–48.

60. Hasan, A. R. and C. S. Kabir, "A Study of Multiphase Flow Behavior in Vertical Wells." *SPE PE*, May 1988: 263–72.

61. Ansari, A. M., N. D. Sylvester, C. Sarica, O. Shoham, and J. P. Brill, "A Comprehensive Mechanistic Model for Upward Two-Phase Flow in Wellbores." *SPE PF*, May 1994: 143–51.

62. Chokshi, R. N., Z. Schmidt, and D. R. Doty, "Experimental Study and the Development of a Mechanistic Model for Two-Phase Flow through Vertical Tubing." SPE 35676 presented at the Western Regional Meeting held in Anchorage, AK, May 22–24, 1996.

63. Gomez, L. E., O. Shoham, Z. Schmidt, R. N. Chokshi, A. Brown, and T. Northug, "A Unified Mechanistic Model for Steady-State Two-Phase Flow in Wellbores and Pipelines." SPE 56520 presented at the SPE Annual Technical Conference and Exhibition held in Houston, TX, October 3–6, 1999.

64. Wallis, G. B. *One-Dimensional Two-Phase Flow*. New York: McGraw-Hill Book Co., 1969.

65. Al-Najjar, H. S. H. and N. B. A. Al-Soof, "Alternative Flow-Pattern Maps Can Improve Pressure-Drop Calculations of the Aziz et al. Multiphase-Flow Correlation." *SPE PE*, August 1989: 327–34.

66. Kabir, C. S. and A. R. Hasan, "Performance of a Two-Phase Gas/Liquid Model in Vertical Wells." *J. Petroleum Science and Engineering* 1990: 4, 273–89.

67. Hasan, A. R. and C. S. Kabir, "Predicting Multiphase Flow Behavior in a Deviated Well." *SPE PE*, November 1988: 474–8.

68. Hasan, A. R. and C. S. Kabir, "Two-Phase Flow in Vertical and Inclined Annuli." *Int. J. Multiphase Flow*. 1992: 18/2, 279–93.

69. Espanol, J. H., C. S. Holmes, and K. E. Brown, "A Comparison of Existing Multiphase Flow Methods for the Calculation of Pressure Drop in Vertical Wells." SPE 2553, presented at the 44th Annual Fall Meeting of SPE in Denver, CO, Sept. 28–Oct. 1, 1969.

70. Gregory, G. A., M. Fogarasi, and K. Aziz, "Analysis of Vertical Two-Phase Flow Calculations: Crude Oil-Gas Flow in Well Tubing." *JCPT*, January–March 1980: 86–92.

71. Lawson, J. D. and J. P. Brill, "A Statistical Evaluation of Methods Used to Predict Pressure Losses for Multiphase Flow in Oilwell Tubing." *JPT*, August 1974: 903–14.

72. Aggour, M. A., H. Y. Al-Yousef, and A. J. Al-Muraikhi, "Vertical Multiphase Flow Correlations for High Production Rates and Large Tubulars." *SPE PF*, February 1996: 41–8.

73. McLeod, W. E., D. L. Anderson, and J. J. Day, "A Comparison of Vertical Two-Phase Computation Techniques." *ASME Paper 72-Pet-38*, 1972.

74. Vohra, I. R., J. R. Robinson, and J. P. Brill, "Evaluation of Three New Methods for Predicting Pressure Losses in Vertical Oilwell Tubing." *JPT*, August 1974: 829–32.

75. Browne, E. J. P., "Practical Aspects of Prediction Errors in Two-Phase Pressure-Loss Calculations." *JPT*, April 1975: 515–22.

76. Ozon, P. M., G. Ferschneider, and A. Chwetzoff, "A New Multiphase Flow Model Predicts Pressure and Temperature Profiles in Wells." SPE 16535 presented at Offshore Europe 87, Aberdeen, September 8–11, 1987.

77. Rai, R., I. Singh, and S. Srini-vasan, "Comparison of Multiphase-Flow Correlations with Measured Field Data of Vertical and Deviated Oil Wells in India." *SPE PE*, August 1989: 341–8.

78. Pucknell, J. K., J. N. E. Mason, and E. G. Vervest, "An Evaluation of Recent 'Mechanistic' Models of Multiphase Flow for Predicting Pressure Drops in Oil and Gas Wells." SPE 26682 presented at the Offshore European Conference, Aberdeen, September 7–10, 1993.

79. Gilbert, W. E., "Flowing and Gas-Lift Well Performance." *API Drilling and Production Practice*, 1954: 126–57.

80. *Handbook of Gas Lift*. U.S. Industries, 1959.

81. Winkler, H. W. and S. S. Smith, *Gas Lift Manual*. CAMCO Inc., 1962.

82. *Gas Lift Manual*. Teledyne MERLA, 1970.

83. Brill, J. P. and H. D. Beggs, *Two-Phase Flow in Pipes*. Tulsa, OK: University of Tulsa, 1978.

84. Aziz, K., J. R. Eickmeier, M. Fogarasi, and G. A. Gregory, "Gradient Curves for Well Analysis and Design." CIM Special Volume 20, Canadian Institute of Mining and Metallurgy, 1979.

85. Brown, K. E. *The Technology of Artificial Lift Methods*. Vol. 3a. Tulsa, OK: Petroleum Publishing Co., 1980.

86. Beggs, H. D. *Production Optimization Using Nodal Analysis*. Tulsa, OK: OGCI Publications, 1991.

87. Schmidt, Z., "Gas Lift Optimization using Nodal Analysis." CEALC Inc., 1994.

88. Mandhane, J. M., G. A. Gregory, and K. Aziz, "A Flow Pattern Map for Gas-Liquid Flow in Horizontal Pipes." *Int. J. Multiphase Flow*, Vol. 1, 1974: 537–53.

89. Lockhart, R. W. and R. C. Martinelli, "Proposed Correlation of Data for Isothermal Two-Phase, Two-Component Flow in Pipes." *Ch. Eng. Progress*, January 1949: 39–48.

90. Love, D. L., "Improve Two-Phase Flow Calculations." *OGJ*, January 7, 1974: 78–82.

91. Dukler, A. E., "Gas-Liquid Flown in Pipelines. I Research Results." AGA-API Project NX-28. May 1969.

92. Ros, N. C. J., "An Analysis of Critical Simultaneous Gas Liquid Flow through a Restriction and its Application to Flowmetering." Applied Scientific Research, Sect. A, Vol. 9: 374–88.

93. Omana, R., C. Houssiere, K. E. Brown, J. P. Brill, and R. E. Thompson, "Multiphase Flow through Chokes." SPE 2682 presented at the 44th Annual Fall Meeting of SPE, Denver, CO, Sept. 28–Oct. 1, 1969.

94. Guzhov, A. I. and V. F. Medvediev, "Critical Gas-Oil Mixture Flow through Wellhead Chokes." (in Russian) *Neftianoe Hoziastvo*. May 1968: 47–52.

95. Ashford, F. E. and P. E. Pierce, "Determining Multiphase Pressure Drops and Flow Capacities in Down-Hole Safety Valves." *JPT*, September 1975: 1145–52.

96. Sachdeva, R. Z. Schmidt, J. P. Brill, and R. M. Blais, "Two-Phase Flow through Chokes." SPE 15657 presented at the 61st Annual Technical Conference and Exhibition of SPE, New Orleans, LA, October 5–8, 1986.

97. Perkins, T. K., "Critical and Subcritical Flow of Multiphase Mixtures through Chokes." *SPE Drilling and Completion*. December 1993: 271–6.

98. Ramey, H. J., "Wellbore Heat Transmission." *JPT*, April 1962: 427–35.

99. Willhite, G. P., "Overall Heat Transfer Coefficients in Steam and Hot Water Injection Wells." *JPT*, May 1967: 607–15.

100. Carslaw, H. S. and J. C. Jaeger. *Conduction of Heat in Solids*. London: Oxford University Press, 1950.

101. Hasan, A. R. and C. S. Kabir, "Aspects of Wellbore Heat Transfer During Two-Phase Flow." *SPE PF*, August 1994: 211–6.

102. Shiu, K. C. and H. D. Beggs, "Predicting Temperatures in Flowing Wells." *J. Energy Resources Technology*, March 1980: 2–11.

103. Sagar, R., D. R. Doty, and Z. Schmidt, "Predicting Temperature Profiles in a Flowing Well." *SPE PE*, November 1991: 441–8.

104. Proano, E. A., J. M. Mach, and K. E. Brown, "Systems Analysis as Applied to Producing Wells." *Congreso Panamericano de Ingeniera del Petroleo*, Mexico City, March 1979.

105. Mach, J. M., E. A. Proano, and K. E. Brown, "A Nodal Approach for Applying Systems Analysis to the Flowing and Artificial Oil and Gas Wells." Society of Petroleum Engineers Paper No. 8025, 1979.

106. Szilas, A. P., "Field-Integrated Production System Cuts Costs." *OGJ*, June 13, 1983: 125–28.

107. Takács, G., A. P. Szilas, and V. A. Sakharov, "Hydraulic Analysis of Producing Wells." (in Hungarian) *Koolaj es Foldgaz*, May 1984: 129–36.

3 | Gas Lift Valves

3.1 Introduction

Gas lifting is a process of lifting fluids from an oil well by the injection of gas (a) continuously in the upward-rising liquid column (in case of continuous flow gas lift), or (b) underneath an accumulated liquid slug in a relatively short period of time to move the slug to the surface (intermittent lift). In both cases, high-pressure gas (or air in early practice) from the surface is led downhole, either through the casing-tubing annulus, the tubing, or some special conduit. This gas is injected into formation fluids at predetermined depth or depths in the well. To achieve an efficient operation and to ensure that the proper amount of gas is injected at all times, gas entry must be controlled by utilizing some kind of a downhole control device. In modern practice, gas lift valves are used for downhole gas injection control.

The gas lift valve is the heart of a gas lift installation since it provides the necessary control of well production rates and its performance determines the technical and economic parameters of fluid lifting. Because of its importance, there have been a great number of varieties of gas lift valves developed over the years. The abundance of the different operating principles and technical features allows the operator to choose the proper type of valve for any case.

Chapter 3 covers all aspects of gas lift valves including their evolution, operating principles, performance, and application. Most of the discussion is related to the family of pressure-operated valves since this is the only class that survived from the many different kinds of operating principles. This is why almost all gas lift valves today belong to this category. After describing the constructional details of modern gas lift valves, operational features of the different available models are discussed. The conditions of opening and closing as well as their dynamic performance will be described by presenting the latest developments. The proper application of gas lift valves, including the selection and the setting of the right valve is detailed as follows. The chapter is concluded with the description of running and retrieving operations of most types of gas lift valves.

3.1.1 Downhole gas injection controls

This section discusses the ways gas injection into the fluid column was controlled before the use of gas lift valves. As shown, control methods continuously evolved to the present day with the aim of increasing the efficiency of fluid lifting.

Open Tubing

In early days of gas lifting, a tubing string was suspended into the well with no packer, see Figure 3–1. The well could be produced by injecting gas

(a) down the casing-tubing annulus and removing fluids out the tubing

(b) down the tubing string and removing fluids out the annulus

In either case, gas had to be blown around the bottom of the tubing string, resulting in excessive and mostly uncontrollable gas usage. This so-called *open installation* had several disadvantages, the most important of them being the necessity to kick (unload) the well every time it is brought back to production. Finding of the correct injection depth was extremely difficult and the installation allowed practically no control over the volume of injected gas.

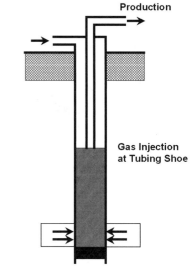

Fig. 3–1 Open tubing installation with gas injection at the tubing shoe.

Foot-pieces

The next step to more efficient gas injection control was the use of so called *foot-pieces*. These devices were run on the bottom of the tubing string (Fig. 3–2) and provided a more-or-less controlled gas injection through a special injection string run parallel to the tubing. The first U.S. patent for an *oil ejector* (foot-piece) was issued to A. Brear in 1865. [1]

There was much controversy over foot-pieces in the technical literature of that time. Some manufacturers claimed that by finely dividing the injected air into small bubbles, the air would do more useful work. Foot-pieces with such features directed the injected gas stream into the tubing in a vertical upward direction. Others found that little difference existed between various makes of foot-pieces since the dispersed air bubbles tended to coalesce immediately above the foot-piece. This was proved by the experiments of Davis and Weidner in 1914, who showed that the type of foot-piece had very little effect on the efficiency of the system.

Both before-mentioned injection control methods had one main disadvantage that the only point of gas injection was around the bottom of the tubing string. For deep wells, this requires excessively high kickoff (unloading) pressures. For example, if a 5,000-ft dead well is standing full of a 0.4 psi/ft gradient fluid, a gas pressure of 5,000 0.4 = 2,000 psi would be required to blow gas around the bottom of the tubing string. In most cases, this much pressure was not available and safety methods in handling these pressures had not been introduced. That is why these early gas lift methods were limited to shallow wells where such high initial unloading pressures were not required.

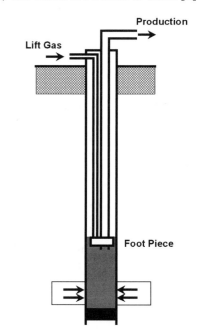

Fig. 3–2 Gas lift installation with a foot-piece as injection control.

Jet Collars and Orifice Inserts

The next development in gas lift equipment was to devise a means for reducing the high initial unloading pressures required in deep wells. This is accomplished by introducing the unloading process, during which liquid is displaced from the annulus (in tubing flow installations) using a stepwise procedure. Gas pressure, applied to the annulus, U-tubes liquid into the tubing and depresses liquid level to a given depth corresponding to injection pressure and static liquid gradient conditions. Then gas is injected into the tubing above the depressed liquid level, aerating the liquid column in the tubing. A drop in tubing pressure due to a reduction of flowing gradient above the gas injection point decreases bottomhole pressure. As annular and tubing spaces communicate at the bottom of the tubing, liquid level in the annulus is further depressed. Now gas injection at a lower depth is already possible, and the previous process is repeated at the next valve. Unloading proceeds until the required injection depth is reached. By utilizing this stepwise unloading method, there is no need for a high kickoff pressure and the usual operating pressure can be used for unloading purposes as well.

Early unloading operations used two procedures. [2] Jet collars (special tubing collars with holes) could be opened and closed from the surface by turning the tubing string. These collars (run at different depths) provided gas injection points during unloading operations. Although (at least in principle) they can be closed from the surface, if some collars do not close completely after the unloading process, then lift gas usage becomes excessive.

The other method used orifice inserts installed in the tubing string. These can be installed by wireline after the tubing is already in place by punching holes in the tubing string at certain depths. The main problem of using orifice inserts is that during normal operation, all the orifice inserts above the required injection point are uncovered and are therefore passing gas. Under these conditions, gas usage is most often high and inefficient. In addition, the backflow and turbulence created at orifice locations cause excessive wear and holes in the casing.

To overcome the disadvantages of orifice inserts left open in the tubing string, several methods were developed to close upper holes while injecting gas through lower ones. This feat was a distinct advantage over previous practice since injection gas volumes were under closer control. These requirements brought about the development of gas lift valves.

Kickoff Valves

Early gas lift valves were called kickoff valves because they were originally used in wells that needed only to be started flowing. After an initial boost of gas, or *kickoff*, such wells would be able to flow naturally for shorter or longer periods.

One of the first valves of this type was the velocity-controlled valve. Its principle of operation is that upper valves are closed by the increased mixture velocity in the tubing while lower valves remain open and inject gas. Other kickoff valve types worked on different principles and all were utilized for continuous flow operations. Because of their generally unreliable operation, all kickoff valves were eventually replaced by more efficient valve types, to be discussed in the following section.

3.1.2 Evolution of gas lift valves

For detailed discussions on early gas lift valves as well as exact historical data, the reader is advised to consult Brown [2] and Shaw [3]. Both books present a chronological list of events and include many drawing figures from the numerous early patents on gas lift valves.

The spring-loaded differential valve first appeared in 1934, was extremely popular, and is still being utilized today. It incorporates a spring that provides the force to hold it open. The valve opens and closes, based on a difference between the tubing and the casing pressures. For example, with a spring force setting equivalent to a pressure of 100 psig and a casing pressure of 700 psig, the valve requires 700 − 100 = 600 psi or greater in the tubing to open. Although widely used in early continuous flow operations, this valve is no longer popular because of its reliability problems and the emergence of better valves.

Mechanically controlled valves were opened and closed from the surface by some mechanical means, most frequently by wireline. This type of valve was used in intermittent lift wells; its biggest disadvantage was the problem of unloading which operation had to be accomplished manually.

The specific gravity differential valve, on the other hand, was used for unloading continuous flow installations. Its main element is a flexible diaphragm that opens and closes the valve based on the pressures acting on its two sides. A control force is provided by a liquid charge of a set specific gravity, and the valve opens based on the actual fluid gradient valid at valve depth. This valve proved to be an excellent one but because of its bulkiness was later replaced by modern types.

Since the early days of gas lifting, an extremely large number of gas lift valves working on different principles had been introduced. Brown [2] states that in the last 100 years, more than 25,000 patents related to gas lift equipment have been issued in the United States alone. Among those, the valve type that has revolutionized the gas lift industry is the pressure-operated gas lift valve. This valve uses for its operation the pressure of the injected gas as a means of control, provides more and better control of gas injection, and operates automatically. Because of its excellent features, it has almost obsoleted all other valve types after the World War II.

After some initial applications of metal bellows in gas lift valves, the first pressure-operated gas lift valve patent was issued to W.R. King [4] in 1944. His valve utilized a metal bellows charged with high-pressure gas and sealed at the surface as the control force to open and close the valve. In the open position (Fig. 3–3), lift gas from the casing annulus is injected into the tubing through the valve's open port. The valve port is closed by a drop in the casing pressure and can be opened again by increasing the casing or the tubing pressures. In summary, valve operation is controlled by downhole pressure conditions which, in turn, can be greatly affected by changing the surface injection pressure. Present-day bellows valves are very similar to this valve and have retained all the principal features of their forerunner.

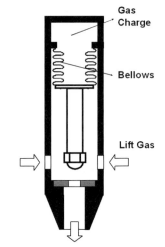

Fig. 3–3 A typical pressure-operated gas lift valve.

Since the King valve, the first major change in gas lift valve design was the flexible sleeve valve, based on a patent issued to L.L. Cummings [5] in 1953. This concentric valve (Fig. 3–4) uses a resilient element as its moving member and is run as a part of the tubing string. The valve's radial dimensions are identical to those of the tubing string it is run in with: its OD and ID being equal to the respective dimensions of the coupling and the tubing. The resilient element opens and closes the gas injection ports, depending on the condition whether casing pressure is higher or lower than the valve's gas charge pressure. This pressure-operated valve is suitable for both continuous flow and intermittent lift operations and was very popular in slim-hole completions.

Developments are being made in gas lift equipment even now, but the pressure-operated valve is the type of valve that is used in practically all gas lift installations worldwide. That is why the present chapter will deal primarily with this valve type in detail.

3.1.3 Overview of valve types

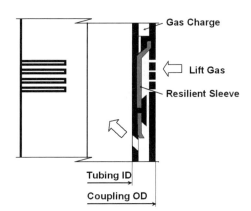

Fig. 3–4 Schematic of a flexible sleeve concentric gas lift valve.

The great variety of different operating principles and features of gas lift valves developed over the years makes it difficult to set up a system of classification covering all possible types. Based on different aspects, valves can be grouped in different classes. The following overview introduces the main classes where current or early gas lift valves belong to and then gives a short description of their main technical features.

Based on their control of operation, all gas lift valves can be classified into one of the following groups:

- Mechanically controlled from the surface (by wireline, drop bar, etc.).

- Other control methods include flow velocity, specific gravity, etc.

- Pressure-operated valves are opened and closed by injection and/or production pressures. They include a huge number of different subtypes to be described in subsequent sections.

According to their application, gas lift valves can be used for unloading or as an operating valve, as follows:

- Unloading valves are used for the startup of gas lift operations only and are usually closed during normal production. A string of unloading valves with the proper setting is activated every time the well is brought back to production after shutdown.

- Operating valves ensure normal gas lift operations and inject the right amount of gas into the well. The valves suited for continuous and intermittent gas lifting may be different because of the different requirements of these two principal types of gas lift.

Gas lift valves can also be classified according to the method they are run in the well, as follows:

- Conventional valves are attached to the outside of the tubing in special mandrels and can be run and retrieved along with the tubing string only.

- Retrievable valves require special mandrels with inside pockets to receive the valve. Such valves are run on wireline tools inside the tubing string and can be retrieved without the need to pull the tubing.

- Concentric valves are special subs in the tubing string and can be run and retrieved with the tubing string only.

- Pack-off valves combine the advantages of concentric and wireline-retrievable valves. They are run and set inside the tubing where holes in the tubing were previously made.

- Pump-down valves are employed in installations where the pump-down technique of well completion is used. They require a way of fluid circulation in the well for setting and retrieval. Special tool strings allow for running and retrieval of downhole equipment by changing the direction of circulation to the well.

- Coiled tubing (CT) valves are required in installations where a CT string is used to produce the well, and lift gas is injected through the tubing—CT annulus.

Constructional features can also be used for distinguishing different valves.

- Differential valves usually contain a single spring, the force of which determines the pressure conditions under which the valve opens and closes.

- Bellows valves contain a metal bellows charged with a predetermined gas pressure. The pressure acting on the bellows area provides a control force that, in relation with the other forces coming from injection and production pressures, determine the valve's operation. Subdivisions of these valves include the following:

  - single element valves having only the bellows as a source of control force.

  - valves utilizing a spring in addition to the gas-charged bellows.

  - valves with a spring as the only control force with an uncharged bellows.

- Pilot valves contain two sections: a pilot and a main valve. The gas pilot controls the operation of the main valve and the cooperation of the two sections provides the desired results.

- Flexible sleeve valves have concentric flexible elements and a gas charge and open or close the valve's injection ports, depending on the injection pressure.

- Other valve types include control mechanisms different from those noted previously.

Finally, gas lift valves can be classed according to the place fluid flow occurs in the well (in the tubing or in the casing annulus), although most valves can be applied in both cases depending on the type of the valve mandrel used.

- Tubing flow valves inject gas from the casing-tubing annulus into the tubing string, whereas

- Casing flow valves allow the injection from the tubing string to the well's annular space.

3.1.4 Supporting calculations

To accurately describe the operation of pressure-operated gas lift valves, one must consider the valve's actual operational conditions: the temperature and the injection pressure at valve setting depth. Both parameters affect the behavior of the valve in the well and their proper knowledge is of prime importance in gas lift design. Temperature at valve setting depth affects the pressure in the valve's gas dome, and in turn, determines the force arising on the bellows cross-sectional area. Injection pressure at valve depth, on the other hand, provides a force trying to open the valve. However, both the previous parameters, *i.e.* dome charge pressure and injection pressure at valve depth, are not readily available and have to be calculated from surface data. The present section describes the ways to solve this problem.

3.1.4.1 Dome charge pressure calculations. When working with pressure-operated gas lift valves, one must consider that valves are (and can only be) charged with high-pressure gas at the surface at a controlled temperature. When run in the well, the temperature of the valve and the gas charge assumes the well's flowing temperature valid at the given depth. Since well temperatures are greater than the valve's surface charging temperature, actual pressure in the valve dome will always be greater than at charging conditions. If we disregard the very small change in dome volume between the two cases, the Engineering Equation of State (Equation 2.15) can be solved as:

$$\frac{p_d}{Z(p_d, T_v) \, T_v} = \frac{p_d'}{Z(p_d', T_{cb}) \, T_{cb}}$$

3.1

This equation allows one to find downhole or surface dome pressures from the knowledge of the other pressure. Solution for the downhole bellows pressure results in the following formula, where the usual workshop temperature of 60 °F is assumed to prevail at charging conditions:

$$p_d = \frac{Z(p_d, T_v)}{Z(p_d', T_{cb})} \, \frac{T_v + 460}{520} \, p_d'$$

3.2

where: $Z(p_d, T_v)$ = deviation factor at downhole conditions, -

  $Z(p_d', T_{cb})$ = deviation factor at charging conditions, -

  T_v = valve temperature at setting depth, F

  T_{cb} = valve temperature at charging conditions, F

  p_d' = dome charge pressure at 60 °F, psia

The use of the previous formula requires an iterative calculation process because the deviation factor at downhole conditions, $Z(p_d, T_v)$, is a function of the downhole dome pressure p_d that is to be calculated. For the first guess, downhole dome pressure is assumed to be equal to surface dome pressure and a better approximation for p_d is calculated from Equation 3.2. Using the value just calculated, another value of p_d is found. If the two values differ considerably from each other, a new iteration step is preformed. After the required accuracy is attained, the final value of p_d is accepted.

The previous procedure relies on the accurate prediction of deviation factors for the gas the valve dome is charged with. Because of its beneficial features and predictability, the universally adopted medium for charging gas lift valves is nitrogen gas. Deviation factors of N_2 gas were experimentally measured by Sage and Lacy [6], their results were fitted by the present author with the following expressions, valid in the ranges of 0–3,000 psia pressure and 60–400 °F temperature:

$$Z = 1 + b\,p + c\,p^2 \qquad\qquad 3.3$$

$$b = (1.207 \times 10^{-7}\,T^3 - 1.302 \times 10^{-4}\,T^2 + 5.122 \times 10^{-2}\,T - 4.781)\,10^{-5} \qquad\qquad 3.4$$

$$c = (-2.461 \times 10^{-8}\,T^3 + 2.640 \times 10^{-5}\,T^2 - 1.058 \times 10^{-2}\,T + 1.880)\,10^{-8} \qquad\qquad 3.5$$

where: T = temperature, F

          p = pressure, psia

The previous formulas give accuracies better than ±0.1% in the indicated ranges of pressure and temperature. Their utilization in the iterative solution discussed previously allowed the development of charts like the one presented in Figure 3–5, where downhole dome pressure p_d is plotted as a function of surface dome charge pressure p_d' for different temperatures. Appendix D contains full-page copies of similar charts with different pressure ranges.

Winkler and Eads [7] pointed out that charts similar to Figure 3–5 and published in different books [8, 9] contain systematic errors for dome pressures more than 800–1,000 psig. They proposed the following formulas to calculate the dome charge pressure at elevated temperatures, which include the effect of the deviation factor for nitrogen gas:

$$p_d = p_d' + m(T_v - 60) \qquad\qquad 3.6$$

$$m = -0.00226 + 0.001934\,p_d' + 3.054 \times 10^{-7}\,(p_d')^2 \qquad\qquad \text{for } p_d' < 1{,}238 \text{ psia} \qquad 3.7$$

$$m = -0.267 + 0.002298\,p_d' + 1.84 \times 10^{-7}\,(p_d')^2 \qquad\qquad \text{for } p_d' > 1{,}238 \text{ psia} \qquad 3.8$$

where: Tv = valve temperature at setting depth, F

          p_d' = dome charge pressure at 60 °F, psia.

The equations given previously permit the calculation of surface dome charge pressures from downhole dome pressures by solving Equation 3.6 for p_d', as given in API Spec 11V [10].

$$p_d' = \frac{-b + \sqrt{b^2 - 4ac}}{2\,a} \qquad\qquad 3.9$$

The constants a, b, and c depend on the value of the surface dome charge pressure p_d' which is to be calculated. A simple approximation is to assume an ideal gas at a constant volume, as given in the following:

	if $p_d \dfrac{520}{460 + T_v} < 1{,}238$ psia	if $p_d \dfrac{520}{460 + T_v} > 1{,}238$ psia
a =	$3.054 \times 10^{-7}\,(T_v - 60)$	$1.84 \times 10^{-7}\,(T_v - 60)$
b =	$1 + 1.934 \times 10^{-3}\,(T_v - 60)$	$1 + 2.298 \times 10^{-3}\,(T_v - 60)$
c =	$-2.226 \times 10^{-3}\,(T_v - 60) - P_d$	$-0.267 \times 10^{-3}\,(T_v - 60) - P_d$

A detailed investigation of published data from Winkler and Eads [7] confirmed that their results are identical to those found from Figure 3–5.

-·

Example 3–1. Find a nitrogen charged gas lift valve's dome pressure at the well temperature of 180°F if its surface charging pressure at 60°F is 600 psig. Use Figure 3–5 as well as Equation 3.6.

Solution

Using Figure 3–5 and starting from 600 psig charging pressure, we can read a downhole dome pressure of 750 psig, at the valve temperature of 180°F.

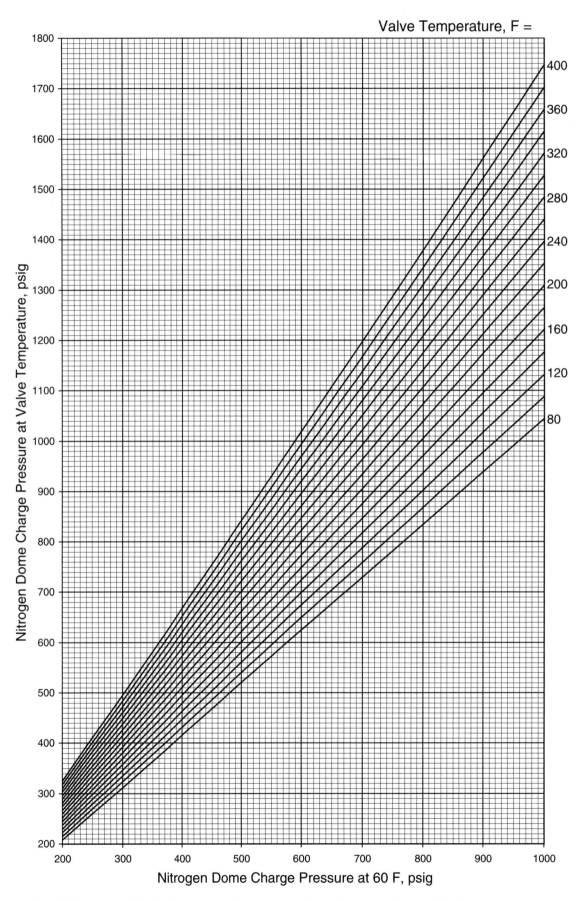

Fig. 3–5 Chart for calculating the change in dome pressures for gas lift valves charged with nitrogen gas.

When using Equation 3.6, the constant m is found from Equation 3.7, since dome charge pressure is below 1,238 psia:

$$m = -0.00226 + 0.001934\,(600 + 14.7) + 3.054\ 10^{-7}\,(600 + 14.7)^2 = 1.3$$

Now downhole dome pressure from Equation 3.6 is

$$p_d = (600 + 14.7) + 1.3\,(180–60) = 770.7 \text{ psia} = 756 \text{ psig.}$$

Example 3–2. Find a nitrogen charged gas lift valve's required surface charge pressure if its dome pressure at 120°F is 820 psig. Use Figure 3–5 as well as Equation 3.9.

Solution

From Figure 3–5 at a downhole dome pressure of 820 psig and a valve temperature of 120°F, we read a required charge pressure of 727 psig.

For the solution with Equation 3.9, the approximate charge pressure is

$$(820 + 14.7)\ 520 / (460 + 120) = 748.3 < 1,238 \text{ psig}$$

The constants in Equation 3.9 are evaluated accordingly:

$$a = 3.054\ 10^{-7}\,(120 – 60) = 1.832\ 10^{-5}$$

$$b = 1 + 1.934\ 10^{-3}\,(120 – 60) = 1.116$$

$$c = -2.226\ 10^{-3}\,(120 – 60) – (820 + 14.7) = -834.8$$

The required surface charging pressure at 60 F is found from Equation 3.9:

$$p'_d = \{-1.116 + [(-1.116)^2 + 4\ 1.832\ 10^{-5}\ 834.8]^{0.5}\,\} / (2\ 1.832\ 10^{-5}) = 739.9 \text{ psia} = 725 \text{ psig.}$$

Some manufacturers use natural gas to charge their gas lift valves. The behavior of natural gas, as discussed in Section 2.2.4.1, heavily depends on its composition and cannot be as accurately predicted as the behavior of nitrogen gas. Therefore, calculation accuracy is usually lower than in the case of nitrogen charge. Using the same principles as for a nitrogen gas-charged bellows, charts depicting the change in dome charge pressure can be developed. Figure 3–6 is an example for a $\gamma = 0.7$ specific gravity natural gas charge.

3.1.4.2 Injection pressure vs. depth. As discussed in Section 2.4.4, the injection gas column in the casing-tubing annulus is usually assumed to be in a static condition and the pressure distribution along the well depth is calculated accordingly. As detailed in that section, the pressure distribution at any temperature is linear with depth. This observation permitted the development of Figure 2–10, which gives gas pressure gradients as a function of surface injection pressure. As already mentioned, some sources [9, 11] presented charts similar to Figure 2–10 but assumed constant deviation factors for the development of the charts. Consequently, gas gradients read from them contain errors, and they predict increasingly lower gradients for higher surface injection pressures.

Use of the previous and similar charts permits the calculation of downhole pressures from known surface injection pressures. However, many times in gas lift calculations, the opposite direction of calculation is required: the surface pressure is sought for a given downhole gas injection pressure. For such cases, either an iterative scheme similar to the one described in the flowchart of Figure 2–9 or a graphical approach can be used. One kind of a graphical solution is given in Figure 3–7, which is valid for a gas specific gravity of $\gamma = 0.7$. All other assumptions are identical to those used for constructing the gas pressure gradient chart presented in Figure 2–10.

Appendix E contains full-page charts similar to Figure 3–7 for different gas gravities.

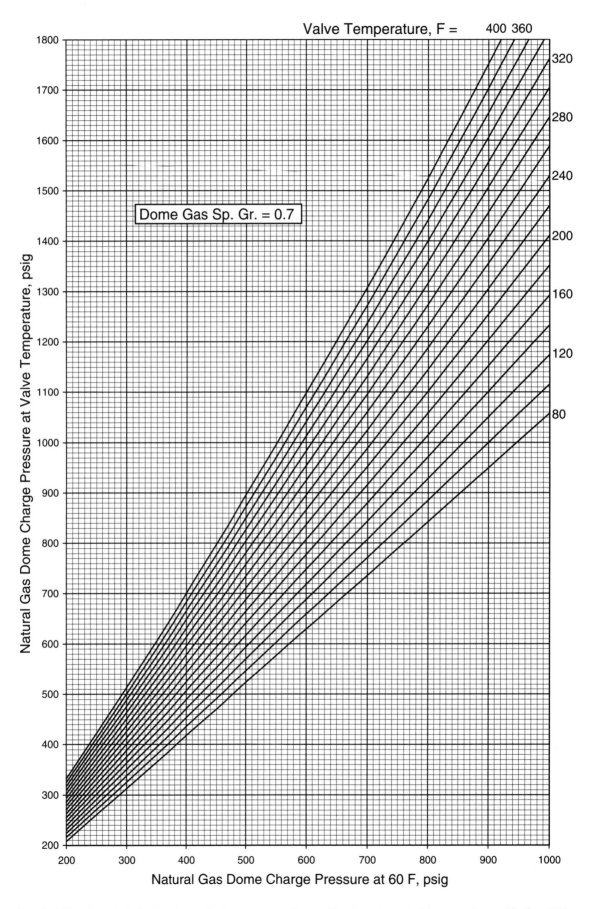

Fig. 3–6 Chart for calculating the change in dome pressures for gas lift valves charged with a natural gas of Sp.Gr. = 0.70.

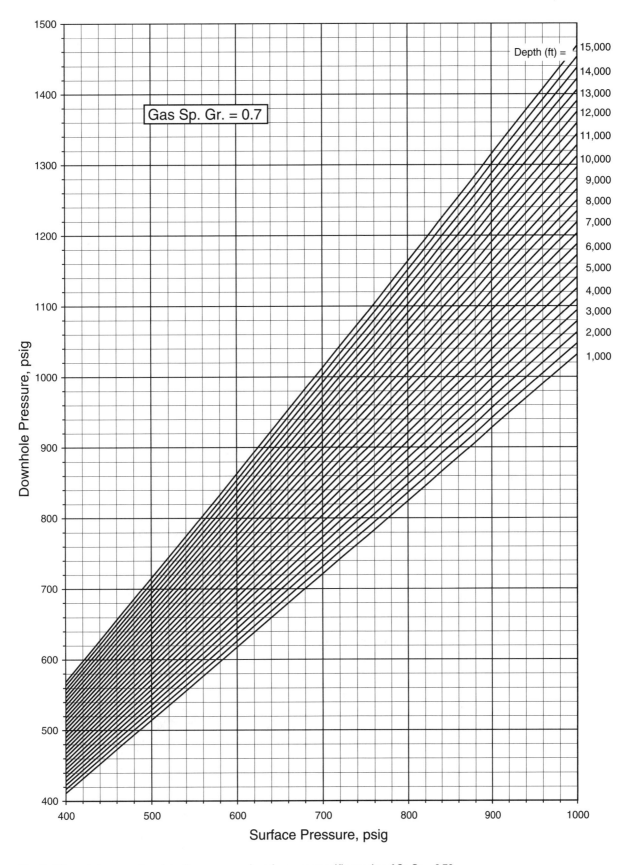

Fig. 3–7 Downhole vs. surface injection pressure chart for a gas specific gravity of Sp.Gr. = 0.70.

––·

Example 3–3. Use Figure 3–7 to find the surface gas pressure corresponding to a gas lift valve's opening pressure of 900 psig, valid at the valve setting depth of 7,500 ft.

Solution

On Figure 3–7 starting from a downhole pressure of 900 psig, a horizontal line is drawn to the intersection with the 7,500 ft depth line. The surface pressure is read as 743 psig.

––·

3.2 Pressure-operated Gas Lift Valves

3.2.1 Introduction

Out of the great many kinds of gas lift valves invented in the last hundred years, only the pressure-operated type survived to these days. Pressure operation means that the valve's behavior is controlled by injection pressure, production pressure, or both. This control mechanism is easily adapted to gas lifting since pressure control is a frequently occurring operation in the oilfield. Using the familiar pressure control devices at the surface, gas lift valves are easily controlled by changing the surface injection pressure, which in turn involves pressure changes at the valve's setting depth. This pressure provides a control signal to the valve and the valve behaves according to its predetermined mechanical features. In order that the valve *senses* the changes in pressure, all pressure-operated valves are provided with a reference pressure or force that comes from a gas-charged dome, a spring, or a combination of both. By properly setting these references at the surface (with the right amount of pressure charge or spring force), the valve can be made to open and close according to the actual requirements of the gas lift installation.

Before the detailed analysis of the different pressure-operated valves, the terminology and construction details of the common types are discussed in the following.

3.2.1.1 Valve parts terminology. Figure 3–8 displays the parts of a single-element gas lift valve equipped with a reverse flow check valve. The valve has a gas dome filled with high-pressure gas (usually nitrogen) connected to a metal bellows. The bellows allows the movement of the valve stem and retains the valve's charged pressure. Its function is similar to a sealed piston on which a force arises that corresponds to its cross-sectional area and the contained pressure. Practice has shown that pistons cannot be used in gas lift valves and the use of a metal bellows is universally accepted. Still, for calculating opening and closing conditions, the bellows is treated analogously to a piston with a cross-sectional area equal to the effective bellows area.

Lift gas injected from the surface enters the valve through the valve inlet ports. Depending on pressure conditions, the valve stem tip opens or closes the flow of injection gas through the valve port. Valves are usually equipped with reverse flow check valves having their own seats and closing devices (dart or disk). The check valve in Figure 3–8 is a normal closed one and prevents flow against the normal gas-injection direction.

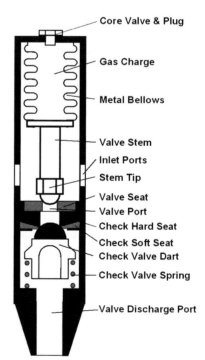

Fig. 3–8 Valve parts terminology of a single element bellows valve.

Many valves have, in addition to the bellows charge, a spring providing a supplemental control force. The spring, as seen in Figure 3–9, is of the compression type, its normal force being set at the surface before the valve is run in the well.

The concentric flexible sleeve valve does not have a metal bellows as its major moving element. As given in Figure 3–10, this valve is made up in the tubing string, hence the name concentric. The dome is charged with a set pressure at the surface, and this pressure keeps the main member, called a *flexible sleeve* or *resilient element*, firmly on the inside of the inlet ports preventing gas injection. In the open position, as seen in Figure 3–10, the resilient element is forced to move off the inlet ports, and injection gas can flow through the finned retainer and through the valve discharge ports into the fluid to be lifted. This valve type's reverse flow check is a flexible sleeve that closes as soon as flow in the wrong direction starts.

3.2.1.2 Valve construction details

3.2.1.2.1 Core valve and tail plug. The dome of gas lift valves is charged to a predetermined gas pressure at the surface. To facilitate charging, core valves and tail plugs are used at the top of the dome body as shown in Figure 3–11. Part A shows a generally accepted solution where a Dill valve core (as used in motorcar tires) provides filling of and release from the dome. Because the Dill valve would permit high pressure to enter the dome and the charged pressure would increase at

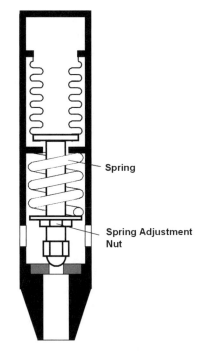

Fig. 3–9 Valve parts terminology of a spring-loaded valve.

downhole conditions, a tail plug with a metal-to-metal seal on a copper gasket is also provided. Most gas lift valves utilize this construction because of its advantages: filling of the dome with gas is easy, the charged pressure is easily decreased by tapping on the core valve, and the Dill valve is readily available everywhere.

Part B of Figure 3–11 shows a different solution for valve charging. Here, two plugs with their respective copper gaskets are used to confine the dome charge pressure. Charging of the valve is more difficult than with the previous assembly because use of a special apparatus is required, which allows a leakproof charging and the simultaneous tightening of the primary plug. The tail plug is used for safety only and does not need any special equipment.

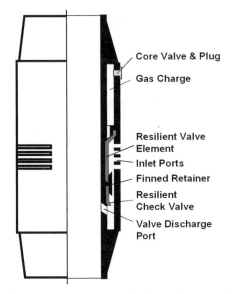

Fig. 3–10 Valve parts terminology of a concentric flexible sleeve valve.

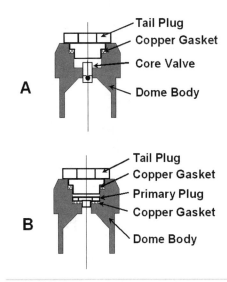

Fig. 3–11 Details of a gas lift valve's core valve and tail plug.

3.2.1.2.2 *Gas charge.* Early gas lift valves used natural gas as a dome charge medium. However, today most gas lift valves are charged with nitrogen gas because of its availability and other beneficial features: N_2 gas is inexpensive, inflammable, and noncorrossive. The properties of N_2 gas are well known and its deviation factor has been determined in a broad range of pressures and temperatures. In summary, gas lift valves charged with nitrogen gas are reliable and their operation at different temperatures is easy to calculate.

In contrast, the performance of a valve charged with natural gas is more difficult to predict because of the inaccuracies in calculating deviation factors that change with gas composition, which is seldom known accurately. In summary, the use of natural gas is not recommended for charging gas lift valves.

3.2.1.2.3 *Bellows assembly.* The bellows is the heart of the gas lift valve, and its proper operation over prolonged periods determines the reliability of the valve itself. The bellows performs the most important function of the gas lift valve by allowing the valve stem tip to move on and off the seat while maintaining the dome-charged pressure. In respect to outside pressure, the bellows compresses and stretches many times during its life span and should therefore be of good quality and well protected against adverse conditions to ensure a long working life.

Metal bellows used in gas lift valves (see Fig. 3–12) are manufactured from multi-ply seamless Monel material. Made in different outside diameters for use in the usual sizes of gas lift valves (ODs of 1 ½ in., 1 in., and ⅝ in.), they are made of 3-ply construction and contain about 20–30 convolutions. The two ends of the bellows are soldered to the valve stem and the valve dome, respectively. During operation, because of the axial loading, the bellows is compressed and its outer and inner convolutions come to lie on each other, and bellows *stacking* occurs, see Figure 3–13. In this case, no more movement of the stem tip is possible and the bellows can be permanently damaged if further compressed by outside pressure. Therefore, good valve design ensures that bellows stacking occurs after the stem tip completed its full travel.

Practice has shown that the weakest points in the bellows are the valleys and peaks of the convolutions. When subjected to high pressures in the well, these points tend to *knife-edge*, *i.e.* their radii are significantly decreased and the high stresses thus produced can rupture one or more of the plies of the bellows. An additional detrimental effect is the change in bellows *stiffness* (spring rate), which causes the valve's operating pressure to deviate from its set pressure. These conditions occur when high pressure differentials across the bellows are present, especially during running the unloading valve string in the well when lower gas lift valves are exposed to high hydrostatic pressures.

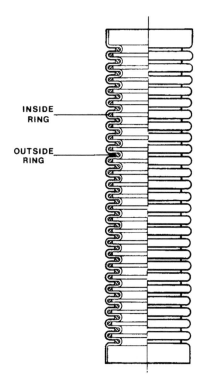

Fig. 3–12 Metal bellows with Teflon rings for bellows protection.

INSIDE RING

OUTSIDE RING

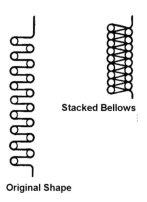

Stacked Bellows

Original Shape

Fig. 3–13 Bellows stacking due to excessive axial load.

To prevent damage to the bellows and eventually the valve itself, bellows protection is applied. Basic protection is built in the valve by external and internal supports preventing excessive lateral movement of the bellows, as well as by mechanical stops limiting its axial travel. Additional protection is provided against the adverse effects of high differential pressures across the bellows in the following forms.

- Bellows support rings as shown in Figure 3–12 are fitted in inside and outside convolutions. These prevent the occurrence of knife-edging, and the spring rate of the assembly is kept at its original value.

- Controlled hydraulic deformation (pre-forming) of the bellows with 2,000–6,000 psi differential pressures during the manufacturing process reduces the radii of convolutions and ensures that the bellows maintains its shape under adverse conditions.

- Trapping of an incompressible liquid (silicone oil) in the bellows after it reaches its full travel prevents a high pressure differential to occur across the bellows wall.

- Even the original King valve [4] used mechanical sealing of the bellows against well fluids activated after full stem travel.

- If the gas charge acts on the outside of the bellows assembly (see Fig. 3–14), then the bellows is not compressed as in conventional valve types but will stretch as the valve opens, and stacking of convolutions is prevented.

A commonly occurring problem with gas lift valves is valve stem chatter, when the stem tip hits the valve seat with a high frequency as the result of resonance. This unpredictable condition usually results in the loss of tightness between the valve tip and seat, as well as in bellows ruptures due to fatigue. To prevent valve chatter, most gas lift valves have some dampening mechanism built-in in the bellows assembly, the most common solution being the presence of a high-viscosity liquid. This liquid is forced to move across restrictions thereby dampening the movement of the valve stem tip.

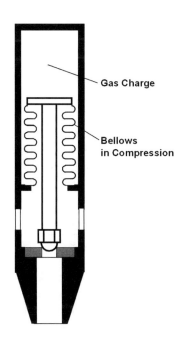

Fig. 3–14 Bellows assembly with the gas charge acting on its outside.

The bellows in a gas lift valve acts very similar to a helical spring: an increase in axial load results in a corresponding reduction of its length. The gas lift valve's stem travel, therefore, depends on the behavior of the bellows, which can be described by the spring rate, conveniently expressed in axial load required per unit travel, in lb/in. units. The spring rate of a bellows assembly consists of the spring rates of the uncharged bellows and the gas charge. If the valve contains a compression spring in addition to the dome charge, its spring rate must properly be considered when calculating the net spring rate of the whole bellows assembly. The magnitude and relative importance of these factors is easily seen if a balance of the forces acting on the valve stem is written. For an infinitesimal stem travel dx measured from the closed position at $x = 0$, dome pressure increases due to the decrease of dome volume and an axial force arises. This force, along with the spring force occurring in the bellows assembly (which may include a compression spring) works against the opening force coming from the injection pressure acting on the full area of the bellows A_b:

$$dp_d A_b + k \, dx = dp_i A_b \qquad\qquad 3.10$$

where: p_d = dome pressure, psi

A_b = effective bellows area, sq in.

k = spring rate of the uncharged bellows assembly, lb/in.

p_i = injection pressure, psi

dx = valve stem travel, in.

Dividing Equation 3.10 by A_b dx, we get:

$$\frac{dp_d}{dx} + \frac{k}{A_b} = \frac{dp_i}{dx}$$

3.11

The terms of this equation are expressed in psi/in, and are called *load rates* in the gas lift literature. As seen, load and spring rates of the individual components are related by the bellows area A_b. The net load rate of the bellows assembly is the sum of the individual load rates, because the *springs* corresponding to the bellows assembly and the dome charge are connected in parallel (their axial movements being identical). The second term on the left-hand side, in addition to the load rate of the bellows itself may contain the effect of a compression spring, often used in gas lift valves.

To find the effect of the compression force arising in the dome charge of a valve, assume an ideal gas charge and isothermal conditions during bellows movement. As shown by Decker [12], these conditions very closely model the actual behavior of nitrogen-gas-charged valves with a maximum dome charge pressure of 1,500 psig and temperatures of up to 300 °F. If the Ideal Gas Law is written for the gas volume contained in the gas dome in the original and in the compressed cases, we get:

$$P_{d_1} V_{d_1} = P_d (V_{d_1} - x A_b)$$

3.12

Solving the previous formula for the dome pressure at any valve travel results in:

$$P_d = \frac{p_{d_1} V_{d_1}}{V_{d_1} - x A_b}$$

3.13

where: p_{d_1} = dome pressure at x = 0 stem travel, psi

V_{d_1} = dome volume at x = 0 stem travel, cu in.

Equation 3.13 was used to create Figure 3–15 where the variation of dome charge pressure with valve stem travel is shown for a usual 1½ in. gas lift valve. In the possible valve stem travel range, p_d is clearly a linear function of the valve travel for any initial dome pressure. Therefore, under normal operating conditions, the dome charge of a gas lift valve behaves just like a spring with a constant spring rate. Its load rate is found after differentiating Equation 3.13 with respect to x:

$$\frac{dp_d}{dx} = \frac{p_{d_1} V_{d_1} A_b}{(V_{d_1} - x A_b)^2}$$

3.14

As seen, load rate depends on the initial dome pressure, initial dome volume, and the cross-sectional area of the bellows. The load rate increases with an increase in the initial dome charge pressure, but the effect of the dome volume is more complex. The decrease in dome volume due to valve stem travel is always negligible in gas lift valves and the valve's net load rate is constant. In the example shown in Figure 3–15, a total valve travel of 0.2 in. means a dome volume change of 0.2 in. times 0.77 sq in. = 0.15 cu in., which is about 3% of the initial dome volume of 5 cu in. If initial dome volume and volume change were comparable, than a nonlinear behavior would occur.

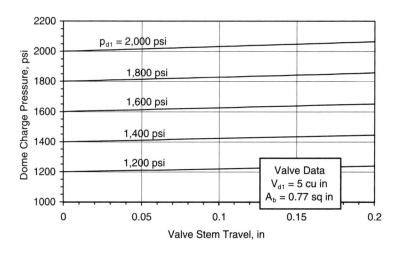

Fig. 3–15 Increase in dome charge pressure due to bellows compression for various initial charge pressures.

The net load rate of bellows assemblies (the right-hand side of Equation 3.11) can be measured with a special probe, following the recommendations in API RP 11V2. [13] The test involves precise measurement of valve stem travels for different injection pressures acting on the full bellows area. If plotted with stem travel on the horizontal axis, a sudden change in slope indicates bellows stacking and the limit of valve stem travel, as seen in Figure 3–16. Before reaching full stroke, measured points lie on a straight line, the slope of which defines the bellows assembly load rate B_{lr} for the given dome charge pressure, in psi/in units. As indicated in Figure 3–16, measurements made with increasing and decreasing injection pressures differ from each other, demonstrating hysteresis due to mechanical and viscous friction in the different parts of the bellows assembly. Such plots permit the determination of not only the valve's load rate but also its maximum possible valve stem travel.

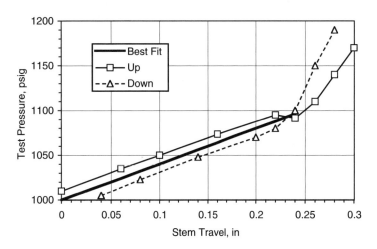

Fig. 3–16 Example plot of injection pressure vs. valve stem travel for the determination of the bellows assembly load rate.

Example 3–4. Find the contribution of the compression of the bellows to the load rate of a 1 ½ in. OD valve with a 5 cu. in. dome volume, a bellows area of 0.77 sq. in., and an initial dome charge of 1,000 psi. The valve does not have a spring and its measured net load rate was 400 psi/in., and its maximum valve stem travel is 0.2 in. Calculate the injection pressure increase required to fully open the valve.

Solution

The load rate of the dome for the given conditions is found from Equation 3.14 as:

dp_d/dx = (1,000 5 0.77) / (5 − 0.2 0.77)² = 164 psi/in.

The mechanical load rate of the uncharged bellows assembly is the difference of the valve's net load rate and the calculated value, according to Equation 3.11:

k/A_b = 400 − 164 = 236 psi/in.

The pressure increase required to fully open the valve is found from the net load rate and the maximum valve travel:

Δp_i = 400 0.2 = 80 psi.

3.2.1.2.4 Spring. For various reasons, many gas lift valves include a compression spring providing an additional force to keep the valve closed, as shown in Figure 3–9. In an unloading valve string, such valves are closed by the spring force even if the bellows fails and can thus prevent uneconomical use of injection gas. Metal springs used in valves usually have a much greater spring rate than the uncharged bellows, thus their spring rate determines the *stiffness* of the valve. The spring force is unaffected by temperature, which is an advantage in installation designs with limited temperature data where the operation of valves with only a gas charge could be unreliable.

There are gas lift valves without a gas charge whose entire closing force is supplied by a spring. The spring rate of the bellows (filled up with some liquid) can be disregarded because of its small value, as compared to that of the spring.

Compression springs used in these valves are very *stiff* and allow very limited valve stem travel. Therefore, the valve can never fully open and will always *throttle*, i.e. restrict the effective area open to injection gas flow to less than the valve port area. As mentioned previously, the operation of spring-loaded valves is unaffected by well temperatures so their opening pressure can be very accurately set at the surface and will not change in the well.

Example 3–5. If the valve in the previous example contains a compression spring to assist the bellows charge, find its total load rate and the pressure needed to fully open it. The load rate of the uncharged bellow assembly with the spring was measured as 1,890 psi/in.

Solution

The load rate of the dome is identical to the value calculated in the previous example:

dp_d/dx = 164 psi/in.

The valve's net load rate, according to Equation 3.11 is the sum of the two load rates:

dp_i/dx = 164 + 1,890 = 2,054 psi/in.

The injection pressure must overcome the bellows charge pressure by the amount equal to the product of the valve's load rate and the maximum stem travel:

Δp_i = 2,054 0.2 = 410 psi.

Comparison of the two examples shows that

(a) spring-loaded valves have much higher load rates than those without a spring

(b) in a spring-loaded valve, the contribution of bellows compression to the net load rate is very small

(c) spring-loaded valves require much higher pressures to fully open

3.2.1.2.5 Ball and seat. Gas lift valves inject lift gas through their main port, and the operation of the valve heavily depends on a proper seal between the valve seat and the stem tip. These parts are manufactured from Monel, Tungsten carbide, or a ceramic material for a long and leak-free operational life and are matched together. When one of them is damaged, both must be replaced with a new, matching pair. As seen in Figure 3–17, valve stem tips are usually metal balls attached interchangeably to the valve stem. The size of the ball must fit the port size and generally, balls 1/16 in. larger in diameter than the port inside diameter are used. The seat is also replaceable and is usually of the floating design enabling the seat to a limited vertical movement for better sealing.

Most gas lift valves have sharp-edged valve seats (Part A in Fig. 3–17) or a very shallow bevel on their seats. Other seats (Part B in Fig. 3–17) may have a greater bevel but the taper angle is not standardized. Finally, valve stem tips of a tapered design are also used (Part C in Fig. 3–17) to increase the throttling behavior of some valves. The amount of bevel or chamfer in the seat affects the ball-

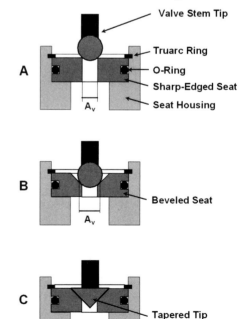

Fig. 3–17 Valve stem tip and valve seat combinations.

seat contact area, the size of which is very important in calculating the forces acting on the valve stem. For a sharp-edged seat, the contact area A_p is equal to the port bore area, but for beveled seats, it is always greater than that. (Part B) It is very important that manufacturers publish valve mechanical data with due regard to this effect.

As the ball moves off the seat, the valve opens a proportionally increasing area through which gas injection can take place. At any valve stem position, the area open to flow equals the lateral area of the frustum of a right circular cone [14, 15] whose base is the edge of the valve seat, its tip being the center of the ball (see Fig. 3–18). The flow area gradually increases with valve stem travel, and at some point, actual throughput area will reach the size of the valve port area. Experimental data given by several sources [8, 16] for sharp-edged seats show that the flow area through a gas lift valve linearly increases with valve stem travel until the maximum area equivalent to port area is reached, as shown in Figure 3–19. As indicated in the figure, very small amounts of valve stem travel are needed to fully open the gas-charged gas lift valve. Due to their much greater B_{lr}, spring-loaded valves have much less stem travel at identical conditions and can never be fully opened. These valves, therefore, will always throttle and restrict the rate of gas injection through the valve.

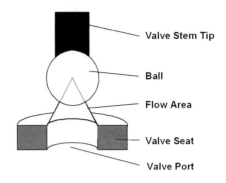

Fig. 3–18 The flow area through a gas lift valve's port at an arbitrary valve stem position.

3.2.1.2.6 Check valve. Gas lift valves, installed in the well, provide communication between the tubing and the casing-tubing annulus at their setting depths. Valves in the unloading valve string as well as the operating valves are opened by injection and/or production pressures and permit injection of lift gas into the wellstream when a sufficient pressure differential exists between injection and production pressures. However, valves can be opened by these pressures at times when the pressure differential points against the desired direction of injection and flow across the valve in the wrong direction can occur. To prevent such situations, reverse flow check valves (*check valves* for short) must be attached to gas lift valves, either separately or as an integral part of the valve assembly.

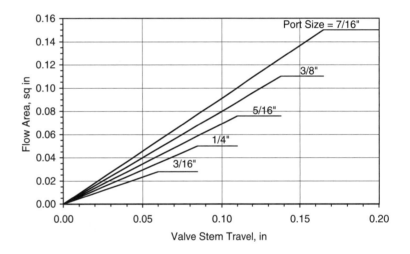

Fig. 3–19 Valve stem travels required to open an area equal to full port area.

The most important functions of reverse flow check valves are the following:

- If all valves in the unloading valve string are equipped with check valves then, after the initial unloading, well fluids cannot re-enter the space reserved for gas injection (either the casing-tubing annulus or the tubing). This ensures that a constant fluid level is maintained above the operating valve and no unloading is necessary after the well is shut down.

- Many times after well completion, removal of drilling mud left in the well is necessary. A normal unloading operation would accomplish this, but all valves would be severely damaged (*cut out*) since the sand-laden mud would have to be passed through the valves. In such cases, circulating against the normal injection direction prevents mud from flowing across valves, provided they have check valves. As soon as a pressure differential due to circulation pressure builds up against the normal direction, the check valves close and prevent flow through the valves.

- If remedial work such as acidizing or hydraulic fracturing is to be performed on a gas lifted well, check valves are a must. These operations involve pressurizing the tubing string and pumping acid or other liquids down the hole, which would be impossible without the use of check valves preventing communication between the tubing and the annulus.

- In wells needing only a *kickoff* by gas lift to start to flow naturally, check valves prevent well fluids from entering the casing-tubing annulus when high flowing pressures in the tubing occur.

Reverse flow check valves used in gas lift valves operate on various principles. As shown in Figure 3–20, velocity-controlled valves (Part A) are closed as soon as flow in the wrong direction starts; addition of a light spring (Part B) makes a more positive closing; a gravity controlled valve (Part C) is assisted by gravity to close. Usually, all check valves have a double sealing mechanism as described in Figure 3–8. The initial seal is provided between the check dart and an elastomeric soft seat after which the dart seals on the metal hard seat. Selection of the right type of check valve depends on the direction of injection gas flow through the gas lift valve as indicated in Figure 3–20.

Flexible sleeve valves also require the use of check valves that, as shown in Figure 3–10, are resilient sleeves installed in the proper direction.

Because of the previous reasons, all gas lift valves include an integral or separate reverse flow check valve. The presence of the check valve does not affect the operation of the valve, *i.e.* its opening characteristics are not influenced. This is easy to understand if a normally open check valve is used (Part A in Fig. 3–20), which permits production pressure to act across the check valve's port to the valve's main port. In case of a normally closed check valve, it interacts with the main valve as shown in Figure 3–21. When the gas lift valve is closed, the check valve dart is kept on its seat by its light spring. Production or tubing pressure acting from below also pushes the dart to close, but the pressure trapped between the main valve port and the check valve port still can attain tubing pressure. This is due to the negligible capacity of this space and the always-present imperfection of seal between the check dart and the seat. In summary, in spite of the presence of the check valve, tubing pressure can act on the main valve stem, and the valve's opening is not affected.

3.2.2 Valve mechanics

3.2.2.1 Introduction. This section discusses the mechanical construction of the most important kinds of pressure-operated gas lift valves and gives an overview of their operational features. All valves will be fully described and their opening and closing conditions will be detailed using force balance equations. Although these balance equations are only valid for static conditions, and the dynamic performance of the valves is covered in a later section, basic understanding of the various valve's performance is ensured.

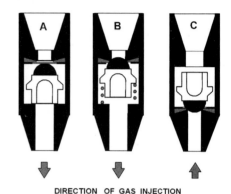

Fig. 3–20 Reverse flow check valve constructions.

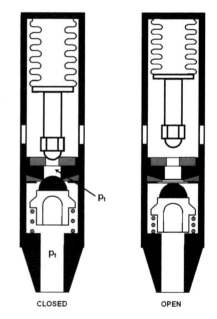

Fig. 3–21 The interaction of a gas lift valve and its reverse flow check valve.

Formerly, when classifying gas lift valves, the industry used the terms *casing pressure operated* and *tubing pressure operated* to denote valves that are more sensitive to casing or tubing pressure, respectively. These terms are correct if a tubing flow installation is used where casing pressure automatically means injection gas pressure and tubing pressure reflects to the pressure of the produced fluid. For casing flow installations, however, the previous terms can be misleading because gas is injected through the tubing string and fluid production occurs in the annulus. This is the reason why the previous terms seem to be outdated and are being gradually replaced by the expressions *injection pressure operated* (IPO) and *production pressure operated* (PPO), which are more general and always refer to the right pressures. In the following section, these terms will be used.

Throughout the discussion of the various valve types, for the sake of simplicity and to provide a uniform treatment, valves will be depicted in their conventional (tubing outside mounted) versions for tubing flow installations, wherever possible. These basic valves, as will be seen later, can be used in tubing or casing flow installations as well as in injection or production pressure operations, depending on the various possible combinations of their versions (conventional or wireline retrievable) and the types of gas lift mandrels used.

Appendix F contains tables with mechanical data on gas lift valves available from some leading manufacturers of gas lift equipment.

3.2.2.2 Unbalanced valves with spread.

A gas lift valve is called unbalanced if its (a) opening or (b) opening and closing pressures are influenced by production pressure. This means that the opening or closing conditions of the valve depend on the prevailing production (or tubing) pressure, whereas balanced valves open and close at the same injection pressure. Valves with variable opening pressures but constant closing pressures exhibit *valve spread*, a term used for the difference between those pressures. In contrast, valves having variable opening as well as closing pressures do not have spread.

The first gas lift valve, the King valve, is an ideal example for an unbalanced one, and it will be used in the following to illustrate the operation and basic concepts of many other valve types.

3.2.2.2.1 *IPO valves.*

Single Element Valves

The original King valve [4] is a classical example of an unbalanced gas lift valve with spread. It uses only a gas charge in its dome to control the valve's operation and is sometimes called a single element valve. As seen in Figure 3–22 in the closed position injection pressure, p_i acts on a much larger area than production pressure p_p. This is why the valve is designated as IPO. In order to examine the behavior of this very common valve type, let us investigate its opening and closing conditions. It is easy to see that the valve stem opens or closes the valve port to gas injection, depending on its axial position, which, in turn, varies according to the net force arising on the valve stem. By writing up the balance of forces on the valve stem, the conditions for opening and closing at downhole conditions can be found.

In the closed position (see Fig. 3–22) dome charge pressure p_d, acting on the net area of the bellows A_b provides a sufficiently large closing force to keep the valve stem on the port. All the other forces acting on the valve stem work in the other direction and try to open the valve. The greater opening force comes from the injection pressure acting on the bellows assembly from below on a *donut* area equal to the net bellows area minus the valve port area, $(A_b - A_v)$. A much smaller force arises on the valve stem tip, coming from production pressure p_p. At the instant the valve opens, the closing force must be in equilibrium with the sum of the opening forces and the following force balance equation can be written:

$$p_d A_b = p_i(A_b - A_v) + p_p A_v$$

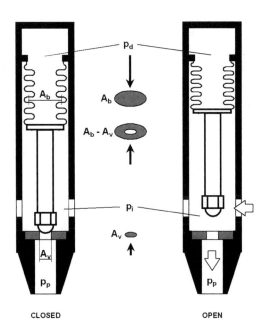

Fig. 3–22 Schematic drawing of an unbalanced, bellows-charged gas lift valve.

3.15

Since the valve is IPO, it is most sensitive to injection pressure and this is the pressure it is usually opened with. This is the reason why the previous equation is solved for injection pressure. The resulting formula is called the valve's opening equation and gives the value of the injection pressure necessary to open the valve.

$$p_{io} = \frac{p_d}{1 - \dfrac{A_v}{A_b}} - p_p \frac{\dfrac{A_v}{A_b}}{1 - \dfrac{A_v}{A_b}} \qquad\qquad 3.16$$

For simplification, introducing the term $R = A_v/A_b$ we get:

$$p_{io} = \frac{p_d}{1 - R} - p_p \frac{R}{1 - R} \qquad\qquad 3.17$$

where: p_d = dome pressure at valve temperature, ps,

        p_{io} = opening injection pressure at valve setting depth, psi

        p_p = production pressure at valve setting depth, psi

        R = geometrical constant, -

As seen from the opening equation, the injection pressure required to open the valve in the closed position depends not only on the dome-charged pressure but on the production pressure as well. The higher the production pressure, the lower the injection pressure necessary for opening the valve. At a constant production pressure, as soon as the injection pressure reaches the value calculated from Equation 3.17, the valve starts to open. For the following discussion, it is assumed that as soon as the valve opens, the valve stem completely lifts off the seat and the full area of the port is opened for injection gas flow.

If the valve is fully open, as in Figure 3–22, a new balance of the forces acting on the valve stem must be written up. When doing so, a general assumption is that the pressure on the valve stem tip equals the injection pressure. This can be justified because the total flow area of the valve inlet ports is much greater than the cross-sectional area of the valve port, so flowing gas pressure will drop across the valve port. This means that the pressure upstream of the port is very close to injection pressure. Thus, injection pressure acts on the total bellows area and tries to keep the valve open. Closing force comes from the dome charge pressure, as before, and the balance of forces results in the following equation:

$$p_d\, A_b = p_i\, A_b \qquad\qquad 3.18$$

Solving for the injection pressure at which the valve closes, we find the valve's closing equation:

$$p_{ic} = p_d \qquad\qquad 3.19$$

where: p_d = dome pressure at valve temperature, psi

        p_{ic} = closing injection pressure at valve setting depth, psi

By comparing the opening and closing equations of the unbalanced gas lift valve (Equations 3.17 and 3.19, respectively), we find that the valve opens when injection pressure exceeds dome charge pressure but closes when injection pressure drops below dome charge pressure. In most gas lift valve design calculations, the surface injection pressure and the production pressure at valve depth are specified and valve dome pressure is to be found. For this case, solving the opening equation for p_d results in the following formula where p_{io} has to be calculated from the surface injection pressure:

$$p_d = p_{io}\,(1 - R) + p_p\, R \qquad\qquad 3.20$$

It should be mentioned that the previous treatment considers static conditions only and the dynamic performance of the valve (subject of a later section) may be different. This is caused by the fact that the pressure below the valve stem may be different from injection pressure when the valve closes. In intermittent gas lift, at the moment the operating valve closes, production and injection pressures are very close to each other and the valve closes at the injection pressure found from its closing equation. In continuous flow applications, however, injection pressure is always greater than production pressure, and the valve closes at an injection pressure different from the value found from Equation 3.19.

A graphical presentation of the opening and closing performance of an unbalanced IPO valve at downhole conditions is given in Figure 3–23. The presence of a reverse flow check valve, always attached to gas lift valves, prevents operation at any production pressure greater than injection pressure, above the $p_p = p_i$ line. At injection pressures lower than dome charge pressure, the required production pressure to open the valve would be more than the actual injection pressure and, because of the action of the reverse flow check, the valve is closed. This area is indicated by the darkest shading in the figure. The opening injection pressure (Equation 3.17) is a linear function of production pressure and is shown in bold line. Opening of the valve occurs along this line, whereas closing occurs along the vertical line $p_i = p_d$. Inside the triangle formed by the line of the opening pressure and the constant dome charge pressure, the state of the valve depends on its previous state: if it was closed before, then it stays closed as long as injection or production pressures (or both) do not cross the bold line. If the valve was opened previously, then it stays open until injection pressure drops below the dome charge pressure.

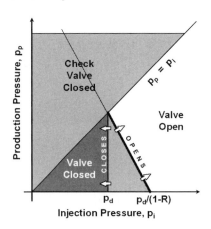

Fig. 3–23 Opening and closing performance of an unbalanced IPO gas lift valve.

When working with gas lift valves, it should be always kept in mind that actual performance of the valve occurs at the pressure and temperature prevailing at its setting depth. As discussed in Section 3.1.4.1, valves are charged on the surface at a standard temperature of 60 °F, and their dome pressure increases with increased well temperatures. The force balance equations, however, have the same form at any temperature because the valve's geometrical data are constant. This is the reason why the slope of the opening equation is the same for any valve temperature, as shown for an example case in Figure 3–24.

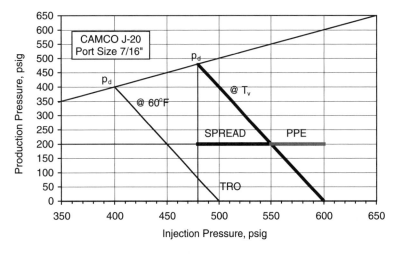

Fig. 3–24 Graphical definition of spread and production pressure effect.

Valves are set in special valve testers, usually at zero production pressure. Applying the opening equation and substituting a production pressure of $p_p = 0$, we can find the injection pressure required to open the valve at these conditions from Equation 3.17. This pressure is called the valve's test rack opening (TRO) pressure value:

$$TRO = \frac{p_d'}{1 - R}$$

3.21

where: p_d' = dome pressure at 60 °F, psi.

Unbalanced gas lift valves are characterized by two widely used parameters: production pressure effect (formerly known as tubing pressure effect), and valve spread. Both are discussed and defined in the following in conjunction with Figure 3–24.

The parameter called production pressure effect (PPE) is the second term on the right-hand side of the opening Equation 3.17. It represents the contribution of production pressure to the valve's opening injection pressure and can be defined as the difference between the valve's opening pressure at zero production pressure and its actual opening pressure:

$$\text{PPE} = p_p \frac{R}{1 - R} \tag{3.22}$$

For a given valve size and port diameter, PPE is zero at zero production pressure and increases linearly with production pressure, see Figure 3–24. The slope of this function is called the production pressure effect factor (PPEF):

$$\text{PPEF} = \frac{R}{1 - R} = \frac{\dfrac{A_v}{A_b}}{1 - \dfrac{A_v}{A_b}} \tag{3.23}$$

As seen, PPEF is a geometrical constant for a given valve with a given port size and represents the drop in the valve's opening injection pressure for a unit increase in production pressure. In a tubing flow installation, production pressure equals tubing pressure, PPEF may therefore be called tubing effect factor (TEF), a formerly widely used term. Equation 3.23 reveals that for a given valve PPEF increases as valve port size increases.

Valve spread is defined as the difference between the opening and closing pressures of a gas lift valve. This difference is caused by the fact that different pressures act on the valve port when the valve is in the open and when in the closed positions. Spread is a function of the valve geometry, dome charge pressure, and production pressure and is defined as:

$$\text{SPREAD} = p_{io} - p_{ic} \tag{3.24}$$

After substitution of the opening pressure from Equation 3.17 and the closing pressure from Equation 3.19, we get:

$$\text{SPREAD} = \frac{p_d - p_p R}{1 - R} - p_d = \frac{R}{1 - R}(p_d - p_p) \tag{3.25}$$

The previous formula contains the PPEF defined by Equation 3.23, and the final formula for valve spread is

$$\text{SPREAD} = \text{PPEF}(p_d - p_p) \tag{3.26}$$

As shown in Figure 3–24 for an example valve, valve spread is zero when injection and production pressures are identical and valve spread is maximum at zero production pressure. For a given valve size (given bellows area) valve spread heavily depends on port size. This is illustrated in Figure 3–25 where valve characteristics for every available port size of an example valve are given for a set dome charge pressure. Spread (the difference of opening and closing pressures) is seen to increase as port diameter increases.

Valve spread is a very important parameter in intermittent gas lift installations utilizing a choke on the surface to inject the lift gas into the annulus. Under such

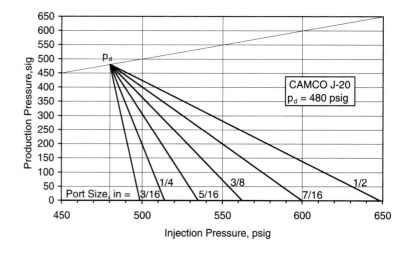

Fig. 3–25 Variation of valve spread with port size for a given gas lift valve.

situations, casing pressure continuously increases until valve opening pressure is reached and the gas lift valve starts to inject gas into the tubing. As soon as the valve opens, casing pressure starts to drop and the gas lift valve closes when casing pressure drops to the dome pressure of the valve. Because of this process, lift gas is stored in the annulus whose pressure extremes are the valve's opening and closing pressures. The amount of gas stored in the annulus and used for fluid lifting can easily be calculated, based on the volume of the annulus and the pressure drop, the latter being equal to valve spread. Therefore, the amount of gas used for one intermittent cycle is a direct function of the actual valve spread. This implies that the selection of valve ports is a critical factor in intermittent gas lift design because valve spread is defined by the port area. An improper port size always results in too low or too high injection volumes per cycle.

Representative Valves

Selected valves from various manufacturers are: CAMCO J-20, J-40, J-50; MERLA N Series; McMurry-Macco C-2, C-1, C-3. The PPEF of these valves lies in the range of PPEF = 0.04 – 0.73.

Example 3–6. Calculate a valve's PPE and spread at a production pressure of 200 psig, and a valve temperature of 160 °F, for a gas lift valve with a bellows area of 0.77 sq. in., and a 7/16 in. port (port area 0.154 sq. in.). The valve's TRO pressure is 500 psig.

Solution

First calculate the geometrical factor R, the ratio of the port and bellows areas:

$R = 0.254 / 0.77 = 0.2$

Dome charge pressure at shop conditions (60 °F) can be found from the definition of TRO, Equation 3.21:

$p'_d = 500 \ (\ 1 - 0.2 \ ) = 400$ psig.

To find the dome charge pressure at downhole conditions, Figure 3–5 is used and a downhole dome pressure of $p_d = 480$ psig is read. Using this value, PPE is calculated from the PPEF, which is easy to find from valve geometrical data (Equation 3.23):

PPEF = 0.2 / (1 – 0.2) = 0.25

PPE at 200 psig production pressure is found from Equation 3.22:

PPE = 200 0.25 = 50 psig, which is the reduction in opening pressure due to the effect of production pressure.

For calculating valve spread, Equation 3.26 is used:

Spread = 0.25 (480 – 200) = 70 psig.

The previous calculations are illustrated in Figure 3–24.

Example 3–7. An operating valve in an intermittent installation is run to 7,500 ft where the temperature is 190 °F and the tubing (production) pressure due to a starting liquid slug is 400 psig. Surface operating gas lift pressure is 900 psig, lift gas is of 0.8 specific gravity, and the valve is a CAMCO J-20 with a ½ in. port (A_b = 0.77 sq in, A_v = 0.2 sq. in.). Find the following parameters:

(a) the opening pressure at valve depth

(b) the closing pressure at setting depth

(c) valve spread at valve depth

(d) the required TRO pressure

Solution

(a) The valve is assumed to be opened by the full injection pressure, its surface opening pressure is thus p_{io}' = 900 psig. To find the downhole opening pressure, the pressure gradient for a surface pressure of 900 psig and a gas specific gravity of 0.8 is found from Fig. B-1 in Appendix B as 33.5 psia/1,000 ft. Opening pressure at valve depth is

p_{io} = 900 + 33.5 7,500 / 1,000 = 1,151 psig.

(b) Since valve opening and production pressures at valve depth are known, Equation 3.20 can be used to find the required dome charge pressure, after substitution of R = 0.2 / 0.77 = 0.26:

p_d = 1,151 (1 – 0.26) + 400 0.26 = 956 psig, which is the valve's closing pressure at depth.

(c) Valve spread at setting depth is the difference of the opening and closing pressures:

Spread = 1,151 – 956 = 195 psig.

(d) For determining the TRO pressure of the valve, the dome charge pressure at 60 °F is found from Figure 3–5 as p_d' = 750 psig, and according to Equation 3.21:

TRO = 750 / (1 – 0.26) = 1,014 psig, which is the setting pressure for the valve.

Double Element Valves

Sometimes, in addition to the bellows charge, a spring is also used to provide a closing force in an unbalanced gas lift valve as shown in Figure 3–26. The spring can ensure that an unloading valve with a ruptured bellows does not stay open and inject gas unnecessarily. The operation of such valves (often called double element valves) is described the same way as was done with those without a spring, by writing up the force balance equations. The spring force is usually represented by an equivalent pressure term p_{sp} (measured in a valve tester) which is assumed to act on the *donut* area of $(A_b–A_v)$. The force thus generated tries to close the valve and helps the dome charge pressure. In contrast to the force caused by the dome pressure, the spring force is insensitive to temperature, so the same value is used for surface and downhole conditions.

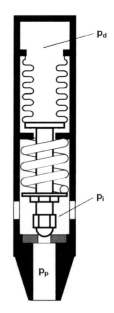

Fig. 3–26 Double element gas lift valve.

The opening conditions for this type of valve are described by the following force balance equation valid at the instant the valve opens:

$$p_d A_b + p_{sp}(A_b - A_v) = p_i(A_b - A_v) + p_p A_v \qquad\qquad 3.27$$

Solving for the opening injection pressure and using the terminology as before:

$$p_{io} = \frac{p_d}{1-R} - \text{PPEF } p_p + p_{sp} \qquad\qquad 3.28$$

where: p_d = dome pressure at valve temperature, psi

p_{io} = opening injection pressure at valve setting depth, psi

p_p = production pressure at valve setting depth, psi

PPEF = production pressure effect factor defined by Equation 3.23

p_{sp} = spring force effect, psi

The valve's closing conditions are described by the following force balance equation where, as was done for the valve without a spring, the valve is assumed to be fully open and injection pressure acts under valve stem.

$$p_d A_b + p_{sp}(A_b - A_v) = p_i A_b \qquad\qquad 3.29$$

From this, the closing injection pressure of the valve is easily found:

$$p_{ic} = p_d + p_{sp}(1 - R) \qquad\qquad 3.30$$

In case opening injection pressure and production pressures are specified, solution of Equation 3.28 permits the calculation of the required dome charge pressure:

$$p_d = (p_{io} - p_{sp})(1 - R) + p_p R \qquad\qquad 3.31$$

The other characteristics of this valve type also reflect the presence of the spring and are given as follows, except the PPEF which is identical to the one given for the previous valve type.

$$\text{TRO} = \frac{p'_d}{1-R} + p_{sp} \qquad\qquad 3.32$$

$$\text{SPREAD} = \text{PPEF}\,[p_d - p_p + p_{sp}(1 - R)] \qquad\qquad 3.33$$

where: p'_d = dome pressure at 60 °F, psi.

Reference is made here to Section 3.2.1.2 where the behavior of the dome charge and a spring was compared. As discussed, metal springs used in gas lift valves have a much higher load rate than the gas-charged bellows assembly. This makes spring-loaded valves to *throttle* because the stem can never completely lift off from the seat and gas injection rate through the valve is limited. The double element valve discussed previously is not considered a throttling valve because the spring is only used to complement the gas charge and to provide positive closing in the event of a bellows failure. The main application of this valve type is in installations where extremely high valve operating pressures are required.

Throttling valves utilizing strong springs with no gas charge in their domes behave very differently from those discussed here and will be described under a different group of valves.

Example 3–8. Two unbalanced bellows-charged valves have the same TRO pressure of 1,200 psig, one has a spring with a spring force effect of p_{sp} = 500 psig. Geometrical parameters are identical with A_b = 0.77 sq in and A_v = 0.154 sq in, find the bellows charge pressure required for each valve.

Solution

For the valve without a spring, dome charge pressure is found from Equation 3.21 as:

$$p'_{d_1} = 1,200 \; 0.8 = 960 \text{ psig, where}$$
$R = A_v/A_b = 0.154 / 0.77 = 0.8.$

The other valve contains a spring and Equation 3.32 must be solved for the dome charge pressure:

$$p'_{d_2} = (1,200 - 500) \; 0.8 = 560 \text{ psig.}$$

Figure 3–27 presents calculation results with the plot of the common opening and closing equations. As seen, the spring-loaded valve could be charged to a much lower gas pressure because of the high spring force effect.

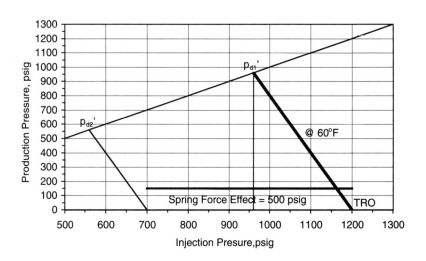

Fig. 3–27 Performance of the gas lift valves given in Example 3–8.

Example 3–9. For the same case as in Example 3–7, assume that the valve is equipped with a spring set to p_{sp} = 400 psig and find the same parameters as before.

Solution

(a) The opening pressure at valve depth is the same as before: p_{io} = 1,151 psig.

(b) Opening and production pressures at valve depth being known, Equation 3.31 is used to find the required dome charge pressure:

$$p_d = (1,151 - 400)(1 - 0.26) + 400 \; 0.26 = 660 \text{ psig.}$$

The valve's closing pressure at depth is evaluated from Equation 3.30:

$$p_{ic} = 660 + 400 (1 - 0.26) = 956 \text{ psig.}$$

(c) Valve spread at setting depth is the difference of the opening and closing pressures:

$$\text{Spread} = 1,151 - 956 = 195 \text{ psig.}$$

(d) For determining the TRO pressure of the valve, the dome charge pressure at 60 °F is found from Figure 3–5 as p'_d = 523 psig, and according to Equation 3.32:

$$\text{TRO} = 523 / (1 - 0.26) + 400 = 1,107 \text{ psig, which is the setting pressure for the valve.}$$

3.2.2.2.2 *PPO valves.*

Valves with Normal Seats

If the unbalanced gas-charged valve with or without a spring is installed so that production pressure can act on the greater area $(A_b - A_v)$, then it becomes a PPO valve. This type of valve was formerly called a tubing-pressure or a fluid-operated valve. As shown in Figure 3–28, the valve is the same as before, but the pressures acting on it have replaced each other. Production pressure (in tubing flow installations tubing pressure) enters the gas inlet ports and provides an opening force on the area $(A_b - A_v)$. Since injection pressure is exerted over the valve port area A_v, its contribution to the net opening force is low. The opening equation for a double element (gas-charged and spring-loaded) valve is very similar to the one written for IPO valve with the pressures exchanged:

$$3.34 \quad p_d A_b + p_{sp}(A_b - A_v) = p_p (A_b - A_v) + p_i A_v$$

Because this valve is much more sensitive to production pressure than injection pressure, the previous equation is solved for production pressure at the instant the valve opens. Using the terminology as before and denoting the group $R/(1 - R)$ by injection pressure effect factor (IPEF), we get:

$$p_{po} = \frac{p_d}{1 - R} - \text{IPEF } p_i + p_{sp} \qquad 3.35$$

where: p_d = dome pressure at valve temperature, psi

p_{po} = opening production pressure at valve setting depth, psi

p_i = injection pressure at valve setting depth, psi

IPEF = injection pressure effect factor

p_{sp} = spring force effect, psi

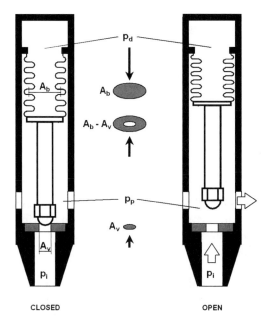

Fig. 3–28 Unbalanced valve with normal seat used in PPO service.

The term IPEF is a geometrical constant for a given valve with a given port size and represents the drop in the valve's opening production pressure for a unit increase in injection pressure. In a tubing flow installation, injection pressure equals casing pressure; IPEF may therefore be called casing effect factor (CEF), a formerly widely used term.

In the open position, the pressure under the valve stem is assumed to equal production pressure because the valve is supposed to open fully. To ensure this kind of operation, small ports are used in PPO valves. If the port is larger the valve must be choked upstream of the port. Through this solution, production rather than injection pressure will act on the total bellows area in the open position making the valve relatively insensitive to injection pressure. The valve's closing conditions are described by the following force balance equation, valid in the valve's open position:

$$p_d A_b + p_{sp}(A_b - A_v) = p_p A_b \qquad 3.36$$

From this, the closing production pressure of the valve is easily found:

$$p_{pc} = p_d + p_{sp}(1 - R) \qquad 3.37$$

If the opening and closing equations are combined, the following formula describes the relationship of the opening and closing production pressures:

$$p_{po} = \frac{p_{pc} - R\,p_i}{1 - R}$$

3.38

The PPO valve is set in the shop with zero injection pressure and its TRO pressure is described by the following equation:

$$TRO = \frac{p'_d}{1 - R} + p_{sp}$$

3.39

where: p'_d = dome pressure at 60 °F, psi.

All performance equations derived previously must be modified if gas charge only or spring force only is used in the valve. For a valve without a spring, spring force effect must be set to zero, whereas for valves without a gas charge, dome pressure is set to zero in the opening (Equation 3.35), the closing (Equation 3.37) equations, as well as in the TRO formula (Equation 3.39).

A graphical presentation of the opening and closing performance of an unbalanced PPO valve at downhole conditions is given in Figure 3–29. The reverse flow check valve closes at any production pressure greater than injection pressure, the area above the $p_p = p_i$ line. The valve's opening pressure (shown in bold line), if compared to the opening pressure of an IPO valve, varies much less with injection pressure because injection pressure acts on the port area of the valve only. The valve stays closed below this line, indicated by the darkest shaded area. Closing production pressures are represented by the horizontal line at a production pressure equal to p_{pc} found from Equation 3.37.

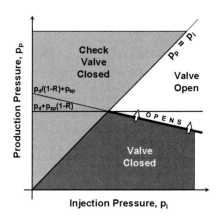

Fig. 3–29 Opening and closing performance of an unbalanced PPO gas lift valve.

If injection pressure is held constant and production pressure is increased, the valve opens as soon as production pressure reaches the opening pressure p_{po}. However, it is obvious that this pressure is below the valve's closing pressure so the valve would immediately close after opening. When closed, the opening forces would make the valve reopen and the process would go on. This type of operation, of course, is impractical and would inhibit the use of such valves in practice. In spite of this, PPO unbalanced valves are still used and give successful operation. The explanation to this seemingly absurd situation lies in the limitations of the static force balance equations, which do not allow the determination of the actual pressures acting on the valve stem tip. The gas injected through the port into the valve body attains a high enough velocity to generate a sufficient dynamic pressure on the valve tip to keep the valve open as long as production pressure does not drop. This is why such valves are used with port diameters less than 1/4 in. or are choked upstream of the port.

Valves with Crossover Seats

Figure 3–30 illustrates a bellows-charged valve with a crossover seat that converts an IPO valve to a PPO valve. Compared to an IPO valve, only the valve housing containing the seat is changed. Injection pressure now acts on the valve port area and production pressure is exerted over the area $(A_b - A_v)$, making the valve operate by production pressure. For proper operation, the total cross-sectional area of the bypass openings that lead production pressure to the bellows must be greater than the port area. To ensure this, the gas inlet ports are often equipped with chokes upstream of the valve seat. The operation of this valve is the same as that of the conventional solution when an IPO valve is used in the proper valve mandrel.

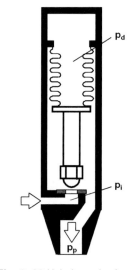

Fig. 3–30 Unbalanced valve with crossover seat for PPO service.

Crossover seats make it possible to use the same valve mandrel for injection and production pressure operation or using a tubing flow mandrel in a casing flow installation. Although these solutions are acceptable if the proper gas lift mandrels are not available, valves with crossover seats are generally not recommended. The main reason is that only relatively small valve ports limiting gas injection rates can be used, and the small channels in the bypass area are prone to plugging.

Representative Valves

In principle, all unbalanced IPO valves can be converted to production pressure operation. Many times, separate valves with crossover seats are offered like the CAMCO JR-20, JR-40, JR-50 valves. The IPEF of these valves has a range of IPEF = 0.02 to 0.1. If compared to the PPEF of IPO valves, IPEF values are much lower, showing that the effect of injection pressure on the opening pressures of these valves is almost negligible.

Example 3–10. Determine whether an unbalanced PPO valve in an unloading valve string stays closed after the unloading operation if tubing pressure during normal operation at the valve's setting depth of 3,000 ft is 550 psig. Surface injection pressure is 800 psig, injection gas gravity is 0.8. The valve is spring loaded only with a 3/16 in. port (R=0.094).

Solution

To find the downhole injection pressure, the pressure gradient for a surface pressure of 800 psig and a gas specific gravity of 0.8 is found from **Fig. B–1** in **Appendix B** as 29 psia/1,000 ft. Injection pressure at valve depth is

p_{io} = 800 + 29 3,000 / 1,000 = 887 psig.

The closing production pressure of the valve is selected according to the manufacturer's recommended design procedure to p_{pc} = 620 psig and the valve's opening pressure is found from Equation 3.38 as:

p_{po} = (620 – 0.094 887) / (1 – 0.094) = 592 psig, which is greater than the flowing tubing pressure, the valve will stay closed.

To find the TRO pressure, the spring force effect is found from the closing equation (Equation 3.37) by substituting zero dome charge pressure:

p_{sp} = 620 / (1 – 0.094) = 684 psig.

Using the previous value and Equation 3.39:

TRO = 684 psig.

Example 3–11. Determine whether an unloading valve stays closed after unloading if production (tubing) pressure at the valve setting depth of 4,600 ft is 1,200 psig. Surface injection pressure is 1,300 psig, injection gas gravity is 0.8. A gas-charged and spring-loaded valve with a 5/32 in. port (R=0.066) is used with a spring force effect of p_{sp} = 800 psig, valve temperature is 140 °F.

Solution

The annulus pressure gradient for a surface pressure of 1,300 psig and a gas specific gravity of 0.8 is found from **Fig. B–1** in **Appendix B** as 53 psia/1,000 ft. Injection pressure at valve depth is

p_{io} = 1,300 + 53 4,600 / 1,000 = 1,544 psig.

The closing production pressure of the valve is assumed as p_{pc} = 1,340 psig, and the valve's opening pressure is found from Equation 3.38:

p_{po} = (1,340 − 0.066 1,544) / (1−0.066) = 1,326 psig, which is greater than the flowing tubing pressure, the valve will stay closed.

The dome charge pressure at valve depth is found from the closing equation (Equation 3.37):

p_d = 1,340 − 800 (1−0.066) = 593 psig.

Dome pressure at surface conditions is found from Figure 3–5 as $p_d^!$ = 510 psig, and the TRO pressure is evaluated from Equation 3.39:

TRO = 510 / (1 − 0.066) + 800 = 1,346 psig.

––·–

3.2.2.3 Unbalanced valves without spread.

The gas lift valves belonging to this category have opening and closing pressures that are identical but vary with production pressure. Their spread, *i.e.* difference between opening and closing pressures, is zero but the PPEF is a nonzero variable. Such valves are often called *throttling valves* since they can never fully open and always restrict the gas injection rate. Due to this throttling action, the pressure below the valve stem tip is always very close to production pressure, making the stem to rise and descend according to variations in tubing pressure. Gas injection rate, therefore, is tightly controlled, making this type of valve to ideally suit the requirements of continuous flow gas lift wells.

3.2.2.3.1 IPO valves.

Valves with a Choke Upstream of the Port

The gas lift valve shown in the closed position in Figure 3–31 is a bellows-charged valve with a spring and two distinct constructional features. The valve stem is sealed just below the bellows, and the gas inlet ports are equipped with relatively small diameter chokes. Because of the seal, injection pressure acts on the $(A_v - A_b)$ area irrespectively of the valve's actual state, *i.e.* both in the open and the closed positions. The function of the small chokes is to keep the pressure under the valve stem tip constant at production pressure. Obviously, this is true for the closed position. When the valve opens, gas flows through the inlet chokes and the valve port with its pressure dropping from injection pressure to production pressure. Since the inlet chokes offer a much more reduced area than the valve port, the pressure drop occurs across the chokes. Consequently, production pressure is maintained in the valve body and, of course, below the valve stem tip. These features make the valve sensitive to production pressure and adjust its gas throughput capacity according to changes in production pressure.

Force balance equations are written with the assumptions that valve stem diameter and port diameter are practically the same, and no friction occurs on the seal ring. Opening and closing forces come from the dome charge pressure, injection, and production pressures and they are identical in the open and the closed positions:

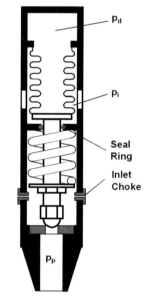

Fig. 3–31 Unbalanced valve with a choke upstream of the port for IPO service.

$$p_d A_b + p_{sp}(A_b - A_v) = p_i (A_b - A_v) + p_p A_v \qquad 3.40$$

Solution of this equation for p_i gives the opening and closing injection pressures:

$$p_{io} = p_{ic} = \frac{p_d}{1 - R} - \text{PPEF } p_p + p_{sp}$$

3.41

where: p_d = dome pressure at valve temperature, psi

   p_{io} = opening injection pressure at valve setting depth, psi

   p_{ic} = closing injection pressure at valve setting depth, psi

   p_p = production pressure at valve setting depth, psi

   PPEF = production pressure effect factor defined by Equation 3.23

   p_{sp} = spring force effect, psi

The valve is set in the shop with zero injection pressure and its TRO pressure is described by the following equation:

$$\text{TRO} = \frac{p'_d}{1 - R} + p_{sp}$$

3.42

where: p'_d = dome charge pressure at 60 °F, psi.

A graphical presentation of the opening and closing performance of the valve at downhole conditions is given in Figure 3–32. The opening and closing injection pressures are represented by the same linear function of production pressure and are shown in bold line. Opening and closing of the valve occurs along this line, and there is no difference between the opening and the closing pressures, *i.e.* the valve's spread is zero. It is unbalanced because the opening and closing pressures depend on production pressure. The valve is very sensitive to production pressure, and any drop in that pressure can make it close.

Spring-loaded Valves without Bellows Charge

As already discussed, metal springs used to load gas lift valves have much higher load rates than a gas-charged bellows assembly. This is why valves whose only closing force comes from the compression of a metal spring cannot be fully opened and will always throttle and restrict gas flow. This feature is utilized in the valve type shown schematically in Figure 3–33. The dome of the valve is not charged with gas, and the bellows only serves to allow the injection pressure to lift the valve stem.

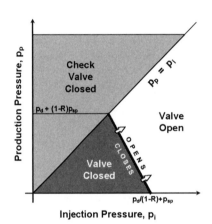

Fig. 3–32 Opening and closing performance of an unbalanced IPO gas lift valve with a choke upstream of the port.

All closing force comes from the metal spring in which a compression force arises as the valve stem lifts above the seat. The spring force, compared to the effect of a bellows charge, is relatively high and it is not affected by temperature. In installation design, this is a definite advantage because all the inaccuracies associated with the improper knowledge of well temperatures are completely eliminated.

The other important feature of the valve is the use of a tapered seat and valve stem tip. This enhances the valve's characteristic feature of restricting or *throttling* gas throughput rates. Since valve travel is very limited due to the high load rate of the compression spring, the area opened to gas flow is always small, as compared to the valve port size. The combined effects of the spring as a loading element as well as that of the tapered seat and valve stem result in a basic feature of this valve: even in the open position, the pressure acting on the valve tip is production pressure. The reason for this is that gas inlet ports have a much larger cross-sectional area than the valve seat contact area, causing a large pressure drop across the seat. (see Fig. 3–33)

It follows from the previous considerations that force balance equations for the open and the closed valve are the same. They include the spring force as a closing component, as well as the opening forces coming from injection pressure acting on the area $(A_b - A_v)$ and from production pressure acting on the valve seat contact area A_v.

$$F_{sp} = p_i (A_b - A_v) + p_p A_v \qquad\qquad 3.43$$

where: F_{sp} =   compression force in spring, lb

p_i =   injection pressure at valve setting depth, psi

p_p =   production pressure at valve setting depth, psi

A_b, A_v =   bellows and seat contact areas, sq in.

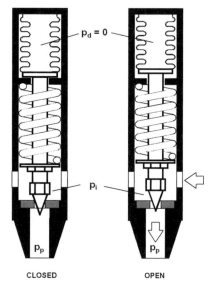

Fig. 3–33 Spring-loaded throttling valve without gas charge used in IPO service.

The amount of spring force is mechanically adjusted during the valve setting procedure. Its direct measurement is not feasible, therefore it must be measured indirectly, by using the appropriate valve tester. Valve testers used for this type of valve measure the pressure applied to the total bellows area just before the instant the valve closes. Although this value, called the spring adjustment pressure, indicates the spring force at closing conditions, it is used for the opening conditions as well, thus the variation of spring force with valve stem travel is neglected. This is justified by the small amount of valve stem travel that makes the variation in spring force to be negligible in the operation range of the valve.

Assuming that the measured spring adjustment pressure acts on the full bellows area, the force balance equation is modified to:

$$p_{sa} A_b = p_i (A_b - A_v) + p_p A_v \qquad\qquad 3.44$$

where: p_{sa} = spring adjustment pressure, psi

Solution of the previous equation for the injection pressure at opening and closing conditions gives the valve's opening and closing equations:

$$p_{io} = p_{ic} = \frac{p_{sa}}{1 - R} - \text{PPEF } p_p \qquad\qquad 3.45$$

where: $\text{PPEF} = R / (1 - R)$

In design calculations, injection and production pressures are known at valve opening conditions and the previous formula, if solved for spring adjustment pressure, allows the calculation of the required valve setting pressure:

$$p_{sa} = p_{io} - R (p_{io} - p_p) \qquad\qquad 3.46$$

The opening and closing performance of the valve is shown in Figure 3–34. Note that, in contrast to bellows-charged valves, the performance curve is the same for surface and downhole conditions, because the spring force is not affected by temperature, a big advantage

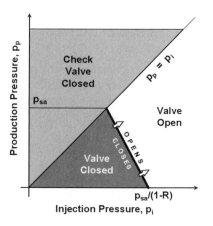

Fig. 3–34 Opening and closing performance of a spring-loaded, IPO gas lift valve.

of this valve category. The bold line represents the opening and closing injection pressures. Since there is no difference between them, the valve's spread is zero. The valve is production pressure sensitive and can be closed by a drop in production pressure, even if injection pressure stays constant.

The IPO throttling valve's gas throughput characteristics (discussed in Section 3.2.3.2) make it an ideal continuous flow valve. Classical example is the MERLA L Series, often called a *proportional response* valve [17] because it properly responds to the injection gas requirements of continuous flow gas lift wells.

Representative Valves

MERLA L Series is the classic valve, other valves are: CAMCO J-46-0; McMurry-Macco CF-1, CF-2, R-2, R-1. The PPEF of these valves lies in the range of PPEF = 0.04–0.67.

-- --

Example 3–12. Determine the setting of one valve in an unloading valve string so that it should stay closed after the unloading operation. Setting depth is 4,450 ft, surface injection pressure is 1,000 psig, gas gravity is 0.65, the flowing production pressure at valve depth is 657 psig. The valve is a MERAL LM-16RA with ¼ in. port and $R = 0.24$.

Solution

Injection pressure at valve depth is calculated from a pressure gradient of 27 psi/1,000 ft, found from Fig. B–1 in Appendix B as:

$$p_{io} = 1,000 + 27 \ 4,450 / 1,000 = 1,120 \text{ psig.}$$

The opening production pressure is taken as flowing production pressure plus 10% of the difference between injection and production pressures at valve depth (according to the manufacturer's recommendation):

$$p_{po} = 657 + 0.1 \ (\ 1120 - 657 \ ) = 703 \text{ psig.}$$

The required spring adjustment pressure is found from Equation 3.46:

$$p_{sa} = 1,120 - 0.24 \ (\ 1,120 - 703 \ ) = 1,020 \text{ psig.}$$

If the valve is set to this pressure, it won't open during normal operations because the flowing production pressure is lower than the valve's opening pressure.

-- --

3.2.2.3.2 *PPO valves.* The spring-loaded throttling valve, if used in the proper valve mandrel, becomes a PPO valve, as shown in Figure 3–35. Production pressure now acts on the greater part of the bellows area $(A_b - A_v)$, while injection pressure acts on the valve seat area A_v. The only closing force comes from the metal spring since the valve dome is not charged. Because of the behavior of the spring and the tapered seat, the valve does not fully open but throttles just like the IPO version. As discussed before, the port area open to flow is always restricted and injection pressure must drop to production pressure across the valve

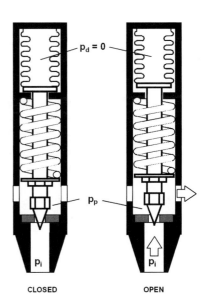

Fig. 3–35 Spring-loaded throttling valve without gas charge used in PPO service.

seat. All these features result in the same pressure conditions in the closed and the open position of the valve and the opening and closing equations will be the same:

$$p_{po} = p_{pc} = \frac{p_{sa}}{1-R} - \text{IPEF } p_i \qquad\qquad 3.47$$

where: p_{po}, p_{pc} = opening and closing production pressures at valve setting depth, psi

$\qquad\qquad p_i$ = injection pressure at valve setting depth, psi

$\qquad\qquad p_{sa}$ = spring adjustment pressure, psi

$\qquad$ IPEF = $R/(1-R)$ = injection pressure effect factor

The previous formula is the same as Equation 3.45, only the pressures are exchanged. If injection and production pressures are known at valve opening conditions, the previous formula can be solved for spring adjustment pressure:

$$p_{sa} = p_{po} - R\left(p_i - p_{po}\right) \qquad\qquad 3.48$$

Figure 3–36 shows the opening and closing performance of the valve, where the performance curve is the same for surface and downhole conditions, because the spring force is not affected by temperature. Opening and closing of the valve occurs along the bold line; no spread exists. Since the valve is more sensitive to production pressure than to injection pressure, it is easily opened by a small increase in production pressure at valve setting depth. This feature is best utilized in intermittent lifting where multipoint gas injection is desired.

The same operational features can be reached if a crossover seat and the proper mandrel are used with an IPO throttling valve.

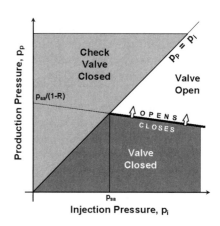

Fig. 3–36 Opening and closing performance of a spring-loaded, PPO gas lift valve.

Representative Valves

Selected valves from various manufacturers are: CAMCO BKF-6; MERLA R Series (with crossover seats); McMurry-Macco RF-1, RF-2. IPEFs lie in a range of IPEF = 0.04 to 0.23. Again, comparison of these values and the PPEFs of IPO valves shows that these valves are almost insensitive to injection pressure.

3.2.2.4 Balanced valves. Balanced valves, in contrast to unbalanced ones, open and close at the same pressure. Their performance is governed by one pressure only, either injection or production pressure. Thus IPO balanced valves work independently of tubing pressure; whereas PPO balanced valves are not affected by injection pressure. This category includes not only the bellows-charged but the flexible sleeve type of valve as well. As before, they are treated in two groups: IPO and PPO valves.

3.2.2.4.1 IPO valves.

Bellows Valves

The two basic construction features of the balanced, bellows-charged gas lift valve presented in Figure 3–37 are as follows:

(a) a seal ring on the valve stem divides the valve body in two sealed parts

(b) injection pressure is exerted on the valve seat area both in the open and the closed position

The force balance equations, as will be seen in following paragraphs, do not contain the production pressure so the valve is absolutely insensitive to that pressure. This only holds if the cross-sectional areas of the valve stem and the port are equal, what is assumed in the following.

The closing force in the valve comes from the bellows charge pressure acting on the full bellows area. In the closed position, two opening forces are present but both are the results of injection pressure: one acting on the area $(A_b - A_v)$, the other one on the valve seat area A_v. Just before opening, the force balance is the following:

$$p_d A_b = p_i (A_b - A_v) + p_i A_v \qquad 3.49$$

After the valve opens, the forces acting on the valve stem do not change because injection pressure is still applied to the valve seat. Consequently, the force balance equation written previously is valid for both the opening and closing conditions. Solving it for the opening and closing pressures we get:

$$p_{io} = p_{ic} = p_d \qquad 3.50$$

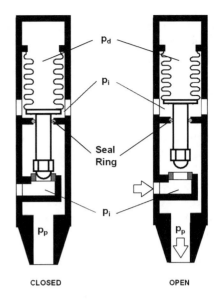

Fig. 3–37 Balanced, bellows-charged gas lift valve for IPO service.

where: p_{io}, p_{ic} = opening and closing injection pressures at valve setting depth, psi

p_d = dome charge pressure at valve temperature, psi

This valve, therefore, opens and closes at the same injection pressure, which equals the dome charge pressure at prevailing conditions. There is no valve spread since opening and closing pressures are identical, and production pressure does not affect the valve's operation. Consequently, in intermittent operations, greater port sizes can be used without excessive valve spread and excessive lift gas consumption. If, on the other hand, unbalanced IPO valves are used, the applicable port size must be limited because of its effect on valve spread.

The balanced and unbalanced valves inject different amounts of gas during the intermittent cycle. The unbalanced one, due to its spread, allows the gas stored in the annulus to be used for fluid lifting, thus decreasing the instantaneous surface injection rate. Balanced valves, on the other hand, do not have spread and cannot utilize the annulus for gas storage. This is why the high instantaneous injection rates required for an efficient fluid lifting must completely be supplied by the surface gas lift system. The balanced valve always injects the same amount of gas to the production string, as much is available on the surface.

Flexible Sleeve Valves

This valve category is different from the valves described so far because of its concentric construction with radial dimensions identical to those of the tubing string. These valves are run as a part of the tubing string and are pulled along with it. They use, instead of a metal bellows, a resilient element (flexible sleeve) as a working part through which gas injection into the tubing string takes place.

The valve schematically shown in Figure 3–38 has a chamber (like the bellows valve's gas dome) charged at the surface with high-pressure gas. Dome pressure can act from the inside on the flexible sleeve, which is fixed in the valve housing to bridge the valve inlet ports. In the closed position, dome pressure is greater than injection pressure acting through the inlet ports from the outside, and the sleeve is forced to seal off gas entry into the

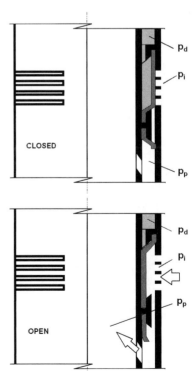

Fig. 3–38 Balanced, flexible sleeve type gas lift valve for IPO service.

valve. As soon as injection pressure reaches and exceeds the valve's dome charge pressure, the flexible sleeve moves off the inlet ports and allows gas to enter the valve. Gas then flows past the finned retainers to the valve discharge ports, and the valve starts to inject gas into the tubing string. Injection continues with the full cross-sectional area until injection pressure drops below the valve's dome pressure. At this moment, dome pressure working against injection pressure forces the flexible sleeve to shut off the inlet ports and gas injection stops.

The valve includes a flexible reverse flow check valve preventing backflow from the tubing string should tubing pressure exceed casing pressure. It is a velocity-activated valve, shutting off the flow path as soon as fluid flow from the tubing is activated.

The flexible sleeve balanced gas lift valve, as described previously, opens and closes at the same injection pressure that equals the dome charge pressure at valve setting depth. Because it does not have spread, the valve cannot be used in conditions where gas storage in the annular volume is required. The gas throughput capacity of this valve is quite high and it works as an expanding orifice, always passing the same amount of gas as injected on the surface.

Representative Valves

The flexible sleeve valve was exclusively manufactured by OTIS Engineering Co., as Series C valves.

3.2.2.4.2 PPO valves. The balanced IPO valve, if installed in the proper mandrel, becomes a PPO valve, as shown in Figure 3–39. The valve is insensitive to injection pressure and opens and closes at the same production pressure equal to the dome charge pressure at valve temperature. The opening and closing equations are:

$$p_{po} = p_{pc} = p_d \qquad\qquad 3.51$$

where: p_{po}, p_{pc} = opening and closing production pressures at valve setting depth, psi

p_d = dome charge pressure at valve temperature, psi

The valve's main features are the same as those of the IPO valve described previously.

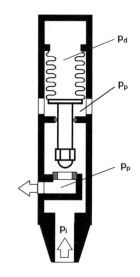

Fig. 3–39 Balanced, bellows-charged gas lift valve for PPO service.

3.2.2.5 Pilot valves. Pilot valves contain two valves in one: (a) a pilot section, and (b) a main (power or slave) section. The pilot section, usually a bellows-charged unbalanced valve with or without a spring, controls the operation of the set. When pressure conditions in the well at valve setting depth satisfy the pilot's opening requirements, the pilot opens and signals to the main section. Depending on the construction of the main valve, it then performs the required operation and starts injecting gas. Most pilot valves are designed for intermittent lift and have large main port sizes for an efficient lifting of the accumulated liquid slug to the surface.

3.2.2.5.1 IPO pilot valves. The pilot-operated valve shown in Figure 3–40 is used in intermittent gas lift installations and will be shown to provide ideal conditions for an efficient lifting of liquid slugs. It consists of a bellows-charged unbalanced valve as the pilot section and a power piston as the main section. The pilot section may contain a spring in addition to the bellows charge to keep the valve closed. The main section is a spring-loaded differential valve with a light compression spring and an exceptionally large diameter (from 3/8 in. up to 3/4 in.) gas discharge port. Note that gas injection can only take place through the main valve's inlet and discharge ports.

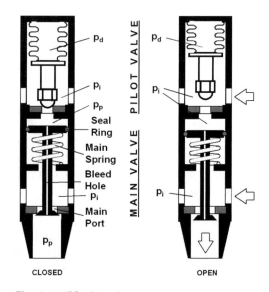

Fig. 3–40 IPO pilot valve.

In the closed position, the pilot did not open yet and the bleed hole in the main piston allows production pressure to reach the area below the pilot port. The pilot section, therefore, functions just like any unbalanced IPO valve in the closed position. The main valve port is kept closed by the spring force and the force arising from the difference of the injection and production pressures acting on the piston area. As injection and/or production pressures increase, the pilot section opens when its opening equation is satisfied. As soon as the pilot opens, gas with injection pressure fills the space above the main piston. Although the bleed hole continuously bleeds from this pressure, its effect is negligible and injection pressure will exist here as long as the pilot is open. This high pressure, acting on the piston area against the spring force depresses the main valve and opens the main discharge port to gas injection into the produced fluid.

After opening, the pilot section will fully open because the pressure on its seat changes from production to injection pressure. At the same time, the main valve also snaps open due to the big pressure differential across its piston. The gas lift valve now injects gas through the main valve's inlet and discharge ports, which can have as large cross-sectional areas as made possible by constructional restrictions. The valve, therefore, provides ideal conditions for liquid slug lifting by: (a) its snap action and (b) allowing very high instantaneous gas flow rates.

While the liquid slug leaves the well to the surface, injection pressure continuously decreases at valve setting depth and at the proper instant the pilot valve closes. Its closing injection pressure equals the dome charge pressure because in the valve's open position, injection pressure acts on the full bellows area. As soon as the pilot section closes, the pressure trapped above the main piston decreases to tubing pressure through the small bleed hole. This greatly reduces the force acting downward on the main valve stem, which held the valve open. The main spring's compression force, along with force coming from the casing pressure acting on the main piston overcome the opening force, and the main discharge port is shut down.

It must be clear from the previous description that the pilot section behaves as an unbalanced IPO gas lift valve. Its opening and closing pressures are found from formulas derived before, and the behavior of the valve can graphically be presented as was done in Figure 3–23. For a double element valve the relevant equations are listed as follows:

$$p_{io} = \frac{p_d}{1-R} - \text{PPEF}\, p_p + p_{sp}$$ 3.52

$$p_{ic} = p_d + p_{sp}(1-R)$$ 3.53

where: p_{io}, p_{ic} = opening and closing injection pressures at valve setting depth, psi

p_d = dome pressure at valve temperature, psi

p_p = production pressure at valve setting depth, psi

PPEF = $R/(1-R)$ = production pressure effect factor

R = Av/Ab = ratio of valve seat and bellows areas

p_{sp} = spring force effect, psi

In design calculations, opening as well as production pressures are specified, and solution of the opening equation permits the calculation of the required dome charge pressure:

$$p_d = (p_{io} - p_{sp})(1-R) + p_p R$$ 3.54

The valve's TRO pressure and spread are:

$$\text{TRO} = \frac{p'_d}{1-R} + p_{sp}$$ 3.55

$$\text{SPREAD} = \text{PPEF}\,[p_d - p_p + p_{sp}(1-R)]$$ 3.56

where: p'_d = dome pressure at 60 °F, psi

The great advantage of using the pilot-operated valve just discussed lies in its spread characteristics. As pointed out previously, the complete valve's spread is determined by the seat area of the pilot section and has nothing to do with the actual area open to gas injection. This is contrary to the behavior of the unbalanced valve where valve port area and valve spread are determined by each other. As shown in Figure 3–25, the larger the port, the greater the valve's spread. This condition results in an often-unresolved conflict in the selection of the proper valve. On one hand, the possible largest port area is desired for an efficient fluid lifting process, on the other hand, valve spread must be matched to the existing annulus volume. This whole problem is eliminated if a pilot valve is used because a large gas passage port can be used while the valve's spread can be selected as necessary.

Representative Valves

Some pilot valve types with bellows charge only are: CAMCO CP-2, BF-2, PK-1, RPB-5, RP-6; McMurry-Macco RPV-1, RPV-2. Their PPEF lie between 0.07 and 0.58. Valves with spring-loaded, no bellows charge pilot sections are: McMurry-Macco RPV-2S, RPV-1S; MERLA WF Series. PPEFs of 0.06 to 1.9 are available.

Example 3–13. The operating valve of an intermittent installation is a CAMCO PK-1 pilot-operated valve without a spring. It is run to 7,500 ft where the valve temperature is 120 °F. Surface injection pressure is 900 psig, gas relative density is 0.8 and gas injection on the surface is controlled by a choke. In order to inject the proper amount of gas per cycle, the required pressure reduction in the annulus is 110 psig. An oil of 0.85 specific gravity is lifted with a starting slug length of 1,600 ft. Select the proper port size for the pilot section of the valve and calculate its setting pressure.

Solution

Casing pressure at valve depth is found from the surface injection pressure, which is also the valve's opening pressure. From Fig. B–1 in Appendix B, a gas gradient of 33.5 psi/1,000 ft is read and the opening pressure is

p_{io} = 900 + 33.5 7,500 / 1,000 = 1,151 psig.

Tubing pressure at valve depth equals the hydrostatic pressure of the oil column:

p_p = 1,600 0.85 .433 = 589 psig.

The closing casing pressure is found from the definition of spread:

p_{ic} = 1,151 − 100 = 1,051 psig, and this is also the valve's dome charge pressure at valve depth.

From Equation 3.56, with p_{sp} = 0, since the valve is only dome charged, PPEF can be found:

PPEF = 110 / (1,051 − 589) = 0.24.

From manufacturer data a pilot port of ¼ in. is selected with a PPEF = 0.23 and A_v = 0.058 sq. in.

To find the TRO pressure, dome charge pressure at 60 °F from **Fig. D–1** in **Appendix D** at the given temperature is read as 930 psig. Using Equation 3.55 with p_{sp} = 0 and R = 0.058 / 0.31 = 0.187 we get:

TRO = 930 / (1 − 0.187) = 1,144 psig.

Example 3–14. A similar installation to the previous example has a McMurry-Macco RPV-2S pilot-operated valve without a gas charge, with A_b = 0.31, run to 8,500 ft. Injection pressure at the surface is 800 psig, gas relative density is 0.8. The required pressure reduction in the annulus is 85 psig. Starting slug length is 1,200 ft, oil specific gravity equals 0.75. Select the proper pilot port size and calculate valve setting pressure.

Solution

Gas gradient in the annulus from Fig. B–1 in Appendix B is 29.5 psi/1,000 ft. The opening casing pressure of the valve is thus:

$$p_{io} = 800 + 29.5 \ 8,500 \ / \ 1,000 = 1,051 \ \text{psig.}$$

The hydrostatic pressure of the starting oil slug equals the tubing pressure at valve depth:

$$p_t = 1,200 \ 0.75 \ .433 = 390 \ \text{psig.}$$

Solving the opening equation (Equation 3.52) for spring force effect p_{sp} and substituting this into Equation 3.56, while setting dome charge pressure to zero, the following second-order equation is received:

$$R^2 \ p_p - R(p_{io} - p_p) + \textbf{SPREAD} = 0$$

Upon substitution of the known variables and solving the previous equation, the value of $R = 0.14$ is found. For the given bellows area, this results in a port area of $A_v = 0.043$ sq in. The closest available port size is 3/16 in. with $A_v = 0.029$, $R = 0.165$, and PPEF = 0.198. This valve is selected and the spring force effect is found from the opening equation (Equation 3.52 with $p_d = 0$) as:

$$p_{sp} = 1,051 - 0.198 \ 390 = 974 \ \text{psig.}$$

Using this value, the valve's actual spread is checked with Equation 3.56:

Spread = 0.198 [974 (1 – 0.165) –390] = 84 psig, which is close to the required 85 psig.

Finally, the valve setting pressure is equal to the spring force effect since the bellows is not charged with gas:

TRO = 974 psig.

Valves with Pilot Port

Although not strictly a pilot-operated valve, but the valve given in Figure 3–41 ensures the same operation as detailed previously. It contains a double valve stem tip and seat; the main seat is usually non-replaceable and can have port sizes up to ¾ in., whereas the area of the control seat A_c is selected according to the desired valve spread. This solution ensures that valve spread and gas throughput area are independent of each other.

When the pressures at valve setting depth approach opening conditions, production pressure acting on the control seat area A_c along with the effect of injection pressure acting on the bellows area start to lift the valve stem. As the stem moves upward, it compresses the light spring contained in the outer stem tip, the spring force then lifts the outer stem tip off the main seat, opening the full cross section of the main port to gas injection. As soon as it is in the open position, the small spring in the double stem tip expands and moves the outer stem tip to its original position on the valve stem. The valve will close at its bellows charge pressure, like any unbalanced valve. This valve, therefore, provides a sufficiently great gas throughput area required for intermittent lifting but allows the designer to select the right amount of valve spread. An added advantage is that it is *snap acting*.

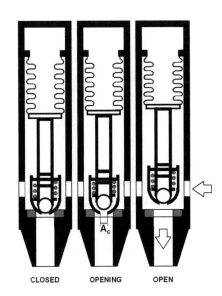

Fig. 3–41 Unbalanced valve with special pilot port for IPO service.

3.2.2.5.2 *Combination valves.* Several gas lift valves containing pilot valves are mainly or completely sensitive to production (tubing) pressure. They can be called *fluid operated* valves and are often designated as *combination valves.*

Fluid Differential Valves

The valve shown in Figure 3–42 is identical to an unbalanced dual element gas lift valve, with the only difference that the bellows is not attached to the valve stem. The bellows assembly works as a pilot and the valve stem with the compression spring as the main valve. The valve stem with the spring is thus converted into a fluid differential valve, which opens and closes, depending on the difference between injection and production pressures.

In the closed position, injection pressure is not sufficient to open the pilot, which keeps the main valve closed. If the bellows were connected to the valve stem, the valve would open at an injection pressure greater than dome charge pressure, just like any unbalanced valve. In this case, however, the bellows will move off the stem before injection pressure could open the valve. The result is the *cocked* position when the pilot is open because injection pressure is higher than its dome charge pressure. This feature of the valve prevents gas injection in situations when injection pressure is lower than designed.

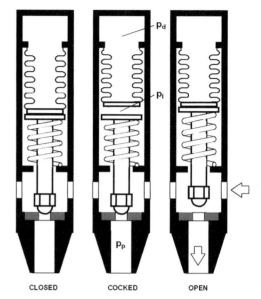

CLOSED   COCKED   OPEN

Fig. 3–42 Fluid differential pilot valve.

Now the main valve, consisting of the valve stem and spring, is able to operate according to the balance of the forces acting on it. The force keeping it closed comes from the injection pressure; opening forces include the effects of production pressure and the spring force. If spring force is substituted with an equivalent pressure that is assumed to act on valve seat area A_v, and considering that the opening and closing pressures both act on the valve seat, we get:

$$p_i A_v = p_p A_v + p_{sp} A_v \qquad\qquad 3.57$$

Solving this equation for the opening production pressure:

$$p_{po} = p_i - p_{sp} \qquad\qquad 3.58$$

where: p_{po} = opening production pressure at valve setting depth, psi

   p_i = injection pressure at valve setting depth, psi

   p_{sp} = spring force effect, psi

The valve's opening equation (Equation 3.58) shows that the valve requires a constant difference, equal to the spring force effect, between the injection and the production pressures to open. Since this difference is provided by a spring force, it is not affected by temperature, a great advantage in installation design. In an intermittent installation, the opening production pressure equals the fluid load (hydrostatic pressure) against the valve. Therefore, starting slug length is adjusted by selecting the right spring for the given injection pressure at valve setting depth. Generally, spring pressures of 100–400 psi are used.

In the open position, injection pressure is assumed to drop to production pressure across the valve seat, like in an unbalanced valve. Therefore, injection pressure will act on the full bellows area, providing the force to keep the valve open. The closing conditions are reached when the pilot forces the main valve to close. Closing force comes from the dome charge pressure, opening forces from injection pressure and the spring. The force balance at the instant the valve closes is the following:

$$p_d A_b = p_{ic} A_b + p_{sp} A_v \qquad\qquad 3.59$$

From the previous equation, the valve's closing injection pressure is constant as follows:

$$p_{ic} = p_d - Rp_{sp} \qquad\qquad 3.60$$

The spread of the valve is the difference between opening and closing injection pressures and is derived as:

$$\textbf{SPREAD} = p_{sp}(1-R) - p_d + p_p \qquad\qquad 3.61$$

where: p_d = dome charge pressure at valve setting depth, psi

$R = A_v/A_b$ = ratio of valve seat area to bellows area

p_p = production pressure at valve setting depth, psi

It is interesting to note that the spread of this valve increases with production pressure. If starting slug length is increased, the valve automatically increases its spread and injects more gas below the liquid lifted. This operation is more desirable than the behavior of the IPO pilot valve (see Section 3.2.2.5.1) that exhibits smaller spreads with increased tubing loads. This is exactly the opposite of the required operation because greater slug lengths, of course, need more injection gas to efficiently lift them to the surface.

This valve is usually tested in a special valve tester, where its closing pressure is determined when injection and production pressures are equal. The test rack closing pressure is given by the following formula:

$$TRC = p_d' - R\,p_{sp} \qquad\qquad 3.62$$

where: p_d' = dome charge pressure at 60 °F, psi.

Figure 3–43 shows the performance of the fluid differential pilot valve at valve setting depth in a graphical presentation. The opening conditions are depicted with a bold line, below and parallel to the $p_p = p_i$ line, and closing is at a constant injection pressure. In intermittent lift applications, the valve opens when production pressure (starting slug pressure) plus spring pressure effect exceed the injection pressure. As the liquid slug travels toward the wellhead, injection and production pressures at valve depth are equalized and start to bleed down to the flow tube. As these pressures simultaneously decrease, the valve will close as soon as its closing injection pressure is reached. In a continuous flow installation, tubing pressure is relatively constant and higher than the opening pressure, so the valve stays open injecting gas continuously.

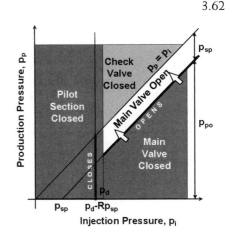

Fig. 3–43 Opening and closing performance of the fluid differential pilot valve.

Fluid Pilot Valves

The *fluid pilot* valve shown in Figure 3–44 has features very similar to the previous valve type. Its pilot section is an unbalanced, bellows-charged valve, and the main section is a differential valve with a spring. The valve stem of the main section has a cross-sectional area equal to the valve seat area. As seen, the top of the stem is sealed against production pressure, allowing injection pressure to act on it at all times. The compression spring of the main valve keeps the valve in an open position, making the valve a normal open one. Pilot port area A_v and main port area A_m are different. In valves designed for intermittent gas lift operation, main port area is smaller than the area of the pilot port; in continuous flow valves this ratio is reversed. For the following discussion assume an intermittent valve and consequently $A_m < A_v$.

In the closed position, the pilot section behaves like an unbalanced gas lift valve and will open when injection pressure reaches its opening pressure. The main section is still closed because the relatively high injection pressure, acting on top of its valve stem forces it to be seated. Although the main valve is closed, tubing pressure from below can still reach to the port of the pilot because of the small space involved and the imminent leaking of the seats.

As soon as the pilot section opens, the valve assumes a *cocked* position with the main valve ready to operate. Injection pressure now reaches below the pilot port, but flow cannot start because of the main stem still resting on its seat. The main valve will open when the force keeping it closed (injection pressure) is equal to the forces trying to open it, which come from the tubing and spring forces. A balance of the forces at the instant the main valve starts to open is the following:

$$p_i A_m = p_p A_m + p_{sp} A_m \qquad 3.63$$

Opening production pressure for the main valve is found from this equation as:

$$p_{po} = p_i - p_{sp} \qquad 3.64$$

where: A_m = port area of the main valve, sq in.

p_{po} = opening production pressure at valve setting depth, psi

p_i = injection pressure at valve setting depth, psi

p_{sp} = spring force effect, psi

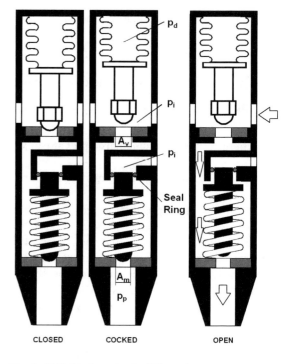

Fig. 3–44 Fluid pilot valve for IPO service.

The opening equation is the same as that of the fluid differential valve (Equation 3.58) and shows that this valve also keeps a constant difference between the injection and production pressures. It fully opens when production pressure combined with the spring effect can lift the valve stem against the injection pressure. The differential between injection and production pressures is set with the spring force that is unaffected by valve temperature making valve setting calculations easier. In intermittent operations, starting slug length above an operating valve is selected at will by using the right spring force. Usual spring force effects are 150–250 psi, but lighter tubing loads require stronger springs.

While the valve is open, injection pressure exists upstream of the pilot port, keeping the pilot fully open. This is ensured by the relation of the port areas, the main port being smaller than the pilot port. In intermittent lift, after the start of gas injection, production and injection pressures at valve depth are very close to each other, and they gradually decrease as the liquid slug surfaces. As soon as injection pressure drops below the pilot's dome charge pressure, the pilot closes and stops gas injection. The force balance at the instant the valve closes:

$$p_d A_b = p_{ic} A_b \qquad 3.65$$

Hence, the valve closing pressure is given as follows:

$$p_{ic} = p_d \qquad 3.66$$

where: p_d = dome charge pressure at valve setting depth, psi

Therefore, this valve is opened by production pressure when the differential between injection and production pressures becomes less than the set spring effect. Its closing, however, occurs at a significant drop in injection or production pressures. This operation provides a very efficient control over the amount of gas used in an intermittent cycle.

The difference between the opening and closing injection pressures is the spread of the valve:

$$\text{SPREAD} = p_{sp} - p_d + p_p \qquad\qquad 3.67$$

where: p_d = dome charge pressure at valve setting depth, psi

   p_{sp} = spring force effect, psi

   p_p = production pressure at valve setting depth, psi

Just like with the previous valve type discussed, valve spread increases with production pressure. This valve, then, injects more gas if starting slug length is increased, an advantageous feature. The IPO pilot valve, in contrast, has smaller spreads with increasing slug lengths, *i.e.* production pressures. This difference in operation becomes important if surface injection pressure fluctuates at the wellsite. The common pilot valve (see Section 3.2.2.5.1) will open for an increase in injection pressure when a sufficient liquid slug length is not yet available, and is shut down by insufficient injection pressure. The fluid pilot valve, on the other hand, works independently of injection pressure and always opens when the right slug length has accumulated above it.

The TRO pressure of the pilot section is identical to that of the unbalanced single element valve and is found from:

$$\text{TRO} = \frac{p_d'}{1 - R} \qquad\qquad 3.68$$

where: p_d' = dome charge pressure at 60 °F, psi

A graphical presentation of the performance of the fluid pilot valve at valve setting depth is given in Figure 3–45 where opening and closing conditions are shown in bold lines. The pilot section opens at pressures greater than its dome charge pressure, and the main valve can only open after the pilot. At any injection pressure, the gas lift valve opens when the sum of production pressure p_{po} and spring effect p_{sp} is greater than injection pressure. Closing of the fluid pilot valve occurs at the dome charge pressure.

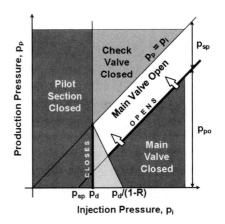

Fig. 3–45 Opening and closing performance of the fluid pilot valve.

The previous discussions were valid for valves designed for intermittent lift. The same valve, however, can be used in continuous flow installations as well. Continuous flow valves have main port areas A_m greater than pilot port area A_p and usually are equipped with tapered stems in the pilot section. The modifications make the pilot sensitive to production pressure and production pressure maintained upstream of the pilot port. Because of this, the main valve stem will always move according to the actual differential between the injection and production pressures. Production pressure in continuous flow installations being relatively high and constant, the valve will stay open and inject gas continuously.

Example 3–15. The operating valve in an intermittent installation is set to 5,000 ft, where the temperature is 140 °F. Surface injection pressure is 700 psig, gas specific gravity 0.6. Use a fluid pilot valve with $A_b = 0.77$ sq in., and a starting slug length of 900 ft of 0.9 gravity oil. Separator pressure is 100 psig, and the annulus pressure reduction required to supply the right amount of gas per cycle is 100 psi. Select the spring effect for the main valve, then calculate the required pilot port size and valve setting data.

Solution

The fluid load on the valve is found from the separator pressure and the hydrostatic head of the accumulated liquid column. This will give the opening tubing (production) pressure:

$$p_{po} = 100 + 900 \cdot 0.9 \cdot 0.433 = 451 \text{ psig.}$$

The main valve should open at the injection pressure at depth, where the gas gradient is found as 16 psig/1,000 ft from Fig. B–1 in Appendix B:

$$p_i = 700 + 16 \cdot 5{,}000 / 1{,}000 = 780 \text{ psig.}$$

The spring force effect is calculated from the main valve's opening equation (Equation 3.64) as follows:

$$p_{sp} = p_i - p_{po} = 780 - 451 = 330 \text{ psig.}$$

Out of the 100 psig required pressure drawdown in the annulus, let 50 psig be the *over-build* pressure, *i.e.* the difference between the opening pressure of the pilot and the main valves. Thus, the pilot will open at:

$$p_{io} = 780 - 50 = 730 \text{ psig.}$$

The rest (50 psig) is the pilot's spread, and the pilot's closing pressure will be:

$$p_{ic} = 730 - 50 = 680 \text{ psig, which equals the dome charge pressure at valve depth.}$$

At this point, the pilot's opening casing, tubing pressures, and dome charge pressure are known, and the required R value can be determined from the opening equation of an unbalanced valve (Equation 3.17):

$$R = (p_{io} - p_d) / (p_{io} - p_p) = (730 - 680) / (730 - 451) = 0.179.$$

From manufacturer data, a pilot port of 7/16 in. is taken with $R = 0.195$.

Dome charge pressure at 60 °F is found from Figure 3–5 as $p'_d = 580$ psig.

TRO pressure is evaluated from Equation 3.21:

$$\text{TRO} = 580 / (1 - 0.195) = 720 \text{ psig.}$$

A check can be made to find the pilot's actual spread for the selected port size by using Equation 3.25:

$$\text{Spread} = 0.195 / (1 - 0.195) [680 - 451] = 55.5 \text{ psig,}$$

which is close to the assumed value.

Flexible Sleeve Pilot Valves

The flexible sleeve pilot valve is a concentric valve whose pilot is a balanced concentric valve, its main valve being a spring-loaded fluid differential valve. The valve shown schematically in Figure 3–46 is run and retrieved with the tubing string. The pilot section works exactly like the balanced flexible sleeve valve (see Section 3.2.2.4) and opens when injection pressure exceeds the dome charge pressure. This allows injection gas to enter the valve but in this *cocked* position, no gas injection takes place if production

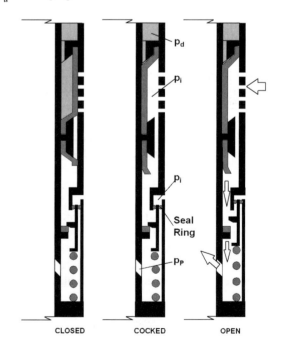

CLOSED COCKED OPEN

Fig. 3–46 Flexible sleeve pilot valve for IPO service.

pressure is still low, like in intermittent installations. The main valve now takes control and will open when the differential between injection and production pressures is less than spring force effect. Springs are replaceable and usually have equivalent pressures of 150–350 psi. In the open position, injection gas enters through the gas inlet ports and flows inside the valve past the main valve into the flow conduit.

As seen, the operation of the flexible sleeve valve is identical with that of the bellows type fluid pilot valve. The only difference is that the pilot now is a balanced valve, whereas in the bellows type valve, it is an unbalanced one. The force balance equations are written identically to the bellows valve and the opening and closing equations are reproduced here:

$$p_{po} = p_i - p_{sp} \qquad\qquad 3.69$$

$$p_{ic} = p_d \qquad\qquad 3.70$$

As shown in Figure 3–47, opening and closing conditions are very similar to those of the bellows-type pilot valve. Valve spread increases with production pressure, too:

$$\mathbf{SPREAD} = p_{sp} - p_d + p_p \qquad\qquad 3.71$$

where: p_{po} = opening production pressure at valve setting depth, psi

           p_{ic} = closing injection pressure at valve setting depth, psi

           p_i = injection pressure at valve setting depth, psi

           p_d = dome charge pressure at valve setting depth, psi

           p_{sp} = spring force effect, psi

           p_p = production pressure at valve setting depth, psi

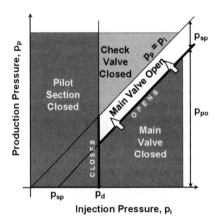

Fig. 3–47 Opening and closing performance of the flexible sleeve pilot valve.

The valve can be used in intermittent as well in continuous flow installations. All advantages are identical with those described previously for the bellows-type fluid pilot valve.

3.2.3 Dynamic performance of gas lift valves

3.2.3.1 Introduction. As already discussed, gas lift valves are used for both well unloading and normal operation. The valves in an unloading valve string are designed to accomplish a stepwise process of displacing well fluids from the annulus and the tubing string. During unloading, gas injection should occur at each successively lower valve, in order to reach the operating depth. Injection of gas from the annulus to the tubing string at an upper valve decreases the tubing pressure at this and at the next lower valve's level. The lower valve can only inject gas if tubing pressure is reduced below the casing pressure valid at its setting depth. It should be obvious that the more gas is injected by the upper valve, the less the tubing pressure will be at the lower valve level. If the gas rate passed by the upper valve is insufficient to achieve the right amount of tubing pressure drop, the lower valve won't be able to inject gas and the unloading process stops.

The injection capacity of an operating valve is also important. Changes in injection and/or production pressures require changes in injection rates for the well to operate efficiently. The greater the well rate, the more important it is to properly design and maintain the right injection volumes. This is why in high-volume offshore producers, small errors in valve capacity calculations may result in great changes in well rates.

Although gas passage characteristics of gas lift valves are of paramount importance, the gas lift industry did not investigate the problem in depth for many years. In the past, the performance of gas lift valves was described by the static force balance equations detailed in Section 3.2.2 and the assumption that an open valve behaves like a square-edged orifice. The valve was assumed to quickly and fully open as soon as the injection and production pressures satisfied the opening conditions. When fully open, the gas flow rate through the valve was computed from the port size and the equation originally proposed by the Thornhill-Craver Co. (see Equation 2.65) This equation was developed for flow of ideal gases through fixed orifices with a constant cross-sectional area open to flow. Gas lift valves, however, do not provide a constant flow area because the valve stem rarely lifts completely off the seat. The position of the stem, in relation to the valve seat is a function of the pressure conditions and the valve, therefore, acts as a variable orifice.

In one of the first papers [18] on the previous problem, the authors state that the gas lift valve investigated (a CAMCO R-20) behaved as a variable-orifice Venturi device and not as a simple orifice. A Venturi device is a converging-diverging nozzle in which the minimum flowing pressure occurs at its throat and there is a considerable pressure recovery downstream of the throat. The cross-sectional area of the throat changes with the position of the valve stem which, in turn, varies with changes in the injection and/or production pressure. In conclusion, the pressure acting on the tip of the valve stem never equals injection pressure, as suggested from the static force balance equations. The unbalanced valve, therefore, does not have a constant closing pressure.

The first practical investigation on the dynamic performance of gas-charged bellows valves was done by L.A. Decker [12] who laid the foundation for present-day valve testing procedures. He derived the formulae to describe the behavior of the bellows assembly and introduced the concept of the bellows load rate. His analytical model allows the determination of the valve stem position as a function of the mean effective pressure acting on the bellows area.

In order to improve gas passage calculations, Winkler and Camp [19] experimentally derived discharge coefficients for a given unbalanced valve and modified the Thornhill-Craver formula (see Equation 2.65) to include the gas deviation factor. Their experimental work was limited to the throttling portion of valve performance and the approach adapted (the discharge coefficient principle) was later proved to be misleading.

3.2.3.2 Early models. In Section 3.2.2.3.1 the spring-loaded valve without bellows charge was shown to have special features markedly different from those of other gas lift valves. In normal operation, this valve never opens fully, and the pressure upstream of the port is very close to production pressure both in the open and in the closed positions. It was clear from their inception that they could not be treated as other valves because they never opened the full port area. Therefore, this *throttling valve* required special considerations and great efforts to fully describe its performance under different well conditions. The original manufacturer, MERLA Tool Corp., realized the need for flow capacity data and performed hundreds of flow tests since the early 1960s. [17, 20]

The typical performance of the throttling valve is illustrated in Figure 3–48, where measured gas throughput capacities of an LM-12 valve are compared to those of a fixed choke having the same port size. In the usual applications, injection pressure is kept constant, and gas injection rate varies with production pressure. When production and injection pressures are equal (800 psig), though it is open, the valve does not pass gas since there is no pressure differential across its seat. At decreasing production pressures, the gas rate increases rapidly up to a maximum value, than declines and diminishes at the valve closing pressure, p_{pc}. The linear portion of the curve is the throttling range of the valve where gas injection rate varies linearly with production pressure. Along this line, gas rate decreases in spite of the increase in differential pressure caused by decreasing production pressures. This controversy is explained by an investigation of the valve stem position which, in turn, is determined by the pressure below the valve stem tip. Since the valve stem tip always restricts gas flow, this pressure is close to production pressure, and the area open to gas flow is proportional to production pressure. The linear relationship between valve stem

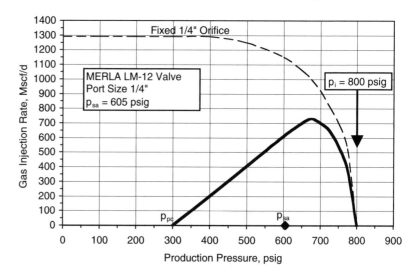

Fig. 3–48 Comparison of the gas injection rates through a fixed orifice and a throttling gas lift valve.

position and production pressure is ensured by the loading element of the valve, *i.e.* the spring, which has a constant spring rate. The net result of these effects is the observed linear decrease of gas injection rate with the decrease in production pressure.

The throttling valve's gas passage characteristics are different from the behavior of a fixed orifice as shown in Figure 3–48. At lower production pressures, flow through the orifice reaches critical conditions, and there is no control over the gas volume injected in the production string. The gas lift valve, in contrast, allows an ideal control of the gas throughput capacity. In continuous flow operations, the throttling portion of the performance curve permits the maintenance of a constant flow gradient in the production string, a basic requirement for an efficient lifting of well fluids. This valve then properly responds to fluctuations in well inflow conditions and is therefore often called a *proportional response* valve.

The results of several hundreds of measurements [20] were published for each throttling valve [21] similar to the example given in Figure 3–49 where basic data for the valves LM-16 and LM-16R are listed along with performance parameters discussed following. The parameter F_e represents the dynamic value of the A_v/A_b ratio, which is calculated from experimental data by solving the valve opening/closing equation (Equation 3.45) for R and substituting $F_e = R$:

$$F_e = \frac{p_i - p_{sa}}{p_i - p_{pc}} \qquad\qquad 3.72$$

The closing production pressure p_{pc} is also expressed from the opening/closing equation where F_e is used instead of R:

$$p_{pc} = p_i - \frac{p_i - p_{sa}}{F_e} \qquad\qquad 3.73$$

where: p_i = injection pressure at valve setting depth, psi

                p_{sa} = spring adjustment pressure, psi

                p_{pc} = closing production pressure at valve setting depth, psi

In the throttling region, where such valves normally operate, calculation of the gas capacity is based on the linear relationship given as follows:

$$q_{sc} = M(p_p - p_{pc}) \qquad\qquad 3.74$$

where: q_{sc} = injection gas rate at standard conditions, Mscf/d

            M = slope of throttling line, found from diagram, Mscf/d/psi

            p_p = production pressure at valve setting depth, psi

Since the gas flow rate cannot reach values higher than a maximum (see Fig. 3–48), values calculated from the previous formula must be limited. The maximum possible rate is found from:

$$q_{max} = K M(p_i - p_{pc}) \qquad\qquad 3.75$$

where: K = correction value read from diagram

– · –

Example 3–16. Calculate the gas injection capacity of an LM-16 valve with a 0.34 in. port if the spring adjustment pressure is p_{sa} = 900 psig, the tubing pressure at valve depth is 800 psig. The injection pressure at valve setting depth is 1,050 psig.

VALVE PERFORMANCE DATA

Valve Type: LM-16 and LM-16R
Bellows Area: $A_b = 0.23$ sq in

Port Size, in	0.25	0.34
Ball Size, in	3/8	1/2
F_e, -	0.25	0.50
K, -	0.68	See K graph

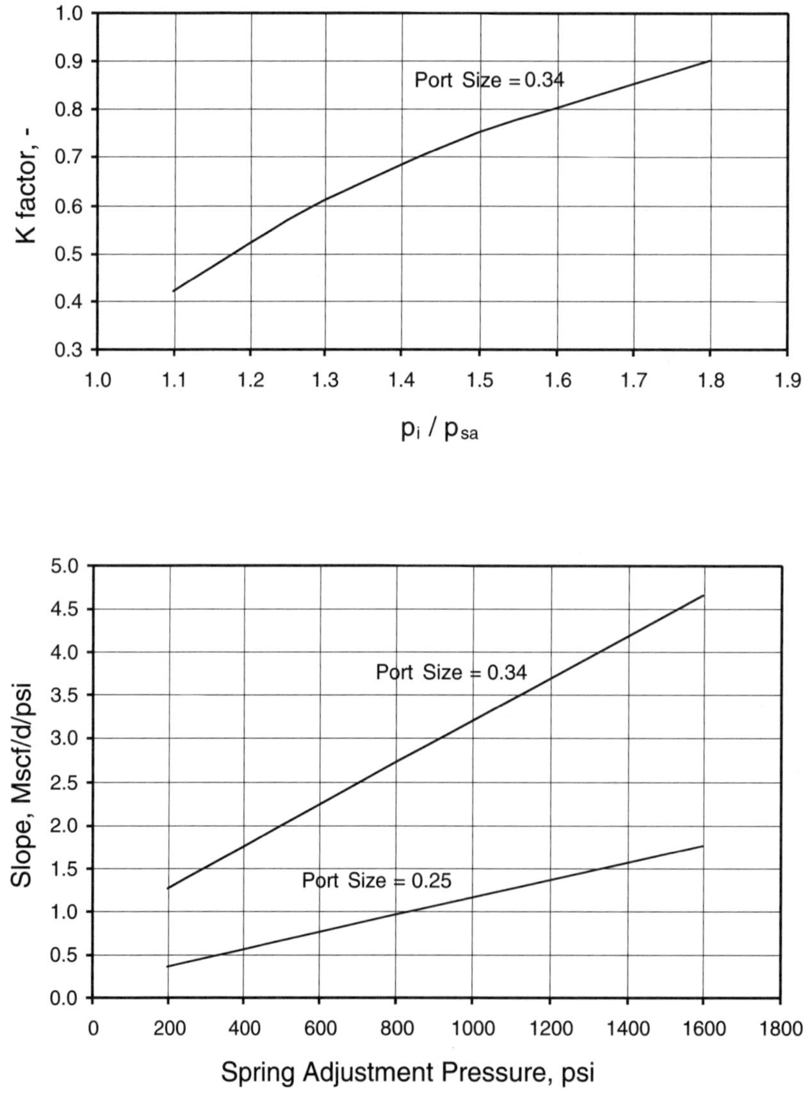

Fig. 3.49
Example valve performance data for a throttling gas lift valve. [21]

Fig. 3–49 Example valve performance data for a throttling gas lift valve. [21]

Solution

The closing production pressure for the given conditions is found from Equation 3.73, by using the value $F_e = 0.5$ read from Figure 3–49:

$$p_{pc} = 1{,}050 - (1{,}050 - 900) / 0.5 = 750 \text{ psig.}$$

Injection rate through the valve is calculated from Equation 3.74 with $M = 2.9$ Mscf/d/psi read from Figure 3–49:

$$q_{sc} = 2.9 \, (\, 800 - 750 \,) = 145 \text{ Mscf/d.}$$

In order to check the calculated rate, K must be found for:

$$p_i/p_{sa} = 1{,}050 / 900 = 1.17 \text{ from Figure 3–49 as } K = 0.5.$$

The maximum possible rate from Equation 3.75:

$$q_{max} = 0.5 \; 2.9 \, (\, 1{,}050 - 750 \,) = 435 \text{ Mscf/d, which is greater than the actual, the calculated rate is valid.}$$

3.2.3.3 General models. Research into the dynamic behavior of gas lift valves received a new impetus in the 1980s, after the founding of TUALP (Tulsa University Artificial Lift Projects), an industry-sponsored research consortium. Several papers were published [22–25], covering the main results of many years' experimental work. The approach adopted by TUALP and most of the present-day investigations can be summed as up as follows:

- for each valve a large database is set up by measuring all significant flow parameters for a variety of closely controlled test conditions

- from the measured gas rates and other relevant parameters, flow coefficients are calculated for each case

- using statistical methods, flow coefficients and other pertinent performance data are correlated

Nieberding et al. [22] were the first to point out that gas passage through gas lift valves can occur under two different flow patterns: orifice and throttling flow. Orifice flow is very similar to gas flow through a fixed choke, whereas throttling flow resembles flow through a variable area Venturi device. Figure 3–50 shows schematic performance curves in the gas rate vs. production pressure coordinate system for these two flow patterns. The orifice flow model can be divided into two regions: subcritical and critical. For a constant injection pressure p_i decreasing production pressures entail an increase of the gas rate until critical conditions are reached at $p_p = p_{p \, cr}$. Gas flow velocity across the valve port now equals the velocity of sound, and injection rate remains at its critical value despite any

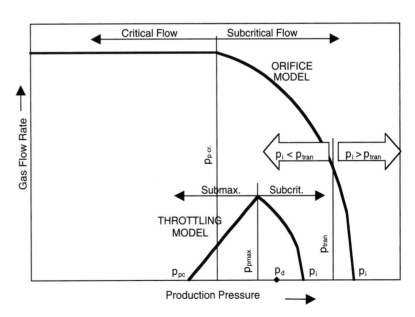

Fig. 3–50 Schematic gas lift valve performance curves for the orifice and the throttling flow patterns.

further decrease in tubing pressure. This flow pattern occurs with the valve stem at its maximum travel when the valve port behaves as a fixed orifice (see Section 2.4.5).

For injection pressures less than a definite transition pressure p_{tran}, the valve behaves differently and is in the throttling flow pattern. When production pressure decreases from $p_p = p_i$, the gas rate increases from zero, due to an increasing pressure differential across the valve seat. After reaching a maximum value, injection rate linearly decreases with production pressure until gas flow ceases at the closing production pressure p_{pc}. This kind of behavior was discussed in detail previously. The throttling type of performance curve is also divided into two regions by the production pressure belonging to the maximum flow rate and denoted by p_{pmax}. The subcritical region is similar to the orifice flow and the sub-maximal region is the usual application range of the valve.

Discussion of the calculation models of Nieberding et al. [22] and others [23–25] seems to be superfluous since the API RP 11V2, to be detailed following, adopted most of their results.

3.2.3.3.1 The API RP 11V2 procedure. The publication API Recommended Practice for Gas-Lift Valve Performance Testing (API RP 11V2) [13] is the product of a long process and the contribution of many individuals and institutions. It covers recommendations on general measurement requirements, valve performance test procedures, and calculation models for the determination of gas injection rates. Adherence to these prescriptions ensures that valves of different makes are investigated identically, and their performance under identical conditions can be compared. In the following, the recommended model for calculating gas injection rates through gas lift valves is detailed.

As described in connection with Figure 3–50, the boundary between orifice and throttling flow patterns is given by the transition injection pressure p_{tran}, for a specific port size and valve closing pressure. The value of the transition pressure is calculated from the valve closing pressure and the dynamic A_p/A_b ratio denoted by F_e and defined by Equation 3.72 as follows:

$$p_{tran} = p_d(1 + F_e)$$
3.76

where: p_d = valve closing (dome) pressure at valve setting depth, psi

F_e = dynamic A_p/A_b ratio

The API RP 11V2 recommends the use of the previous formula, but some investigators use the next formula:

$$p_{tran} = \frac{p_d}{(1-R)}$$
3.77

where: R = the ratio of port and bellows areas

The Orifice Flow Model

The orifice model is to be used if actual injection pressure p_i is greater than the transition pressure defined by Equation 3.76. The calculation of the gas injection rate depends on the condition whether actual production pressure p_p is below or above the critical value. The critical pressure ratio $(p_2/p_1)_{cr}$ is found from the flow performance tests conducted on the given valve and is usually constant for a given port size. The critical production pressure is found from:

$$p_{pcr} = p_i \left(\frac{p_2}{p_1}\right)_{cr}$$
3.78

For production pressures greater or equal than the critical value calculated from the previous formula, gas flow is subcritical. The gas throughput capacity of the valve is calculated from a modified flow meter formula including the effects of gas expansion, non-ideal gas deviation factor, and discharge coefficient. Note that the Thornhill-Craver formula (Equation 2.65) is a very simplified version of the flow meter formula given as follows:

$$q_{sc} = 1241 \, A_v \, C_d \, Y \sqrt{\frac{p_i \, (p_i - p_p)}{T_v \, Z_v \, \gamma_g}}$$
3.79

where: q_{sc} = gas flow rate at standard conditions, Mscf/d

A_v = port area, sq in.

C_d = discharge coefficient including the ratio of areas

Y = expansion factor

p_i = injection pressure at valve setting depth, psia

p_p = production pressure at valve setting depth, psia

T_v = valve temperature R

Z_v = gas deviation factor at valve setting depth

γ_g = specific gravity of injected gas

It was shown by several investigators that the term $C_d Y$ is a linear function of the pressure ratio defined as follows. Performing a linear regression analysis on the flow test data, the coefficients of this function are found for the given valve and the given port size. Therefore, the $C_d Y$ term is calculated from the formula:

$$C_d Y = a \, \frac{p_i - p_p}{p_i \, \kappa} + c$$
3.80

where: p_i = injection pressure at valve setting depth, psia

p_p = production pressure at valve setting depth, psia

κ = ratio of specific heats for the gas, -

a, c = coefficients determined for the given valve and port size

If gas flow is at critical conditions, *i.e.* for production pressures below the critical value found from Equation 3.78, the injection rate is constant and equals the value calculated from the flow meter formula (Equation 3.79) for $p_p = p_{p \, cr}$.

The Throttling Flow Model

Modeling of throttling flow is based on a normalization of the measured injection rate vs. production pressure curves [22, 24]. The original curve (see Fig. 3–50) is re-plotted in a coordinate system q/q_{max} vs. a dimensionless pressure defined as follows:

$$N = \frac{p_p - p_{pc}}{p_i - p_{pc}}$$
3.81

where: p_i = injection pressure at valve setting depth, psia

p_p = production pressure at valve setting depth, psia

p_{pc} = production closing pressure at valve setting depth, psia

After this transformation, measurement data are easier to fit with correlating functions. The two important correlations developed for use in gas rate calculations are

(a) the dimensionless pressures belonging to the maximum flow rate

(b) the slope of the throttling portion of the curve.

These correlating functions are developed for each port diameter of the tested valve by regression analysis of the measurement data.

The calculation for the throttling model starts with finding where the production pressure falls: in the sub-maximal or in the subcritical range. As seen in Figure 3–50, the boundary of those is the production pressure belonging to the maximum rate, denoted by p_{pmax}. To find this value, closing production pressure has to be found from a solution of the static opening equation of an unbalanced valve (see Equation 3.73) where the experimentally measured F_e is used instead of R:

$$p_{pc} = p_i \frac{p_i - p_d}{F_e}$$

3.82

Now, using the regression results for the maximum value of dimensionless pressure N_{max} the production pressure where maximum gas flow rate occurs is found:

$$p_{pmax} = p_{pc} + N_{max}(p_i - p_{pc})$$

3.83

where: p_i = injection pressure at valve setting depth, psia

p_{pc} = production closing pressure at valve setting depth, psia

N_{max} = dimensionless pressure at maximum flow rate

Comparing p_{pmax} with the actual production pressure p_p, the actual range in the throttling flow pattern can be found. For production pressures between the closing and maximum values, flow is in the sub-maximal range, for other cases it is in the subcritical range:

IF $p_{pc} < p_p < p_{pmax}$ Flow is sub - maximal

IF $p_p > p_{pmax}$ Flow is sub - critical   3.84

If flow is in the sub-maximal range, the slope of the injection rate vs. production pressure function is found from regression data valid for the given port size of the given valve:

$$\text{slope} = m\, p_d + b$$

3.85

where: p_d = closing (dome) pressure at valve setting depth, psig

m, b = experimentally determined coefficients

The gas injection rate at measuring conditions is found from the slope and the production pressure as follows:

$$q_{scm} = \text{slope}\,(p_p - p_{pc})$$

3.86

Since the experiments are usually performed with air as a medium at shop temperature, all the regression coefficients and the calculated injection rate are valid at test conditions only. The calculated gas rates must therefore be corrected for temperature, gas density, and gas deviation factor valid at valve setting depth:

$$q_{sc} = q_{scm} \sqrt{\frac{T_m \, Z_m \, \gamma_m}{T_v \, Z_v \, \gamma_g}}$$ 3.87

where: q_{scm} = calculated gas flow rate at test conditions, Mscf/d

T_m = test temperature R

Z_m = deviation factor of the test gas at test temperature

γ_m = specific gravity of test gas

T_v = valve temperature at valve setting depth, R

Z_v = gas deviation factor at valve setting depth

γ_g = specific gravity of injected gas

The deviation factors of air can be calculated from the correlation developed by the present author, valid in the ranges $T = 0$ - 300 °F and $p = 0 - 2{,}000$ psia:

$$Z_a = 1.0 - 3.926 \; 10^{-6} \, T + B \, p + C \, p^2$$ 3.88

where: $B = -9.545 \; 10^{-10} \, T^2 + 5.434 \; 10^{-7} \, T - 6.099 \; 10^{-5}$

$C = 1.515 \; 10^{-13} \, T^2 - 9.458 \; 10^{-11} \, T + 1.847 \; 10^{-8}$

T = air temperature, F

p = air pressure, psia

For $p_p > p_{pmax}$, production pressures, flow is in the subcritical range of the throttling model. Since the shape of this range is similar to the subcritical range of the orifice model (see Fig. 3–50), a corrected orifice equation is used to find the gas rate. The steps of the calculation procedure are listed here.

1.   Assume sub-maximal conditions in the throttling flow pattern and calculate accordingly the gas rate with $p_p = p_{pmax}$.

2.   Assume orifice flow conditions to prevail and find the gas rate for $p_p = p_{pmax}$.

3.   Calculate the correction factor defined as follows:

$$C = \frac{q_{throttling}}{q_{orifice}}$$ 3.89

4.   Calculate the group $C_d \, Y$ from Equation 3.80.

5.   Find the gas flow rate from Equation 3.79.

6.   Apply the following correction to the value calculated in Step 5 find the actual gas flow rate through the gas lift valve:

$$q_{sc} = C \, q_{sc} \text{ (Step 5)}$$ 3.90

As seen from the detailed calculation procedure, application of the API RP 11V2 model necessitates the knowledge of several regression parameters for the valve at hand. Table 3–1 contains the required values for some gas lift valve types tested by TUALP. [26]

Valve Manuf.	Valve Type	Port Size	F_e	(p_2/p_1) crit.	a	c	N_{max}	m	b	T_m	γ_m
		in	-	-	-	-	-	-	-	F	-
CAMCO	BK	1/8	0.02	0.58	-0.394	0.64	0.72	7.20E-05	0.07	57.8	1
	BK	3/16	0.09	0.58	-0.416	0.466	0.70	2.80E-05	0.25	70.1	1
	BK	1/4	0.15	0.58	-0.347	0.407	0.64	2.90E-04	0.44	61.5	1
	BK	5/16	0.23	0.58	-0.255	0.346	0.60	-4.00E-04	1.53	66.2	1
CAMCO	BK1	3/16	0.063	0.52	-0.964	0.904	0.74	2.00E-04	0.5	54.9	1
	BK1	1/4	0.13	0.50	-0.701	0.774	0.73	2.70E-04	0.562	53.2	1
	BK1	5/16	0.19	0.58	-0.885	0.765	0.72	-7.60E-04	1.905	71.5	1
	BK1	3/8	0.26	0.54	-0.356	0.499	0.71	1.50E-03	0.45	57.3	1
McMurry	Jr STD	1/8	0.03	0.53	-0.773	0.846	0.73	1.30E-04	0.08	66.9	1
	Jr STD	3/16	0.11	0.53	-0.972	0.864	0.74	4.30E-04	0.25	66.5	1
	Jr STD	1/4	0.18	0.53	-0.746	0.767	0.70	7.50E-04	0.4	62.9	1
	Jr STD	5/16	0.24	0.53	-0.362	0.541	0.64	2.70E-04	1.53	61.6	1
MERLA	NM16R	3/16	0.08	0.53	-0.541	0.680	0.74	2.50E-04	0.12	62.6	1
	NM16R	1/4	0.12	0.53	-0.595	0.706	0.71	6.50E-04	0.18	62.0	1
	NM16R	5/16	0.22	0.53	-0.339	0.582	0.62	7.20E-04	0.96	60.1	1
CAMCO	R20	3/16	0.12	0.63	-0.943	0.945	0.65	2.68E-03	-1.1681	75.1	1
	R20	1/4	0.15	0.64	-0.778	0.882	0.63	3.00E-03	-0.9859	81.9	1
	R20	5/16	0.18	0.60	-0.655	0.769	0.62	2.11E-03	0.4560	74.6	1
	R20	3/8	0.19	0.58	-0.522	0.606	0.61	1.84E-03	1.5588	72.5	1

Table 3–1 Performance parameters of selected gas lift valves according to Schmidt [26].

Example 3–17. Calculate the gas flow rate through a CAMCO BK valve with a 5/16 in. port if injection and production pressures are 1,100 psig and 900 psig, respectively. The dome charge pressure at the valve temperature of T_v = 140 °F is 800 psig, injection gas specific gravity is 0.7.

Solution

To determine the actual flow model to be used, the transition pressure is calculated first. From Equation 3.76:

p_{tran} = 800 (1 + 0.23) = 984 psig; where F_e was found from Table 3–1.

Since actual injection pressure is greater than this value, the orifice model is to be used. Now a check is made to decide on the proper range by calculating the critical production pressure from Equation 3.78:

$p_{p\,cr}$ = 1114.7 0.58 = 646.5 psia; where the critical pressure ratio 0.58 was read from Table 3–1.

Actual production pressure being greater than the critical value, the subcritical range of the orifice model is to be used. The group C_dY is evaluated from Equation 3.80 with the coefficients a and c taken from Table 3–1.

$C_d Y$ = -0.255 (1,114.7 − 914.7) / 1,114.7 / 1.25 + 0.346 = 0.31; where κ = 1.25 was used.

The gas flow rate formula contains the gas deviation factor at valve conditions. For the 0.7 specific gravity gas, critical parameters are evaluated from Equation 2.23 and 2.24:

p_{pc} = 709.6 − 58.7 0.7 = 668.5 psia, and

T_{pc} = 170.5 + 307.3 0.7 = 385.6 °R.

Reduced pressure and temperature are: p_{pr} = 1114.7 / 668.5 = 1.67 and T_{pr} = (140 + 460) / 385.6 = 1.56.

With these values and using the Papay formula (Equation 2.28) we get for the deviation factor:

Z_v = 1 - 3.52 1.67 / ($10^{0.9813\ 1.56}$) + 0.274 1.67^2 / ($10^{0.8157\ 1.56}$) = 0.87.

The gas flow rate is found from Equation 3.79:

$$q_{sc} = 1241 \frac{\left(\frac{5}{16}\right)^2 \pi}{4} 0.31 \sqrt{\frac{1114.7(1114.7 - 914.7)}{(140 + 460)\ 0.87\ 0.7}} = 729 \text{ Mscf/d}$$

Example 3–18. Find the injection rate for the same valve as in Example 3–17 if the production pressure is 400 psig.

Solution

Transition and critical production pressures are identical to those of the previous example. The orifice model in critical flow is used, since actual production pressure is lower than the critical value.

The group $C_d Y$ is evaluated from Equation 3.80 with the coefficients a and c from Table 3–1. Instead of the actual production pressure the critical value is substituted in the formula:

$C_d Y$ = -0.255 (1,114.7 − 646.5) / 1,114.7 / 1.25 + 0.346 = 0.26; where κ = 1.25 was used.

The gas deviation factor is the same as before and the rate is found from Equation 3.79 with the substitution of $p_p = p_{p\ cr}$:

$$q_{sc} = 1241 \frac{\left(\frac{5}{16}\right)^2 \pi}{4} 0.26 \sqrt{\frac{1114.7(1114.7 - 646.5)}{(140 + 460)\ 0.87\ 0.7}} = 935 \text{ Mscf/d}$$

Example 3–19. Find the injection rate for the same valve as in Example 3–17 if injection and production pressures are 950 psig and 600 psig, respectively.

Solution

Transition pressure is the same as before (984 psig), but now the actual injection pressure is below this value and the throttling model must be used. In order to find which is the range—sub-maximal or subcritical— production pressure belonging to the maximum flow rate must be calculated first.

To find p_{pmax}, production closing pressure is evaluated from Equation 3.82:

$p_{pc} = 950 - (950 - 800) / 0.23 = 298$ psig; where F_e was found from Table 3–1.

$p_{p\ max}$ is found from Equation 3.83 with N_{max} read from Table 3–1.

$p_{p\ max} = 298 + 0.6 (950 - 298) = 689$ psig.

Since the actual production pressure is less than the previous value, the flow pattern is sub-maximal. The slope of the injection rate vs. production pressure function is found from Equation 3.85 with the coefficients read from Table 3–1:

slope = -0.0004 800 + 1.53 = 1.21 Mscf/d/psi.

The injection rate at test conditions is calculated from Equation 3.86:

$q_{sc\ m} = 1.21 (600 - 298) = 365$ Mscf/d.

Test measurements were performed with air at $T_m = 66.2$ °F (from Table 3–1). The deviation factor of air at this temperature and at the injection pressure is found from Equation 3.88, where factors B and C are calculated first:

$B = -9.545\ 10^{-10}\ 66.2^2 + 5.434\ 10^{-7}\ 66.2 - 6.099\ 10^{-5} = -2.92\ 10^{-5}$,

$C = 1.515\ 10^{-13}\ 66.2^2 - 9.458\ 10^{-11}\ 66.2 + 1.847\ 10^{-8} = 1.29\ 10^{-8}$.

$Z_m = 1.0 - 3.926\ 10^{-6}\ 66.2 - 2.92\ 10^{-5}\ 964.7 + 1.29\ 10^{-8}\ 964.7^2 = 0.984$.

The gas deviation factor at injection pressure and injection temperature is calculated from the reduced thermodynamic parameters, where the pseudocritical parameters are identical to those in Example 3–17:

$p_{pr} = 964.7 / 668.5 = 1.44$ and $T_{pr} = (140 + 460) / 385.6 = 1.56$.

Using the Papay formula (Equation 2.28) for the deviation factor:

$Z_v = 1 - 3.52\ 1.44 / (10^{0.9813\ 1.56}) + 0.274\ 1.44^2 / (10^{0.8157\ 1.56}) = 0.88$.

The corrected gas injection rate can now be calculated from Equation 3.87:

$$q_{sc} = 365 \sqrt{\frac{(66.2 + 460)0.9841.0}{(140 + 460)\ 0.880.7}} = 432 \text{ Mscf/d}$$

3.2.3.3.2 *Valve Performance Curves.* The calculation model described in API RP 11V2 allows the determination of gas lift valve performance curves, which show the gas throughput rates for different conditions. Figure 3–51 contains relevant curves for the CAMCO BK type valve with a 5/16 in. port, a dome charge pressure of 800 psig, and a valve temperature of 140 °F. The transition pressure, as calculated from Equation 3.76 and shown in bold line, divides the two characteristic flow patterns of the valve. Injection pressure is held constant for each curve, and gas rate varies with production pressure. Use of such performance curves is helpful when analyzing the injection conditions of gas lift valves.

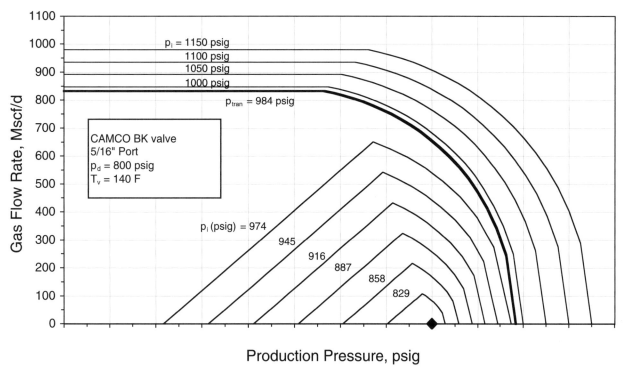

Fig. 3–51 Gas throughput performance of a CAMCO BK gas lift valve, as calculated from the API RP 11V2.

Figure 3–52 schematically compares the static and dynamic performance of an unbalanced gas lift valve at valve temperature. According to the force balance equation, the valve opens along the *static* line, but actual opening occurs along the *dynamic* line. As discussed before, the actually measured F_e ratio (see Equation 3.72) is always less than the A_v/A_b ratio calculated from the valve's geometrical data. This means that for a given production pressure, the valve always opens at a lower injection pressure than the value received from the force balance equation. Gas injection rates through the valve are superimposed on the figure for several injection pressures, showing the throughput capacity of the valve. For injection pressures greater than the transition pressure, orifice flow takes place, whereas for lower injection pressures throttling occurs.

The behavior of the unbalanced valve can be analyzed in Figure 3–52 for continuous or intermittent operations. In continuous flow gas lifting, the valve is designed to be open at a relatively constant production pressure in its throttling range.

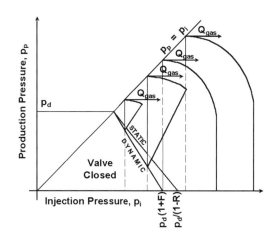

Fig. 3–52 Comparison of the static and dynamic performance of an unbalanced gas lift valve.

Injection and production pressures being constant, the valve injects a constant gas rate. Small fluctuations in production pressure result in gas injection rate changes: a heavier fluid load increases; a lighter one decreases the injection rate. The valve, therefore, operates exactly as required and maintains a constant pressure gradient in the production string, a basic requirement for an efficient continuous flow operation.

In a well placed on intermittent lift, after the operating valve opens, injection and production pressures tend to equalize and then, as the liquid slug in the tubing string travels to the surfaces, both pressures bleed down simultaneously. As seen in Figure 3–52, as production pressure approaches injection pressure, the gas rate through the valve decreases, independently of the flow pattern (orifice or throttling). As the two, almost identical pressures progressively bleed down, the valve finally closes at its dome charge pressure. This behavior compares well to the results of the static force balance equations that state that the valve always closes at a constant pressure.

3.2.4 Setting of gas lift valves

3.2.4.1 Introduction. Before they are run in the well, gas lift valves must be set to the accurate specifications determined during the installation design calculations. The proper setting of each valve in the installation ensures that the well will operate according to the original design. Installation design calculations prescribe the required valve type, port size, dome charge pressure, spring adjustment, differential spring setting, etc. When the required valve with the necessary port is selected, the gas lift expert needs to correctly set the valve's pressure setting. As will be seen later, direct setting of the dome charge pressure is seldom possible so opening or closing pressures are set in special valve testers. The recommended pressure setting procedure is detailed in Appendix B of API Spec. 11V1 [10].

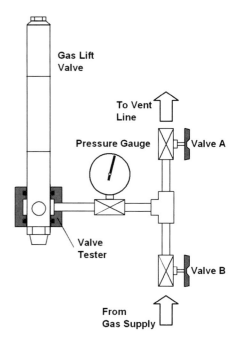

Fig. 3–53 Schematic drawing of a sleeve-type gas lift valve tester.

For bellows-type gas lift valves, the most important phase of valve setting is the checking of the dome charge pressure because the nitrogen gas contained in the valve dome is subject to great changes in temperature. Charging is done at a controlled shop temperature but the valve, after run in the well, operates at a much higher temperature valid at valve setting depth. The resultant dome pressure increase can accurately be calculated (see Section 3.1.4.1), making it possible to properly account for the temperature effect. If the valve does not contain a gas charge but a spring only, valve setting procedures are simpler since the behavior of metal springs is not affected by temperature. Therefore, the valve can be set to downhole specifications in the shop.

Valve setting procedures necessitate the use of several pieces of special equipment:

- **Valve Testers.** Depending on valve type, such testers allow the determination and adjustment of the opening or the closing pressure of the valve.

- **Water Baths.** Sufficiently sized water containers with temperature controls are essential to keep the valves under the proper temperature during setting operations.

- **Agers.** Water-filled pressure chambers are used to *age* gas lift valves. During the aging process, valves are subjected to as high as 5,000 psig pressure for several cycles to find out possible problems.

In the following, pressure setting of different gas lift valves is detailed according to the valve testers generally available.

3.2.4.2 Sleeve-type valve tester. The sleeve (or ring) type valve tester is shown in Figure 3–53. The gas lift valve is inserted in a properly sized sleeve with its gas inlet ports sealed by O-rings. Atmospheric pressure acts on the valve port and test pressure can be applied to the dome minus port area, $(A_b - A_p)$. If high-pressure supply gas is injected into the tester, it tries to open the valve by working against the dome charge pressure that acts on the full bellows area A_b. Since supply pressure acts on a smaller area than the dome charge does, the valve opens at a pressure higher than dome charge pressure. The pressure required to open the valve equals the TRO pressure derived for several gas lift valves in the section on valve mechanics. (Section 3.2.2)

Setting of an unbalanced, bellows-charged gas lift valve (without a spring) with the sleeve type valve tester means adjusting the valve's actual TRO pressure to the value dictated by installation design calculations. The main steps of the usual setting procedure are as follows:

1. While the valve is installed in the tester, use the proper charging equipment to charge the valve dome with nitrogen (or natural) gas to a pressure close to the desired pressure. If the original dome pressure is too high, use a *de-airing* tool to lower it to a value close to the desired pressure.

2. Remove the valve from the tester and put it in the water bath whose temperature is closely controlled to the desired charging temperature, usually 60 °F. Allow the valve to reach charging temperature.

3. Put the valve back in the tester and, while Valve A is closed, open Valve B thereby slowly opening the valve. (see Fig. 3–53)

4. Using the de-airing tool, release nitrogen pressure through the valve core until the required valve opening pressure is reached.

5. After installing the tail plug, put the valve in the ager and apply the recommended pressure (about 5,000 psig) for the recommended time.

6. Take the valve out of the ager and put it back into the tester to check the opening pressure.

7. Repeat Steps 4 to 6 until a constant opening pressure is observed.

8. Put a small amount of silicon fluid on top of the core valve and properly tighten the tail plug.

In case of double-element gas lift valves, the previous procedure is slightly modified because of the effect of the spring. The typical TRO equation of such valves takes the form:

$$\text{TRO} = \frac{p_d'}{1 - R} + p_{sp} \qquad\qquad 3.91$$

where: p_d' = dome pressure at 60 °F, psi

p_{sp} = spring force effect, psi

R = A_v/A_b

In this case, two parameters must be set: (a) the spring force effect, and (b) the TRO of the valve. In order to properly set these, first the valve is installed in the tester without any dome charge. The opening pressure measured under these conditions will be equal to the spring force effect p_{sp}. By making appropriate adjustments to the spring, p_{sp} can be set to the required value. Since spring force is insensitive to temperature, this part of the setting procedure does not involve putting the valve into the water bath. After spring adjustment is finished, the dome is charged with nitrogen gas and the TRO pressure is determined according the previous procedure while the valve temperature is closely controlled.

The sleeve type valve tester can be used to set the following types of gas lift valves: unbalanced IPO and PPO valves and the pilot section of pilot valves. Balanced and flexible sleeve valves require special testers.

3.2.4.3 Encapsulated valve tester.

As discussed previously, sleeve-type valve testers allow the measurement and setting of valve opening pressures at shop conditions. The encapsulated valve tester, shown in Figure 3–54, is used to find the gas lift valve's closing pressure, which is very difficult to determine with the sleeve-type tester. As seen, this tester simulates actual well conditions since pressure can be applied not only to the bellows but also to the valve port area. The two pressure gauges measure injection (upstream) and production (downstream) pressures acting on the bellows and the port areas, respectively.

Almost any type of gas lift valve can be set and adjusted with the use of this valve tester, but its main application is for spring-loaded throttling valves without any dome charge pressure. The usual setting procedure involves the following steps:

1. The valve is installed in the tester with an initial spring setting.

2. With Valve A closed, open Valve B and apply supply pressure to the inlet ports of the gas lift valve. As pressure increases, at one point the gas lift valve opens and injection pressure reaches to Valve A. Both pressure gauges read the same value.

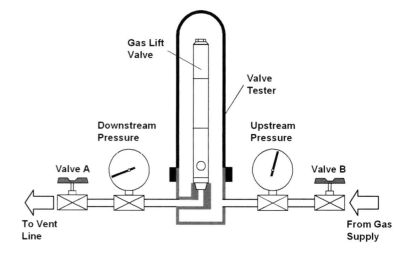

Fig. 3–54 Schematic drawing of an encapsulated gas lift valve tester.

3. Close Valve B to cut off gas supply. The gas lift valve stays open and the gauges show identical upstream and downstream pressures.

4. Slowly open Valve A and let the gas trapped in the tester to bleed down to the atmosphere. Note that the pressures acting on the bellows and the valve port are identical and they balance the spring force. Since the gas lift valve is still open and upstream and downstream pressures are equal, the continuous pressure reduction slowly reduces the valve stem travel of the gas lift valve. Finally, the stem tip seats on the valve port and the gas lift valve closes. When closed, downstream pressure reduces to zero but upstream pressure is trapped in the tester and can be read on the upstream gauge. This is the valve's closing or spring adjustment pressure p_{sa}, defined in Figure 3–34.

5. If the measured closing pressure is different from the prescribed value, the tester is disassembled and the spring setting of the gas lift valve is adjusted accordingly.

6. Steps 1–5 are repeated until the desired valve closing pressure is reached.

If a gas lift valve with a bellows charge is to be set, the procedure is modified to ensure that the gas charge is always under the specified valve setting temperature. This involves frequent immersion of the valve in the controlled temperature water bath.

3.2.5 Application of gas lift valves

3.2.5.1 Valve requirements for different services

3.2.5.1.1 Continuous flow. In continuous flow gas lift operations, high-pressure gas is injected into the fluid column to decrease the flowing density and the pressure gradient from the injection point upward. The decrease of flowing pressure in the flow conduit (either tubing or casing) reduces flowing bottomhole pressure to such a level that reservoir energy, insufficient to maintain natural flow, is now able to move reservoir fluids to the surface. In order to minimize energy requirements, the amount of injected gas must be adjusted to provide optimum performance since either excessive or insufficient injection rates cause the well's fluid rate to change. For an ideal continuous flow operation, therefore, a constant amount of gas is to be injected at the proper depth, provided well fluid rates (oil, water, and gas) and pressures are constant.

Under actual conditions, well inflow rates and well stream compositions are not steady but smaller or greater fluctuations about the daily averages occur. For continuous flow gas lift, the most important changes are those in the liquid rate and the gas content of the well stream. If lift gas is injected at the proper depth and in the right amount for the daily average conditions, an optimum performance can be maintained for those conditions. Fluctuations in well liquid rate and produced GLR, however, make the production pressure at valve setting depth to deviate from its design value as shown in

Figure 3–55. An increase in liquid rate or a drop in GLR entails an increase in production pressure, whereas a decreased liquid rate or an increased GLR works vice versa. Since optimum lift gas usage is maintained at design conditions only, gas injection into the flow conduit must be controlled accordingly. This task automatically is accomplished if an operating valve of the proper type is selected.

It follows from the previous discussion that a proper operating valve for continuous flow gas lift should adjust the volume of injected gas according to the fluctuations in producing pressure at valve setting depth. This kind of *proportional response* operation can only be ensured if the valve is sensitive to production pressure and injects more gas when production pressure increases and injects less when it decreases. The behavior of the operating valve maintains a constant producing pressure at valve setting depth and, at the same time, achieves an optimum usage of lift gas for maximum energy efficiency.

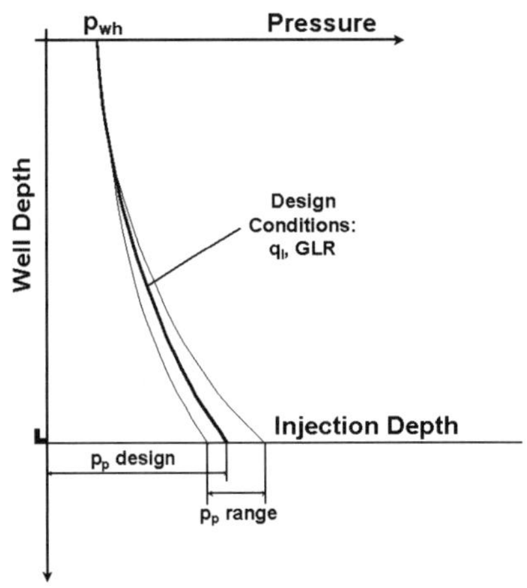

The ideal operating valve for continuous flow operations, as seen previously, must have gas throughput characteristics identical to those of throttling valves. This requires the use of IPO or PPO spring-loaded valves without bellows charge. Unbalanced single or double element valves, as shown in Section 3.2.3 also provide this kind of performance and are also used in continuous flow installations.

Fig. 3–55 Fluctuations in production pressure at the operating valve due to changes in liquid rate and/or formation GLR.

3.2.5.1.2 Intermittent lift. The mechanism of intermittent gas lifting is absolutely different from that of the continuous flow gas lift process. Since production is in cycles, lift gas is only periodically injected into the well at high instantaneous rates to ensure a piston-like displacement of the liquid column accumulated in the tubing string just above the operating valve. In order to use the energy of the lift gas at a maximum efficiency, the proper amount of gas is to be injected rapidly. Injection must cease as soon as the gas volume sufficient to lift the liquid slug to the surface has entered the tubing string. Depending on design conditions, the full amount of the necessary gas volume may come from the annulus (used as an accumulation chamber between cycles), or directly from the surface, or from a combination of both sources.

Based on the previous description of the intermittent gas lift cycle, the main requirements for an operating valve in intermittent installations are as follows:

- The operating valve must be *snap-acting, i.e.* open and close immediately. When operating conditions dictate, it must open or close fully to provide optimum lifting of the liquid column. If the valve is not snap-opening, its slow opening may cause lift gas to simply bubble through the liquid slug without lifting it. Snap-closing, on the other hand, prevents the injection of more-than-sufficient gas volumes and ensures a proper use of gas energy.

- Since the valve should inject relatively high instantaneous gas volumes, it must use a large enough port. Ports of minimum ¼ in. and up to 1 in. are required. Use of smaller ports does not provide the necessary throughput capacity, and the rising velocity of the liquid slug may not be maintained at the required level. Efficiency of lifting may become low, in extreme cases the liquid slug on its travel to the wellhead may even fall back before reaching the surface.

- Depending on the type of surface gas injection control, the operating valve must have the proper amount of spread in order to inject the right amount of gas into the tubing string. If a surface intermitter is used and no gas is stored in the annulus during the cycle, spread is irrelevant and even a balanced valve can be used. In all other cases, gas storage in the annulus is required and the valve should provide the proper amount of spread which, combined with the annular volume available for gas storage, determines the minimum volume of gas injected during the cycle.

It should be noted that the last two requirements contradict each other since, according to the definition of spread, valve port size and spread are interrelated (see Equation 3.26). If the maximum port size is selected to ensure a sufficiently large instantaneous injection rate, the spread of the valve may become too big, as compared to the actual annular volume of the well. The problem can be solved by the use of a pilot-operated gas lift valve.

One of the ideal valves for intermittent lift, as shown previously, is the pilot-operated valve that provides the great port diameters required for high gas flow rates and, at the same time, allows the selection of the right valve spread. Practically all other valve types, except the throttling valves, can also be used in intermittent gas lift.

3.2.5.1.3 Well unloading. Unloading involves the progressive removal of liquids from the annulus of the well in order to reach the operating depth. During this process, gas lift valves in the unloading valve string successively inject gas into the flow conduit and close as soon as the next valve below starts to inject gas. The basic requirement for a gas lift valve to be used in well unloading is, therefore, the capability to transfer the point of gas injection downward. Successive closing of upper valves can be ensured in two ways: either by decreasing the injection pressure for each valve or by maintaining the same injection pressure and having the valves to close with some other mechanism. As will be seen in later sections on continuous and intermittent gas lifting, use of almost every type of valve is possible in an unloading valve string.

3.2.5.2 Advantages and limitations of different valves. In the following, the advantages and disadvantages of the most important types of gas lift valves are detailed. [8, 27]

Unbalanced IPO Valves with Spread

- This valve type can be used in continuous and intermittent operations with or without a spring and has the advantages as follows.

- The most universally known valve type, most operators are familiar with its performance and design.

- Rugged, simple construction, probably the least expensive of all valve types. Easier to repair than other gas lift valves.

- Its setting is easy and simple; it does not require special equipment.

- In borderline wells, it can intermit or produce continuously, depending on actual well conditions.

- In continuous flow operations, a large port size can be used for lifting large liquid volumes.

- If used for intermittent lift, it is less expensive than other applicable valve types, *e.g.* pilot valves.

- In intermittent lift, the annulus volume can be used to store injection gas between cycles. This reduces the instantaneous gas demand of the well and can reduce the size of the compressor required.

The disadvantages of this valve type are

- The dome charge of the bellows is sensitive to well temperature at valve setting depth. For proper design, operating temperatures must accurately be predicted.

- In continuous flow operations, a large port can cause heading conditions if production pressure at valve setting depth fluctuates. To prevent this situation, either a smaller port must be used or the valve must be choked upstream of the port, both solutions restrict the maximum gas injection rate through the valve.

- In intermittent lift installations with surface choke control, the effects of valve port size and valve spread are contradicting. The port size required for an efficient fluid lifting may cause an excessive gas usage per cycle because of the valve's increased spread.

- For unloading valve string design, surface opening pressures of the valves must be decreased to enable upper valves to close. In deeper wells, the full surface injection pressure cannot be utilized and the operating valve cannot be run deep enough.

Throttling IPO Valves

These valves do not have a gas-charged dome; the control force comes from a spring. Because of their special behavior, they can only be used in continuous flow operations with the following advantages:

- A simple, rugged construction.

- Since there is no gas charge, the valve is unaffected by well temperature. This makes well installation design simpler and errors due to improper knowledge of flowing temperature are eliminated.

- The valve's proportional response to production pressure fluctuations makes it automatically maintain the proper pressure gradient and the right injection rate in the flow conduit.

- Since surface opening pressure is kept constant for the unloading valve string, gas injection at the maximum depth for achieving a minimum injection GLR can be ensured. These valves, therefore, allow gas injection at deeper points than unbalanced IPO valves.

Disadvantages include

- Due to their throttling action, gas passage through the valve is limited. In high-capacity wells, several valves may have to be run to the same setting depth.

- Valve setting is more difficult and necessitates the use of special equipment.

- Because of the limited gas capacity of these valves, the total number of unloading valves may be greater than for the unbalanced bellows-charged valves.

Throttling PPO Valves

Used in continuous or intermittent installations, these valves are spring loaded without any dome charge. Advantages include

- Valve operation and setting is unaffected by temperature.

- Since the valve is practically insensitive to injection pressure, it can be used for installations where the surface injection pressure fluctuates.

- Can be used in dual installations since valve operation is controlled by production pressure.

Disadvantages of using this valve type are

- Because of the limited gas passage capacity of this gas lift valve, multipoint gas injection may be needed in intermittent lift operations.

- The versions of this valve type with crossover seats are generally not recommended because their port sizes are limited and the bypass channels are prone to plugging by sand and debris.

Balanced Bellows-type Valves

In general, these valves include balancing mechanisms and seals on the valve stem, parts making the valve complicated and more expensive than an unbalanced one. The fine channels and seal rings are ideal sand traps and may cause frequent valve failures. Because of these disadvantages, manufacturers no longer offer gas lift valves of this type.

Balanced Flexible Sleeve Valves

These valves can be used in continuous as well as in intermittent installations. Some advantages include the following:

- In continuous flow, the valve works as a variable area choke and offers extremely large gas passage areas. Since it injects whatever amount of gas as injected at the surface, it can be used for high liquid rates.

- In intermittent lift installations with a surface intermittent controlling gas injection, the valve opens as soon as the intermitter and, due to its large total port area injects the proper amount of gas per cycle.

- In borderline wells, the same valve can be used for continuous or intermittent operations, depending on actual well conditions.

- Disadvantages of flexible sleeve balanced valves are

- Because of the gas charge, they are sensitive to well temperature, and their design heavily depends on the proper estimation of flowing temperatures.

- Since these valves are run as a part of the tubing string, retrieving of faulty valves necessitates the pulling of the tubing string and the use of a workover rig.

Pilot Valves

Although some types can also be used in continuous flow operations, pilot valves excel in intermittent lift with the following advantages:

- The valve has a controlled spread combined with a large injection port. Spread is adjusted in the pilot section independently of size of the main gas injection port.

- With a pressure regulator and a fixed choke at the surface, this valve allows the control of the starting slug length. This feature ensures that each intermittent cycle uses the ideal volume of lift gas.

Disadvantages of this valve type are as follows:

- More complicated and more expensive than single element valves.

- Sand or debris may easily cause bleed holes and other small channels to plug and this may lead to valve failure.

- Setting procedures require special valve testers.

3.3 Running and Retrieving of Gas Lift Valves

3.3.1 Gas lift valve mandrel types

Installation design calculations specify the depths the operating and the unloading gas lift valves must be run in the well. The valves are positioned at the proper depths in gas lift valve mandrels, *i.e.* special subs in the tubing string allowing communication between the annulus and the tubing string through the valves. In early gas lifting, valve mandrels accepted outside mounted valves, necessitating their running and retrieval along with the tubing string.

Because of the need for a workover rig and the great time demand involved, these conventional mandrels were later replaced with mandrels accepting inside mounted gas lift valves. This solution has many advantages, but the most important one is the opportunity to install gas lift mandrels (equipped with dummy valves) at the initial completion of the well. The following sections discuss the different mandrel types available today.

3.3.1.1 Conventional mandrels.

A conventional or outside mounted gas lift valve mandrel is a short tube with tubing connections at its two ends and a ported lug where the gas lift valve is installed. The inside diameter of the mandrel is identical to that of the tubing string, in order to provide a full cross-sectional area for fluid flow and to prevent any choking of the wellstream. The gas lift valve itself is of the conventional, or outside mounted variant, described in the illustrations of Section 3.2. One version of the conventional gas lift mandrel is depicted in Figure 3–56 where the gas lift valve is mounted with the port area connected to the tubing string. If an unbalanced gas lift valve is used then, depending on the valve seat configuration (normal or crossover), the valve becomes an IPO or PPO valve for tubing or casing flow installations. The check valves to be used for these combinations vary with the direction of gas injection, as previously given in Figure 3–20. The gas lift mandrel given in Figure 3–57, if used in a tubing flow

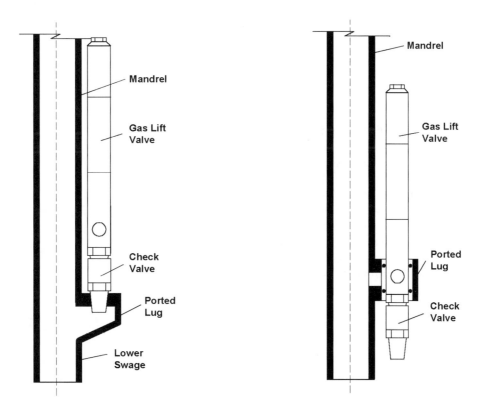

Fig. 3–56 Conventional, outside mounted gas lift mandrel with connection to the gas lift valve port.

Fig. 3–57 Conventional, outside mounted gas lift mandrel with connection to the gas lift valve's gas inlet ports.

installation, allows PPO service with any unbalanced gas lift valve. As seen, the use of gas lift valves for injection or production pressure operation is established with the selection of the proper gas lift mandrel.

The conventional mandrels have a basic disadvantage in practice: they can be run and retrieved with the tubing string only. Therefore, after a well dies, it is necessary to pull the tubing string and to run it back into the well with the mandrels in place. In addition, repair of faulty valves is only possible by pulling the tubing string with all the costs and downtime associated with such operations. This is the main reason why retrievable gas lift valves and mandrels were developed.

3.3.1.2 Wireline retrievable mandrels. Wireline operations have been used for many decades for special operations like running downhole pressure and temperature surveys, and running and pulling special equipment into the well. Instead of a workover rig, they require a small surface unit only and allow operations to be carried out under pressure without the need to kill the well. All these advantages can be utilized in gas lifting if wireline retrievable mandrels and valves are used. The retrievable mandrels are special subs installed in the tubing string and have an elliptical cross section allowing an unobstructed fluid passage area throughout their length. They contain a side pocket (hence the other name: *side pocket mandrel*) with special seating surfaces where gas lift valves can be installed.

The two basic versions of retrievable gas lift mandrels are those with and without a *snorkel*, a piece of pipe protruding from the mandrel body. Figure 3–58 shows a gas lift mandrel, designated as Type S in API Spec 11V1 [10], without a snorkel, which, depending on the type of the gas lift valve's seat (normal or crossover) can be used for any kind of operation: tubing or casing flow, injection or production pressure operation. Type E mandrels (Fig. 3–59) are used for injecting gas on top of a downhole chamber in a chamber lift installation. The Type C mandrel given in Figure 3–60 has ports inside of the valve pocket and has a snorkel as well. It can provide the same kinds of operation as Type S. Finally, Type W mandrels (Fig. 3–61) have a flow path from the top of the valve pocket into the snorkel.

Wireline retrievable gas lift mandrels require the use of retrievable valves, which only differ from the conventional ones in the way they are installed. Figure 3–62 shows a retrievable gas lift valve installed in the proper mandrel and is used in the following discussion. The figure also shows the details of the mandrel: the two polished bores that allow sealing of the valve's packings, the undercut with the ports for gas entry into the valve, and the locking recess for keeping the valve in place. The gas lift valve has two packings, an upper and a lower one that seal against the polished surfaces of the polished bores in the mandrel. The valve is kept in place by a latch installed on its top. When the valve is being installed, the ring in the latch is forced to enter the locking recess of the mandrel and ensures a precise and safe vertical position. When retrieval is necessary, the valve is

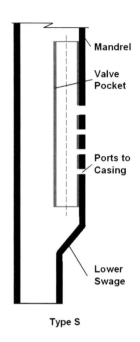

Fig. 3–58 Type S wireline retrievable gas lift mandrel with ports to casing.

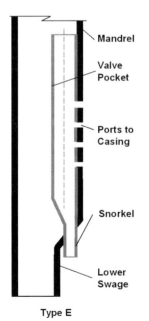

Fig. 3–59 Type E wireline retrievable gas lift mandrel with ports to casing and snorkel.

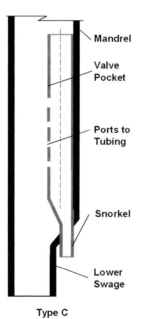

Fig. 3–60 Type C wireline retrievable gas lift mandrel with ports to tubing and snorkel.

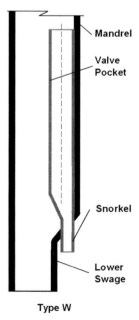

Fig. 3–61 Type W wireline retrievable gas lift mandrel with flow path from the top of the valve pocket into the snorkel.

caught by the fishing neck on the latch and, after the shear pin is broken by a sufficient pull force applied to the wireline, the latch ring collapses and the valve is ready to be pulled to the surface.

The advantages of using wireline retrievable mandrels include the following:

- There is no need to pull the tubing when running or retrieving of gas lift valves is necessary.

- Valves can be installed and retrieved selectively, making repair of faulty valves easy.

- Mandrels with dummy valves in place may be installed at the time of initial well completion while the well is still flowing. When later in the life of the well gas lifting is required the dummies are pulled and gas lift valves are installed with little costs incurred.

3.3.2 Running and retrieving operations

Conventional gas lift valves require pulling the tubing string for valve installing and retrieving operations because valves are installed on the outside of the mandrels. Every time a valve failure occurs, the tubing string must be pulled until the defective valve reaches the surface. This necessitates the use of a workover rig with all the complications involved. Retrievable valves, on the other hand can be pulled and run through the tubing string with wireline operations. The wireline is lowered into the well through a lubricator under pressure with a wireline tool assembly at its bottom. The wireline tool assembly comprises many components: sinker bars, jars, running or pulling tools, etc., and provides a firm contact with the valve or other piece of equipment run in or pulled out of the well.

The wireline tool designed for working in gas lift mandrels performs three basic functions:

(1) it locates the mandrel

(2) it orients the running or pulling tool in the proper azimuth

(3) it provides the lateral offset required to position the tool over the side pocket

Early solutions included a *kick-over tool* with three arms that expanded as soon as the tool entered the mandrel where it properly oriented the running or pulling tool. The evolution of wireline techniques brought about the emergence of modern *positioning tools*, ensuring a simpler and more reliable operation. Figure 3–63 presents the various phases of valve retrieval with a generally available type of the wireline positioning tool.

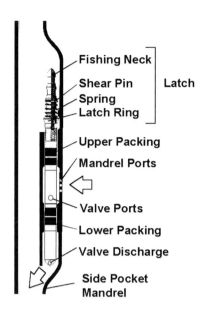

Fig. 3–62 Retrievable gas lift valve installed in its mandrel.

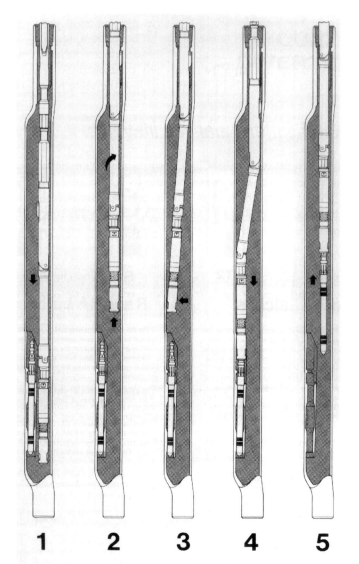

Fig. 3–63 Phases of pulling a retrievable gas lift valve.

The wireline tool assembly with a pulling tool at its bottom is lowered into the well to the depth of the mandrel from which a gas lift valve is to be pulled. When it enters the side pocket mandrel, it is still in a rigid, locked position, as shown in Phase 1. A slow pull on the wireline raises the positioning tool and the key of the tool engages the orienting sleeve situated at the top of the mandrel. The sleeve, while the tool is slowly moved upward, forces the key to rotate the positioning tool. Finally, the key is stopped in its vertical movement at the top of the sleeve's slot, as seen in Phase 2. The positioning tool is now stuck in the mandrel, and this is indicated on the surface by a rapid increase in the wireline load. An additional pull on the wireline (about 200 lb above the free weight of the positioning tool and the wireline in the well) forces the positioning tool's pivot arm to swing out and lock in position. The pulling tool is now located just above the valve pocket. (Phase 3) By slowly lowering the positioning tool, and after some hammering the pulling tool engages the valve's fishing neck (Phase 4) while special guides in the mandrel ensure its proper position. After the shear pin in the valve latch (see Fig. 3–62) is broken by an upward pull on the wireline, the valve is pulled out of the side pocket. The positioning tool is further raised with the wireline as long as its key engages the orienting sleeve again. Sufficient pull breaks a shear pin in the tool, and the pivot arm leaves the mandrel locked in a vertical position as shown in Phase 5. The positioning tool can now be brought to the surface with the gas lift valve attached to it.

Running of gas lift valves involves the same phases as discussed previously with the difference that a running tool with the gas lift valve connected to it (instead of a pulling tool) is situated at the bottom of the positioning tool.

3.4 Other Valve Types

During the evolution of gas lift equipment, many different valve types working on different principles had been developed. Recent advances include a surface controlled electric valve [28] in which the size of the valve port and the gas throughput capacity is remotely controlled. As discussed previously, in today's gas lift technology, the pressure-operated valve is the dominant type but the valve types presented in this section deserve introduction.

3.4.1 Differential gas lift valve

The first widely used gas lift valve was the differential valve that first made unloading with the operating gas lift pressure possible. It was very popular at the early days of gas lift but the emergence of the pressure-operated valve practically eliminated its application. The differential valve is not a pressure-operated valve; its operation is based on the difference between the pressures acting on it: the injection and the production pressures.

The differential valve depicted in Figure 3–64 is connected to the tubing string at the top and is run in the well as shown. It contains a differential spring the compression force of which is set by adjusting the nut on the valve stem. The spring force keeps the valve in the open position at surface conditions, and the valve is equipped with small replaceable orifices as shown.

If the valve is in the closed position in the well, then the opening forces on the valve stem are the force coming from production pressure and the force of the spring, while the only closing force is the result of the injection pressure. Since both pressures act on the same valve stem area A_v, the balance of forces reduces to the following simple equation describing the opening conditions for the differential valve:

$$p_p + p_{sp} = p_i \qquad 3.92$$

where: p_p = production pressure at valve conditions, psi

p_{sp} spring force effect, psi

p_i = injection pressure at valve conditions, psi

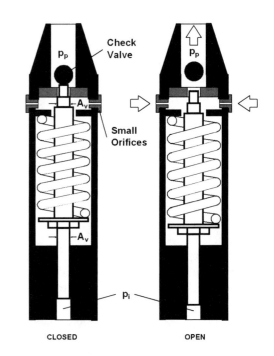

Fig. 3–64 Operation of the differential valve.

When the valve is open, gas injection into the tubing takes place across the small orifices. (see Fig. 3–64) Since the combined area of these orifices is much less than the valve port area A_v, the pressure exerted on the valve port is very close to production pressure. Compared to the closed position, the same pressures act on the same areas, making the valve's opening and closing conditions identical. The valve, therefore, opens and closes when the difference of injection and production pressures overcomes the spring force. For example, if the spring is adjusted to a spring force effect of 150 psi, and the injection pressure at valve setting depth is 900 psi, the valve opens when production pressure is more than 750 psi. As soon as production pressure drops below this value, the valve will close.

The differential valve operates as described previously only if sufficiently small orifices are used. If the orifices are large enough to allow injection pressure to exist inside the valve during gas injection, the valve cannot close and stays open. This is because the opening and closing forces coming from injection pressure cancel each other out, and the spring force keeps the valve in the open position. The need for small orifices in the valve severely restricts the gas injection rates the valve can achieve, which limits the application of differential valves. On the other hand, valve operation is not affected by temperature, an advantageous feature of this gas lift valve type.

3.4.2 Orifice valve

In many continuous flow gas lift wells, the operating valve is a simple choke (square-edged orifice) installed in the gas lift valve mandrel and often called an orifice valve. These *valves* come in the same housing and main dimensions as gas lift valves and can be installed in standard conventional or wireline retrievable mandrels. For preventing backflow into the flow conduit (tubing or casing annulus), they are equipped with standard reverse flow check valves.

The operation of an orifice valve in a well placed on continuous flow gas lift is analyzed in Figure 3–65 where the pressure distributions in the tubing and the annulus are given. The bottom part of the figure represents the performance of a simple orifice and shows the gas injection rates across the orifice as a function of the injection and production pressures valid at valve setting depth. Continuous flow gas lift design requires a pressure differential of about 100–200 psi between the annulus and tubing pressures at the depth of the operating valve. Most of the time, this implies that gas flow across the orifice under design conditions is in the subcritical range, as shown in Figure 3–65. Since tubing pressure at the operating valve is never constant but fluctuates with smaller or greater changes in the liquid rate and/or formation GLR, the injection rate across the orifice will change accordingly.

If operating tubing pressure at valve setting depth increases due to an increase in liquid rate or a decrease in formation GLR, the orifice injects less gas because gas flow through it is in the subcritical range. Similarly, if production pressure decreases, the gas injection rate across the orifice increases. This behavior is, however, contrary to the basic principle of continuous flow gas lifting, *i.e.* the maintenance of a constant tubing pressure at the operating valve. When more gas is needed, the orifice injects less, and vice versa, while the ideal gas lift valve for continuous flow works exactly the opposite way (see Section 3.2.5.1.1). In conclusion, orifice valves used as an operating gas lift valve in a continuous flow gas lift installation are usually not recommended. Their use implies flow stability problems and the occurrence of casing or tubing heading, as detailed in a later section of this book.

3.4.3 Nozzle-Venturi valve

The nozzle-Venturi gas lift valve invented by Schmidt [29] replaces the orifice valve and prevents the heading problems associated with subcritical flow through orifices. It is similar to an orifice valve in construction but instead of a fixed choke, the flow control element is a converging-diverging Venturi device. The nozzle-Venturi and the orifice valves are compared in Figure 3–66, where

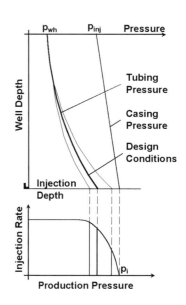

Fig. 3–65 Operation of an orifice valve in a continuous flow gas lift installation.

cross sections and pressure distributions along the flow path are shown. It is assumed that critical flow with sonic velocity occurs in the throat section of both devices. Flowing pressure in the orifice valve (identical to a square-edged orifice) is shown to decrease to the critical pressure downstream of the throat and stay relatively constant along the flow path. In the nozzle-Venturi valve, in contrast, a great amount of pressure recovery takes place and the outflow pressure attains about 90% of the inflow pressure. This means that a pressure differential of about 10% is enough to achieve critical flow conditions in the throat of the nozzle-Venturi valve, which equals to a critical pressure ratio of about $(p_2/p_1)_{cr} = 0.9$. As discussed in Section 2.4.5, the critical pressure ratio for natural gases through fixed orifices is about $(p_2/p_1)_{cr} = 0.554$, a much lower value. Therefore, critical flow through a nozzle-Venturi valve is much easier to achieve than through the simple orifice valve.

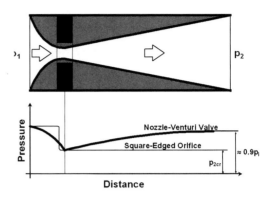

Fig. 3–66 Cross section and pressure distributions in an orifice and a nozzle-Venturi valve.

The gas passage characteristics of the orifice and the nozzle-Venturi valves are compared in Figure 3–67 where gas injection rates are plotted vs. production pressure for a constant injection pressure. Both devices behave similarly: flow rate increases with a decrease in production pressure in the subcritical range and remains constant for pressure ratios below the critical value. The difference between the two gas lift valves is the critical pressure ratio: the orifice valve (square-edged orifice) reaches critical flow conditions if production pressure is approximately 60% of the injection pressure, whereas the nozzle-Venturi valve needs a much lower pressure decrease.

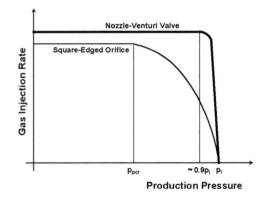

Fig. 3–67 Comparison of gas passage characteristics of orifice and nozzle-Venturi gas lift valves.

The advantages of using the nozzle-Venturi valve as the operating valve in a continuous flow installation are related to its behavior discussed previously. Pressure differentials across operating valves are in the range of 100–200 psi which, as was shown in Figure 3–65, often result in subcritical flow through orifice valves. The same differential across a nozzle-Venturi valve, however, can usually achieve critical flow conditions (see Fig. 3–67). Since gas injection rate under these conditions is constant and independent of any fluctuations in production pressure, the nozzle-Venturi valve very effectively prevents heading and gas lift instability. It can also be used in dual installations to ensure proper gas injection rates for the individual zones. [30, 31]

3.5 API Designation of Gas Lift Valves and Mandrels

The *API Specification for Gas Lift Equipment* [10] contains a system of designation for gas lift valves and wireline retrievable valve mandrels. The use of these designations ensures a fast and efficient specification of gas lift equipment for purchasing or other purposes. The following sections discuss the application of these standardized designations.

3.5.1 Gas lift valves

The API designation for gas lift valves [10] utilizes a six-term alphanumeric code system as detailed in Figure 3–68. Most of the codes in the specification require no explanation. Valves are classified under the *Valve Type* heading into the broad categories of injection, production pressure, and pilot-operated groups. As detailed in Section 3.2, these categories refer more to the application than the construction of the valve and can be meaningless in many cases.

The flow configuration code used in API Spec. 11V1 is explained in Figure 3–69. The four types distinguished are permutations of the normal and crossover valve seats (see Section 3.2.2.2.2) and the two possible directions of gas injection through the valve. The final code of the API designation refers to the service class of the application: standard, non-sour corrosion service, or stress corrosion cracking service in sour environments.

3.5.2 Gas lift mandrels

The standard designation for wireline retrievable gas lift mandrels according to API Spec. 11V1 is given in Figure 3–70. The designation contains alphanumeric codes in three groups. The first group contains codes for the tubing connection, the material/service (a special letter code for standard and stress corrosion cracking service), and the size of the side pocket. The second group gives constructional details of the mandrel: whether a guard and/or an orienting sleeve are provided (see Section 3.3.5) and the drift diameter. The final group specifies the flow configuration (see Figs. 3–58 to 3–61) of the mandrel and the internal and external test pressures applied.

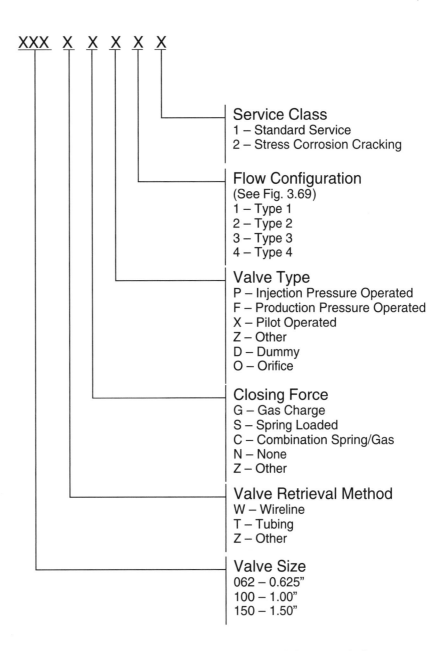

Service Class
1 – Standard Service
2 – Stress Corrosion Cracking

Flow Configuration
(See Fig. 3.69)
1 – Type 1
2 – Type 2
3 – Type 3
4 – Type 4

Valve Type
P – Injection Pressure Operated
F – Production Pressure Operated
X – Pilot Operated
Z – Other
D – Dummy
O – Orifice

Closing Force
G – Gas Charge
S – Spring Loaded
C – Combination Spring/Gas
N – None
Z – Other

Valve Retrieval Method
W – Wireline
T – Tubing
Z – Other

Valve Size
062 – 0.625"
100 – 1.00"
150 – 1.50"

Fig. 3–68 API designation of gas lift valves according to API Spec. 11V1. [10]

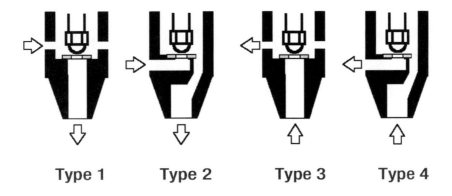

Type 1   Type 2   Type 3   Type 4

Fig. 3–69 Possible flow configurations in gas lift valves.

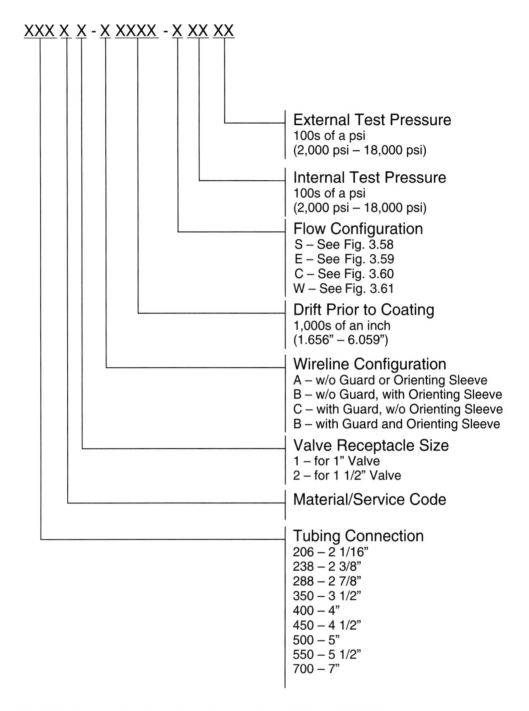

XXX X X - X XXXX - X XX XX

External Test Pressure
100s of a psi
(2,000 psi – 18,000 psi)

Internal Test Pressure
100s of a psi
(2,000 psi – 18,000 psi)

Flow Configuration
S – See Fig. 3.58
E – See Fig. 3.59
C – See Fig. 3.60
W – See Fig. 3.61

Drift Prior to Coating
1,000s of an inch
(1.656" – 6.059")

Wireline Configuration
A – w/o Guard or Orienting Sleeve
B – w/o Guard, with Orienting Sleeve
C – with Guard, w/o Orienting Sleeve
B – with Guard and Orienting Sleeve

Valve Receptacle Size
1 – for 1" Valve
2 – for 1 1/2" Valve

Material/Service Code

Tubing Connection
206 – 2 1/16"
238 – 2 3/8"
288 – 2 7/8"
350 – 3 1/2"
400 – 4"
450 – 4 1/2"
500 – 5"
550 – 5 1/2"
700 – 7"

Fig. 3–70 API designation of gas lift mandrels according to API Spec. 11V1. [10]

References

1. Brown, K. E., C. Canalizo, and W. Robertson, "Evolution of Gas Lift." Proceedings of the 8th West Texas Oil Lifting Short Course, Lubbock, TX 1961: 13–25

2. Brown, K. E. *Gas Lift Theory and Practice*. Tulsa, OK: Petroleum Publishing Co., 1967.

3. Shaw, S. F. *Gas-Lift Principles and Practices*. Houston, TX: Gulf Publishing Co., 1939.

4. King, W. R., "Time and Volume Control for Gas Intermitters." U.S. Patent 2,339,487, 1944.

5. Cummings, L. L., "Gas Lift Valve." U.S. Patent 2,642,889, 1953.

6. Sage, B. H. and W. N. Lacey, "Thermodynamic Properties of the Lighter Paraffin Hydrocarbons and Nitrogen." *API Monograph No. 37*. American Petroleum Institute, 1950.

7. Winkler, H. W. and P. T. Eads, "Algorithm for More Accurately Predicting Nitrogen-Charged Gas-Lift Valve Operation at High pressures and Temperatures." Paper SPE 18871 presented at the Production Operations Symposium held in Oklahoma City, OK, March 13–14, 1989.

8. Winkler, H. W. and S. S. Smith. *Gas Lift Manual*. CAMCO Inc., 1962.

9. Brown, K. E. *The Technology of Artificial Lift Methods*. Vol. 2a. Tulsa, OK: Petroleum Publishing Co., 1980.

10. "Specification for Gas Lift Equipment." *API Spec. 11V1*, 2nd Ed., American Petroleum Institute, 1995.

11. *Field Operation Handbook for Gas Lift*. Revised Edition, OTIS Eng. Co., 1979.

12. Decker, L. A., "Analytical Methods for Determining Pressure Response of Bellows Operated Valves." SPE 6215, available from the Society of Petroleum Engineers, 1976.

13. "Recommended Practice for Gas Lift Valve Performance Testing." *API RP 11V2*, 2nd Ed., American Petroleum Institute, 2001.

14. Hepguler, G., Z. Schmidt, R. N. Blais, and D. R. Doty. "Dynamic Model of Gas-Lift Valve Performance." *JPT* June 1993: 576–83.

15. Bradley, H. B. (Ed.) *Petroleum Engineering Handbook*. Chapter 5. Society of Petroleum Engineers, 1987.

16. Decker, K. L., "Gas Lift Valve Performance Testing." Paper SPE 25444 presented at the Production Operations Symposium held in Oklahoma City, OK March 21–23, 1993.

17. Orris, P. W., L. J. Bicking, E. E. DeMoss, and W. B. Ayres. *Practical Gas Lift*. MERLA Tool Corp. Garland, TX, 1963.

18. Neely, A. B., J. W. Montgomery, and J. V. Vogel, "A Field Test and Analytical Study of Intermittent Gas Lift." *SPE Journal*, October 1974: 502–12.

19. Winkler, H. W. and G. F. Camp, "Dynamic Performance Testing of Single-Element Unbalanced Gas-Lift Valves." *SPE PE*, August 1987: 183–90.

20. Decker, K. L., "Computer Modeling of Gas-Lift Valve Performance." Paper OTC 5246 presented at the 18th Annual Offshore Technology Conference held in Houston, TX, May 5–8, 1986.

21. *Gas Lift Manual*. Section 5. Teledyne MERLA, 1979.

22. Nieberding, M. A., Z. Schmidt, R. N. Blais, and D. R. Doty, "Normalization of Nitrogen-Loaded Gas-Lift Valve Performance Data." *SPE PF*, August 1993: 203–10.

23. Hepguler, G. Z. Schmidt, R. N. Blais, and D. R. Doty, "Dynamic Model of Gas-Lift Valve Performance." *JPT*, June 1993: 576–83.

24. Acuna, H. G., Z. Schmidt, and D. R. Doty, "Modeling of Gas Rates through 1-in., Nitrogen-Charged Gas-Lift Valves." Paper SPE 24839 presented at the 67th Annual Technical Conference and Exhibition, Washington D.C., October 4–7, 1992.

25. Sagar, R. K., Z. Schmidt, D. R. Doty, and K. C. Weston, "A Mechanistic Model of a Nitrogen-Charged, Pressure Operated Gas Lift Valve." Paper SPE 24838 presented at the 67th Annual Technical Conference and Exhibition, Washington D.C., October 4–7, 1992.

26. Schmidt, Z. *Gas Lift Optimization Using Nodal Analysis*. CEALC Inc., 1994.

27. Davis, J. B., P. J. Thrash, and C. Canalizo. *Guidelines To Gas Lift Design and Control*. 4th Edition, OTIS Engineering Co., Dallas, 1970.

28. Schnatzmeyer, M. A., C. J. H. Yonker, C. M. Pool, and J. J. Goiffon, "Development of a Surface-Controlled Electric Gas-Lift Valve." *JPT*, May 1994: 436–41.

29. Schmidt, Z., "Nozzle-Venturi Gas Lift Flow Control Device." U.S. Patent 5,743,717, 1998.

30. Tokar, T., Z. Schmidt, and C. Tuckness, "New Gas Lift Valve Design Stabilizes Injection Rates: Case Studies." Paper SPE 36597 presented at the Annual Technical Conference and Exhibition held in Denver, CO, October 6–9, 1996.

31. Faustinelli, J., G. Bermudez, and A. Cuauro, "A Solution to Instability Problems in Continuous Gas-Lift Wells Offshore Lake Maracaibo." Paper SPE 53959 presented at the Latin American and Caribbean Petroleum Engineering Conference held in Caracas, Venezuela, April 21–23, 1999.

4 | Gas Lift Installation Types

4.1 Introduction

Different gas lift installations contain different combinations of downhole equipment: gas lift valves, packers, mandrels, nipples and subs, etc. The type of gas lift installation is mainly governed by the type of lift used: continuous flow or intermittent lift. Equipment selection is based on original well conditions but should provide for the necessary flexibility to reduce the number of future workover operations. A proper gas lift installation design, therefore, assures trouble-free operation for the entire *productive life* of a well. [1, 2]

The usual gas lift installation types are classified in two broad categories: *tubing flow* and *casing flow* installations. In a tubing flow installation, lift gas is injected in the casing-tubing annulus, and production occurs through the tubing string; whereas a casing flow installation allows gas injection through the tubing string, and the well is produced from the annulus. The following sections present detailed discussions of the most common versions of gas lift installations.

4.2 Tubing Flow Installations

In these cases, lift gas from the surface is injected down the casing-tubing annulus, and liquid production takes place up the tubing string. Except in wells with extremely large liquid production rates, tubing flow installations are generally recommended because of their many advantages: no corrosive and/or abrasive liquids flow in the casing string, well killing is easily accomplished, etc.

4.2.1 The open installation

In the *open gas lift installation*, the tubing is simply hung inside the casing string, and no packer is run in the well (Fig. 4–1). This is the original installation type used in the early days of gas lifting when no gas lift valves were installed on the tubing and lift gas had to be injected at the *tubing shoe*. This resulted in poor economics of gas usage that is why today's open installations contain gas lift valves.

In continuous flow gas lift, an open installation usually means that lift gas is injected at the bottom of the tubing string. The selection of the proper setting depth for the tubing as a function of the surface injection pressure is extremely difficult. After the tubing is run to a predetermined depth, any changes in surface injection pressure or in the well inflow rate cause wide fluctuations in the injected gas rate. The well produces in *heads*, and the liquid rate fluctuates accordingly. Therefore, the basic requirement of continuous flow gas lifting, *i.e.* the injection of a constant gas volume ensuring an optimum flowing gradient, cannot be achieved. Lift gas usage is excessive, and the economic parameters of production are unfavorable. This is why an open installation without the use of gas lift valves is *not* recommended.

If gas lift valves are also run in an open installation, a substantial length of tubing should extend below the operating valve as shown in Figure 4–2 to prevent gas to be *blown around* the tubing shoe. The fluid *seal* thus created permits a constant flowing bottomhole pressure independent of injection pressure fluctuations. Bottomhole pressure in such cases can attain the same values as if a packer was installed, and the well can be lifted with optimum gas usage.

There are several drawbacks of open installations in continuous flow:

- Every time the well is shut down, well fluids rise in the annulus and the well should be *unloaded* to the depth of the operating valve at the next startup. During the repeated unloading operations, gas lift valves are exposed to liquid flow that can gradually damage them.

- During normal operation, because of fluctuations in the gas lift pressure, the annulus fluid level rises and falls exposing lower gas lift valves to severe *fluid erosion*.

All of these effects usually result in the leaking and eventual failure of the valves below the operating fluid level. In conclusion, an open installation for continuous flow gas lifting is not generally recommended even if gas lift valves are also used. However, in certain cases when running of a packer is undesirable or even impossible, the open installation with the proper *fluid seal* in the well can provide a satisfactory solution. As already mentioned, the *fluid seal* ensures that the same flowing bottomhole pressures can be reached as with a packer.

In intermittent wells, flowing bottomhole pressures are usually much lower than in continuous flow operations, and it is very likely that injection gas has to be blown around the bottom of the tubing string. In addition to the previous disadvantages, an open installation has some more drawbacks for intermittent lift wells:

- If no gas lift valves are run in the well, tubing pressure must blow down near to separator pressure after the liquid slug has been produced to the surface. This increases the intermittent cycle's gas requirement and usually leads to *excessive gas usage*.

- When gas lift valves are also used, it must be assured that gas injection takes place always through the *operating valve* and not around the tubing shoe. This can only be achieved if the tubing string is run to a sufficient depth below the deepest possible annular liquid level. As wells produced by intermittent lift usually have low bottomhole pressures, this condition can seldom be met.

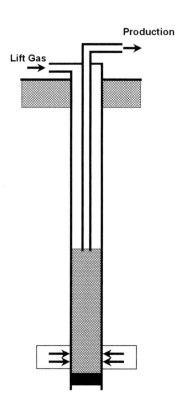

Fig. 4–1 Open tubing flow installation with gas injection at the tubing shoe.

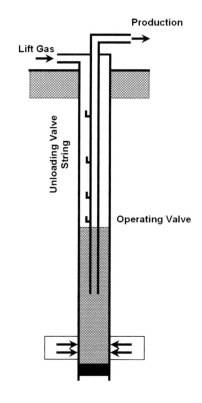

Fig. 4–2 Open tubing flow installation with a *fluid seal* and gas lift valves.

- During the period when gas is injected into the annulus, injection pressure can directly act on the formation and thus prevent the required reduction in bottomhole pressure. Since intermittent wells usually exhibit low flowing bottomhole pressures, the effects of this condition are more pronounced than for continuous flow wells. An open installation, therefore, restricts the increase of liquid production; that is why wells near *abandonment* cannot be produced with this installation type.

In conclusion, use of an open installation in intermittent gas lift operations is *not* recommended. For continuous flow wells, although the open installation has its drawbacks, when conditions do not permit the use of other solutions, it does not necessarily result in poor production economics.

4.2.2 The semi-closed installation

This type of installation (Fig. 4–3) differs from the open installation by a *packer* set at a deep enough point in the well. The packer isolates the tubing from the annulus and eliminates most of the disadvantages of an open installation. Very often, in order to limit the damage done by well fluids passing through unloading valves, a *sliding side door* is run just above the packer. This is opened during *initial unloading* and provides communication between the annulus and the tubing string. Most of the liquid flow from the annulus to the tubing takes place through the open side door, reducing the amount passed by the open unloading valves. After unloading, the door is closed by a wireline operation, and the well is put on gas lift. The semi-closed installation offers these *advantageous features*:

- Well fluids from the formation cannot enter the annulus during shutdowns because of the packer.

- Flow of fluids from the tubing string is prevented by the check valves on each gas lift valve.

- After an initial unloading of the installation, *startup* of production is easy since unloading of the annulus is not necessary.

- The formation is *sealed* from injection pressure by the packer; thus, injection pressure cannot prevent the reduction of the flowing bottomhole pressure.

The semi-closed installation is the *ideal* and most common type of installation for continuous flow gas lift wells because it provides for a constant bottomhole pressure required for optimum conditions.

In intermittent lift, this installation allows the reduction of bottomhole pressures below the levels attainable with an open installation. Due to the injection pressure having no effect on bottomhole pressure, more liquid can be lifted from the well especially in wells exhibiting low bottomhole pressures. A great number of intermittent wells therefore are produced with semi-closed installations.

A semi-closed installation has its *drawbacks* in some intermittent wells, which comes from the effect of the injection pressure acting on the formation during the time the operating valve is open. This condition limits the attainable decrease in the bottomhole pressure or, especially in weak formations, can even result in *backflow* into the formation. In wells having very low reservoir pressures, these effects are more pronounced and a closed installation is recommended.

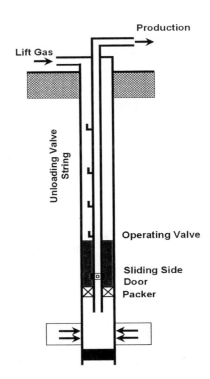

Fig. 4–3 Semi-closed gas lift installation.

4.2.3 The closed installation

A *closed gas lift installation* differs from a semi-closed one by the application of a *standing* (check) valve at the tubing shoe, see Figure 4–4. Most wells on intermittent gas lift employ this type of downhole construction. The standing valve eliminates the effects of gas lift pressure on the formation since during injection gas pressure in the tubing is isolated from the formation by the standing valve, while in the annulus the packer seals it. The average flowing bottomhole pressure during an intermittent cycle can thus be kept to a minimum, and it is not affected by injection pressure. A sliding side door to reduce the damage of the unloading valve string is often run above the packer.

Figure 4–5 shows the schematic variation of bottomhole pressure in the function of time for several injection cycles in a well placed on intermittent lift. The bold line denotes conditions for a closed installation and the normal line holds for a semi-closed installation. In order to achieve an efficient fluid lifting process, a gas injection pressure of two to three times as much as the hydrostatic pressure of the accumulated liquid slug is required. In a semi-closed installation, this high pressure directly acts on the formation during the gas injection period, as shown in the figure. The high pressure involved may force well fluids to *flow back* into the formation and, at the same time, increases the average backpressure on the formation. In the closed installation, average bottomhole pressure is always lower, and daily production rates can be consequently higher. The installation of the *standing valve* in the well, therefore, usually allows higher production rates to be achieved from an intermittent gas lift well.

4.2.4 Chamber installations

Chamber installations are used in intermittent gas lift wells with low formation pressures near *abandonment*, because such installations enable the lowest possible flowing bottomhole pressures to be attained in gas lift operations. The explanation for this feature is that well fluids in chamber installations accumulate in special chambers having substantially larger diameters than the tubing. If the same liquid column height is allowed to accumulate, chambers will contain much greater liquid volumes due to their bigger capacities. Since bottomhole pressure is proportional to the hydrostatic pressure of the liquid column (which, in turn, depends on column height only), pressure buildup follows the same pattern in both the closed and the chamber installations. This means that during the same period, the accumulated liquid volume is *greater* in a chamber installation than in a closed one. Assuming the same daily cycle number, the chamber installation will therefore achieve a much greater total daily liquid production rate.

If, on the other hand, the same liquid volume per cycle is allowed to accumulate, the *hydrostatic head* against the formation will be much lower in a chamber as compared to the tubing, due to the higher liquid capacity and the lower liquid column height required. At the same time, the shorter liquid column accumulates faster in the chamber, making the number of daily cycles increase, which entails a proportional increase of the well's production rate. In conclusion, chamber lift installations can highly enhance the liquid production from wells with *low* to *extremely low* bottomhole pressures and relatively high productivities.

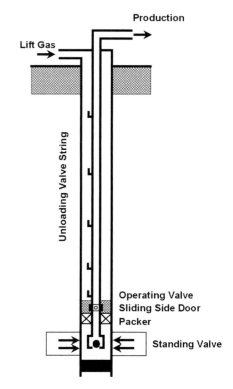

Fig. 4–4 Closed gas lift installation.

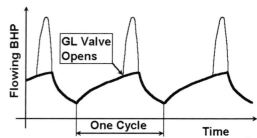

Fig. 4–5 Variation of bottomhole pressure in the open and semi-closed installations.

Chamber installations fall in two broad categories. *Two-packer chamber* installations have their chambers formed by a casing section between two packers, while *insert chamber* installations employ a large diameter pipe section at the bottom of the tubing string. Other chamber installations for special conditions (open-hole completion, sand removal, etc.) are also known. [3]

Detailed design procedures for chamber lift are presented in chapter 6 of this book.

4.2.4.1 Two-packer chamber installations.

A typical two-packer chamber installation is depicted in Figure 4–6. The accumulation chamber comprises the casing-tubing annular space between two packers where the upper packer contains a bypass for gas injection. The advantages of this installation are illustrated by presenting the sequence of operation in an intermittent cycle.

Just after the previous slug has been removed to the surface, tubing pressure bleeds down and fluid inflow through the open standing valve starts. Well fluids simultaneously enter both the chamber space and the tubing string thanks to a *perforated nipple* in the tubing string situated just above the bottom packer. The accumulating fluid column continuously increases the well's flowing bottomhole pressure, and the inflow rate proportionally decreases with time. As the fluid level rises, produced gas is trapped at the top of the chamber. This gas volume would be compressed and would further increase the flowing bottomhole pressure causing fluid inflow to cease before the chamber fills completely. To eliminate this problem, a *bleed valve* is installed in the tubing string near the top of the chamber that allows produced gas to escape from the chamber.

After the necessary amount of well fluid has accumulated, lift gas from the surface is injected into the annulus, and the operating gas lift valve opens. Gas is injected at the top of the chamber with a sufficiently high pressure forcing the bleed and the standing valves to close. Liquid from the chamber space is continuously *U-tubed* into the tubing and gas injection below the liquid slug takes place only after all liquid has been transferred to the tubing. High-pressure injection gas then displaces the liquid slug to the surface and the intermittent lift cycle is repeated.

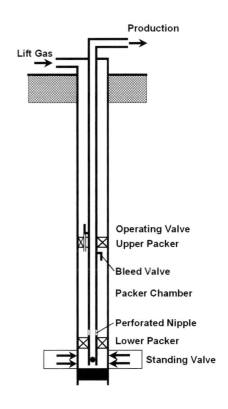

Fig. 4–6 The two-packer chamber gas lift installation.

In conclusion, the most important advantages of chamber lift installations are

(a) *very low* flowing bottomhole pressures can be achieved

(b) the *efficiency* of lifting the liquid slug is improved over the conventional intermittent installations, as will be shown later [4]

4.2.4.2 Insert chamber installations.

In cases where two packers cannot be set in the well (open-hole completions, long perforated intervals, bad casing, etc.) the insert chamber installation is used. In this case, the chamber is constructed from a larger pipe section at the bottom of the tubing string. Since it has to be run through the casing string, the diameter of an insert chamber is considerably less than that of the casing. In spite of this disadvantage, all benefits of chamber lift operation can be realized with insert chambers as well. The main components and the operation of an intermittent cycle are identical to those described previously for the two-packer chamber installation.

Two basic versions of insert chambers can be classified based on the *type of packer* used. In cases when the well's production rate justifies the use of a more expensive *bypass packer*, the installation schematically depicted in Figure 4–7 is used. As shown, the tubing is run all the way down to the bottom of the chamber and gas injection takes place through a perforated nipple at the bottom of the tubing string. This arrangement is advantageous because, after the accumulated liquid is U-tubed into the tubing, the liquid slug rises in a constant diameter conduit. Lift gas breakthrough, due to the constant flow velocity, is thus reduced.

An inexpensive solution, mostly used in shallow, low capacity *stripper wells*, utilizes a simple hookwall packer and a *dip tube* of a smaller diameter than that of the tubing. As shown in Figure 4–8, the dip tube is inserted in the tubing string using a *hanger nipple*, and gas injection takes place at the top of the annulus formed by the tubing and the dip tube. Since the diameter of the dip tube is much smaller than that of the tubing, *gas breakthrough* is more pronounced, causing a larger amount of the starting slug length to be lost during its trip to the surface.

4.2.4.3 Special chamber types.

There are several different chamber installations developed for special conditions such as for wells with extra long open-hole sections, bad casing, etc. [5, 6] In wells with very tight formations, the solution given in Figure 4–9 can be used where the casing or the open-hole section forms the chamber and the tubing run below the packer acts as the dip tube. This installation, often called the *open-hole chamber*, can be used in tight formations or in low producers with sand problems. Variants of this installation type include one with a bypass packer as shown in Figure 4–9 or with a simple hookwall packer and a smaller-diameter dip tube.

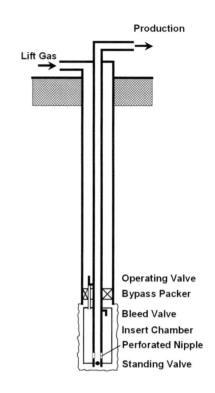

Fig. 4–7 The insert chamber gas lift installation using a bypass packer.

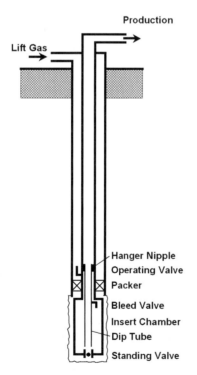

Fig. 4–8 Insert chamber gas lift installation using a dip tube.

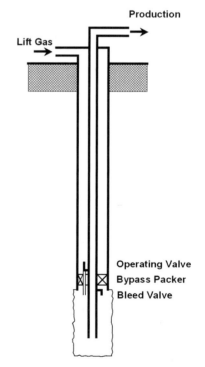

Fig. 4–9 Special chamber gas lift installations for tight formations.

4.2.5 Multiple Installations

Multiple installations involve production of more than one formation through the same well. Although production by gas lift of several *zones* is possible, *dual installations* with two formations being simultaneously produced are predominantly used. As in case of any dual or multiple oil well installation, the main principle to be followed is that different zones must be produced *independently* of each other. The reason for this is that accountability of the liquid rates coming from the different formations cannot be ensured if *commingling* of the wellstreams is allowed.

It follows from the previous principle that dual gas lift installations are equipped with two tubing strings reaching to the depth of the two zones opened in the well. Since the two tubing strings share a *common* casing annulus and must therefore utilize the same injection pressure, interference of the two zones' operation is a basic problem. One rarely used solution involves a separate conduit that supplies gas to one of the zones, the other zone being supplied from the casing annulus.

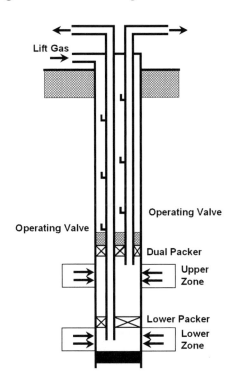

The generally used dual gas lift installation utilizes the well's annulus to supply lift gas to each zone, as shown in Figure 4–10. Two packers are used to isolate the productive formations, the upper one being a *dual packer* with two bores for the two tubing strings. Both the *short string* (the tubing string producing the upper zone) and the *long string* (through which the lower zone is produced) have their individual operating valves and unloading valve strings. Due to the possible difference in productivities of the formations, they may be placed on different kinds of lift: either continuous flow or intermittent lift.

The possible combinations of lift types (continuous-continuous, continuous-intermittent, intermittent-intermittent) require different operation of their respective operating gas lift valves and need consequently different operating gas lift pressures and pressure patterns. In order to prevent interference of the two operating valves, it has to be ensured that they operate independently of each other. One usual solution is to use IPO valves on one string and PPO valves on the other.

More details and design procedures for dual gas lifting are presented in a separate chapter of the present book.

Fig. 4–10 Dual gas lift installation with a common annulus.

4.2.6 Miscellaneous Installation Types

4.2.6.1 Pack-off installations.
In oilfield operations, it is standard practice to install gas lift mandrels with dummy valves in a well that is still flowing. When the time comes to place the well on gas lift, the dummies are retrieved, and gas lifting can commence. This solution saves the cost and the time associated with a workover to convert the well to the new conditions. Many times, however, wells are originally completed for flowing production only, and it is necessary to convert them to gas lifting after they die. In order to reduce the re-completion costs involving pulling the tubing and running it back with the gas lift mandrels in place, the pack-off technique of well completion was developed.

A pack-off gas lift valve installed in the tubing string is shown in Figure 4–11. Before running the valve or valves, the tubing is *perforated* with a mechanical or gun perforator at the proper depth(s) where gas lift valve(s) are to be installed. Now the pack-off gas lift valve assemblies are run in the well with standard wireline operations, bottommost first. The valve assembly is made up of a *collar lock*, a lower and an upper *pack-off*, the *gas lift valve*, and a *hold down*. The two pack-offs straddle the perforated hole in the tubing and prevent injection gas to directly enter the tubing. Injection gas from the annulus enters the gas lift valve and is directed to the tubing to lift well fluids to the surface. The valve used in the pack-off assembly may be of the concentric or the conventional outside-mounted type.

Spacing of pack-off gas lift valves follows the same procedure as for other types of valves. Experience has shown that the pressure drop occurring across pack-off assemblies is usually negligible [7], and pressure conditions in the tubing are practically identical with and without the pack-offs. The main advantages of using the pack-off technique can be summarized as

- Substantial equipment and workover *costs* can be saved, especially in an offshore environment where well servicing is expensive.

- The well does not have to be *killed*—an additional and many times decisive cost saving argument.

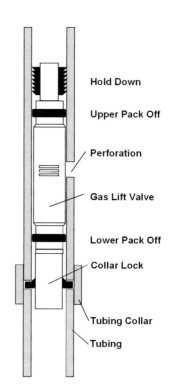

Fig. 4–11 Schematic drawing of a pack-off gas lift valve installed in the tubing string.

- *Downtime* and production loss are minimized because of the low time demand of the required wireline operation.

In addition to a very efficient and low-cost conversion of flowing wells to gas lifting, pack-off assemblies may very effectively be used also in special conditions, such as replacing leaking gas lift valves, repairing of tubing leaks, running additional valves below existing ones, etc.

4.2.6.2 Macaroni installations.

Tubingless or ultra-slim hole completions are used in small-diameter holes and are cased with a normal tubing string of 2 $^7/_8$ in. or 3 $^1/_2$ in. size. Although application of such installations is limited to medium to low oil producers, the reduced drilling and completion costs make them very competitive. The tubing sizes that can be run in these wells are $^3/_4$ in., 1 in., 1 $^1/_4$ in., or 1 $^1/_2$ in., often called *macaroni* tubing, hence the name of the installation type. Macaroni installations only differ from normal-size gas lift installations by the size of their tubulars. In a well producing a single zone a packer and a semi-closed installation is used for continuous flow gas lifting. The limitation of macaroni installations is the flow capacity of the small tubing sizes, although maximum liquid production with a 1 $^1/_2$-in. macaroni tubing can reach 900 bpd. [8]

The utilization of macaroni tubing is not restricted to tubingless completions. Many times, macaroni tubing is run inside the tubing string of a single or multiple completion to convert the well to gas lifting. In dual installations, the use of a macaroni tubing in one of the tubing strings eliminates the problems associated with lifting two strings from a *common annulus*. The zone with the macaroni string is supplied with injection gas from the tubing-macaroni annulus, while the other zone gets gas from the casing annulus.

4.2.6.3 CT installations. Conversion of flowing wells to gas lift can also be accomplished with the use of a CT string. Using conventional gas lift mandrels installed at the proper depths in the coiled tubing string, the CT string is run inside the well's tubing and is hung at the wellhead. As seen in Figure 4–12, lift gas is injected into the annulus formed by the tubing and the coiled tubing string, and the well is produced through the CT string. Compared to the pack-off installation, the need to perforate the production tubing is eliminated, which is a definite advantage. The use of the relatively small diameter CT string, however, restricts application of this installation type to medium to low producers.

4.3 Casing Flow Installations

In a *casing flow* installation, gas is injected down the tubing, and production rises in the casing-tubing annulus (see Fig. 4–13). It is generally used in continuous flow gas lift wells producing *extremely large* liquid rates that exceed the capacity of the tubing string run in the well. Its main drawback is that the casing is exposed to well fluids restricting application to noncorrosive liquids.

No packer is run in the well, and the gas lift valves can be of the conventional or wireline retrievable type with check valves reversed. As is the case for tubing flow, it is not recommended to blow gas around the tubing shoe. That is why the tubing is *bull-plugged* and injection gas can only enter the well through the gas lift valves. Use of the bull plug also ensures that the tubing must not be unloaded after every shut-in of the well.

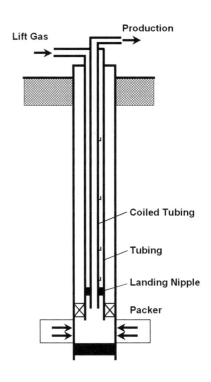

Fig. 4–12 CT gas lift installation.

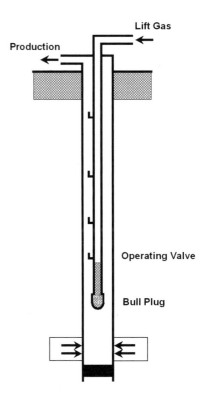

Fig. 4–13 Casing flow gas lift installation with bull-plugged tubing.

References

1. *Gas Lift*. Book 6 of the Vocational Training Series. Dallas, TX: American Petroleum Institute, 1965.

2. Winkler, H. W. and S. S. Smith. *Gas Lift Manual*. CAMCO Inc., 1962.

3. Brown, K. E. *The Technology of Artificial Lift Methods*. Vol. 2a. Petroleum Publishing Co., 1980.

4. *Field Operation Handbook for Gas Lift*. Revised Edition, OTIS Eng. Co., 1979.

5. Winkler, H. W. and G. F. Camp, "Down-Hole Chambers Increase Gas-Lift Efficiency." Part One, *PE*, June 1956: B-87–100.

6. Winkler, H. W. and C. F. Camp, "Down-Hole Chambers Increase Gas-Lift Efficiency." Part Two, *PE*, August 1956: B-91–103.

7. Brown, K. E., R. W. Donaldson, and C. Canalizo, "Pack-off assembly cuts gas-lift installation costs." *PE*. May 1963: B-42–B-50.

8. Thrash, P. J., "What about gas lifting tubingless completions." *PE*. September 1960: B-33–9.

5 | Continuous Flow Gas Lift

5.1 Introduction

Continuous flow gas lift is a continuation of natural flow with the only difference that lift gas from the surface is artificially injected into the flow conduit. The vast majority (about 90%) of gas-lifted wells are placed on this type of operation. The great popularity of this type of gas lifting comes from its wide range of applicability: liquid production rates from a few hundreds to many tens of thousands of barrels are possible.

Chapter 5 presents a full coverage of the topics related to continuous flow gas lifting. After a detailed description of the basic principles and the basic design of continuous flow installations, the description of system performance is discussed. Since continuous flow gas lifting is a prime application for multiphase flow theories and practice, the description of the well's producing system is based on Systems Analysis (NODAL) principles. The different typical cases are reviewed and analyzed, and the stability of a gas-lifted well's operation is investigated.

Due to its great practical importance, optimization of continuous flow installations is covered in detail. Since the definition of optimum conditions can vary in different situations, all possible cases (prescribed or unlimited liquid or gas injection rates, existing compressor, etc.) of optimizing a single well are discussed, and the proper optimization procedures are fully described. Optimization for a group of wells involves the correct distribution of the available injection gas with different possible objectives in mind, and the proper allocation of lift gas between a group of wells of different behavior is also discussed with worked examples.

In most of the wells placed on continuous flow gas lift, the startup of production is facilitated by the operation of a string of unloading gas lift valves. The proper design and operation of these valves allows a stepwise removal of the kill fluid from the annulus down to the operating valve's setting depth. Valve spacing procedures for the use of the most popular types of gas-lift valves are fully described with illustrated examples.

The last section of chapter 5 discusses the different ways surface gas injection control are applied on continuous flow wells.

5.2 Basic Features

5.2.1 The mechanism of continuous flow gas lift

Continuous flow gas lifting involves a continuous injection of lift gas into the wellstream at a predetermined depth, usually from the casing-tubing annulus to the tubing string. The injection of a proper amount of gas significantly decreases the flowing mixture's average density above the injection point. Since the major part of pressure drop (PD) in vertical multiphase flow is generally due to the change in potential energy, a marked reduction of pressure gradient occurs in the affected tubing section. The whole tubing string's resistance to flow being thus significantly reduced, the pressure available at the well bottom will now become sufficient to move well fluids to the surface. Therefore, with all other conditions unchanged, the well that was dead before will start to produce, thanks to the injection of lift gas into the wellstream. This is the *basic mechanism* of continuous flow gas lifting: due to the continuous injection of lift gas from an outside source, the flow resistance of the well tubing is reduced, allowing reservoir pressure to move well fluids to the surface.

The previous mechanism ensures that formation gas (produced from the well) is fully utilized for fluid lifting. In most other types of artificial lift, especially those utilizing some kind of a downhole pump, formation gas is either harmful or completely detrimental to the operation of the production equipment. Since this is the case for intermittent gas lift as well, continuous flow gas lifting can be considered the sole kind of artificial lift completely using the formation's natural energy stored in the form of dissolved gas.

5.2.2 Applications, advantages, limitations

In addition to the general applications of gas lift techniques detailed in chapter 1, continuous flow gas lift is used to

- Produce *high* to *extremely high* liquid rates from any depth.

- Lift sand or solid-laden liquids because sand settling problems in the flow string are negligible.

- Increase the liquid rate of otherwise flowing wells.

- Produce large volumes of water for waterflood operations.

- Back-flow water injection wells.

Because of the relatively high average well temperature in continuous flow installations, flowing viscosity of highly viscous crudes can be kept low, which many times is the only feasible way to produce such wells.

Basic advantages and limitations of continuous flow gas lifting can be summarized as follows, after [1, 2]:

Advantages

- Continuous flow gas lift, in contrast to intermittent gas lifting, fully utilizes the energy of the formation gas.

- Since gas injection and production rates are relatively constant, neither the gas supply, nor the gathering systems are overloaded, if properly designed.

- Within its application ranges, continuous flow gas lifting is *relatively flexible* to accommodate varying well conditions.

- The fairly constant bottomhole pressure means ideal inflow conditions at the sandface.

- Surface injection gas control is simple; usually a choke on the injection line is sufficient.

Limitations

- The primary limitation of continuous flow gas lifting stems from its operating mechanism and is the *available formation pressure* at well bottom. Since a flowing multiphase mixture column is to be maintained in the flow string at all times, the required *FBHP* to lift well fluids to the surface is considerably high. As described in Section 2.5.3, even increased *GLRs* cannot decrease this pressure below a minimum value, which is determined by the liquid rate, well depth, flow string size, and other parameters. If the actual *FBHP*, as calculated from the well's inflow performance data and reflecting the formation pressure and the required pressure drawdown, is less than the minimum pressure required at the bottom of the flow string, the well cannot be produced by continuous flow. Therefore, in contrast with artificial lift methods utilizing a subsurface pump, *FBHPs* cannot be decreased sufficiently to produce the well. Thus, a sufficiently *high reservoir pressure* is a basic requirement for the application of continuous flow gas lifting.

- As the well's liquid rate decreases due to the depletion of the formation, continuous flow gas lifting cannot be applied until the abandonment of the well.

5.3 Principles of Continuous Flow Gas Lifting

5.3.1 Dead wells vs. gas lifting

In order to fully understand the mechanism of continuous flow gas lifting, the various reasons why flowing wells cease to flow or die have to be discussed first. Although there is a great variety of conditions that may lead to a well's dying, most of them can be reduced to two basic causes. These involve (a) the reduction of reservoir or *formation pressure*, and (b) the reduction of *gas content* in the flowing wellstream.

Let us investigate a hypothetical case shown in Figure 5–1. In the original conditions, the well flowed naturally with a gas-liquid ratio of GLR_p while its static and flowing bottomhole pressures were $SBHP_1$ and $FBHP_1$, respectively. Formation pressure then dropped to $SBHP_2$ and, in order to maintain the same liquid rate, flowing bottomhole pressure had to be reduced to $FBHP_2$, to keep the drawdown at the original level DP. Since it is assumed that the well's liquid and gas production rates did not change, flow with the same production gas-liquid ratio GLR_p must take place in the tubing string. After calculating or plotting the corresponding pressure traverse curve shown in dashed line, it turns out that the well *cannot flow* against the required wellhead pressure (*WHP*). All other parameters being constant, the drop in formation pressure, therefore, can lead to a well's dying.

The second basic case, as shown in Figure 5–2, involves a reduction of the gas content of the flowing mixture in the tubing string. Even with the original formation and *FBHPs*, the reduction of the gas-liquid ratio from GLR_{p1} to GLR_{p2} can increase the flowing pressure gradient to such an extent that the

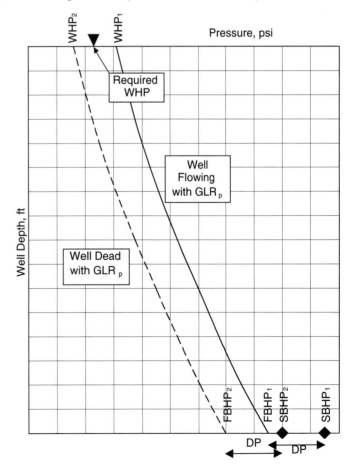

Fig. 5–1 How a well dies due to a significant drop in formation pressure.

well *can no longer flow* against the required *WHP*. Reduction of the production *GLR* can be caused, for example, by an increase in water cut.

It must be obvious that the wells in the previous two examples would return to flowing production if production *GLRs* could be increased sufficiently. This is what happens in continuous flow gas lifting when gas from an outside source is injected into the well. In both examples, if a sufficient amount of lift gas is injected at the well bottom into the flowing wellstream, the total *GLR* can be increased to such an extent that tubing pressure at the surface attains or exceeds the required *WHP*. Under these conditions, the well resumes production at the previous liquid rate.

In summary, continuous flow gas lifting can be considered an *extension* of the well's flowing life. It is accomplished by continuously injecting a proper amount of lift gas into the upward-moving wellstream to supplement the well's production *GLR*. The artificially increased *GLR* reduces the flowing gradient in the tubing string and allows existing formation pressure to lift well fluids to the surface against the *WHP* required to move them into the gathering system.

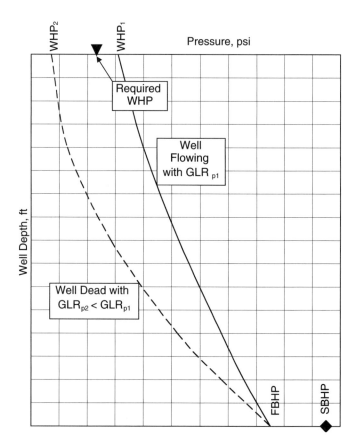

Fig. 5–2 How a well dies: the effect of decreasing formation GLRs.

5.3.2 Basic design of a continuous flow installation

There are two basic operational parameters to be determined when designing a continuous flow gas lift installation: (a) the depth of the *operating valve, i.e.* the point of gas injection, and (b) the *amount of lift gas* to be injected at that point. The exact determination of these quantities varies with several given or assumed parameters detailed as follows, like surface injection pressure, required *WHP*, etc. In the following, a very basic case is considered where all relevant parameters are held constant.

Figure 5–3 describes the basic design of a continuous flow gas lift installation for the following conditions:

- the well's desired liquid rate is given

- the *WHP* required to move well fluids to the surface gathering system is known

- well inflow parameters are available

- surface injection gas lift pressure is specified

- the well's tubing size is given

With the previous parameters given, the two main tasks of the installation design can be accomplished by the use of a graphical procedure. Design calculations are done in a rectangular coordinate system with the ordinate representing well depth (with zero depth at the top) and the abscissa representing pressure. The use of this coordinate system provides a simple and easily understandable way to depict well pressures.

1. First the well's static bottomhole pressure (*SBHP*) is plotted at total well depth. The *kill fluid* gradient is started from this pressure and is extended toward the surface until it intersects the ordinate axis. The intersection denotes the static liquid level in the well.

2. Since inflow parameters are known, the required *FBHP* is found from the desired rate and the *IPR* (as discussed in Section 2.3) and is plotted at total well depth.

3. The next task is to plot the injection pressure distribution in the well's annulus starting from the surface operating gas lift pressure. This pressure should be available at the wellsite at all times and must be selected based on compressor discharge pressure, line losses, and an allowance for fluctuating line pressures. Starting from the surface injection pressure p_{inj}, the increase in annulus pressure can be calculated with the procedures detailed in Section 2.4.4.

4. Since liquid rate and formation gas-liquid ratio GLR_p are known, a pressure traverse starting from the *FBHP* can be established. This curve may be calculated from any multiphase flow correlation or may be traced from available flowing gradient curves.

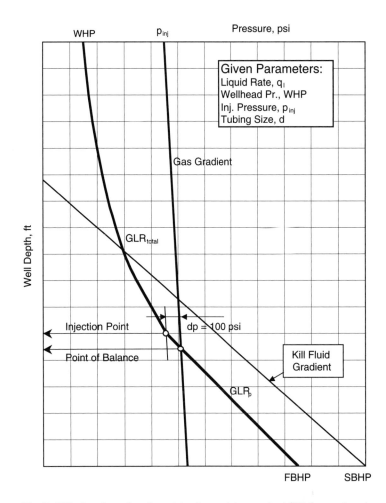

Fig. 5–3 Finding the point of gas injection and the required GLR for continuous flow gas lifting.

5. The intersection of this curve with the annulus gas gradient curve represents the *point of balance* between the tubing and annulus pressure. At this depth, injection pressure in the casing-tubing annulus equals the flowing pressure in the tubing string.

6. The *operating gas lift valve* should be run above the point of balance to ensure a pressure differential across the valve during operation. For this reason, a parallel to the gas gradient line is drawn at a distance of about 100 psi, the intersection of which with the flowing tubing pressure defines the depth of the required gas injection point.

7. Now the gas injection requirement is found. For this, the pressure traverse curve that fits between the point of injection and the *WHP* should be found, if gradient curves are used. In computer calculations, a trial-and-error procedure can be used to find the proper gas-liquid ratio GLR_{total}.

8. Injection gas requirement is found by subtracting the amount of gas supplied by the formation GLP_p from the total ratio:

$$GLR_{inj} = GLR_{total} - GLR_p \qquad\qquad 5.1$$

where: GLR_{total} = calculated GLR above the point of gas injection, scf/bbl

GLR_p = formation GLR, scf/bbl

Design Considerations

The previous procedure results in reliable designs as long as all required parameters are well known. In practice, however, this is not always the case, and the following considerations should be taken into account:

- The surface gas injection pressure should be based on the pressure always available at the wellsite. In case of a *fluctuating line pressure*, its minimum value must be taken as the design pressure. This ensures that the required surface injection pressure is available at all times.

- The *pressure differential* across the operating gas lift valve, usually taken as 100 psi, severely affects the well's FBHP and, consequently, its rate. This is especially true for high-capacity wells with high PIs, where a considerable increase of liquid rates can be achieved if the pressure differential is reduced. [4]

- Due to the well-known calculation errors in vertical multiphase pressure drop (*PD*) predictions, as well as due to fluctuations in well inflow performance, actual injection depths can differ from those found from the previous design procedure. In high-capacity wells, this can result in production losses, and *bracketing* of gas lift valves is recommended. This means that several closely spaced gas lift valves are run close to the design injection depth, facilitating gas injection at a deep enough point.

- In high-volume wells, the required gas injection volume can be more than that available at the wellhead. In such cases, lower injection *GLRs* must be used, and the restriction on gas availability may severely limit the liquid rate lifted.

- In continuous flow installations, the operating valve must be open at all times and its opening pressure should be set accordingly. Valve setting, therefore, is very critical and should be based on the minimum flowing tubing pressure and, in the case of gas-charged gas lift valves, on a proper estimate of valve temperature.

- If well inflow performance data are not reliable or are unknown, the basic design procedure detailed previously cannot be used to find the depth of the operating valve with the proper accuracy.

Example 5–1. Find the point of gas injection and the required gas volume for a well with the following data. Perforation depth is 7,000 ft, where formation pressure is 2,500 psi, the well's *PI* is 2.5 bpd/psi. The well produces 600 bpd of oil with a formation GLR_p of 100 scf/bbl through a tubing with a 2 3/8 in. OD and a *WHP* of 200 psi. Lift gas of 0.75 relative gravity is injected with a surface injection pressure of 800 psi.

Solution

The well's *FBHP* is found from the productivity index principle as:

$$p_{wf} = p_{ws} - q / PI = 2,500 - 600 / 2.5 = 2,260 \text{ psi.}$$

All graphical constructions will be accomplished with the help of Figure 2–46, a gradient curve sheet valid for the previous conditions. First, the pressure traverse below the point of gas injection is plotted starting from the *FBHP*, see Figure 5–4.

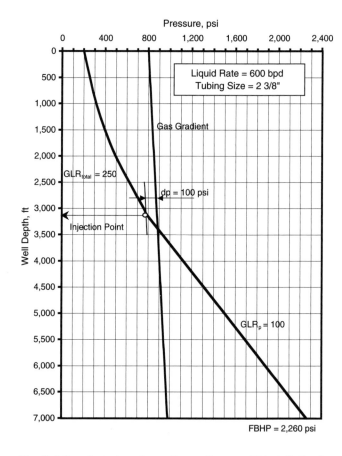

Fig. 5–4 Sample design of a continuous flow gas lift installation for Example 5–1.

Next, calculate and plot the distribution of gas pressure in the annulus. From Fig. B–1 in Appendix B, the static gas gradient for the given injection pressure and gas gravity is found as 26 psi/1,000 ft. Gas pressure at the perforation level is found as

$$800 + (26 / 1,000) 7,000 = 982 \text{ psi}$$

A parallel with the gas pressure line, drawn with a pressure differential of $dp = 100$ psi, intersects the pressure traverse valid below the gas injection at the depth of 3,100 ft, which specifies the required injection point. Using the gradient curve sheet, the pressure traverse connecting the injection point with the specified *WHP* can be found. The gas requirement belonging to the proper curve equals 250 scf/bbl, and the injection requirement is found from Equation 5.1 as

$$GLR_{inj} = 250 - 100 = 150 \text{ scf/bbl}$$

5.3.3 Basic considerations

There are two very basic considerations affecting continuous flow gas lift design that must be mentioned before any detailed discussion. Both relate to the effectiveness of the gas lift process and can be easily understood by using basic multiphase flow principles only. One shows the effect of injection depth, the other shows the effect of multipoint injection on the performance of a well placed on continuous flow gas lift.

5.3.3.1 The effect of injection depth.
The effect of gas injection depth on the performance of a continuous flow gas lift well will be shown for two basic cases: for constant and variable liquid rates. If the well is assumed to produce at a constant liquid rate, the effect of the depth of gas injection is schematically shown in Figure 5–5. Liquid rate being constant, the well's *FBHP* is also constant. In the knowledge of the formation gas-liquid ratio GLR_b, flowing tubing pressure starts from *FBHP* at well depth and changes along the solid curve drawn upward. Assume that three gas lift valves at different depths are installed in the well and investigate the required injection gas volumes if injection occurs through each of them in turn.

If, in each case, the well is produced against the same surface backpressure, *WHP*, then using the basic design procedure described before, the *GLRs* belonging to each injection point can be determined. In Figure 5–5, these are denoted as GLR_1, GLR_2, and GLR_3 for the injection points 1, 2, and 3, respectively. If the pressure traverse curves above each injection point are compared to the individual curves of a gradient curve sheet, then it must be obvious that the steeper the curve the higher the *GLR*, which means that in this case $GLR_1 > GLR_2 > GLR_3$. Thus, without performing any detailed calculations, a basic conclusion can be drawn: *deeper gas injection points* result in lower injection gas requirements.

Let us now investigate the case with variable liquid rates and assume that the well produces against a constant *WHP*. Further, assume that there is a sufficient volume of injection gas available to achieve the *minimum*

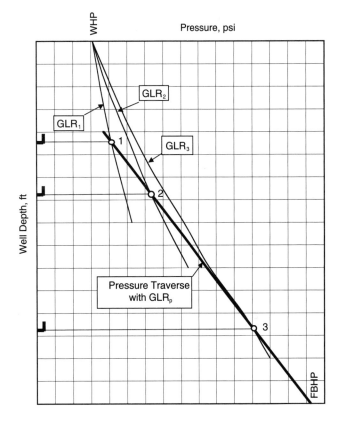

Fig. 5–5 Schematic drawing showing the effect of injection depth on the required GLR for a well producing a constant liquid flow rate.

pressure gradient above the point of gas injection. As shown in Figure 5–6, a pressure traverse curve belonging to the minimum pressure gradient can be extended toward the well bottom. Again, gas lift valves run to three different depths are assumed. Starting from points 1, 2, and 3, pressure traverses representing the pressure distribution below the gas injection points and belonging to the well's production gas-liquid ratio, GLR_p, can be constructed. Calculation of these curves requires an iterative solution because the liquid rates to be used depend on the values of the flowing bottomhole pressures $FBHP_1$, $FBHP_2$, and $FHBP_3$. If the rates thus found are q_{l1}, q_{l2}, and q_{l3}, then it is easy to see that they increase in the same order, i.e. $q_{l1} < q_{l2} < q_{l3}$. Therefore, gas injection at deeper depths results in higher liquid production rates from the same well.

From the previous section, a generally applicable conclusion can be drawn on the effect of gas injection depth on the performance of a well placed on continuous flow gas lift: deeper gas injection points are always preferable. This is the reason why usually the injection of gas at the deepest possible point is pursued in continuous flow gas lift design.

5.3.3.2 Multipoint vs. single-point gas injection.
Multipoint gas injection or *multipointing* means that lift gas is simultaneously injected at several points along the tubing string. In contrast with intermittent operations, this practice is usually not recommended for continuous flow gas lifting for the reasons highlighted in the following example.

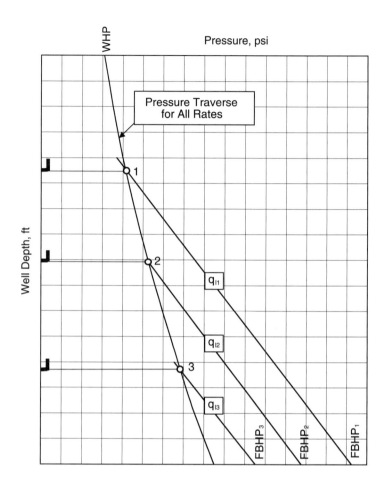

Fig. 5–6 The effect of gas injection depth on the well's liquid rate with unlimited lift gas availability.

Using the gradient curve sheet presented in Figure 2–47, gas injection at a single and two different depths is compared in Figure 5–7 for an example well. If lift gas is injected at the well bottom (Point 1) at 8,000 ft, a total *GLR* of 1,000 scf/bbl is required to lift the well's liquid production of 600 bpd against the *WHP* of 300 psi, as indicated by the solid curve. Now assume that an additional valve is located at 4,500 ft and the same total gas volume is injected through the two gas lift valves and, for simplicity, let the individual injection rates through the valves be equal. Since the bottom valve injects half of the amount as before, the pressure above it will follow the pressure traverse curve belonging to a *GLR* of 500 scf/bbl, as shown by the curve connecting points 1 and 2. At point 2, the second gas lift valve injects the same amount of gas as the lower one, increasing the *GLR* to the total value of 1,000 scf/bbl. After calculating or plotting the *WHP* developing under these conditions, we find that the well is unable to flow against the required *WHP* of 300 psi since it develops a significantly lower *WHP*.

To correct the previous situation and to enable the well to produce by continuous flow gas lift, an additional volume of gas (more than the assumed 500 scf/bbl) must be injected at the upper valve. As shown in Figure 5–7, a total *GLR* of 2,000 scf/bbl is needed to produce the well at the same conditions as with the single gas lift valve at well bottom, see the dashed line. Thus, as compared to single-point gas injection, lift gas requirement has increased twofold for the two-point gas injection case.

Even this simple example can prove that multipointing in continuous flow gas lifting can result in very inefficient operations as compared to gas injection at a single depth. Therefore, in normal operations, the required gas volume is usually injected through a single operating gas lift valve only, which is usually run to the *deepest possible depth*.

Multipointing (multipoint gas injection) can be advantageous in intermittent installations where individual valves in the valve string open and pass gas as the upward-rising liquid slug passes them, thereby ensuring a constant rising velocity for the slug. In continuous flow operations, efficient multipoint gas injection is only feasible under strictly controlled conditions, as given by Raggio [3], using a special valve string design and utilizing throttling-type PPO gas lift valves.

5.3.4 The effects of operational parameters

Even the elementary treatment of continuous flow gas lifting presented so far allows one to study the effects of the most important parameters (wellhead pressure, injection pressure, etc.) on the efficiency of fluid lifting. It is easy to understand that the efficiency of continuous flow gas lifting is related to the amount of lift gas that is used for lifting the given amount of liquid to the surface. The less the gas requirement, the more efficient is the use of its energy, and, on the other hand, high gas requirements mean poor energy utilization. The amount of lift gas needed, however, is calculated from the injection GLR found from the basic design calculations. Therefore, injection gas-liquid ratio GLR_{inj} is a good indicator of the effectiveness of continuous flow gas lifting.

A more thorough analysis of lift efficiency must include the calculation of lifting costs and has to be based on the knowledge of the power needed to compress the given amount of gas. Compressor power, however, is determined by the gas flow rate, although the pressure conditions (suction and discharge pressures) and other parameters must be considered as well. In spite of these facts, a basic evaluation of the most important operational parameters on lift efficiency of continuous flow gas lifting can be accomplished by comparing the $GLRs$ calculated under different conditions. In the following, the effects of WHP, injection pressure, and flow pipe diameter will be described. [5, 6]

5.3.4.1 Wellhead pressure. Using the basic design calculations, as detailed in conjunction with Figure 5–3, a diagram schematically depicted in Figure 5–8 can be developed if several $WHPs$ are assumed. Well inflow data and formation GLR being constant, lift gas is injected at the same depth for all cases. After finding the gas requirements for all wellhead pressures, the values GLR_1, GLR_2, and GLR_3 are found, corresponding to the wellhead pressures WHP_1, WHP_2, and

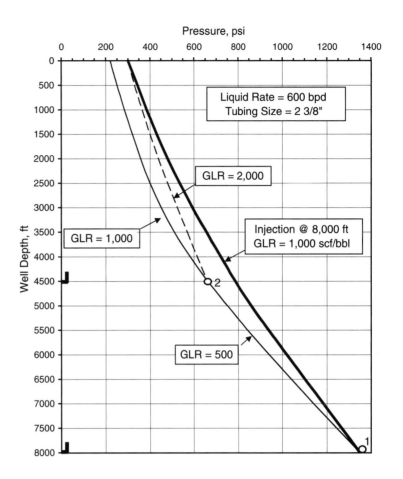

Fig. 5–7 Comparison of single-point and multipoint gas injections in a continuous flow gas lifted well.

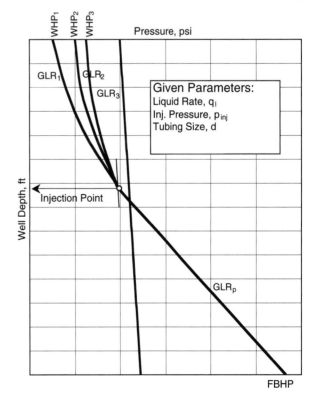

Fig. 5–8 Comparison of lift gas requirements for varying WHPs.

WHP_3. It follows from the basic concepts of vertical multiphase flow that the gas requirements increase as the WHPs increase, i.e.: $GLR_1 < GLR_2 < GLR_3$. The conclusion is, therefore, quite obvious: higher *WHPs* require more gas to be injected for producing the same liquid rate. This is the reason why, in general, in efficient continuous flow gas lift installations, the lowest possible *WHP* is desired. The use of wellhead chokes (standard in flowing wells) is usually not recommended and streamlining of the wellhead–flowline connection may be advantageous, especially for high-capacity wells. [6]

_ . _ . _ . _ . _ . _ . _ . _ . _ . _ . _ . _ . _ .

Example 5–2. Using the same well as in Example 5–1, find the gas requirement of continuous flow gas lift operations for a *WHP* of 400 psi.

Solution

As shown in Figure 5–9, the point of gas injection is at the same depth as before. Using the gradient curve sheet in Figure 2–46, the pressure traverse connecting the injection point with the new *WHP* is found. The gas requirement, as read from the proper curve, equals 1,100 scf/bbl. Injection requirement is found from Equation 5.1 as:

$$GLR_{inj} = 1,100 - 100 = 1,000 \text{ scf/bbl.}$$

As shown, an increase in the wellhead pressure from 200 psi to 400 psi entails an almost tenfold increase in lift gas requirement for the example case.

_ . _ . _ . _ . _ . _ . _ . _ . _ . _ . _ . _ . _ .

5.3.4.2 Gas injection pressure. The available surface injection pressure has a definite impact on the efficiency of continuous flow gas lifting, as schematically shown in Figure 5–10. Greater injection pressures mean that gas injection into the flow pipe can take place at greater depths, and, as presented in conjunction with Figure 5–5, deeper points of gas injection result in lower gas requirements. This is easy to explain if the average flowing gradients of the different cases are compared, since higher injection pressures mean lower gradients and, consequently, less gas is needed to decrease the flowing gradient. As a result, lift gas requirements belonging to increased injection pressures decrease, i.e. $GLR_1 > GLR_2 > GLR_3$.

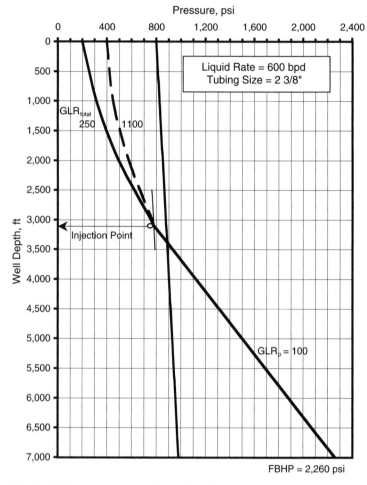

Fig. 5–9 Calculation results for Example 5–2.

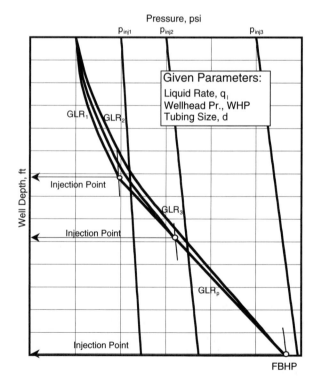

Fig. 5–10 Comparison of lift gas requirements for different surface gas injection pressures.

It can be concluded from the previous discussion that surface gas injection pressure should be set as high as possible, if a continuous flow installation with a sufficiently high efficiency is desired. In other words, gas injection at the *depth of the perforations* is ideal. Increasing the surface gas pressure, however, in addition to the implications caused by the higher gas compression requirements, has another limiting factor. It must not reach a pressure higher than the well's *FBHP* at perforation depth, since greater pressures would be unnecessary for ensuring gas injection at the bottom.

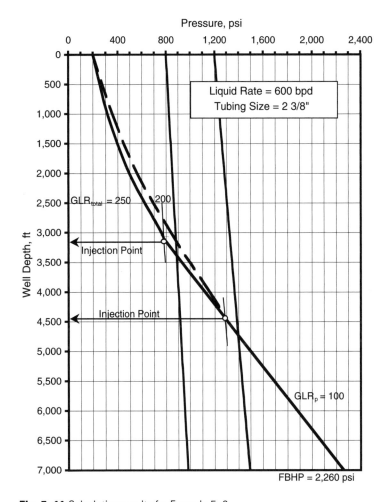

Example 5–3. Demonstrate the effect of an increased surface injection pressure of 1,200 psi on the performance of the well described in the previous examples, if *WHP* is kept at 200 psi.

Solution

The pressure traverse below the point of gas injection, as well as the gas requirement for the injection pressure of 800 psi are identical to those in the previous example and are shown in Figure 5–11.

The distribution of gas pressure in the annulus for the 1,200 psi case is calculated from Figure B–1 in Appendix B, where the static gas gradient is found as 42 psi/1,000 ft. Gas pressure at the perforation level is found from

Fig. 5–11 Calculation results for Example 5–3.

$$1{,}200 + (42 / 1{,}000)\,7{,}000 = 1{,}494 \text{ psi}$$

The line drawn parallel with the gas pressure line intersects the pressure traverse at the required injection point at a depth of 4,450 ft. The pressure traverse connecting this injection point with the *WHP* is found with the help of the gradient curve sheet in Figure 2–46. Gas requirement belonging to the proper curve equals 200 scf/bbl, and Equation 5.1 gives the injection requirement as

$$GLR_{inj} = 200 - 100 = 100 \text{ scf/bbl}$$

5.3.4.3 Tubing size. The size of the flow pipe (*i.e.* the tubing size for tubing flow installations) markedly influences the performance of vertical multiphase flow, as it was shown in chapter 2. Since continuous flow gas lifting involves flow of a multiphase mixture with varying gas content, the injection gas requirement must reflect the size of the flow pipe. The effect of pipe size on the multiphase pressure drop, however, is not as simple as that of the *WHP* and can vary in different ranges of the flow parameters. A smaller pipe may develop less pressure drop than a bigger one, provided mixture flow rates are low to medium. For higher liquid flow rates, bigger pipes become more and more favorable as the mixture rate increases because friction losses become the governing factor in the total pressure drop.

As briefly discussed previously, contrary to the effects of wellhead and injection pressures, no decisive conclusion can be made on the effect of flow pipe size on the lift gas requirements. This is the reason why this topic is discussed in more detail in the next sections, where systems analysis methods are applied to the description of continuous flow gas lift.

5.4 Description of System Performance

5.4.1 Introduction

The previous sections dealt with quite simple cases where only the tubing string's performance was studied for a known liquid flow rate. Not included in the evaluation were the performance of the formation and the surface system. However, as discussed in Section 2.7, for a proper analysis of well performance, the complete *production system* consisting of the formation, the well, the surface flowline, and the separator must be simultaneously considered. This can be accomplished by the application of systems analysis methodology. [8, 9]

The production system of a continuous flow gas lift well is shown with the node points (see Section 2.7) in Figure 5–12. The tubing string is divided at the depth of the operating gas lift valve (where the injection of lift gas takes place) in two sections. Although the liquid flow rate q_l is identical in both sections, the section below the point of gas injection contains the gas produced from the formation only, GLR_p, whereas the section above the injection point contains the injected gas volume as well, GLR_{total}. The flowline contains the same liquid and gas rates as those reaching the wellhead.

In short, the systems analysis procedure consists of the following basic steps.

- A solution node is selected first. This can be any node point in the system the proper selection of which facilitates the evaluation of different assumed conditions.

- A range of liquid flow rates is selected for subsequent calculations.

- Starting from the two endpoints of the system (Node 1 and Node 5 in Figure 5–12) and working toward the solution node, the flowing pressures at every node point are calculated for the first assumed liquid rate.

- After repeating the calculations in Step 3 for every assumed liquid flow rate, two sets of pressure–rate values will be available at the solution node. These values represent the performance curves for the two subsystems created by the solution node.

- According to a basic rule of systems analysis (see Section 2.7) inflowing and outflowing pressures at the solution node must be equal. Thus, the intersection of the two performance curves defines the well's liquid flow rate under the given conditions.

Continuous flow gas lift wells are ideally suited for systems analysis; this is why their performance is described with the use of that methodology in the next sections.

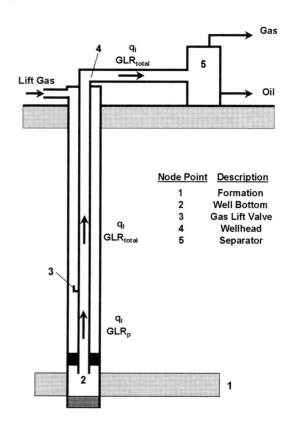

Node Point	Description
1	Formation
2	Well Bottom
3	Gas Lift Valve
4	Wellhead
5	Separator

Fig. 5–12 The production system of a well placed on continuous flow gas lift.

5.4.2 Constant WHP cases

In some cases in continuous flow gas lift operations, the operating *WHP* is assumed to stay constant, usually justified by the existence of a very short flowline. The production system in such cases is comprised of the well's tubing and the formation, and the surface flowline's performance is disregarded. This simplification, of course, reduces the amount of the required calculations but can give very misleading results, especially if an improperly sized flowline (with a too big or a too small diameter) is used.

5.4.2.1 Injection pressure given.

Many times, the available surface injection gas pressure is given. In such cases, the solution node for systems analysis is selected at the well bottom (Node 2 in Figure 5–12), resulting in two subsystems: the tubing string and the formation. The inflow pressure at Node 2, represented by the *FBHP*, must be set equal to the outflow pressure from the same node, represented by the tubing intake pressure, to find the well's liquid flow rate.

When describing the performance of a continuous flow gas lift well, the previous procedure can be applied to two different cases depending on the available amount of injection gas. These are (a) unlimited and (b) limited lift gas availability.

Unlimited Gas Availability

If the lift gas volumetric rate available at the wellsite is not limited by compressor capacity or other restrictions, total *GLR*s for the different liquid rates can be set equal to their optimum values. Optimum *GLR* means, as described in detail in Section 2.5.3.7, that value for which the tubing pressure traverse curve exhibits the lowest gradient. This *GLR*, therefore, ensures the best use of the lift gas because the pressure drop in the tubing string is at a minimum. Although optimum *GLR*s are usually high, the unlimited availability of lift gas makes it possible to use them for the calculation of well performance.

The required steps of the calculation model are detailed as follows in conjunction with the schematic drawing in Figure 5–13.

1. Plot the surface gas injection pressure p_{inj} at zero depth.

2. Starting from the surface injection pressure, calculate the gas pressure distribution in the well's annulus and plot it with well depth.

3. Assume a liquid production rate q_l and find the optimum total *GLR* belonging to that rate and the tubing size used.

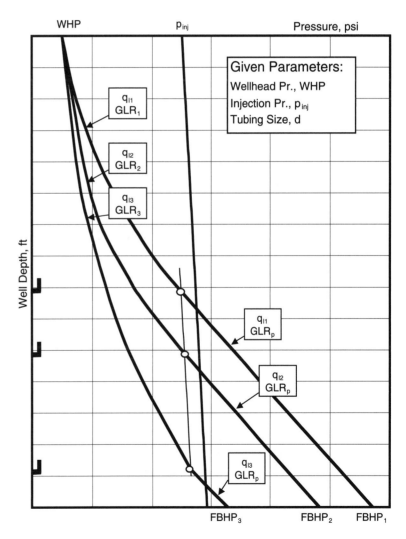

Fig. 5–13 Graphical determination of a well's tubing performance curves for known wellhead and injection pressure cases.

4. Starting from the set *WHP* and using the previous rate and *GLR*, calculate the pressure distribution in the tubing string.

5. Find the *depth of gas injection* at the intersection of the above curve and a parallel drawn to the annulus gas pressure at depth. Use of a pressure differential of $\Delta p = 100$ psi is generally accepted.

6. From the point of gas injection determined previously, calculate the pressure traverse in the tubing string below the gas injection using the formation gas-liquid ratio GLR_p.

7. Find the *FHBP* at the perforation depth from the previous pressure traverse curve.

8. Repeat Steps 4–7 with properly selected liquid rates and their optimum *GLRs*. The liquid flow rate q_l and *FBHP* pairs constitute the points of the tubing performance curve, *i.e.* the outflow pressures at Node 2 in Figure 5–12.

9. Plot the tubing performance curve in a liquid rate–bottomhole pressure coordinate system, as shown in Figure 5–14.

10. Plot the well's inflow performance curve, representing the inflow pressures at Node 2. Depending on reservoir conditions, either a constant PI model or some other IPR method can be used.

11. Find the well's liquid rate at the intersection of the inflow and outflow pressures.

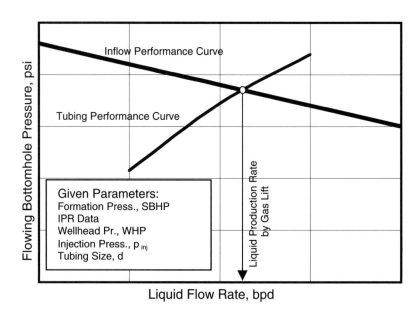

Fig. 5–14 Schematic determination of a gas lifted well's liquid rate by systems analysis with the solution node at the well bottom.

Example 5–4. Find the well's liquid production rate if *WHP* is constant at 200 psi, and there is an unlimited supply of lift gas available at the wellsite with an injection pressure of 1,000 psi. Well data follows.

Depth of perforations = 7,500 ft	Water cut = 0%
Tubing flow	Oil gravity = 30 API
Tubing ID = 2.441 in.	Water gravity = 1.0
Avg. formation pressure = 2,500 psi	Gas gravity = 0.65
Productivity Index = 2.5 bpd/psi	Inj. gas gravity = 0.6
Formation GOR = 150 scf/bbl	

Solution

The problem can be solved either with the use of gradient curve sheets or with computer calculations, main results are presented in Table 5–1 and Figure 5–15.

q_l	Above injection				Below injection		Formation
	p_4	GLR	p_3	L_{inj}	GLR	$p_{2\,OUT}$	$p_{2\,IN}$
bpd	psi	scf/bbl	psi	ft	scf/bbl	psi	psi
1,000	200	1,500	1,072	7,500	150	1,073	2,100
1,500	200	1,200	1,030	5,666	150	1,721	1,900
2,000	200	1,200	999	4,323	150	2,194	1,700

Table 5–1 System performance calculation results for Example 5–4, unlimited gas supply.

Assumed liquid flow rates and the respective optimum *GLR* values (found from gradient curve sheets) are as follows:

q_l = 1,000 bbl and *GLR* = 1,500 scf/bbl

q_l = 1,500 bbl and *GLR* = 1,200 scf/bbl

q_l = 2,000 bbl and *GLR* = 1,200 scf/bbl

Pressure traverses valid above the point of gas injection and started from the *WHP* are plotted first, and their intersections with the annulus gas pressure give the three gas injection points, see Table 5–1. Starting from the injection points, pressure traverses using the formation *GLR* of 150 scf/bbl are constructed, and the *FBHPs* corresponding to the three liquid rates are found. These values, denoted $p_{2\,OUT}$ in Table 5–1, are the outflow pressures at Node 4 (see Fig. 5–12). Plotted in function of the liquid rate in Figure 5–16, they represent the tubing performance curve. Now *FBHPs* belonging to the assumed liquid rates are calculated from the well's IPR, denoted p_{2IN} in Table 5–1. These values also are plotted in Figure 5–16 as the inflow performance curve. Intersection of the two curves satisfies the requirement $p_{2\,IN} = p_{2\,OUT}$, *i.e.* the inflow and outflow pressures at Node 2 (the well bottom) are equal, and the well produces a liquid rate of about 1,620 bbl.

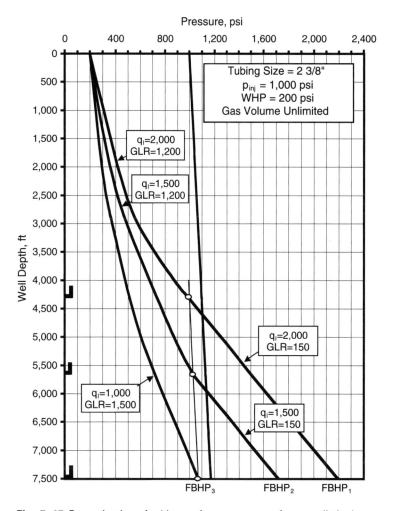

Fig. 5–15 Determination of tubing performance curves for an unlimited gas supply case in Example 5–4.

Limited Gas Availability

Many times the volumetric rate of the lift gas available at the wellsite is limited by compressor capacity or other restraints. In these cases, the available gas injection rate may not be sufficient to achieve the optimum *GLR* in the tubing string above the point of gas injection. System performance calculations, therefore, cannot follow the procedure described for the unlimited gas volume case. Although the main logic of the calculation process is similar to that presented in conjunction with Figure 5–13, the steps should be modified as follows.

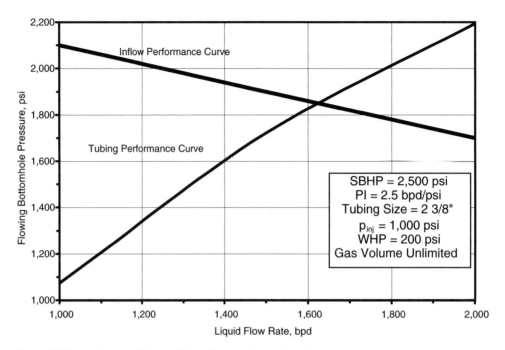

Fig. 5–16 Determination of the well's liquid rate for Example 5–4.

1. Plot the surface gas injection pressure p_{inj} at zero depth.

2. Starting from the surface injection pressure, calculate the gas pressure distribution in the well's annulus and plot it with well depth.

3. Take the first value of the available lift gas rate Q_g.

4. Assume a liquid production rate q_l and find the total GLR valid above the point of gas injection and calculated from the available injection gas rate and the well's formation GLR:

$$GLR_{total} = \frac{Q_g}{q_l} + GLR_p$$

5. Starting from the set WHP and using the previous rate and GLR, calculate the pressure distribution in the tubing string.

6. Find the depth of gas injection at the intersection of the previous curve and a parallel drawn to the annulus gas pressure at depth. Use of a pressure differential of $\Delta p = 100$ psi is generally accepted.

7. From the point of gas injection determined previously, calculate the pressure traverse in the tubing string below the gas injection using the formation gas-liquid ratio GLR_p.

8. Find the $FHBP$ at the perforation depth from the previous pressure traverse curve.

9. Repeat Steps 4–8 with selected liquid rates and total $GLRs$ calculated from the formula in Step 4. Plot the liquid flow rate q_l and $FBHP$ pairs in a liquid rate—bottomhole pressure coordinate system. This constitutes the tubing performance curve belonging to the available injection gas rate Q_g.

10. Select the next value of the available lift gas rate Q_g and repeat Steps 4–8. Plot the points of the tubing performance curve belonging to the actual injection gas rate Q_g.

11. Plot the well's inflow performance curve, representing the inflow pressures at Node 2, on the same chart. Use the proper IPR, either with a constant PI or using the proper model.

12. Find the well's possible liquid flow rates at the intersection of the inflow and outflow pressures.

—··—··—··—··—··—··—··—··—

Example 5–5. Using the same data as in the previous example, find the well's performance if gas rates of 1.0, 1.5, and 1.8 MMscf/d are available for gas lifting.

Solution

Calculation results for the Q_g = 1 MMscf/d case are shown in Figure 5–17. As seen from Table 5–2, available *GLRs* above the point of injection are much less than the optimum values used for the unlimited gas supply case. Tubing performance curves for the three gas rates along with the well's inflow performance curve are depicted in Figure 5–18, from which the well's possible liquid rates can be found. As seen, the well's liquid flow rate decreases as the available injection gas volume decreases.

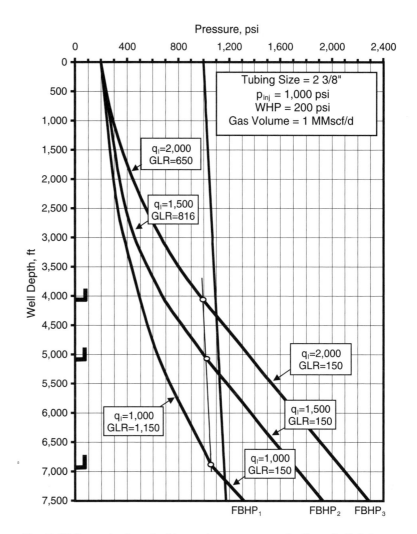

Fig. 5–17 Determination of tubing performance curves for Example 5–5 for an available gas injection rate of 1 MMscf/d.

Q_g	q_l	Above injection				Below injection		Formation
		p_4	GLR	p_3	L_{inj}	GLR	$p_{2\,OUT}$	$p_{2\,IN}$
Mscf/d	bpd	psi	scf/bbl	psi	ft	scf/bbl	psi	psi
1,000	1,000	200	1,150	1,058	6,892	150	1,310	2,100
	1,500	200	816	1,016	5,053	150	1,929	1,900
	2,000	200	650	993	4,061	150	2,286	1,700
1,500	1,000	200	1,650	1,072	7,500	150	1,073	2,100
	1,500	200	1,150	1,027	5,542	150	1,763	1,900
	2,000	200	900	999	4,324	150	2,194	1,700
1,800	1,000	200	1,950	1,072	7,500	150	1,047	2,100
	1,500	200	1,350	1,030	5,666	150	1,721	1,900
	2,000	200	1,050	1,000	4,371	150	2,177	1,700

Table 5–2 System performance calculation results for Example 5–5, limited gas supply.

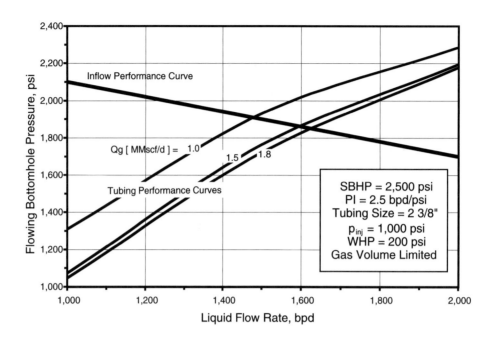

Fig. 5–18 Determination of the well's possible liquid rates for Example 5–5, using different available gas injection rates.

5.4.2.2 The Equilibrium Curve method.

The use of the Equilibrium Curve in continuous flow gas lift design and analysis was introduced by the MERLA Company. [7] The Equilibrium Curve was developed for cases with known wellhead but unknown injection pressures, and the solution node is now selected at the gas injection point. (Node 3 in Figure 5–12) In order to describe the performance of the well, inflow and outflow pressures in Node 3 are to be found. For this, pressure calculations for several *GLRs* are performed starting from the wellhead, moving downward and, at the same time, starting from the well bottom and going up. The intersection of the pressure traverses specifies the actual gas injection point. This procedure, performed for a sufficient number of liquid flow rates, results in several possible injection points and the curve connecting these points specifies the Equilibrium Curve for the given well.

As before, lift gas availability may be restricted at the surface, or an unlimited amount of gas can be used for gas lifting.

Unlimited Gas Availability

The construction of the Equilibrium Curve is best described by a graphical procedure in conjunction with Figure 5–19.

1. Plot the surface gas injection pressure p_{inj} at zero depth.

2. Starting from the surface injection pressure, calculate the gas pressure distribution in the well's annulus and plot it with well depth.

3. A liquid flow rate q_l and the *optimum total GLR*, belonging to that rate and the tubing size, are selected.

4. Starting from the prescribed *WHP* at the surface, a pressure traverse in the tubing string is established with the given liquid rate and *GLR*. This curve represents the flowing pressure distribution in the tubing string above the point of gas injection.

5. Based on the well's inflow performance data, the *FBHP* valid at the given liquid flow rate is calculated from the well's IPR equation. Either a constant PI model or some other IPR method can be used, depending on reservoir conditions.

6. Starting from the *FBHP* at the depth of the perforations, a pressure traverse is extended upward, using the liquid rate q_l and the formation gas-liquid ratio GLR_p.

7. The intersection of the two pressure traverses is determined and marked. This is the *point of gas injection* for the given case.

8. Repeat the above steps with new liquid rates and their optimum *GLR* values.

9. Connect the injection points found in Step 7 to construct the well's Equilibrium Curve.

The use of the Equilibrium Curve is also shown in Figure 5–19. For any surface gas injection pressure p_{inj} the intersection of the gas pressure curve (corrected with the necessary pressure differential Δp) and the Equilibrium Curve determines the required *point of gas injection*. The depth and pressure of the gas injection being known, the well's *FBHP* is determined by extending a pressure traverse down to the perforation depth, using the formation *GLR*. Based on the *FBHP*, the well's liquid rate is found from the IPR equation.

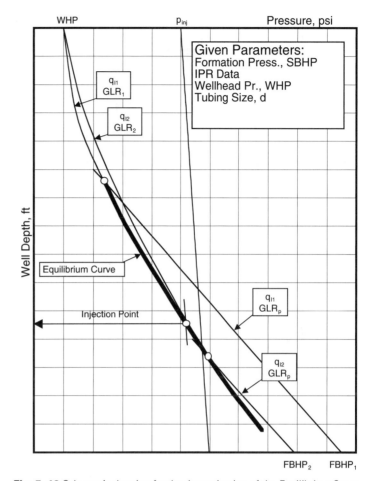

Fig. 5–19 Schematic drawing for the determination of the Equilibrium Curve.

Example 5–6. For the same data as in the previous examples, construct the well's Equilibrium Curve if the available gas injection rate is unlimited.

Solution

Calculation data are contained in Table 5–3 and the results are presented in Figure 5–20. Liquid flow rates of 1,000, 1,500, 1,800, and 2,000 bpd were assumed, based on the data of previous examples. The case for $q_l = 1,000$ bpd is detailed in the following.

The optimum *GLR* for the given rate was found as 1,500 scf/bbl from gradient curve sheets. The pressure traverse valid above the gas injection point is started from the *WHP* of 200 psi and is characterized by the pairs of $p_{3\,OUT}$ and L values, contained in Table 5–3, and plotted in Figure 5–20. These values can be found from the proper gradient curve sheet or from computer calculations. The pressure traverse representing the tubing section below the injection point is calculated with the formation *GLR* of 150 scf/bbl. It is started from the *FBHP* valid at the given 1,000 bpd liquid rate, denoted as p_2 in Table 5–3. (Compare with Fig. 5–12.) Coordinates of the pressure traverse are shown in the columns $p_{3\,IN}$ and L and are plotted in the figure. The intersection of the two pressure traverses defines the required point of gas injection for the given case. Its pressure and depth are denoted as p_{inj} and L_{inj}, respectively and indicated by a small circle in Figure 5–20.

After repeating the previous calculations for the rest of liquid rates, four points of gas injection are found, as shown in Figure 5–20. The curve connecting these points defines the Equilibrium Curve of the well for the conditions of unlimited lift gas availability. As already discussed, this curve allows the determination of the well's rate for different surface injection pressures. If an injection pressure of $p_{inj} = 1,000$ psi, as used in the

q_I	Above Injection				Below Injection				p_{inj}	L_{inj}
	p_4	GLR	$p_{3\,OUT}$	L	p_2	GLR	$p_{3\,IN}$	L		
bpd	psi	scf/bbl	psi	ft	psi	scf/bbl	psi	ft	psi	ft
1,000	200	1,500	200	0	2,100	150	2,100	7,500	475	2,650
			250	1,000			1,100	4,500		
			350	2,000			400	2,500		
			520	3,000						
1,500	200	1,200	200	0	1,900	150	1,900	7,500	667	3,697
			280	1,000			920	4,500		
			370	2,000			600	3,500		
			530	3,000						
			710	4,000						
1,800	200	1,200	200	0	1,780	150	1,780	7,500	1,200	5,868
			410	2,000			1,420	6,500		
			780	4,000			1,080	5,500		
			1,230	6,000						
2,000	200	1,200	200	0	1,700	150	1,700	7,500	1,564	7,111
			440	2,000			1,350	6,500		
			800	4,000						
			1,300	6,000						
			1,630	7,500						

Table 5–3 Equilibrium curve calculation results for Example 5–6, unlimited gas supply.

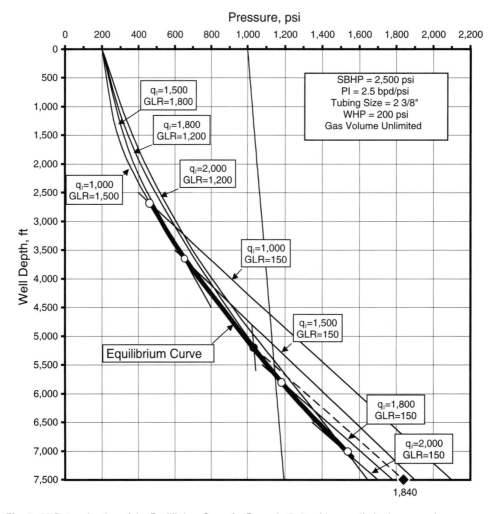

Fig. 5–20 Determination of the Equilibrium Curve for Example 5–6, with an unlimited gas supply.

previous examples, is taken then the *FBHP* is found at 1,840 psi as shown in the dashed line. The well's liquid rate for this pressure is calculated from the PI equation as q_l = 2.5 (2,500 − 1,840) = 1,650 bpd. This compares well with the result (1,629 bpd) of Example 5–4, proving one important feature of systems analysis, *i.e.* that results are independent of the selection of the solution node.

Limited Gas Availability

If the volumetric flow rate of lift gas is limited at the wellsite, an Equilibrium Curve for that gas rate is determined very similar to the procedure described in Figure 5–19. The following steps describe the calculation procedure for constructing Equilibrium Curves for several different available gas rates.

1. Plot the surface gas injection pressure p_{inj} at zero depth.

2. Starting from the surface injection pressure, calculate the gas pressure distribution in the well's annulus and plot it with well depth.

3. Take the first available gas rate Q_g.

4. Assume a liquid production rate q_l and find the total *GLR* for the tubing string above the gas injection point from the available injection gas rate and the well's formation *GLR*:

$$GLR_{total} = \frac{Q_g}{q_l} + GLR_p$$

5. Start from the *WHP* and plot a pressure traverse in the tubing string for the given liquid rate and *GLR*, representing the flowing pressure above the point of gas injection.

6. The *FBHP* valid at the given liquid flow rate is calculated from the well's IPR equation.

7. Starting from the *FBHP* at the perforations, a pressure traverse is extended upward, using the liquid rate q_l and the formation gas-liquid ratio GLR_p.

8. The intersection of the two pressure traverses is determined and marked. This is the *point of gas injection* for the given case.

9. Repeat Steps 4–8 with new assumed liquid rates.

10. Connect the injection points found in Step 9 to construct the Equilibrium Curve belonging to the available gas rate.

11. Take the next available lift gas rate and repeat the previous procedure to find the next Equilibrium Curve.

As seen previously, several Equilibrium Curves will be determined, one for each available lift gas rate. These curves, as before, can be used to find the well's liquid rate for different injection gas pressures.

Example 5–7. For the conditions of the previous examples, construct the well's Equilibrium Curves for the next available gas injection rates: 1.0, 1.5, and 1.8 MMscf/d.

Solution

Calculation data are contained in Table 5–4. Basically, the calculations described in the previous example are repeated for each gas rate. Liquid rates of 1,000, 1,500, and 1,800 bpd are assumed for each case, the *GLR*

Q$_g$	q$_l$	Above Injection				Below Injection				p$_{inj}$	L$_{inj}$
		p$_4$	GLR	p$_{3\,OUT}$	L	p$_2$	GLR	p$_{3\,IN}$	L		
Mscf/d	bpd	psi	scf/bbl	psi	ft	psi	scf/bbl	psi	ft	psi	ft
1,000	1,000	200	1,150	200	0	2,100	150	2,100	7,500	345	2,347
				240	1,000			1,080	4,500		
				310	2,000			400	2,500		
				415	3,000			230	2,000		
	1,500	200	816	200	0	1,900	150	1,900	7,500	1,117	5,225
				310	1,000			1,038	5,000		
				440	2,000			520	3,500		
				700	3,500			175	2,500		
				1,200	5,500						
	1,800	200	705	200	0	1,700	150	1,700	7,500	N/A	N/A
				550	3,000						
				1,070	5,000						
				2,024	7,500						
1,500	1,000	200	1,650	200	0	2,100	150	2,100	7,500	308	2,249
				230	1,000			1,080	4,500		
				280	2,000			400	2,500		
				390	3,000			230	2,000		
	1,500	200	1,150	200	0	1,900	150	1,900	7,500	654	3,886
				270	1,000			1,038	5,000		
				350	2,000			520	3,500		
				500	3,000			175	2,500		
				670	4,000						
	1,800	200	983	200	0	1,700	150	1,700	7,500	N/A	N/A
				275	3,000						
				795	5,000						
				1,749	7,500						
1,800	1,000	200	1,950	200	0	2,100	150	2,100	7,500	289	2,177
				210	1,000			1,080	4,500		
				280	2,000			400	2,500		
				370	3,000			230	2,000		
	1,500	200	1,350	200	0	1,900	150	1,900	7,500	588	3,700
				250	1,000			1,038	5,000		
				320	2,000			520	3,500		
				460	3,000			175	2,500		
				630	4,000						
	1,800	200	1,150	200	0	1,700	150	1,700	7,500	1,574	7,128
				370	2,000			1,350	6,500		
				960	5,000						
				1,670	7,500						

Table 5–4 Equilibrium curve calculation results for Example 5–7, limited gas supply.

values calculated from the gas and liquid rates for the pressure traverses above the point of gas injection are listed in the table. As seen in some cases, no intersection between the pressure traverses exists, making gas injection impossible. Equilibrium Curves belonging to the three cases are shown in Figure 5–21, where description of individual pressure traverses is omitted for clarity. Individual points of resulting curves are indicated by different markers.

Again, if the injection pressure of p_{inj} = 1,000 psi and an available gas rate of 1.8 MMscf/d is assumed, then the point of gas injection is indicated by a solid circle on the appropriate Equilibrium Curve. The *FBHP* belonging to this injection point, as shown by the dashed line, equals 1,830 psi. The liquid rate calculated from the PI equation is q_l = 2.5 (2,500 – 1,830) = 1,675 bpd, a value different from previous results due to tolerances in graphical construction only.

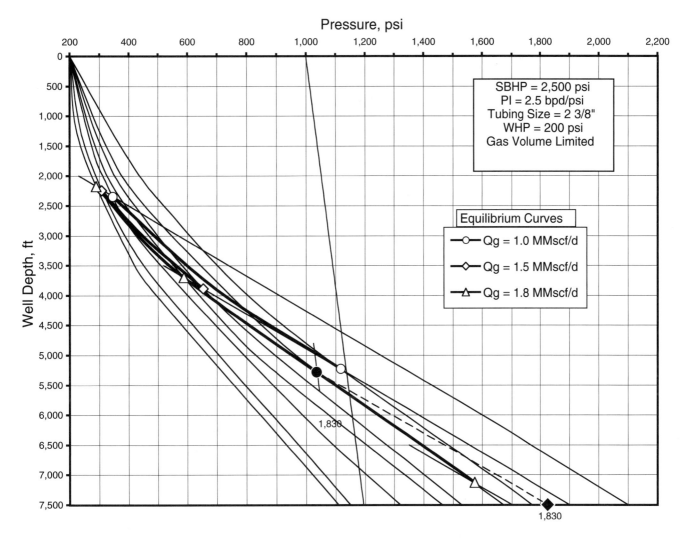

Fig. 5–21 Equilibrium Curves for different available gas injection rates for Example 5–7.

5.4.3 Variable WHP cases

When evaluating the performance of a continuous flow gas lifted well, the effects of all system components must be taken into account, and disregarding the flowline (as done in the previous sections) is not recommended. If the performance of the flowline is also included, then the whole production system of the well, detailed in Figure 5–12, must be studied at the same time. Such cases could be considered as full *Systems Analyses* for which some important applications will be shown in the following.

As already discussed in chapter 2 and in conjunction with Figure 5–12, the selection of the solution node is of prime importance for a meaningful systems analysis. The most common nodes used for this purpose are at the well bottom and at the wellhead, designated as Node 2 and Node 4 in Figure 5–12, respectively. Both selections have their advantages, as detailed later.

5.4.3.1 Solution at the well bottom. By selecting the well bottom (Node 2 in Figure 5–12) as the solution node, the well's production system is divided into the subsystems formed, on one hand, by the formation, and on the other hand by the separator, the flowline, and the production conduit. According to systems analysis principles, performance curves of the two subsystems must be established in order to find the well's liquid production rate.

The required steps for a complete systems analysis are described and can be followed on Figure 5–22.

1. The gas volumetric flow rate available at the wellsite for gas lifting is to be found.

2. A proper range of liquid production rates is selected.

3. Take the first liquid production rate q_l and find the WHP based on the given separator pressure and the pressure drop arising in the flowline. For the calculation of the multiphase pressure drop in the flowline, use the total GLR, i.e. the sum of the formation and injection GLRs.

4. Starting from the WHP, calculate a pressure traverse in the tubing string (representing the flowing pressure above the point of gas injection) for the given liquid rate and total GLR.

5. At the intersection of the above curve and a parallel drawn to the annulus gas pressure at depth, (found from the given or assumed surface gas injection pressure p_{inj}) find the depth of gas injection. Use of a pressure differential of Δp =100 psi is generally accepted.

6. From the point of gas injection determined previously, calculate the pressure traverse in the tubing string below the gas injection using the formation gas-liquid ratio GLR_p.

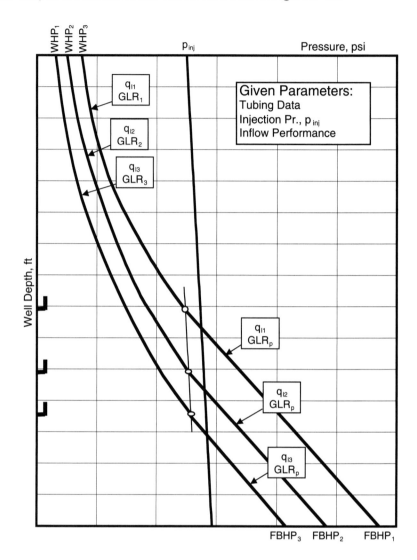

Fig. 5–22 Schematic determination of tubing performance curves for a full systems analysis with the solution node at the well bottom.

7. Find the FHBP at the perforation depth from the previous pressure traverse curve.

8. Repeat Steps 3–7 for the rest of the selected liquid rates and plot the liquid flow rate q_l and FBHP pairs in a liquid rate–bottomhole pressure coordinate system. This constitutes the tubing performance curve belonging to the available injection gas rate Q_g.

9. Using the coordinate system detailed previously, plot the well's IPR curve.

10. The intersection of the two curves determines the well's liquid rate for continuous flow gas lifting, because it is here where inflow and outflow pressures at the solution node (Node 2 in Figure 5–12) are equal.

11. If more than one gas injection rate is considered, the calculations described in Steps 3–8 are repeated and well liquid rates for the injection rates at hand can be found.

Example 5–8. Using well data from the previous examples, evaluate the performance of the well with due consideration for the conditions of the flowline. Separator pressure is 100 psi, flowline length is 3,000 ft, and its inside diameter is 2.5 in. Use gas injection rates of 200, 400, 700, 1,000, and 1,500 Mscf/d.

Solution

Calculations were performed on the computer, and main results are given in Table 5–5. Based on the results in previous examples, assumed liquid rates were selected as follows: 700, 900, 1,100, and 1,300 bpd. The table lists, for each available injection rate Q_g, the calculated pressures at the wellhead p_4 at the gas injection point p_3, and at the outflow side of Node 2, $p_{2\ OUT}$, as well as the *FBHP* found from the well's IPR equation, representing the inflow pressure to Node 2, $p_{2\ IN}$. The final columns contain the well's *FBHP* and liquid rate, valid for the given lift gas injection rate. Performance curves are plotted in Figure 5–23. It can be seen that an increase in gas volume increases the well's liquid production to a point, from where the well's rate decreases.

Q_g	q_l	Flowline		Above Injection		Below Injection		Formation	Solution	
		p_5	p_4	L_{inj}	p_3	GLR	$p_{2\ OUT}$	$p_{2\ IN}$	p_{wf}	q_l
Mscf/d	bpd	psi	psi	ft	psi	scf/bbl	psi	psi	psi	ft
200	700	100	129	5,294	1,022	150	1813	2216	2,114	955
	900	100	139	4,576	1,005	150	2059	2136		
	1,100	100	149	4,050	993	150	2243	2056		
	1,300	100		3,674	985	150	2379	1976		
400	700	100	145	6,297	1,045	150	1477	2216	2,050	1,115
	900	100	157	5,367	1,023	150	1791	2136		
	1,100	100	170	4,660	1,007	150	2034	2056		
	1,300	100	183	4,098	994	150	2232	1976		
700	700	100	168	6,728	1,055	150	1334	2216	2,020	1,192
	900	100	183	5,741	1,032	150	1665	2136		
	1,100	100	199	5,000	1,015	150	1919	2056		
	1,300	100	214	4,383	1,001	150	2134	1976		
1,000	700	100	190	6,985	1,061	150	1249	2216	2,020	1,192
	900	100	208	5,729	1,032	150	1669	2136		
	1,100	100	225	4,998	1,015	150	1919	2056		
	1,300	100	242	4,396	1,001	150	2130	1976		
1,500	700	100	225	6,919	1,059	150	1271	2216	2,042	1,137
	900	100	246	5,519	1,027	150	1740	2136		
	1,100	100	266	4,751	1,009	150	2003	2056		
	1,300	100	286	4,187	996	150	2201	1976		

Table 5–5 System performance calculation results for Example 5–8, solution node at well bottom.

It is interesting to compare the results of this example with those of Example 5–5, where the *WHP* was kept constant at 200 psi. The possible maximum liquid rate achieved by gas lifting, as seen in Figure 5–18 was around 1,620 bpd, at a total *GLR* of about 1,120 scf/bbl. This *GLR* is very close to the optimum *GLR* valid for the given liquid rate (see Example 5–4). In the present case, maximum liquid rate is about 1,200 bpd at a gas injection rate somewhere between 0.7 and 1.0 MMscf/d, and the approximate total *GLR* is around 900 scf/bbl. This clearly shows the effect of a common design mistake when it is assumed that the optimum *GLR* exhibiting the minimum flowing gradient can be achieved in a gas lifted well. As discussed by Brown [10] and also proved by the previous example, the injection gas volume required for maximizing the liquid rate is always less than the one belonging to the *minimum flowing gradient* in the tubing string. The reason for this is that pressure drop in the flowline can significantly increase the flowing *WHP* and hence reduce the well's liquid rate.

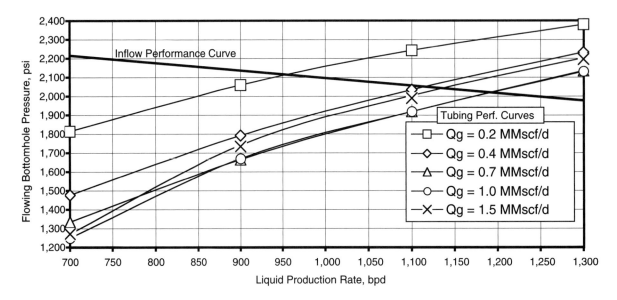

Fig. 5–23 Determination of the well's possible liquid rates for Example 5–8, using different available gas injection rates.

5.4.3.2 Solution at the wellhead.
The solution node at the wellhead (Node 4 in Figure 5–12) creates the subsystems separator + flowline and formation + tubing. Establishing the performance curves for the two subsystems follows the procedure described following and is illustrated in Figure 5–24.

1. Available gas volumetric flow rates are determined, and a properly selected range of liquid production rates is set up.

2. Take the first liquid production rate q_l and calculate the well's *FBHP* from the IPR equation.

3. Starting from the *FBHP* at the depth of the perforations, extend a pressure traverse valid for the given liquid rate, tubing diameter, and the formation GLR_p.

4. At the intersection of the pressure traverse and a parallel drawn to the annulus gas pressure at depth, (found from the given or assumed surface gas injection pressure p_{inj}) find the *depth of gas injection*. Use a pressure differential of $\Delta p = 100$ psi.

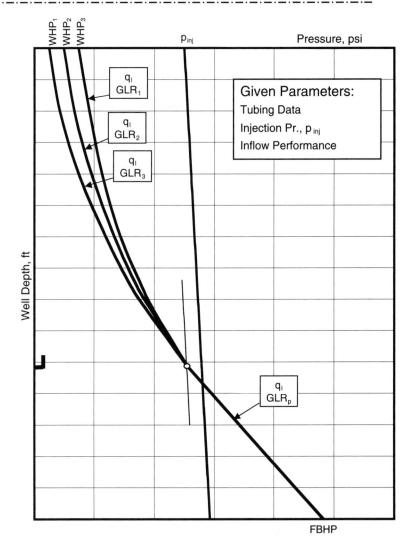

Fig. 5–24. Schematic determination of WHPs for a full systems analysis with the solution node at the wellhead and one assumed liquid rate.

5. Take an injection gas rate and calculate the pressure traverse above the point of gas injection, starting from the injection point found in the previous step. The *GLR* to be used equals the sum of the injection and formation *GLR*s.

6. Find the *WHP* at the surface, from the pressure traverse just calculated.

7. Repeat Steps 5–6 for the remaining gas injection rates and determine the *WHP*s achieved by the different gas injection rates (see Fig. 5–24).

8. Plot the liquid flow rate q_l and flowing *WHP* pairs in a liquid rate—*WHP* coordinate system for each gas injection rate. These constitute the tubing performance curves belonging to the available injection gas rates.

9. Next, calculate the performance curves for the flowline at the available injection rates. For this, first take one gas injection rate.

10. Take the first liquid rate and find the pressure at the intake end of the flowline by adding the pressure drop in the flowline to the separator pressure p_{sep}.

11. Repeat Step 10 for all assumed liquid rates and available injection rates.

12. The pressures at the intake end of the flowline, plotted in the function of liquid rate for the different injection rates (using the same coordinate system as in Step 8), define the performance curves for the separator–flowline subsystem.

13. The inflow and outflow pressures at the solution node (Node 4 in Figure 5–12) must be set equal to find the possible flow rates by gas lifting. Therefore, intersections of the two performance curves determine the well's possible liquid rates

— · —

Example 5–9. Solve the previous problem with the solution node selected at the wellhead.

Solution

Details of computer calculations are contained in Table 5–6. Liquid flow rates of 500, 700, 900, 1,100, and 1,300 bpd were assumed. For each combination of available gas rates Q_g and liquid rates q_l, first the pressure at the inlet end of the flowline $p_{4\ OUT}$ is calculated from the separator pressure p_s and the pressure drop due to multiphase flow. Next, *FBHP*s p_2 are found from the well's IPR equation. As seen in the table, data of the injection points depend on liquid rate only (compare with Fig. 5–24). Pressure calculations above the injection point give possible *WHP*s $p_{4\ IN}$. The final two columns of the table present the pressures and rates of the solution points for each gas injection rate.

Results are plotted in Figure 5–25, where the curves denoted *Outflow* show the performance of the flowline, whereas the *Inflow* curves represent the performance of the rest of the production system. Since at any node, inflow and outflow pressures must be equal, the solid curve (connecting the intersections of the individual curves of the two sets with the same gas injection rate) presents the possible liquid rates from the well.

The maximum possible liquid rate with continuous flow gas lifting, as read from Figure 5–25, is around 1,200 bpd at a gas injection rate of 1.0 MMscf/d and the total *GLR* for this operating point is about 990 scf/bbl. If compared to the results of the previous example, practically identical results were reached, showing the systems analysis results are independent of the selection of the solution node.

Q_g	q_l	Flowline		Below Injection			Above	Solution	
		p_5	$p_{4\,OUT}$	p_2	L_{inj}	p_3	$p_{4\,IN}$	p_{wh}	q_l
Mscf/d	bpd	psi	psi	psi	ft	psi	psi	psi	ft
200	500	100	119	2,300	3,858	989	354	141	954
	700	100	128	2,216	4,103	994	251		
	900	100	139	2,136	4,349	1,000	163		
	1,100	100	149	2,056	4,596	1,006	86		
	1,300	100		1,976	4,843	1,011			
400	500	100	132	2,300	3,858	989	440	171	1,115
	700	100	145	2,216	4,103	994	350		
	900	100	157	2,136	4,349	1,000	262		
	1,100	100	170	2,056	4,596	1,006	177		
	1,300	100	183	1,976	4,843	1,011	94		
700	500	100	152	2,300	3,858	989	479	206	1,197
	700	100	168	2,216	4,103	994	400		
	900	100	183	2,136	4,349	1,000	322		
	1,100	100	199	2,056	4,596	1,006	242		
	1,300	100	214	1,976	4,843	1,011	160		
1,000	500	100	171	2,300	3,858	989	513	234	1,197
	700	100	190	2,216	4,103	994	422		
	900	100	208	2,136	4,349	1,000	348		
	1,100	100	225	2,056	4,596	1,006	270		
	1,300	100	242	1,976	4,843	1,011	187		
1,500	500	100	203	2,300	3,858	989	541	270	1,136
	700	100	225	2,216	4,103	994	450		
	900	100	246	2,136	4,349	1,000	363		
	1,100	100	266	2,056	4,596	1,006	285		
	1,300	100	286	1,976	4,843	1,011	195		

Table 5–6 System performance calculation results for Example 5–9, solution node at the wellhead, variable gas injection rates.

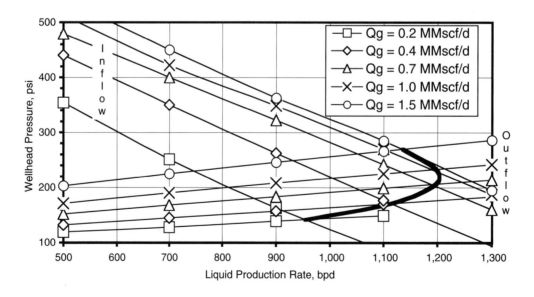

Fig. 5–25 Determination of the well's possible liquid rates for Example 5–9, using different available gas injection rates.

The previous example shows the logic of a computerized solution applied in most computer packages offering systems analysis calculations for gas lifted wells. Although this approach is almost universal in modern practice, hand calculations using gradient curve sheets can also be applied. The next example presents the same problem as in the previous case, but the solution is now oriented to the use of gradient curves.

Example 5–10. Solve the previous problem with the help of gradient curve sheets.

Solution

Compared to the previous example, the only difference is that during calculations, injection *GLRs* are assumed instead of gas injection rates. Care should be taken to use the proper *GLR* in each component of the production system, because the sum of the injected and formation *GLRs* is valid in the flowline and in the tubing section above the gas injection, whereas formation *GLR* must be used below the injection point. Calculation results contained in Table 5–7 are easy to interpret, and the solid curve shown in Figure 5–26 gives the same possible liquid rates as found in the previous example.

GLR$_{inj}$	q$_l$	Flowline		Below Injection			Above	Solution	
		p$_5$	p$_{4\,OUT}$	p$_2$	L$_{inj}$	p$_3$	p$_{4\,IN}$	p$_{wh}$	q$_l$
scf/bbl	bpd	psi	psi	psi	ft	psi	psi	psi	ft
300	500	100	116	2,300	3,849	989	310	160	1,071
	700	100	129	2,216	4,094	994	259		
	900	100	145	2,136	4,340	1,000	207		
	1,100	100	163	2,056	4,587	1,006	152		
	1,300	100	182	1,976	4,834	1,011	91		
600	500	100	126	2,300	3,849	989	412	208	1,199
	700	100	146	2,216	4,094	994	356		
	900	100	170	2,136	4,340	1,000	299		
	1,100	100	195	2,056	4,587	1,006	238		
	1,300	100	222	1,976	4,834	1,011	170		
900	500	100	136	2,300	3,849	989	450	241	1,199
	700	100	162	2,216	4,094	994	393		
	900	100	192	2,136	4,340	1,000	335		
	1,100	100	224	2,056	4,587	1,006	271		
	1,300	100	258	1,976	4,834	1,011	195		
1,200	500	100	145	2,300	3,849	989	469	262	1,154
	700	100	178	2,216	4,094	994	413		
	900	100	214	2,136	4,340	1,000	353		
	1,100	100	252	2,056	4,587	1,006	284		
	1,300	100	291	1,976	4,834	1,011	495		
1,500	500	100	155	2,300	3,849	989	486	280	1,112
	700	100	193	2,216	4,094	994	425		
	900	100	235	2,136	4,340	1,000	362		
	1,100	100	278	2,056	4,587	1,006	286		
	1,300	100	322	1,976	4,834	1,011	179		

Table 5–7 System performance calculation results for Example 5–10, solution node at the wellhead, variable injection GLRs.

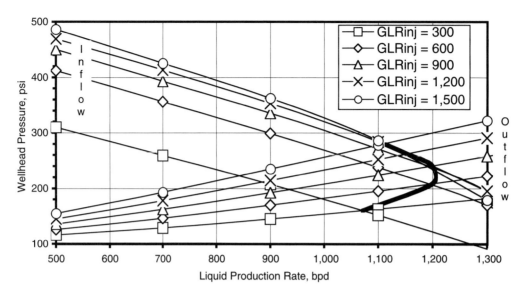

Fig. 5–26 Determination of the well's possible liquid rates for Example 5–10, using different injection GLRs.

5.4.4 System stability

In the preceding discussions, it was tacitly assumed that *steady-state flow conditions* existed in every component of the gas lifted well's production system at all times. In the practice, however, gas lifted wells, just like flowing wells can operate under unstable conditions, called *heading*, when regular and irregular changes in flow parameters (pressures and fluid rates) occur at any point in the system. Heading in flowing wells was first studied by Gilbert [11], the father of production engineering, who offered the following explanation to this undesirable behavior.

As discussed in Section 2.5, the pressure drop in vertical multiphase flow consists of three components: the hydrostatic, frictional, and acceleration terms. If the bottomhole pressure, assuming a constant *WHP*, is plotted in the function of liquid flow rate for a constant *GLR*, then a diagram schematically shown in Figure 5–27 will result. The *hydrostatic term*, represented by the flowing mixture's density, continuously decreases as the gas flow rate (the product of the increased liquid rate and the given *GLR*) increases, whereas *friction and acceleration losses* monotonically grow with gas flow rate. The sum of these terms is the total pressure drop, which, if added to the *WHP*, gives the bottomhole pressure depicted in heavy line and exhibiting a minimum. The minimum point divides the flow conditions into a *stable* and an *unstable region*, as shown in the figure. In the stable region, pressure losses are dominated by friction, and a small increase in gas rate induces an increase in the total pressure drop and bottomhole

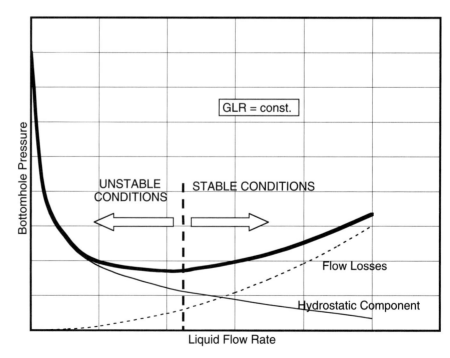

Fig. 5–27 Schematic diagram depicting the variation of pressure losses in vertical multiphase flow with the liquid flow rate.

pressure. In unstable cases, the bottomhole pressure decreases for the same increase in gas rate, because the flow is *slippage-dominated* where the flowing density decreases more than friction losses increase, resulting in a net drop in pressure losses.

Under *stable conditions*, any temporary increase in gas or liquid flow rates tends to increase the pressure drop in the tubing string, which, in turn, increases the well's *FBHP*. This gives rise to a reduction of *drawdown* across the formation and to a resultant decrease of liquid and gas inflow to the well. Thus, the well's flow conditions are automatically stabilized. A temporary decrease in flow rates, on the other hand, decreases the total pressure drop allowing the bottomhole pressure to drop and the fluid rate to increase. The well returns to its original flow conditions and stabilizes. It can be easily shown that similar small changes in gas or liquid flow rates in the unstable range (see Fig. 5–27) may lead to the well's dying or its operating point being shifted to a stable point.

Instability problems in the gas lifted well of Example 5–8 are illustrated in Figure 5–28. Compared to Figure 5–23, tubing performance curves for very low gas injection rates are plotted here, for which the existence of unstable conditions are apparent. Unstable and stable conditions are divided by the dashed curve connecting the points belonging to the minimum bottomhole pressure on every tubing performance curve. As used in most systems analysis computer program packages, only intersections on the right-hand side of the boundary curve are assumed stable.

Grupping et al. [12, 13] have shown that heading in continuous flow gas installations results in inefficient operations because injected gas is not used to best advantage. The main disadvantages of heading are as follows:

- Severe *slugging* at the wellhead occurs, and the big liquid and gas slugs can overload the surface facilities.

- Excessive gas usage is common with a consequently low efficiency of fluid lifting.

- Production control and gas allocation become difficult because of the great pressure fluctuations.

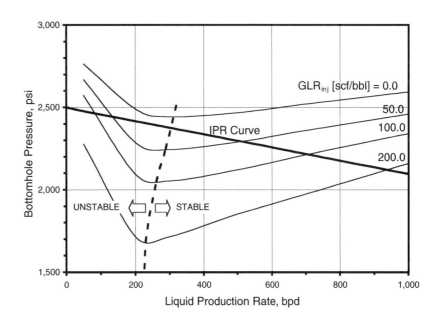

Fig. 5–28 Determination of the well's possible liquid rates for Example 5–8, using low gas injection rates.

The usual ways to control or completely eliminate heading and stabilize gas lifted wells placed on continuous flow include the following:

- Increasing the amount of injected gas, accomplished by installing a surface choke of a greater size. Due to the increased injection volume, the well's liquid rate increases and stable conditions may be reached. This solution, however, can mean *over-injection* when more-than-optimum gas rates are used, and can not be applied if lift gas availability is limited.

- Reducing the port size of the gas lift valve or orifice. However, due to the increased pressure drop through the valve/orifice, a higher injection pressure may be needed.

- Installing a surface *production choke* or decreasing the size of an existing one entails an increase in bottomhole pressure and a reduction of liquid rate but this rate, if stable, is usually greater than the one under heading conditions.

- Moving the injection point deeper can also stabilize the well.

Comprehensive analyses of continuous flow gas lift stability [14–16] utilize a different approach from the ordinary systems analysis procedures described in this section. They proved that the procedures used for stability analysis of flowing wells are inadequate for gas lifted wells because the total GLR in the tubing very much fluctuates with the changes in tubing pressure. Conventional methods, however, do not take this into account and completely disregard the effects of the gas injection system. A proper systems analysis for the purpose of evaluating the stability of gas-lifted wells, therefore, must investigate the system composed of the following:

1. The surface gas *injection choke*. The pressure upstream of the choke is usually constant and is determined by the capacity of the gas supply system. Depending on its downstream pressure, which equals the surface gas injection pressure, flow through the choke can be critical (with a constant gas rate) or subcritical. In practice, however, surface injection chokes operate in the subcritical regime because a large pressure drop across the choke is not desirable.

2. The casing annulus or gas injection string. The physical volume of the casing-tubing annulus or the gas injection string acts as a *buffer* that stores/releases the gas injected into it at the surface, depending on the balance of the inflowing and outflowing instantaneous gas rates. Lift gas leaves this volume at the gas injection point through the gas lift valve.

3. The gas lift valve or orifice. Installed at the injection depth, the gas lift valve, or alternatively a simple orifice, allows the injection of lift gas into the tubing. The gas rate passed by the valve/orifice depends on its size and on the casing and tubing pressures at the injection point. (The operation of simple chokes was described in chapter 2 and that of gas lift valves in chapter 3.) Tubing pressure at the setting depth of the valve/orifice varies with the liquid rate, consequently with the inflow to the well.

4. The tubing string. The tubing pressure at the gas injection point is determined from the separator pressure and the flow losses in the flowline and in the tubing. Total GLR varies with the gas rate injected by the valve/orifice; the liquid rate is governed by fluid inflow from the formation.

5. The productive formation. Inflow from the formation can be assumed *steady-state* or *transient*. In either case, the liquid rate varies with the smaller and/or greater changes in tubing pressure.

The gas injection subsystem (the surface choke, the annulus, and the gas lift valve), the tubing, and the formation are connected to each other at the gas injection point, and their common operating point is a result of their interaction. The stability of this operating point is determined by the well's response to small fluctuations in operating parameters. Tubing pressure at the injection point usually exhibits temporary variations due to variations in inflow rates, gas injection rates, etc. The stability of the system depends on the response of its components to these pressure changes. If, due to an increase in injection gas rate, the pressure difference between the annulus and the tubing increases, then gas injection rate will further increase, and well rates do not stabilize. Flow conditions are in the *slippage-dominated* region; consequently the operating point is *unstable*. On the other hand, if an increased injection rate results in an increase of tubing pressure due to the increased total pressure drop, then flow conditions are friction-dominated. In such cases, pressure differential across the valve/orifice decreases, bringing about a decrease in gas injection rate, and the well's liquid rate will stabilize.

It was shown by Grupping et al. [12, 13] as well as other investigators that system stability heavily relies on the operating modes of the surface gas injection choke and the gas lift valve or orifice. Since both devices can operate in critical or subcritical flow, one of four combinations can exist in a given well. It is easy to understand that the two cases with critical flow in the valve/orifice result in stable flow conditions. Critical flow through the gas lift valve or orifice means that any pressure fluctuations downstream of the valve/orifice do not change the injection rate or the upstream pressure. Therefore, small fluctuations in tubing pressure at the injection depth cannot result in variation of the well's rate, the system is *stable*.

Stability conditions for cases with critical flow in the surface choke and subcritical flow in the gas lift valve or orifice were studied by Asheim [14]. As already mentioned, this is not a very common case, since a great pressure drop through the surface choke is disadvantageous. Asheim developed two criteria on which system stability can be evaluated. The remaining case with subcritical flow both in the gas injection choke and in the gas lift valve/orifice was investigated by Alhanati et al. [16], who corrected Asheim's formulae for such cases.

In oilfields with a great number of heading wells on continuous flow gas lift, it may be advantageous to install *automatic control systems* that ensure stable operating conditions. Such systems utilize variable chokes on the flowline as well as on the gas injection line, controlled by special computer programs to eliminate unstable operation. [17]

Stability problems are eliminated if the gas lift valve operates in the *critical flow* region. Conventional valves, however, require a great pressure drop across their chokes to operate under such conditions. The Nozzle-Ventury (NOVA) Valve, as discussed in chapter 3, needs a very limited pressure differential (about 10%) to enter critical flow, and offers a simple solution to gas lift instability problems, as shown by Tokar et al. [18].

5.4.5 Conclusions

Section 5.4 introduced the use of systems analysis principles to the description of the operation of continuous flow gas lift wells. As shown, most of the conditions occurring in practice can be analyzed with the use of these principles. This makes systems analysis a powerful and universal tool, with the help of which production engineers can tackle everyday problems. As will be shown in later sections, this tool provides a sound foundation for the optimization of gas lift installations.

The performance of a continuous flow gas lift well, in general, can be characterized by the possible operating points of the production system. Depending on the selection of the solution node, these can be plotted on a wellhead pressure–liquid rate (see Fig. 5–25), or on a bottomhole pressure–liquid rate diagram (see Fig. 5–23). In both cases, the required gas injection volumes belonging to each operating point being known makes it possible to prepare liquid rate—gas injection rate diagrams, frequently called *gas lift performance curves*. The performance curve for the well in the previous examples is given in Figure 5–29, as derived from any of Figure 5–23 or Figure 5–25.

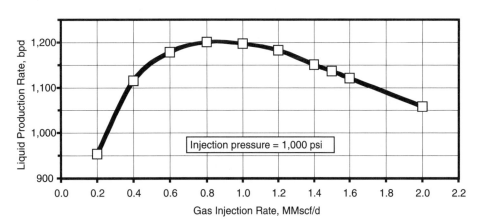

Fig. 5–29 Gas lift performance curve for the example well.

The gas lift performance curve is a plot of the well's liquid rate vs. the gas injection rate for a given surface gas injection pressure and shows the producing system's response to continuous flow gas lifting. Such performance curves have a very characteristic shape, and they have been known and actually measured since the very early days of gas lift history, as demonstrated by Shaw [19]. As seen in Figure 5–29, the well's response to gas injection is different for low and high injection rates. At low injection rates, any increase in the gas volume increases the well's liquid output because multiphase flow is dominated by slippage here, and any additional gas injection decreases the flow resistance of the system. As injection rates increase, the rate of liquid volume increase falls off, and the well's maximum possible liquid rate is reached. After this maximum, any additional gas injection decreases the well's liquid production. In this region of high gas injection rates, multiphase flow in the tubing is dominated by frictional effects meaning that higher mixture rates increase the total pressure drop in the system. Consequently, bottomhole pressure starts to increase and liquid inflow to the well diminishes.

Construction of *performance curves* is usually accomplished with the help of systems analysis computer packages, widely available today. Although such packages offer an easy and quick way to analyze and optimize gas lifted wells, their use requires prudent alertness from the production engineer. Perhaps the most important point is the selection of the proper *multiphase correlations* for the calculation of vertical and horizontal pressure drops in system components. As a warning, consider Figure 5–30 where performance curves calculated with the use of three correlations are shown for the previous examples. Curves with widely different liquid flow rates were received showing the importance of the proper selection. Reference is made here to chapter 2 of this book, where the proper attitude of production engineers to multiphase flow correlations was discussed. Since accurate performance curves cannot be constructed without the knowledge of the best flow correlation for the given field, it is always advantageous to select it beforehand. This is why some computer program packages offering systems analysis calculations provide an option for this process.

Gas lift performance curves have a multitude of uses in the design and analysis of wells placed on continuous flow gas lift. The conditions for attaining maximum and optimum liquid production rates for individual wells or groups of wells are easily found, etc. Such applications are detailed in subsequent sections.

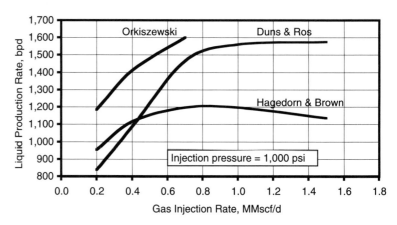

Fig. 5–30 The effect of the selected vertical multiphase flow correlation on the example well's gas lift performance curve.

5.5 Optimization of Continuous Flow Installations

5.5.1 Introduction

For many years in the petroleum industry, natural gas had a low price, and energy costs were also much lower than recently, so it did not matter what amount of injected gas was used to lift a given amount of oil to the surface. Optimum conditions, then, always meant achieving the maximum possible oil rate, irrespective of the injected gas volume. [20] The increased demand and the rising cost of natural gas, however, has completely changed this situation and today optimum conditions for continuous flow gas lifting are based on the economic parameters of fluid lifting.

The most general definition of the *optimum conditions* for continuous flow gas lifting of a single well or a group of wells may be stated as setting up a schedule of gas injection and liquid production rates over time that ensures the maximum *present-value profit* over the life of the wells. Since this definition implies that the wells' production rates are not limited, those cases require a different treatment when, for reservoir or other purposes, well rates are prescribed. Then liquid rates, and consequently, total cash flow is constant, and the objective of optimization is to ensure a *minimum of operating costs* over the life of the field. Both of the previous scenarios have their importance in practice, as will be shown later in this section.

When seeking the optimum conditions for continuous flow gas lifting of a single well or a group of wells alike, the designer must select the right combination of the system parameters listed as follows:

- Tubing size

- Surface gas injection pressure

- Separator pressure

- Flowline size

The order of importance of the previous parameters can vary, depending on actual well and field conditions. Their proper combination (which is also a function of the well's production conditions) satisfies the objective of the optimization process. In practice, due to the great number of possible cases, different scenarios are assumed, based on local conditions, then lift gas requirements and economic parameters are calculated, and finally the one meeting the criteria of optimization is selected.

As mentioned previously, the conditions selected for gas lifting a single well or a group of wells must ensure optimum production over the total life of the well(s). This means that calculations have to be performed with *predicted well data* for future conditions. Also, because of the many possible (and many times unpredictable) changes in well conditions over time (*e.g.* well inflow performance), calculations may have to be repeated from time to time. Although the importance of production time in the process of optimization as well its possible effects on the outcome of optimization are fully recognized, the following discussions treat wells and groups of wells with time-independent production parameters. This is done for the sake of clarity and for demonstrating basic principles that can be easily adapted to account for future conditions.

5.5.2 Optimization of a single well

When dealing with a single well placed on continuous flow gas lift, there may be several typical cases requiring different treatments for finding the well's optimum producing conditions. The well's liquid rate may be prescribed by reservoir considerations, or the capacity of the well can be fully utilized. A compressor delivering lift gas at a given discharge pressure may be available in the field or one has to be selected by the designer. If available, compressor capacity may be unlimited or only a limited amount of lift gas can be used.

In the following section, calculation models for each of the previous cases are detailed.

5.5.2.1 Prescribed liquid rate. Many times, reservoir engineering or other considerations limit the liquid rate to be lifted from the well. The general criteria for optimum gas lift conditions has to be modified to that of securing the least amount of production costs. Since cash inflow is determined by the given liquid rate, maximum profit is ensured by the minimum of costs. In the following, for simplicity, production costs will be defined by the operating costs—and in particular by compression costs, the largest operating cost component in gas lifting. This approach is fully justified for cases when a compressor is available; but for more detailed analyses, considerations for capital expenditures may also be required.

Basically, the operating cost of gas compression can be found from the cost of power and the required brake horsepower of the unit. Brake horsepower calculations, however, may differ for the different compressor types, compression ratios (CR), numbers of stages, etc. [20] This is the reason why, in the following sections, compressor power is approximated by the required *adiabatic power*. Although this is usually lower than the actually observed power, many investigators [5, 19, 21] rely on its use in comparative studies.

The adiabatic power requirement (in HPs) for the compression of an ideal gas is found from

$$HP_a = 8.57 \times 10^{-8} \frac{\kappa}{\kappa - 1} \ T_1 Q_g \left[\left(\frac{p_2}{p_1} \right)^{\frac{\kappa}{\kappa - 1}} - 1 \right]$$

5.2

where: κ = ratio of specific heats of the injected gas

T_1 = suction temperature of the compressor, R

Q_g = gas volumetric rate at standard conditions, scf/d

p_1 = suction pressure of the compressor, psia

p_2 = discharge pressure of the compressor, psia

Substituting κ = 1.3, valid for average natural gases, assuming a suction temperature of T_1 = 520 R, and expressing gas rate in Mscf/d we get

$$HP_a = 0.193 Q_g \left[\left(\frac{p_2}{p_1} \right)^{0.2308} - 1 \right]$$

5.3

where: Q_g = gas volumetric rate at standard conditions, Mscf/d

5.5.2.1.1 Existing compressor. For an existing compressor, the suction and discharge pressures are given, and these must be taken into account in gas lift calculations. Gas injection pressure at the wellsite is always less than compressor discharge pressure; the difference depends on the magnitude of pressure losses in the gas distribution lines. Similarly, WHP is always higher than suction pressure, allowing for pressure losses in the gathering line. Injection and WHPs thus being set, only the tubing and flowline sizes remain to be selected for an optimum operation. The calculation model resulting in the minimum of compression costs is very similar to a systems analysis performed with the wellhead as the solution node. It is described in conjunction with Figure 5–31 in the following:

1. Appropriate values for tubing and flowline sizes are selected.

2. Based on the prescribed liquid production rate q_l, the well's FBHP is calculated from the IPR equation.

3. The distribution of injection pressure in the annulus is determined from the given surface injection pressure.

4. The first tubing size is selected and a pressure traverse valid for the given liquid rate, tubing diameter, and formation GLR_p is extended up the hole from the FBHP at the depth of perforations.

5. The depth of gas injection is found at the intersection of the pressure traverse and a parallel drawn to the annulus gas pressure using a pressure differential of Δp = 100 psi.

6. Assume an injection gas-liquid ratio GLR_{inj} and, starting from the injection point found in the previous step, calculate the pressure traverse above the point of gas injection using the total GLR.

7. Find the WHP from the pressure traverse just calculated.

8. Repeat Steps 6–7 for the other assumed GLRs and determine the WHPs achieved (see Fig. 5–31).

9. Repeat Steps 4–8 for the remaining tubing sizes.

10. Plot the flowing WHP and gas-liquid ratio GLR_{inj} pairs in a wellhead pressure–GLR_{inj} coordinate system to establish the tubing performance curves.

11. Next, calculate the performance curves for the flowline sizes. For this, take the first flowline size.

12. Using the same GLR_{total} values as before, find the pressures at the intake end of the flowline by adding the pressure drop in the flowline to the separator pressure p_{sep}.

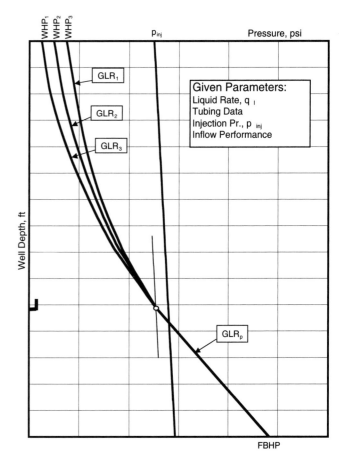

Fig. 5–31 Schematic determination of WHPs for a prescribed liquid rate and different assumed GLRs.

13. Repeat Step 12 for all assumed flowline sizes. The calculated intake pressures, plotted in the function of the *GLR* for the different flowline sizes (using the coordinate system described before), define the performance curves for the separator + flowline subsystem.

14. Intersections of the two performance curve sets determine the well's possible *operating points* for which the required injection *GLRs* are determined.

15. Injection gas requirements for the operating points are calculated from the GLR_{inj} values and the well's prescribed liquid flow rate.

16. Adiabatic powers needed for compressing the lift gas are found from Equation 5.3, and the case with the least power requirement is selected. This defines the tubing and flowline sizes to be used to reach optimum conditions.

The procedure described previously tacitly assumes that lift gas rates are not limited by the capacity of the compressor. If this is not the case and the available lift gas rate is specified then the GLRs used during the calculation process must be limited as follows:

$$GLR_{inj\ max} = \frac{Q_g}{q_l}$$

5.4

where: Q_g = available lift gas volumetric rate, scf/d

q_l = prescribed liquid production rate, bpd

—··—

Example 5–11. Find the optimum conditions for gas lifting the well with the data detailed as follows. Liquid production rate is prescribed as 750 bpd, an unlimited supply of lift gas is available with a compressor discharging at 1,050 psi and a suction pressure of 80 psi. Use tubing sizes of 2 3/8 in. and 2 7/8 in., and flowline sizes of 2 in. and 2.5 in.

Depth of perforations = 6,500 ft	Water cut = 25%
Tubing flow	Oil gravity = 35 API
Flowline length = 4,000 ft	Water gravity = 1.0
Avg. formation pressure = 1,500 psi	Gas gravity = 0.65
Productivity Index = 2 bpd/psi	Inj. gas gravity = 0.6
Formation GOR = 100 scf/bbl	

Solution

Surface injection and separator pressures were assumed as 1,000 psi and 100 psi, respectively, corresponding to the compressor's pressure data. The well's *FBHP* is found from the inflow performance equation as $FBHP = SBHP - q_l/PI = 1.500 - 750/2 = 1,150$ psi. The point of gas injection was found to be identical for both tubing sizes at 6,300 ft.

Calculations were performed on a computer and the Orkiszewski correlation was used for vertical, and the modified Beggs-Brill correlation for horizontal multiphase flow calculations. Results of wellhead and flowline inlet pressure calculations are listed in Table 5–8. Final results are given in Table 5–9, where injection gas requirements and adiabatic horsepower values are displayed for the four combinations of tubing/flowline sizes. Figure 5–32 shows tubing and flowline performance curves in the function of injection *GLRs*. Of the four possible operating points, the case with 2 7/8 in. tubing, and 2.5 in. flowline is selected as the optimum solution because it satisfies the optimization criteria of minimum energy requirement for continuous flow gas lifting of the example well.

GLR$_{inj}$	GLR$_{total}$	WHP for Tubing		WHP for Flowline	
		2 3/8"	2 7/8"	2"	2.5"
scf/bbl	scf/bbl	psi	psi	psi	psi
280	355	106	130	192	134
300	375	117	144	197	137
320	395	129	157	200	139
340	415	139	170	203	140
360	435	150	183	207	142
400	475	169	206	214	145
450	525	192	234	223	149
500	575	212	256	232	153
600	675	247	294	247	161
700	775	276	325	263	169
800	875	300	350	277	176

Table 5–8 Calculated data of tubing and flowline performance curves for Example 5–11.

Tubing Size	Flowline ID	GLR$_{inj}$	Q$_g$	HP$_a$
in	in	scf/bbl	Mscf/d	HP
2 3/8	2	600	450.0	65.0
	2.5	340	255.0	36.8
2 7/8	2	420	315.0	45.5
	2.5	286	214.5	31.0

Table 5–9 Calculated operating points for Example 5–11, assuming an existing compressor.

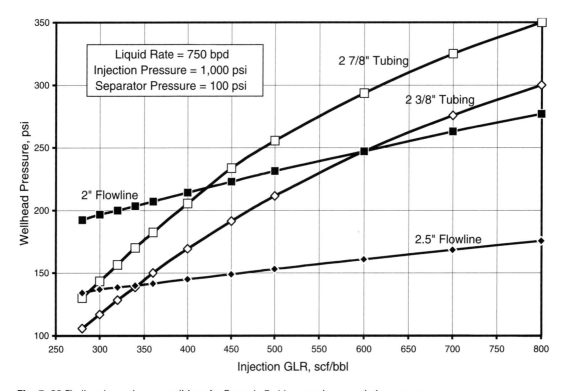

Fig. 5–32 Finding the optimum conditions for Example 5–11, assuming an existing compressor.

5.5.2.1.2 Compressor to be selected. In this case, all system variables affecting the economy of gas lifting must be properly selected: tubing and flowline sizes, compressor suction, and discharge pressures. The proper combination of these variables ensures the minimum compressor power and consequently the minimum of operating costs for lifting the prescribed amount of liquid from the given well. Since the previous variables can have discrete values only and many are constrained by field conditions, the optimization process should start with the selection of their possible values.

Based on local constraints, preferences, and other considerations, a set of viable values for tubing size, flowline size, injection and separator pressures is constructed and calculations are performed for all possible combinations of these. The logic of optimization is similar to the one described for an existing compressor and is basically a systems analysis performed on the gas lifted well with the solution node at the wellhead. The main steps of the optimization process are detailed as follows.

1. Based on field conditions, appropriate values for tubing and flowline sizes, compressor suction and discharge pressures are selected.

2. Calculate the well's *FBHP* based on the prescribed liquid production rate q_l and the well's IPR equation.

3. The first tubing size is selected.

4. Take the first compressor discharge pressure and find the gas injection pressure available at the wellhead, based on pressure loss estimations in the gas distribution line.

5. The distribution of injection pressure in the annulus is determined from the given surface injection pressure.

6. A pressure traverse valid for the given liquid rate, tubing diameter, and formation GLR_p is extended up the hole from the *FBHP* at the depth of perforations.

7. The intersection of the pressure traverse and a parallel drawn to the annulus gas pressure at depth defines the *depth of gas injection*. A pressure differential of $\Delta p = 100$ psi is used.

8. Assume an injection gas-liquid ratio GLR_{inj} and, starting from the injection point found in the previous step, calculate the pressure traverse above the point of gas injection using a gas-liquid ratio of $GLR_{total} = GLR_{inj} + GLR_p$.

9. Find the *WHP* from the pressure traverse just calculated.

10. Repeat Steps 8–9 for the other assumed *GLRs* and determine the *WHPs* achieved.

11. Repeat Steps 4–10 for the rest of compressor discharge pressures.

12. Take the next available tubing size and repeat Steps 4–11.

13. Plot the calculated flowing *WHP* and the assumed gas-liquid ratio GLR_{inj} pairs in a wellhead pressure–GLR_{inj} coordinate system to establish the tubing performance curves.

14. Now establish the performance curves for the available flowlines. For this, take the first flowline size.

15. Based on the first assumed compressor suction pressure, determine the appropriate separator pressure by estimating the pressure drop in the compressor's suction line.

16. Using the same GLR_{total} values as before, find the pressures at the intake end of the flowline for each *GLR* value by adding the multiphase pressure drop in the flowline to the separator pressure p_{sep}.

17. Repeat Steps 15–16 for the rest of available compressor suction pressures.

18. Repeat Steps 15–16 for all assumed flowline sizes. The calculated wellhead (flowline intake) pressures, plotted in the function of the *GLR* for the different flowline sizes and separator pressures (using the coordinate system described before) define the performance curves for the separator–flowline subsystem.

19. Intersections of the two performance curve sets determine the well's *possible operation points* for which the required injection *GLR*s are determined.

20. Injection gas requirements for the operating points are calculated from the GLR_{inj} values and the well's prescribed liquid flow rate.

21. Adiabatic powers needed for compressing the lift gas are found from Equation 5.3, and the operating point with the *least power requirement* is selected. This defines the proper combination of system parameters (tubing and flowline sizes, compressor suction and discharge pressures) that have to be used to reach optimum conditions of gas lifting.

As before, the previous process must be modified for cases when the compressor capacity is limited.

Example 5–12. Find the optimum conditions for gas lifting the well in the previous example including the calculation of compressor parameters.

Solution

Assuming tubing and flowline sizes are identical to those in the previous example, based on the well's liquid production rate. System pressures are set up as follows, where the pressure drops arising in the gas distribution and in the gas gathering lines are accounted for.

Separator Pressure	Suction Pressure
100 psi	80 psi
150 psi	130 psi
200 psi	180 psi
Injection Pressure	**Discharge Pressure**
800 psi	850 psi
1,000 psi	1,050 psi
1,100 psi	1,150 psi

Computer-calculated results are displayed in Table 5–10, which contains calculated *WHP*s for the formation + tubing string and the separator + flowline subsystems, in the function of appropriately selected injection *GLR*s. Tubing and flowline performance curves are plotted in Figures 5–33 and 5–34. Figure 5–33 depicts the cases for a tubing size of 2 3/8 in., where the heavy lines represent tubing performance curves for the three assumed gas injection pressures. Higher injection pressures result in deeper points of gas injection and, consequently, *WHP*s are higher for the same *GLR*. The depths of gas injection for the different injection pressures were found as given next, the values being identical for both tubing sizes. As seen, the highest injection pressure ensures gas injection at the bottom of the well.

GLR_inj	WHPs for Tubing Sizes and Injection Pressures						WHPs for Flowline Sizes and Separator Pressures					
	2 3/8" Tubing			2 7/8" Tubing			2" Flowline			2.5" Flowline		
	1,000 psi	1,100 psi	800 psi	1,000 psi	1,100 psi	800 psi	100 psi	150 psi	200 psi	100 psi	150 psi	200 psi
scf/bbl	psi	psi	psi	psi	psi	psi	psi	psi	psi	psi	psi	psi
280	106	116	67	130	137	103	192	224	261	134	176	221
300	117			144			197	227	264	137	178	222
320	129	141	84	157	165	123	200	231	266	139	179	223
340	139			170			203	234	269	140	180	224
360	150	163	102	183	193	141	207	237	272	142	181	225
400	169	184	117	206	220	157	214	243	277	145	184	228
450	192			234			223	250	284	149	187	230
500	212	229	149	256	272	193	232	257	290	153	190	232
600	247	268	174	294	313	227	247	273	303	161	196	237
700	276	296	196	325	348	249	263	287	316	169	202	242
800	300	323	213	350	374	266	277	300	329	176	209	247
900		344							340			

Table 5–10 Calculated WHPs for the two subsystems in Example 5–12.

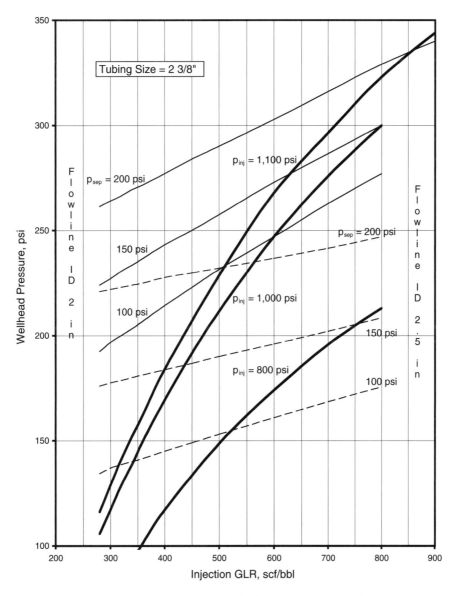

Fig. 5–33 Tubing and flowline performance curves for Example 5–12 and a tubing size of 2 3/8 in.

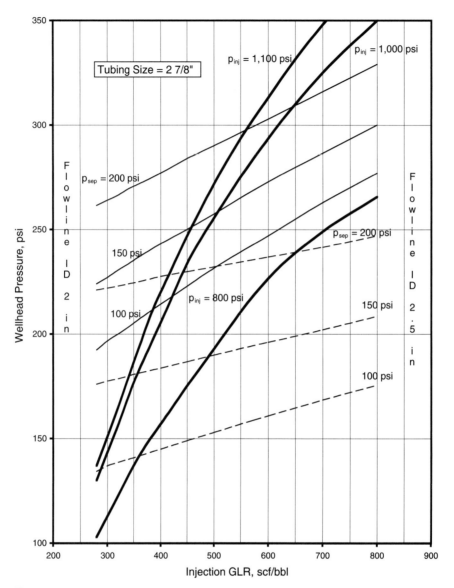

Fig. 5–34 Tubing and flowline performance curves for Example 5–12 and a tubing size of 2 7/8 in.

Injection Pressure	Depth of Injection
800 psi	5,630 ft
1,000 psi	6,300 ft
1,100 psi	6,500 ft

Two other sets of performance curves represent the operation of the separator and the flowline subsystem; the dashed and the solid lines belong to the flowline diameters of 2.5 in., and 2 in., respectively. The three individual curves in each set are valid for the *WHPs* corresponding to the three different separator pressures. As seen, the bigger flowline size involves lower pressure losses and, as a result, lower *WHPs* for the same *GLRs*.

All intersections between the tubing performance curves and the flowline curves constitute system operating points, *i.e.* valid combinations of gas injection pressure, separator pressure, and flowline size, that ensure continuous flow gas lifting of the given well at the given rate. The required injection *GLRs* can be read from the figure. Figure 5–34 contains the same type of data for the tubing size of 2 ⁷/₈ in..

Tubing Size	Flowline ID	p_{inj}	p_{sep}	GLR_{inj}	Q_g	HP_a	CR
in	in	psi	psi	scf/bbl	Mscf/d	HP	-
2 3/8	2	800	100	-	-	-	-
			150	-	-	-	-
			200	-	-	-	-
		1,000	100	600	450.0	65.0	11.2
			150	800	600.0	67.8	7.4
			200	-	-	-	-
		1,100	100	510	382.5	57.9	12.3
			150	630	472.5	56.4	8.0
			200	860	645.0	63.6	6.0
	2.5	800	100	520	390.0	50.1	9.1
			150	755	566.3	55.8	6.0
			200	-	-	-	-
		1,000	100	340	255.0	36.8	11.2
			150	435	326.3	36.8	7.4
			200	565	423.8	39.3	5.5
		1,100	100	310	232.5	35.2	12.3
			150	400	300.0	35.8	8.0
			200	510	382.5	37.7	6.0
2 7/8	2	800	100	-	-	-	-
			150	-	-	-	-
			200	-	-	-	-
		1,000	100	420	315.0	45.5	11.2
			150	505	378.8	42.8	7.4
			200	645	483.8	44.8	5.5
		1,100	100	380	285.0	43.2	12.3
			150	460	345.0	41.2	8.0
			200	565	423.8	41.8	6.0
	2.5	800	100	360	270.0	34.7	9.1
			150	490	367.5	36.2	6.0
			200	650	487.5	38.6	4.4
		1,000	100	286	214.5	31.0	11.2
			150	360	270.0	30.5	7.4
			200	440	330.0	30.6	5.5
		1,100	100	275	206.3	31.2	12.3
			150	340	255.0	30.4	8.0
			200	415	311.3	30.7	6.0

Table 5–11 Calculated operating points for Example 5–12, for the selection of the proper compressor.

Calculated data of the operating points are displayed in Table 5–11 that contains the required gas-liquid ratios GLR_{inj}, daily gas requirements Q_g, and adiabatic compressor powers HP_a. As seen, the use of the greater tubing and/or flowline size is more advantageous due to the lower compressor powers required. In order to facilitate the final selection, the table contains calculated CR values as well. Considering the HP_a and CR values, the optimum solution is found with the following combination of system parameters:

Tubing Size = 2 7/8 in.; Flowline ID = 2.5 in.

Compressor Discharge Pressure = 1,050 psig; Compressor Suction Pressure = 180 psig.

Based on the compression ratio of CR = 5.5, valid for this case, a single-stage compressor can be selected. Although capital costs are not included in the optimization model used, a more rigorous compressor selection would probably result in the same decision.

5.5.2.2 Unlimited liquid rate.

5.5.2.2.1 Economic considerations. If the liquid rate from the well is not limited by reservoir engineering or other constraints, the objective of optimization is to attain a maximum of profit. To reach this goal, the well's liquid rate is selected in such a way that all costs of gas lifting are lower than the revenue from the sale of oil. Since the greatest component of the total operational cost is spent on the compression of the lift gas, the gas lift installation should be designed to ensure an optimum utilization of gas.

It was first shown by Simmons [22, 23] that economic studies of continuous flow gas lifting must be based on the gas lift performance curve concept, discussed in a previous section (see Fig. 5–29). The performance curve is obtained from a complete systems analysis of the gas lifted well's production system and shows the liquid production rate for different values of injected lift gas volumes. A schematic diagram is given in Figure 5–35, where the curve in heavy line shows the producing system's response to continuous flow gas lifting. Knowledge of a well's performance curve allows one to analyze the economic conditions of gas lifting, because both revenues and lifting costs can readily be determined from this curve.

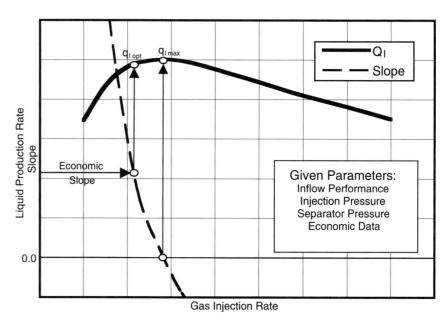

Fig. 5–35 Schematic depiction of a typical gas lift performance curve.

Let us discuss the economics of gas lifting, based on the well's performance curve and assume that the following parameters are known:

(a) the profit made on each barrel of oil produced

(b) the total (capital and operational) cost of gas treatment and compression for a unit volume of injected gas

(c) the specific cost of water treatment and disposal

If we start with a low amount of gas injection, the well's response is a relatively low liquid rate. Additional injection of a small gas volume, however, results in a relatively high increment of liquid rate. One can easily calculate the *increment of revenue* and the *increment of costs* involved, the balance of which is positive in this case. If gas injection rates are further increased by the same increment, liquid production rates increase further but the increments will be less and less, because of the shape of the performance curve, as seen in Figure 5–35. Eventually, a point will be reached where the increments of revenue and production cost are just equal, and from which on any further increase in gas injection rate would result in net economic losses. The conditions belonging to this operating point, therefore, define the well's optimum production by continuous flow gas lifting.

For the optimum point defined previously, the *balance of revenues and costs* can be written as follows, where the left-hand side represents the income gained and the right-hand side is the sum of gas compression and water treatment costs:

$$\Delta q_l (1 - f_w)P = \Delta q_g C_g + \Delta q_l f_w C_w \qquad \qquad 5.5$$

where: Δq_l = liquid production rate increment, bpd

Δq_g = gas injection rate increment, MMscf/d

f_w = water cut

P = profit on oil, $/bbl

C_g = injection gas cost, $/MMscf

C_w = cost of water treatment and disposal, $/bbl

The previous equation can be solved for the difference quotient $\Delta q_l/\Delta q_g$, which is equivalent to the gradient or slope of the gas lift performance curve:

$$\frac{\Delta q_l}{\Delta q_g} = \frac{C_g}{(1 - f_w)P - f_w\, C_w} \qquad\qquad 5.6$$

Based on previous discussions, this is the slope belonging to the *optimum operating point* of gas lifting, introduced by Kanu et al. [24], and called the Economic Slope. Its use in optimizing continuous flow gas lift installations is detailed as follows.

Figure 5–35 shows a sample gas lift performance curve (heavy line) and its derivative or slope (dashed line). At a gas injection rate belonging to the maximum of liquid production rate $q_{l\,max}$, the slope is zero by definition. It must be clear that no profit can be made on the right of this point because costs are higher than revenues. On the left-hand side, however, profits overcome costs, up to the point belonging to the Economic Slope where, as shown previously, they are equal. Therefore, the liquid production rate belonging to this slope represents the *optimum rate* $q_{l\,opt}$ that ensures the most favorable economic conditions of gas lifting for the given well.

The fact that in continuous flow gas lifting the optimum liquid rate is always lower than the possible maximum was proved by many investigators before. [19–24] The difference between the maximum and optimum rates can be investigated by evaluating the formula of the economic slope. (Equation 5.6) If, for some reason, gas compression costs C_g are negligible then, all other conditions unchanged, the right-hand side of Equation 5.6 approaches zero and the optimum rate shifts to the maximum (see Fig. 5–35). A sufficient increase in the profit on oil P acts the same way, whereas decreasing profits decrease the optimum liquid rate. Finally, increased water cuts f_w also increase the well's optimum gas lifted liquid rate.

In conclusion, the determination of the optimum liquid rate for a continuous flow gas lift installation depends on the proper knowledge of the slope of the gas lift performance curve. The curve itself is usually established with the help of systems analysis computer program packages, and its slope can be derived numerically from the data thus received. A common solution involves curve-fitting of the points of the performance curve and calculating the derivative of the resulting function, usually a polynomial or a spline function. [20]

All inaccuracies introduced to gas lift performance curves by improper assumptions in systems analysis calculations are eliminated if performance curves are actually measured at the wellsite. Tokar and Smith [25] developed a microprocessor-controlled setup that, by measuring and automatically modifying gas injection rates to the well, establishes the well's performance curve and automatically determines the economically optimum gas injection rate. The rate of gas injection is set by a flow control valve, liquid and gas rates being measured simultaneously, while all data of these devices are collected by the microprocessor. The operator's only task is to input the economic data (oil profit, compression costs, etc.) to the computer that, by automatic operation, determines and sets the gas injection rate belonging to the economic slope.

5.5.2.2.2 Existing compressor. If an existing compressor is to be used, then the separator and the gas injection pressures are determined from the compressor's suction and discharge pressures, respectively. Thus, only the optimum sizes of the tubing string as well as the flowline have to be found to establish the optimum conditions for gas lifting. The optimization process heavily relies on the knowledge of the proper gas lift performance curves, detailed as follows:

1. Based on field conditions, appropriate values for tubing and flowline sizes are selected.

2. From the given compressor discharge pressure, the gas injection pressure available at the wellhead is calculated, based on pressure loss estimations in the gas distribution line.

3. Based on the given compressor suction pressure, the appropriate separator pressure is found by estimating the pressure drop in the compressor's suction line.

4. The first tubing size is selected.

5. The first flowline size is selected.

6. Using a program package, points of the gas lift performance curve are determined for the actual tubing and flowline sizes, using the injection and separator pressures found in Steps 2 and 3. When performing these calculations, injection GLRs are to be selected so as to provide a sufficient number of points on the performance curve, especially left of the maximum liquid rate.

7. Repeat Step 6 for the rest of flowline sizes.

8. Repeat Steps 5–6 for the rest of tubing sizes.

9. Perform curve-fitting calculations on each performance curve. The author recommends the use of a third-order polynomial of the gas injection rate, which is denoted as x in the following formula:

$$q_l = a_3\, x^3 + a_2\, x^2 + a_1\, x + a_0$$

10. Determine the derivative of each performance curve. For the previous polynomial, this function is defined as

$$q_l = 3\, a_3\, x^2 + 2\, a_2\, x + a_1$$

11. Plot the performance curves along with their derivatives in a coordinate system of liquid rate vs. gas injection rate.

12. Based on field data, find the Economic Slope from Equation 5.6.

13. On the plot of the derivative of each performance curve, find the gas injection rate belonging to the Economic Slope.

14. Find the liquid rates on each performance curve corresponding to the gas injection rates determined in Step 13.

15. Select the maximum of the liquid production rates found in Step 14. The tubing and flowline sizes providing the maximum liquid rate determine the *optimum conditions* of gas lifting for the given well.

The previous procedure is used if the lift gas rate available from the compressor is not limited. It can also be applied to cases with a limited gas availability when Step 15 is modified so as to exclude those liquid rates where the necessary gas injection rate is higher than the available.

– –

Example 5–13. Find the optimum liquid and gas injection rates for the well given in the previous examples if liquid production rate is not limited and an existing compressor is to be used. Compressor discharge and suction pressures are 1,050 psi and 80 psi, respectively. Use a profit on oil of $1/bbl, a water disposal cost of $0.01/bbl of water, and a total of $400/MMscf of gas compression cost.

Solution

As before, based on production data, tubing sizes of 2 3/8 in. and 2 7/8 in.; and flowline IDs of 2 in. and 2.5 in. are selected. From compressor pressures, an injection pressure of 1,000 psi and a separator pressure of 100 psi are established.

The points of the gas lift performance curves, as a result of computer calculations are presented in Table 5–12 for the four combinations of tubing and flowline sizes. Regression calculations gave the following polynomials, where the quality of fitting is defined by the parameter R^2 (R-squared) with $R^2 = 1$ as a perfect fit. As seen, the recommended third-order polynomials give excellent approximation of the original data.

Q_g	Liquid Production Rates			
	2 3/8" Tubing		2 7/8" Tubing	
	2" Flowl.	2.5" Flowl.	2" Flowl.	2.5" Flowl.
MMscf/d	bpd	bpd	bpd	bpd
0.20	595	676	-	-
0.25	-	-	700	823
0.30	681	784	-	-
0.35	707	824	781	927
0.40	729	857	-	-
0.45	745	881	829	1,010
0.50	757	902	846	1,035
0.60	773	930	869	1,073
0.70	779	945	884	1,099
0.80	781	951	892	1,118
0.90	-	-	895	1,131
1.00	776	957	894	1,139
1.10	-	-	891	1,145
1.20	764	947	886	1,148
1.50	-	-	866	1,140

Table 5–12 Calculated points of gas lift performance curves for the cases in Example 5–13.

Tubing Size	Flowline ID	Equation	R_2
2 3/8 in.	2 in.	$y = 592.97\,x^3 - 1643\,x^2 + 1445.7\,x + 372.95$	0.994
	2.5 in.	$y = 693.06\,x^3 - 1990.6\,x^2 + 1864.1\,x + 381.96$	0.998
2 7/8 in.	2 in.	$y = 331.85\,x^3 - 1154.8\,x^2 + 1256.9\,x + 460.76$	0.993
	2.5 in.	$y = 363.92\,x^3 - 1348.7\,x^2 + 1620.7\,x + 507.01$	0.995

The slopes of the performance curves were found from differentiation of the previous functions and are displayed in Table 5–13. The plots of the performance curves along with their derivatives are presented in Figure 5–36 for the 2 3/8 in. and in Figure 5–37 for the 2 7/8 in. tubing sizes.

The Economic Slope is calculated from Equation 5.6 as

Economic Slope = 400/ [(1 − 0.25) 1.0 − 0.25 0.01] = 400 / 0.748 = 535 bbl/MMscf.

Q_g	Calculated Slopes			
	2 3/8" Tubing		2 7/8" Tubing	
	2" Flowl.	2.5" Flowl.	2" Flowl.	2.5" Flowl.
MMscf/d	bbl/MMscf	bbl/MMscf	bbl/MMscf	bbl/MMscf
0.20	860	1151	-	-
0.25	-	-	742	1,015
0.30	620	857	-	-
0.35	514	725	570	810
0.40	416	604	-	-
0.45	327	494	419	628
0.50	247	393	351	545
0.60	115	224	230	395
0.70	17	96	128	267
0.80	-45	10	46	162
0.90	-	-	-15	77
1.00	-61	-38	-57	15
1.10	-	-	-79	-25
1.20	-	-	-81	-44
1.50	-	-	-	-

Table 5–13 Calculated slopes of gas lift performance curves for the cases in Example 5–13.

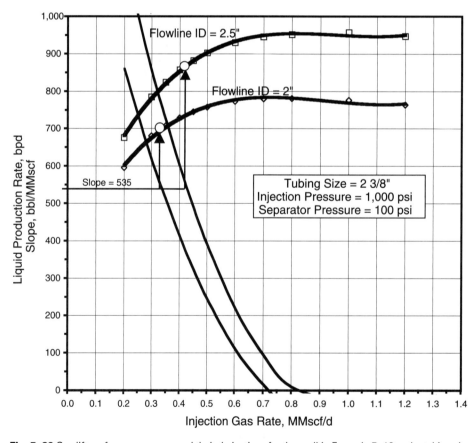

Fig. 5–36 Gas lift performance curves and their derivatives for the well in Example 5–13 and a tubing size of 2 3/8 in.

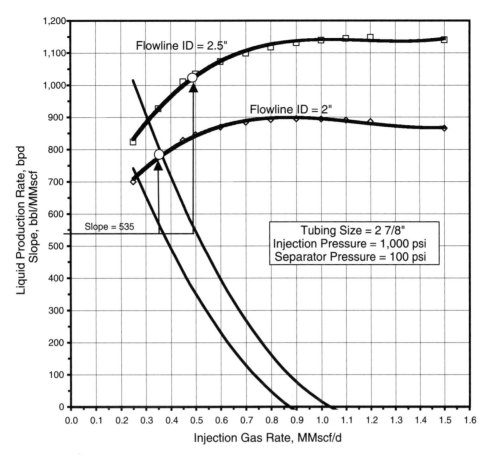

Fig. 5–37 Gas lift performance curves and their derivatives for the well in Example 5–13 and a tubing size of 2 7/8 in.

The parameters of the optimum operating points were read from Figures 5–36 and 5–37 and are:

Tubing Size	Flowline ID	Liquid Rate bpd	Gas Inj. Rate MMscf/d
2 3/8 in.	2 in.	700	0.34
	2.5 in.	870	0.43
2 7/8 in.	2 in.	785	0.37
	2.5 in.	1,030	0.51

From the previous possible cases, the one with the maximum liquid rate is selected as the optimum.

Since this example assumed unlimited lift gas availability, if the gas rate is limited for some reason, then the first case with a required injection rate less than the available volume must be chosen.

5.5.2.2.3 Compressor to be selected. If there is no compressor available for gas lift service or the possible best one is sought for a proposed project, then the calculations detailed in the previous section are repeated for many possible compressor discharge and suction pressures. Even if only a few versions for tubing and flowline sizes are assumed, the theoretically feasible combinations of tubing, flowline sizes, compressor discharge, and suction pressures are numerous. This is one of the typical cases where the use of computer program packages utilizing systems analysis is a must.

Theoretically, the following steps are required to find the optimum gas lift conditions for a well if, in addition to the selection of tubing and flowline sizes, the compressor must also be chosen, and available lift gas volume is unlimited.

1. Appropriate values for tubing and flowline sizes are selected.

2. Several feasible values for compressor discharge and suction pressures are assumed. Care should be taken to include a discharge pressure that allows gas injection at the well bottom for the whole range of expected liquid rates. [26] It is advantageous if the possible discharge/suction pressure combinations correspond to a set of CR values.

3. Based on the pressures selected previously, the gas injection pressures available at the wellhead as well as possible separator pressures are calculated with due respect to flowing pressure losses in the gas distribution line and in the compressor's suction line, respectively.

4. The first values for tubing size, flowline size, gas injection pressure, and separator pressure are selected.

5. Using a computer program package, the gas lift performance curve corresponding to the actual combination of tubing and flowline sizes, injection and separator pressures is established. During calculations, injection gas rates have to be selected so as to provide a sufficient number of points on the performance curve, especially on the left-hand side of the well's maximum liquid rate.

6. Step 5 is repeated for the remaining separator pressures, then for the rest of injection pressures, then for the remaining flowline sizes, and finally for the rest of tubing sizes. As a result, performance curves for all combinations of the four system variables will be available.

7. Perform curve-fitting calculations on each performance curve using a third-order polynomial of the gas injection rate, denoted as x in the following formula.

$$q_l = a_3\, x^3 + a_2\, x^2 + a_1\, x + a_0$$

8. Determine the derivative of each performance curve. For the previous polynomial, this function is defined as

$$q_l = 3\, a_3\, x^2 + 2\, a_2\, x + a_1$$

9. Plot the performance curves along with their derivatives in a coordinate system of liquid rate vs. gas injection rate.

10. Based on field data, find the Economic Slope from Equation 5.6.

11. On the plot of the derivative of each performance curve, find the gas injection rate belonging to the Economic Slope.

12. On each performance curve, determine the liquid production rate corresponding to the injection rate found in Step 11.

13. Select the maximum of the liquid production rates found in Step 12. The combination of system parameters providing the maximum liquid rate determines the optimum conditions of gas lifting for the given well: tubing and flowline sizes, compressor discharge and suction pressures are selected.

As seen previously, a great number of computer runs has to be performed for the solution. Although this does not pose difficulties, calculation efforts may be reduced by observing the practical recommendations given as follows. [27]

Generally, a single tubing size can be selected on the basis of the well's predicted range of liquid production rates. This is because tubing size has a very marked effect on multiphase flow pressure drop and lift gas requirements closely correlate with pressure losses in the tubing. The effect of flowline size on lift gas requirements is also substantial, and a size about the same as the tubing's is usually recommended. The validity of these general rules, of course, can be checked by occasional control calculations.

If tubing and flowline sizes are selected as suggested previously, the total number of cases to be considered is reduced to the number of compressor discharge pressures times the number of suction pressures.

Example 5–14. Find the optimum liquid and gas injection rates and select the proper compressor discharge and suction pressures for the well given in the previous examples, if the well's liquid rate and the available gas injection rate are unlimited. Use the same data as before, and assume that gas compression costs are independent of the number of compressor stages required.

Solution

In order to limit the number of necessary computer runs, a tubing size of 2 ⅞ in. and a flowline ID of 2.5 in. was selected. The selected tubing size was shown in the previous example to produce much greater rates than the size of 2 ⅜ in., and the well's inflow data probably does not necessitate the use of a 3 ½ in. size tubing.

After investigating the well's probable liquid production rates and the corresponding *FBHPs*, it was determined that an injection pressure of 1,000 psi would ensure gas injection at the well bottom. Another injection pressure of 800 psi and several separator pressures were assumed as follows.

Injection Pressure	Discharge Pressure
1,000 psi	1,050 psi
800 psi	850 psi

Separator Pressure	Suction Pressure
100 psi	80 psi
150 psi	130 psi
200 psi	180 psi

Calculated points of the performance curves are contained in Table 5–14, regression analysis resulted in the following polynomials for the six possible cases. R-squared values, as before, demonstrate that input data are properly approximated. (Data for the 1,000 psi injection and 100 psi separator pressure case were calculated in the previous example.)

Q_g	Liquid Production Rates				
	p_{inj} = 1,000 psi		p_{inj} = 800 psi		
	p_{sep} = 150	p_{sep} = 200	p_{sep} = 100	p_{sep} = 150	p_{sep} = 200
MMscf/d	bpd	bpd	bpd	bpd	bpd
0.25	716	601	881	787	673
0.35	867	775	938	854	750
0.40	910	826	977	902	805
0.50	974	899	1004	935	845
0.60	1019	951	1023	959	875
0.70	1051	990	1036	976	897
0.80	1075	1019	1045	989	914
0.90	1091	1039	1053	1004	937
1.00	1103	1054	1049	1004	944
1.20	1116	1097	1035	995	930
1.40	1121	-	-	-	-
1.60	1110	-	-	-	-
1.70	-	1033	-	-	-

Table 5–14 Calculated points of gas lift performance curves for the cases in Example 5–14.

Injection Pressure	Separator Pressure	Equation	R^2
1,000 psi	100 psi	See Example 5–13	
	150 psi	$y = 425.23\,x^3 - 1558.3\,x^2 + 1877.8\,x + 362.99$	0.987
	200 psi	$y = 366.98\,x^3 - 1506.6\,x^2 + 1976.5\,x + 228.91$	0.981
800 psi	100 psi	$y = 185.2\,x^3 - 796.64\,x^2 + 1097.9\,x + 561.49$	0.997
	150 psi	$y = 213.26\,x^3 - 912.17\,x^2 + 1279\,x + 412.18$	0.998
	200 psi	$y = 199.32\,x^3 - 911.53\,x^2 + 1365.2\,x + 265.85$	0.998

Differentiation of the previous functions gave the slopes of the performance curves that are displayed in Table 5–15. The performance curves along with their derivatives are plotted in Figure 5–38 for the 1,000 psi, and in Figure 5–39 for the 800 psi injection pressures.

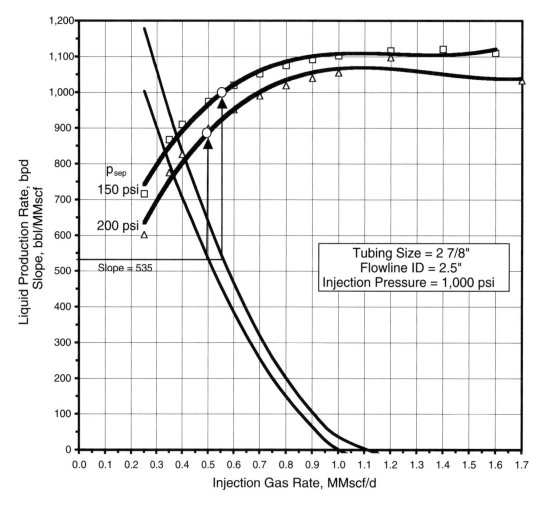

Fig. 5–38 Gas lift performance curves and their derivatives for the well in Example 5–14 and an injection pressure of 1,000 psi.

307

Q_g	Calculated Slopes				
	p_inj = 1,000 psi		p_inj = 800 psi		
	p_sep = 150	p_sep = 200	p_sep = 100	p_sep = 150	p_sep = 200
MMscf/d	bbl/MMscf	bbl/MMscf	bbl/MMscf	bbl/MMscf	bbl/MMscf
0.25	1178	1292			
0.35	943	1057			
0.40	835	947	549	652	732
0.50	638	745	440	527	603
0.60	467	565	342	415	487
0.70	321	407	255	315	382
0.80	201	271	179	229	289
0.90	106	156	114	155	209
1.00	37	64	60	94	140
1.20	-25	-54	-14	11	39
1.40			-44	-21	-15

Table 5–15 Calculated slopes of gas lift performance curves for the cases in Example 5–14.

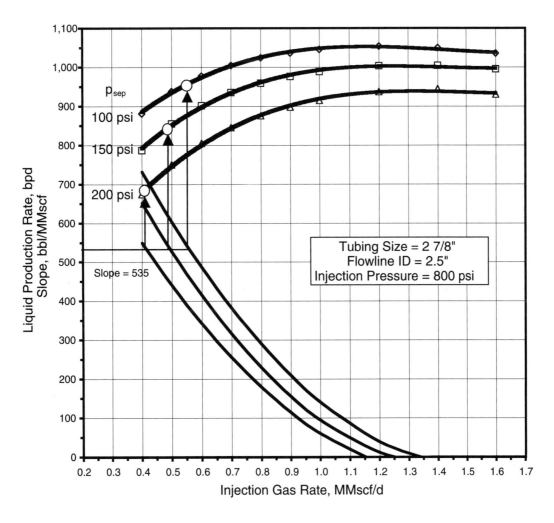

Fig. 5–39 Gas lift performance curves and their derivatives for the well in Example 5–14 and an injection pressure of 800 psi.

Using the Economic Slope calculated in the previous example, the following optimum operating points were read from the previous figures.

Tubing Size	Flowline ID	Injection Pressure	Separator Pressure	Liquid Rate Bpd	Gas Inj. Rate MMscf/d	CR
2 7/8 in.	2.5 in.	1,000 psi	100 psi	1,030	0.51	11.2
			150 pis	1,005	0.55	7.4
			200 psi	890	0.49	5.5
		800 psi	100 psi	955	0.55	9.1
			150 psi	845	0.49	6.0
			200 psi	690	0.42	4.4

From the previous possible cases, the one providing the maximum liquid rate (1,030 bpd) is selected. It should be noted, however, that the compression ratio of $CR = 11.2$ indicates that a compressor with two stages would be required. If the simplifying assumption of identical compression costs for any number of stages is not justified, then different Economic Slopes have to be used for compressors with different numbers of stages and the final results of optimization would change.

As in the previous example, a restriction on lift gas rates eliminates those possible cases where the required gas injection rate is higher than that available for gas lifting.

5.5.3 Allocation of lift gas to a group of wells

5.5.3.1 Introduction. When dealing with several wells placed on continuous flow gas lift, the objectives of optimization must be modified, as compared to the principles discussed so far. In such cases, the system parameters used in the optimization procedures described previously (tubing and flowline sizes, compressor pressures) have already been selected and cannot be changed easily. With these parameters being set (hopefully at their optimum values), the operator's aim is now to reach optimum utilization of the injection gas volume at his/her disposal. As discussed before in conjunction with the gas lift performance curves, different wells respond differently to the injection of the same amount of lift gas. It is now the designer's responsibility to allocate the total available gas volume to the individual wells in a fashion to achieve the maximum possible profit that comes from the sale of the oil produced.

As before, two situations can occur in the oilfield: (a) an unlimited supply of lift gas may be available, or (b) lift gas availability may be limited by some constraint. The two cases require different approaches, and the results of the optimization will be different as well.

For unlimited gas availability, the total gas volume used for gas lifting is not constrained so every well can receive the amount of gas ensuring the *maximum of profit* derived from it. As discussed before, this situation occurs at a gas injection rate where the slope of the gas lift performance curve equals the Economic Slope, calculated from the actual values of the specific incomes and costs. Therefore, optimization of continuous flow gas lifting for a number of wells with an unlimited gas availability is defined as producing every well at its most economic liquid rate. In such cases, field-wide total gas requirement equals the sum of each well's injection rate belonging to its economic liquid production rate.

In case the amount of gas available for gas lifting is limited, the objective of optimization is modified. Because total gas injection volume is known, compression costs are thus defined and fixed so optimization should aim at providing the *maximum of oil production* from the wells. The background theory of the optimization process to be followed is illustrated as follows.

Take two wells, the performance curves of which are given in Figure 5–40. The points belonging to their Economic Slopes are indicated by the points labeled *Unlimited* and define the optimum liquid and gas rates for an unlimited supply of lift gas. The two wells' combined liquid production is 1,559 bpd, total gas injection volume is 1.89 MMscf/d. Now assume that lift gas volume is limited to 1.5 MMscf/d and an optimum allocation of that total volume is desired. Suppose that the wells currently operate as indicated by the points designated *Initial* on their performance curves. Their combined injection rate, of course, equals the amount available (1.5 MMscf/d), and they produce a total of 1,382 bpd liquid. Considering the slopes of the two performance curves, it can be seen that, for the same incremental gas injection volume, Well #1 would respond with a greater increment of liquid rate than Well #2. Therefore, let us inject a greater volume to the first well, by taking a small increment above the current rate. Unfortunately, the gas injection volume to the second well must be decreased by the same increment, in order to satisfy the constraint on the total volume of lift gas. However, the total liquid production of the two wells may have increased.

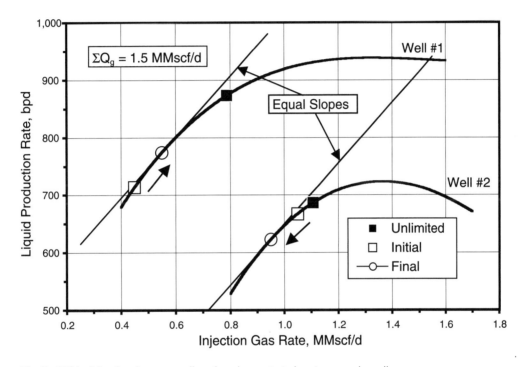

Fig. 5–40 Principle of optimum gas allocation, demonstrated on two sample wells.

If the gas injection rate to Well #1 is gradually increased by the same increment, while the total gas requirement is limited to the available volume, the total fluid rate varies with the calculation steps as shown in Figure 5–41. The figure also contains the variation of slopes (derivatives) of the performance curves. These indicate the individual liquid rates since liquid rate is found from the slope multiplied by the constant increment of injection rate. As seen, the total liquid rate from the two wells has a maximum that is reached when the two *slopes are equal* to each other.

The effect of optimization is shown in Figure 5–40, where the operating points on the two performance curves have moved in the direction of the arrows to their final, optimum points. The optimum total liquid rate is 1,397 bpd, and injection rates to the wells are 0.55 MMscf/d and 0.95 MMscf/d, respectively. The slopes of the performance curves at these two points are shown in the figure and are parallel, as demonstrated previously. It is at these points where an additional increase of gas injection to Well #1 results in an increase of liquid rate identical to the decrease of liquid rate experienced by Well #2, due to the necessary decrease in its injection rate. Consequently, the total liquid rate from the two wells is at a maximum here.

The previous logic, if applied to more than two wells, will give similar final results. This leads to a general conclusion on how lift gas should be allocated to a group of wells if the total gas volume is limited. The rule, first given by Simmons [22, 23], is called the *Equal Slopes* method and says that optimum gas lifting requires two conditions to be met: the slopes at the operating point on each well's gas lift performance curve must be identical while the total gas requirement of the wells must equal the available gas volume.

5.5.3.2 Conventional calculation models.

Gas allocation in continuous flow gas lifting is a heavily discussed topic in the professional literature. Most of the early publications dealt with the wells' performance curves and developed various calculation models to find the best solution. Redden et al. [28] started from the economic optimum of each well and developed a stepwise solution that ensured a maximum utilization of the available gas. Kanu et al. [24] used the Equal Slope model, while Nishikiori et al. [29] developed a nonlinear optimization scheme.

Unlimited Lift Gas Availability

The steps of finding the optimum gas lifting conditions for a group of wells for unlimited gas availability are as follows:

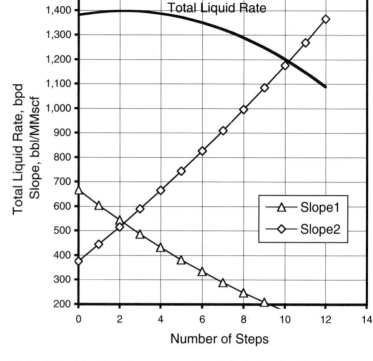

Fig. 5–41 Total liquid rate from two sample wells in the function of iteration steps.

1. Based on existing field conditions, gas lift performance curves for all wells should be established. A computer program package performing systems analysis calculations must be used for this task.

2. Perform curve-fitting calculations on each performance curve using a third-order polynomial of the gas injection rate, denoted as x in the following formula.

$$q_l = a_3 x^3 + a_2 x^2 + a_1 x + a_0$$

3. Determine the derivative of each performance curve. For the previous polynomial, this function is defined as

$$q_l = 3 a_3 x^2 + 2 a_2 x + a_1$$

4. Plot the performance curves and their derivatives in a coordinate system of liquid rate vs. gas injection rate.

5. Based on field data, find the Economic Slope from Equation 5.6 for each well.

6. On the plot of the derivative of each performance curve, find the gas injection rate belonging to the Economic Slope of the given well.

7. For each well, determine the liquid production rate corresponding to the injection rate found in Step 6.

8. Sum the liquid production and gas injection rates for the whole group of wells.

Example 5–15. Find the optimum gas allocation schedule for the five wells defined by the points of their performance curves, given in Table 5–16, if an unlimited supply of lift gas is available in the field. The wells make no water, the profit on oil is $1/bbl, and compression cost is $500/MMscf.

Well #1		Well #2		Well #3		Well #4		Well #5	
Q_g	q_l	Q_g	q_l	Q_g	q_l	Q_g	q_l	Q_g	q_l
MMscf/d	bpd	MMscf/d	bpd	MMscf/d	bpd	MMscf/d	bpd	MMscf/d	bpd
0.40	881	0.40	673	0.10	681	0.20	400	0.80	523
0.50	938	0.50	750	0.20	738	0.30	455	0.90	600
0.60	977	0.60	805	0.30	777	0.40	495	1.00	655
0.70	1,004	0.70	845	0.40	804	0.50	525	1.20	695
0.80	1,023	0.80	875	0.50	823	0.60	547	1.40	725
0.90	1,036	0.90	897	0.60	836	0.70	564	1.50	716
1.00	1,045	1.00	914	0.70	845	0.80	587	1.60	702
1.20	1,053	1.20	937	0.80	853	0.90	594	1.70	666
1.40	1,049	1.40	944	0.90	849	1.00	580		
1.60	1,035	1.60	930	1.00	835				
Coefficients of the Best Fitting Polynomials									
a_3	185.2		199.3		166.2		0.9		165.6
a_2	-796.6		-911.5		-632.0		-362.9		-1198.1
a_1	1097.9		1365.2		682.5		660.9		2343.1
a_0	561.5		265.9		621.9		285.5		-664.0

Table 5–16 Data of gas lift performance curves for the five wells in Example 5–15.

Solution

The coefficients of the best fitting polynomials are contained in Table 5–16, and performance curves along with their derivatives (slopes) are plotted in Figure 5–42.

The Economic Slope, as seen from Equation 5.6, is identical for all wells since they have the same water cut of zero:

Economic Slope = 500/ [(1 – 0) 1.0 – 0.0 0.0] = 500 / 1.0 = 500 bbl/MMscf.

Injection gas and liquid rates for the five wells, read off from Figure 5–42 at the Economic Slope, are listed in Table 5–17. As seen, a total oil rate of 3,457 bpd is achieved while a total gas volume of 2.37 MMscf/d is required for the optimum economic conditions. The table also contains the maximum possible liquid production rates for the case when compression costs are disregarded. It can be observed that achieving maximum liquid production would require more than twice as high an injection rate, but total liquid production would only increase by about 20%.

If water cuts for the different wells are different, Economic Slopes vary for each well, but the logic of the optimization process is the same.

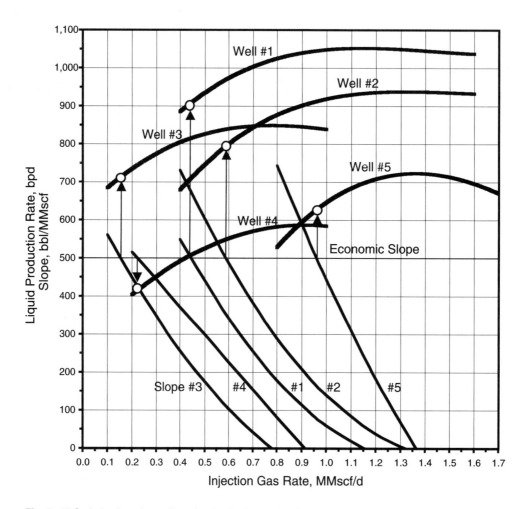

Fig. 5–42 Optimization of gas allocation for the five wells of Example 5–15, unlimited gas availability.

Limited Lift Gas Availability

If the amount of lift gas is limited for the group of wells, two cases are possible: (1) the amount is greater or equal to the sum of gas requirements of the individual wells valid at their Economic Slopes, or (2) there is less gas available than required for the economic optimum. The first case is identical to the unlimited lift gas condition and every well should receive its economic gas injection rate. In the second case, it is customary to apply the method of Equal Slopes described previously. The main steps of the optimization process are given as follows:

1. Using a computer program package performing systems analysis calculations, gas lift performance curves for all wells are established.

2. Perform *curve-fitting* calculations on each performance curve using a third-order polynomial of the gas injection rate, denoted as x in the following formula.

$$q_l = a_3 x^3 + a_2 x^2 + a_1 x + a_0$$

3. Set up a series of increasing slope values (the derivative of the performance curves), starting from zero.

4. Take the first slope value.

Well #	Maximum Liquid Production		Unlimited Gas Availability	
	Q_g	q_l	Q_g	q_l
	MMscf/d	bpd	MMscf/d	bpd
1	1.15	1,052	0.44	908
2	1.32	938	0.59	794
3	0.78	849	0.15	712
4	0.91	587	0.22	414
5	1.36	723	0.96	628
Total	5.53	4,149	2.37	3,457
Avg. GLR [scf/bbl]		1333		685

Table 5–17 Possible injection gas and liquid rates for the five wells in Example 5–15.

5. Calculate, for each well, the gas injection rate belonging to the actual slope. If the performance curves are fitted with the third-order polynomials given is Step 2, gas injection rates are found from the equation Slope = $3 a_3 x^2 + 2 a_2 x + a_1$. This yields a solution as follows, where x stands for the injection gas rate sought:

$$x = \frac{-2a_2 - \sqrt{4a_2^2 - 12a_3 (a_1 - Slope)}}{6a_3}$$

6. For each well, find the liquid production rate corresponding to the gas injection rate found in Step 5. This can be accomplished by substituting each injection rate into the proper equation given in Step 2.

7. Repeat Steps 5 and 6 for the rest of the assumed slope values.

8. Calculate the total gas injection requirement and the total liquid production rate for the group of wells for each assumed slope.

9. Plot the total gas and liquid rates in the function of the slope.

10. Using the available total gas injection rate, ΣQ_g, find the slope corresponding to this value from the plot defined in Step 9.

11. By interpolation, or by using the formula in Step 5, find for each well the gas injection and liquid production rates belonging to the slope determined in Step 10. These values define the optimum conditions and the required allocation of lift gas to each well.

The previous optimization procedure tacitly assumes that water cuts of the individual wells are identical, thus maximum oil production is achieved by maximizing the liquid production. For variable water cuts, the calculation model is modified accordingly.

Example 5–16. Find the optimum gas allocation schedule for the wells defined in the previous example if gas volumes of 2 MMscf/d and 1.7 MMscf/d are available for gas lifting.

Solution

It is known from the previous example that a gas injection rate of 2.37 MMscf/d is required for the economic optimum conditions. This is greater than any of the gas volumes given previously, therefore gas allocation has to be performed with the Equal Slopes model. Assumed slope values are given in Table 5–18, along with the corresponding gas injection and liquid production rates for each well.

Slope	Well #1		Well #2		Well #3		Well #4		Well #5		Field Totals	
	Q_g	q_l	Q_g	q_l	Q_g	q_l	Q_g	q_l	Q_g	q_l	Q_g	q_l
bbl/MMscf	MMscf/d	bpd	MMscf/d	bpd	MMscf/d	bpd	MMscf/d	bpd	MMscf/d	bpd	MMscf/d	bpd
0	1.15	1,052	1.32	938	0.78	849	0.91	587	1.36	723	5.53	4,149
50	1.02	1,049	1.17	935	0.69	846	0.84	585	1.32	722	5.04	4,138
100	0.92	1,042	1.07	927	0.61	840	0.78	580	1.27	719	4.64	4,108
150	0.84	1,032	0.98	917	0.53	831	0.71	571	1.23	713	4.29	4,065
200	0.77	1,019	0.91	904	0.47	820	0.64	559	1.19	706	3.97	4,009
250	0.71	1,005	0.85	889	0.41	806	0.57	544	1.15	697	3.67	3,941
300	0.65	988	0.79	873	0.35	791	0.50	525	1.11	686	3.39	3,863
350	0.59	970	0.73	855	0.30	774	0.43	502	1.07	674	3.12	3,776
400	0.54	951	0.68	836	0.25	755	0.36	476	1.03	660	2.86	3,678
450	0.49	930	0.63	816	0.20	734	0.29	447	1.00	645	2.61	3,572
500	0.44	908	0.59	794	0.15	712	0.22	414	0.96	628	2.37	3,457
550	0.40	885	0.54	771	0.11	689	0.15	378	0.93	610	2.13	3,333
600	0.36	860	0.50	747	0.07	665	0.08	338	0.89	591	1.90	3,201
650	0.32	835	0.46	722	0.03	639	0.02	295	0.86	570	1.68	3,061

Table 5–18 Injection gas and liquid rates in the function of slope for the five wells in Example 5–16.

Total gas injection requirement and total liquid production values are plotted in Figure 5–43, in the function of the slope of the performance curves. The slopes corresponding to the available gas injection rates can be read from the figure and are 578 bbl/MMscf and 645 bbl/MMscf for the total gas rates of 2 MMscf/d and 1.7 MMscf/d, respectively.

The next problem is to find each well's operating point for the previous two slopes. For this, either an interpolation scheme is used to retrieve data from Table 5–18, or the formula in Step 5 can be applied. The results of the calculations are contained in Table 5–19. The table contains all data required for the optimum allocation of the available list gas volumes. It can be seen that, compared to the $\Sigma Q_g = 1.7$ MMscf/d case, an injection volume of $\Sigma Q_g = 2$ MMscf/d (an increase of about 20%) could increase the field's liquid rate by 6% only.

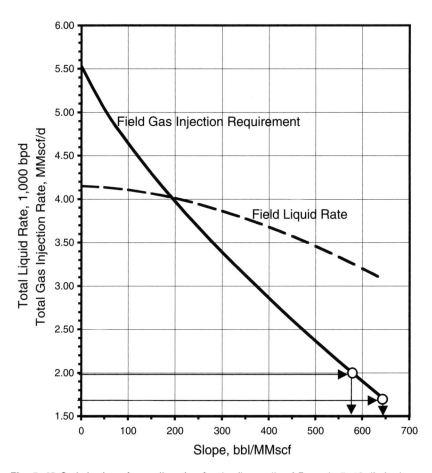

Fig. 5–43 Optimization of gas allocation for the five wells of Example 5–16, limited gas availability.

The results of the previous two examples are summarized in Figure 5–44, which shows the five wells' operating points for the following optimization cases: maximum liquid production, unlimited and limited gas availability.

5.5.3.3 Field-wide optimization.
The lift gas allocation models described in the previous section dealt with individual wells and did not include the effects of their interactions or the effects of other components in the field's network on the performance of the individual wells. The gas lifted well's performance, as seen before, was derived from systems analysis calculations that describe only the cooperation of the production system's components. Several components of the field's network, *e.g.* gathering stations, trunk lines, gas distribution lines, etc. were completely ignored or their effects were highly simplified. This is the reason why the latest gas allocation models treat the field's complete network (including the wells, the gathering and gas distribution systems, etc.) and optimize the distribution of lift gas by correctly accounting for all possible effects. [30] Such allocation models form a part of a field-wide optimization of production operations, and, consequently, require a much greater mathematical apparatus than the procedures described so far.

Well #	Limited Gas Injection Rate Available			
	Σ Qg = 2 MMscf/d		Σ Qg = 1.7 MMscf/d	
	Qg	qₗ	Qg	qₗ
	MMscf/d	bpd	MMscf/d	bpd
1	0.38	871	0.32	837
2	0.52	758	0.47	725
3	0.09	676	0.03	642
4	0.11	356	0.02	300
5	0.91	599	0.86	572
Total	**2.00**	**3,260**	**1.70**	**3,076**
Avg. GLR [scf/bbl]		**614**		**553**

Table 5–19 Parameters of the optimum gas allocation for the five wells in Example 5–16.

In general, lift gas allocation constitutes a complex *optimization problem*, the main features of which can be described as follows.

The *objective function* (*i.e.* the parameter to be maximized) is typically the field's overall profitability or the total oil production rate. The optimization process is always constrained by one or more operational *constraints* that may include the following:

- Availability of lift gas volume (usually the most important constraint).

- Surface gas handling capacity.

- Liquid (oil and water) and gas handling limitations of the gathering system.

- Compressor operational limits (pressures, etc.).

- Flow rates or flow velocities in different network components can be limited.

The *decision variables* of the field-wide lift gas allocation process are the wells' gas injection rates. They are adjusted and optimized to ensure the maximum of the objective function, *i.e.* the maximum profit, or the maximum oil production from the gas lifted wells.

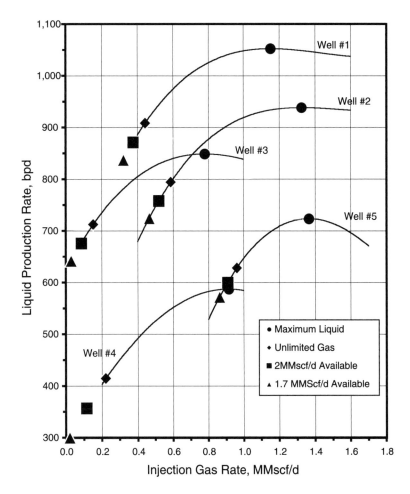

Fig. 5–44 Summary results of gas allocation optimization for Examples 5–15 and 5–16.

The optimization problem roughly described previously requires the use of complex mathematical models, most investigators, like Wang et al. [31, 32] utilize the sequential quadratic programming (SQL) algorithm. Several commercial computer program packages are available for field-wide production optimization that enable the practicing engineer to solve lift gas allocation problems in fields with a great number of wells while satisfying different scenarios.

5.6 Unloading of Continuous Flow Installations

5.6.1 Introduction

Throughout this chapter, it was tacitly assumed that injection of lift gas was possible at all times at the depth of the designed *point of gas injection*. In practice, however, this can only happen if lift gas can reach down to the operating valve with no liquid column existing above the valve in the well's annulus after the application of gas lift pressure. After the initial completion of the well or after any workover operation, this requirement is seldom met and the conditions illustrated in Figure 5–45 can occur. Here a *dead well*, killed before a completion or workover operation, is shown with the operating gas lift valve and all required downhole equipment in place. The kill fluid level, both in the tubing string and in the annulus, is usually quite close to the surface because, as shown in previous sections, continuous flow gas lift can only be applied to wells with sufficiently high bottomhole pressures. The operating valve must be open due to the great hydrostatic pressure of the liquid column above it, allowing communication between the tubing and the annular volume.

In order to start up production, gas lift pressure is applied to the casinghead depressing the liquid level in the annulus and displacing the kill fluid to the tubing string through the open gas lift valve. In most cases, however, the *operating gas lift pressure* is insufficient to depress the annulus liquid level below the gas lift valve (see Fig. 5–45). As shown in the figure, gas lift pressure may force the tubing level to rise up to the wellhead and may also produce some liquid to the surface, but it is usually not high enough to depress the annular level down to the operating valve's setting depth. Since gas cannot reach the valve, no injection is possible and continuous flow gas lifting cannot commence. It must be clear that this situation occurs whenever the gas lift pressure cannot balance the hydrostatic pressure of the liquid column present in the tubing string above the valve setting depth.

Consider the same well being shut down for some reason after a period of production by gas lift in Figure 5–46. Since it was produced before, the annulus liquid level is situated just below the operating valve's level while the liquid level in the tubing string is at the same depth as in the previous example. This is because the liquid level rose in the tubing during the *shut-in period*, but it could not enter the annulus due to the action of the gas lift valve's *check valve*. Now if production of the well is desired and gas lift pressure is applied to the casinghead, lift gas can reach the operating valve, which opens and injects gas into the tubing string. Well startup, in contrast to the case described in Figure 5–45, was possible and easily done.

The two examples showed that under similar conditions, well startup is impossible if the annulus contains kill fluid but is easily accomplished if no liquid is present in the annulus above the operating valve. Starting up well production, therefore, requires the removal of kill fluid from the annulus. This may be accomplished in several ways, the most common procedure being *swabbing* when liquid is removed from the tubing string, thus allowing the annulus liquid level to drop. Instead of swabbing, mostly used on flowing wells, liquid removal, called *well unloading* in the gas lift vocabulary, is generally done by the utilization of the available gas lift pressure. The unloading operation in continuous flow gas lifted wells is facilitated by the use of an *unloading valve string*, i.e. several gas lift valves run to properly designed depths above the operating valve. These valves allow a stepwise removal of the kill fluid from the annulus, and it is thanks to their operation that annulus liquid level can be depressed below the operating valve's depth, allowing the startup of normal gas lift operations.

5.6.2 The unloading process

The basic mechanism of unloading a continuous flow gas lift well is illustrated in Figures 5–47 and 5–48. These figures show a well with a *semi-closed installation* and three gas lift valves, out of which the two upper ones are *unloading valves*, whereas the bottom one is the *operating valve*. Due to the usually high formation pressure, the well's tubing and annulus is almost completely filled up with a kill fluid. As soon as injection gas pressure is applied to the casinghead, a gradual transfer of the fluids from the annulus to the tubing string is taking place through the open gas lift valves. As seen in Part A in Figure 5–47, all valves are open because of the high pressure acting on them, consisting of the gas pressure plus the hydrostatic pressure of the liquid column. Kill fluid is U-tubed by the injection pressure into the tubing and up to the surface. Because of the high pressure exerted on the well bottom, no pressure drawdown across the formation occurs and all the fluid produced to the surface comes from the annulus.

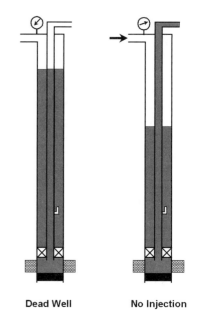

Dead Well   **No Injection**

Fig. 5–45 Unsuccessful startup of a dead well by gas lift.

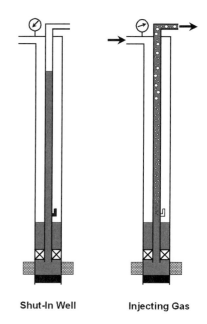

Shut-In Well   **Injecting Gas**

Fig. 5–46 Startup of a previously unloaded dead well by gas lift.

Injection pressure continuously depresses the liquid level in the annulus until the first valve is *uncovered* when it starts to inject gas, since the valve is still open. Part B of Figure 5–47 shows Valve #1 injecting gas into the tubing string and aerating the liquid column above the valve setting depth. Due to the injection of gas through the valve, tubing pressure at the first valve level starts to decrease and reaches a stabilized low value corresponding to the injected *GLR*. Since the tubing and the annulus volumes of the well are interconnected through the still-open lower valves, the pressure and consequently the liquid level must drop in the annulus as well. As a result, the annulus liquid level continuously drops while gas is injected through Valve #1.

Gas injection through Valve #1, as described previously, decreases the tubing pressure to a stabilized value and the annulus liquid level drops to a corresponding stable depth. Valve #2 was run just above this stabilized level, allowing it to be uncovered and to start injecting gas. (Part C) This is a critical moment in the unloading process because two valves admit gas to the tubing string and the upper one has to be closed to meet the two crucial objectives of the unloading procedure. These are (a) to move the point of gas injection down to the operating valve, and (b) to ensure a single-point gas injection. The proper design and setting of the unloading valves guarantees that Valve #1 will close just after injection through Valve #2 has begun. With Valve #1 closed, gas is injected through the second valve alone, as shown in Part D in Figure 5–48. This kind of injection transfer to a lower valve can be ascertained by a number of ways, most of which will be detailed later.

Valve #2, being the sole point of gas injection, brings about a decrease in tubing pressure at its setting depth which, in turn, forces the liquid level in the annulus to drop further. As before, liquid level in the annulus stabilizes and, assuming a correct design of the unloading valve string, Valve #3 is situated just above this stable liquid level. Since it was open from the beginning of the unloading operation, as soon as it is uncovered, Valve #3 starts to inject gas into the tubing (Part E). Note that this is the operating gas lift valve, and it is imperative that Valve #2 be forced to close just after injection has started through the third valve. By this time the objectives of the unloading process have been met, as gas injection takes place through Valve #3 only (see Part F).

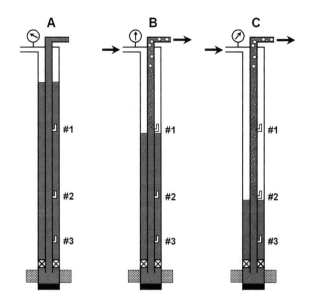

Fig. 5–47 Unloading a continuous flow installation with a string of unloading valves.

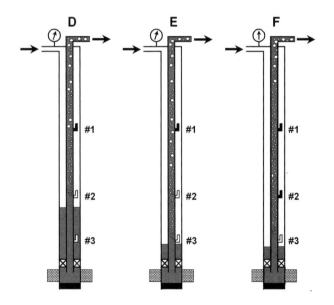

Fig. 5–48 Unloading a continuous flow installation with a string of unloading valves (continued).

At one point during the unloading process, *FBHP* drops below the formation pressure and inflow to the well from the formation starts. From that moment on, the well starts to produce well fluids to the surface, in addition to the kill fluid from the annulus. Due to the transfer of the injection point to the lower valves, *FBHP* decreases and well inflow increases. At the end of the unloading sequence, no more kill fluid is transferred from the annulus, and the well produces a *stable liquid rate* from the formation with gas injection stabilized at Valve #3, which is the *operating gas lift valve*.

5.6.3 General design considerations

The main objectives of unloading valve string design, also called *valve spacing*, are as follows:

- The consecutive transfer of the point of gas injection from the first gas lift valve set at a shallow depth to lower valves and finally to the operating valve.

- To ensure a *single-point gas injection* which, as discussed in Section 5.3.3.2, provides a more economic utilization of injection gas for continuous flow gas lift wells than *multipoint injection*.

The basic goals set forth previously can be met by several different approaches and design procedures. Before the detailed description of the most important valve spacing methods, some basic considerations influencing valve string designs are discussed in the following section.

Injection Pressure and Rate

The operating injection pressure has a great effect on valve spacing since unloading valves can be spaced further apart for higher pressures. In order to facilitate unloading at any conditions, the pressure used for valve spacing calculations must always be available at the wellhead. If line pressure at the well fluctuates due to insufficient system capacity, a value below the minimum available pressure must be used for valve spacing calculations. Pressure differentials between line pressure and design injection pressure of about 50 psi are commonly used [33].

In some cases, a pressure higher than the normal injection pressure, called *kickoff pressure*, is also available at the wellsite. This pressure is used for unloading only and can reduce the total number of unloading valves required. For wells with known gas injection depths, the kickoff pressure is used to space all the valves. If the depth of gas injection is not known or a valve string design is desired where any valve in the string can become the operating one, the setting depth of the first valve only is based on the kickoff pressure. The advantages of using a special kickoff pressure are mostly offset by the difficulties and costs of providing two levels of injection pressure at the well.

The *volume of injection gas* available at the wellsite can also affect unloading operations. For an unlimited gas supply, unloading valve string design is based on the minimum pressure gradient in the tubing string, and the total number of valves is usually reduced. In other cases, the tubing gradient is found from the available gas injection and the desired liquid production rates, and the maximum depth of gas injection will be limited.

Static Liquid Level

Although most wells on continuous flow gas lift have a static fluid level very close to the surface because of their high formation pressures, there are cases when the static liquid level is situated further downhole. It is normal practice to set the first unloading valve just above this static level and to start valve spacing calculations below this depth only.

Unloading to a Pit

Wells are usually unloaded into the flowline against the normal separator pressure. If a lower pressure system is available or the well is unloaded to a *pit*, valve spacing calculations result in deeper and fewer unloading valves. After the operating valve depth is reached, the well production is returned to the flowline and normal gas lift operations commence.

Valve Types

Valve spacing procedures for different gas lift valve types are different. Balanced and many unbalanced valve types require a drop in injection pressure for every valve, in order to properly operate as unloading valves. Throttling valves, on the other hand, do not necessitate any pressure drop to work properly. The valve type selected for the unloading valve string, therefore, has a great impact on valve spacing calculations, and different manufacturers recommend different spacing methods, as will be shown later in this chapter.

Well Temperature

Most unloading gas lift valves include a *bellows* charged with nitrogen gas that provides the closing force at operating conditions. Since bellows charge pressure changes with the temperature existing at valve setting depth, unloading valve string design procedures must correctly account for valve temperatures. The temperature of a *conventional*, outside mounted gas lift valve is very close to the temperature of the gas column present in the annulus. The temperature distribution in the gas column, in turn, is usually approximated by the geothermal gradient. Wireline retrievable valves, on the other hand, are surrounded by the flowing wellstream and are at the temperature of the flowing fluid in the tubing string.

In most valve spacing procedures, valve temperatures are assumed to be linear with well depth. Starting from the bottomhole temperature (equal to formation temperature), a linear distribution is set up to the surface flowing wellhead temperature. For safety reasons, wellhead temperature is usually calculated at the maximum predicted liquid production rate. The assumption of a linear temperature distribution gives lower-than-actual unloading temperatures because of the curvature of temperature traverses observed in gas lifted wells (see Section 2.6.5). Due to the lower temperatures used in the unloading valve string design, actual bellows charge pressures of upper valves will be higher, and this helps to keep the valves closed while injecting gas from a valve below.

Safety Factors

Valve spacing calculations involve many sources for input data and calculation errors that could make the final design's reliability questionable. In order to compensate for input data errors, approximations, and assumptions, gas lift designers apply different safety factors in their valve string design. These include modifications to the expected tubing and injection pressures. As will be shown later, the most common solution is to decrease the injection pressure at each valve's setting depth. This entails, for IPO valves, a decrease in valve closing pressure (see the valve opening equation, Equation 3.20) and ensures that, in spite of the data and calculation errors, the upper valves will close as required. [34]

Flag Valve

If IPO valves are used in the unloading valve string, a positive indication of injection through the operating valve is ensured by *flagging* the bottom valve. A flag valve is set at a substantially lower pressure than those above and its lower operating surface pressure (observed on the surface) indicates that unloading was successful down to the operating valve. The setting pressure of a flag valve is found by assuming an arbitrarily low tubing pressure at its setting depth, usually equal to the pressure valid at an expected minimum liquid production rate. Use of an *orifice* as the operating valve can provide the same operation.

Mandrels Present

In some cases, gas lift valve mandrels were previously installed in the well, and an unloading valve string design without the need for pulling the tubing and re-spacing of valve mandrels is desired. Possible valve setting depths are thus specified, and the objective of valve spacing is to use or skip individual mandrels along with valve setting calculations.

Valve Bracketing

Valve bracketing means that gas lift valves near the expected operating injection point are spaced with a constant vertical distance between them. This solution is required if

- Correction for the inherent inaccuracies in multiphase pressure predictions and in the determination of the injection point is desired

- The normal spacing procedure gives unloading valves *too closely spaced* near the injection point

- The maximum possible liquid rate from a high productivity well is desired by moving the injection point as deep as possible

The use of valve bracketing to compensate for errors in multiphase flow calculations is illustrated in Figure 5–49, after Mach et al. [4]. The schematic diagram contains the basic design calculations for continuous flow gas lifting with the annulus gas pressure line, and the two pressure traverses above and below the point of gas injection, as described in detail in Section 5.3.2. The original design (compare to Fig. 5–3) is shown in bold lines where the depth of the operating valve is a function of the operating pressure differential Δp_{op}. If calculated flowing pressures above the injection point are assumed to be in error, then two new pressure traverses (one for a positive, and one for the same negative error) can be plotted (see the dashed lines). It is customary to assume an error range of ±10% around the calculated pressure traverse, but a different value can also be used. The two extreme pressure traverses define two new injection points at their intersections with the $(p_{inj} - \Delta p_{op})$ vs. depth line. The depths of these points give the limits of the bracketing envelope, the well depth range where additional unloading valves with a uniform spacing should be run.

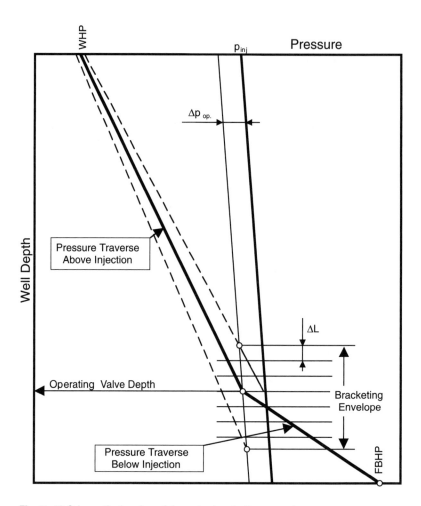

Fig. 5–49 Schematic drawing of the valve bracketing procedure.

The spacing of the unloading valves in the bracketing envelope depends on two parameters: (a) the operating pressure differential Δp_{op} and (b) the flowing fluid gradient below the operating valve. Since the flowing gradient is constant and depends on the formation GLR only, it is easily understood that valve spacing in the bracketing envelope is uniform and is found from

$$\Delta L = \frac{\Delta p_{op}}{grad_f}$$
5.7

where: Δp_{op} = operating pressure differential, psi

$grad_f$ = pressure gradient below the point of gas injection, psi/ft

When valve bracketing for the two other reasons given previously is necessary, additional unloading valves with the spacing defined in Equation 5.7 are used above and below the anticipated *point of gas injection*. [35]

5.6.4 Unloading valve string design procedures

The design of an unloading valve string involves two broad phases: (1) first the running depths of the individual valves are determined; (2) then the *settings* (port size, bellows charge pressure, spring force, etc.) of each valve in the string are calculated so that the valves will operate properly. This order is dictated by the fact that valve settings are affected by the pressures and the temperature valid at the depth of each valve. Proper design results in a perfect unloading of the installation, which means that

- Gas injection is consecutively *transferred* from the top valve to those below

- Each unloading valve closes and ceases to inject gas after the valve below has started to inject

- The point of gas injection is successfully transferred to the operating valve that ensures continuous operation

- All unloading valves remain *closed* while normal production through the operating valve takes place

There is a great multitude of valve spacing procedures developed over the years by different manufacturers of gas lift equipment. Most of them were specifically elaborated for a given type of gas lift valve, but all have a common objective of providing an undisturbed unloading sequence. One way of their classification may be based on the reliability of input data that may be complete and reliable or limited and unreliable. In case all necessary well data are available, the required point of gas injection can very accurately be determined and an unloading valve string reaching down to the depth of gas injection can properly be designed. The main requirements are the knowledge of inflow performance data, the use of accurate multiphase vertical and horizontal PD correlations, and the exact prediction of flowing temperatures, to name the most important items only.

In most situations, however, a complete data set required for a proper unloading valve string design is either unavailable or some important pieces of data (formation pressure, water cut, etc) are changing rapidly in the life of the well. All these conditions lead to an unreliable and/or undefined point of gas injection and to severe difficulties in valve spacing calculations. Since most gas lift installations fall into this category, special valve spacing procedures had to be developed to accommodate such conditions. These designs aim at providing an unloading valve string that can lift the desired liquid rate from *any valve* in the unloading string. An unloading valve string designed this way provides the flexibility needed for moving the actual gas injection point up or down the hole to suit changing well conditions.

In the following, representative valve spacing procedures will be discussed using the two basic design approaches described previously. The discussions of the individual design procedures are grouped according to the type of the gas lift valves used: IPO, PPO, and throttling valves.

5.6.4.1 IPO valves. One of the basic requirements of a successful unloading sequence is that any valve should close as the next lower valve starts to inject gas. Differently stated, the closing pressures of the valves in the string must decrease with well depth so that upper valves close successively. As described in the section on gas lift valve mechanics (Section 3.2.2), the closing injection pressure of an unbalanced IPO valve at valve setting depth is equal to its dome charge pressure and is found from the known actual injection and production pressures as given as follows (see Equation 3.20):

$$p_d = p_{io}(1 - R) + p_p R \hspace{3cm} 5.8$$

where: p_{io} = opening injection pressure at valve setting depth, psi

p_p = production pressure at valve setting depth, psi

R = A_v/A_b the valve's geometrical constant

In order to meet the requirement of decreasing closing pressures, it follows from the previous formula that the *opening injection pressures* of the valves should be decreased downward in the string. The surface injection pressure must therefore be decreased for each lower valve during unloading. As a consequence, the available injection pressure at the bottom (operating) valve is substantially less than full surface pressure would allow. In deep wells, therefore, the maximum attainable injection depth is limited, resulting in *production losses*, especially in high-productivity wells.

5.6.4.1.1 Variable pressure drop per valve. In one of the earliest valve spacing methods, originally developed by the CAMCO Company [2], the closing of the upper valves is ensured by successively dropping the injection pressure for each valve down the string. This solution works for valves with a small PPEF only, because they are sensitive mostly to injection pressure, so their closing pressures can be lowered by decreasing the injection pressure.

The valve spacing procedure described as follows is for wells with an unknown gas injection point and ensures stable gas injection from any valve in the string. The first phase of unloading valve string design is the determination of the *setting depths* of the valves, the basic steps of which are given as follows and are illustrated in Figure 5–50.

1. Starting from the surface injection pressure p_{inj}, draw the distribution of gas pressure in the annulus with depth.

2. Establish the flowing pressure traverse above the point of gas injection for the given tubing size, liquid rate, water cut, and temperature conditions.

3. The traverse should start at the surface from the expected *WHP* and should be selected from a gradient curve sheet or computer-calculated, based on the following considerations. If there is an unlimited gas supply available, then the pressure traverse with the minimum gradient is selected. For limited gas rates, the traverse with a *GLR* equal to the available gas rate divided by the desired liquid rate has to be used.

4. Find the depth of the first unloading valve and select its port size.

   a. The top valve is located at a depth where a line drawn parallel to the unloading gradient, started from the *WHP* intersects the gas pressure line.

   b. The depth of the top valve may also be calculated from

   $$L_1 = \frac{p_{inj} - WHP}{grad_u - grad_g}$$

   5.9

   where: p_{inj} = surface injection pressure, psi

   WHP = wellhead pressure during unloading, psi

   $grad_u$ = unloading gradient, psi/ft

   $grad_g$ = gas gradient in annulus, psi/ft

   c. The unloading gradient to be used is based on the *kill fluid* present in the well's annulus, the desired liquid production rate, and a gas-liquid ratio of *GLR* = 0.

   d. If the previous valve depth is less than the *static fluid level* present in the annulus, the top valve is located at the static fluid level.

   e. Find the minimum tubing pressure at the top valve's depth during unloading, p_{pmin1}.

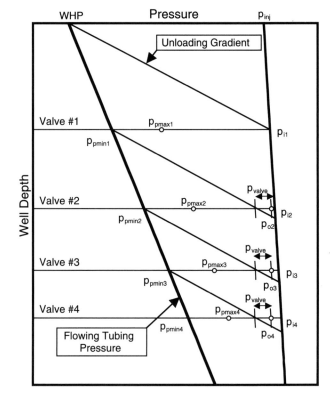

Fig. 5–50 Schematic drawing of the graphical valve spacing procedure for IPO valves using variable PDs per valve.

f. Estimate the injection *GLR* required to lift the desired liquid rate from the depth of the top valve and calculate the daily gas injection requirement.

g. Select a choke size that passes the required gas rate with an upstream pressure equal to injection pressure at valve depth p_{i_1} and a downstream pressure equal to the minimum tubing pressure p_{pmin1}.

5. Find the depth of the second unloading valve and select its port size.

a. The valve is located where the pressure on a line drawn parallel to the unloading gradient and started from the minimum tubing pressure at the top valve depth, p_{pmin1}, is less than the casing pressure by $\Delta p_{valve} = 50$ psi. This pressure differential is assumed to exist when the valve starts to inject gas and is meant to ensure that gas injection through the valve can occur.

b. The distance between the top and the second valve can also be calculated from

$$\Delta L_2 = \frac{p_{inj} - p_{pminl} + L_1\, grad_g - \Delta p_{valve}}{grad_u - grad_g}$$ 5.10

c. Determine the maximum flowing tubing pressure at the top valve p_{pmax1}, while lifting from the second valve. This is done either by using gradient curves or computer calculations.

d. Calculate the decrease, PPE_1, of the top valve's opening pressure caused by the maximum tubing pressure found above and prevailing at the top valve while lifting from the second valve. This decrease is found from the *PPEF* of the top valve as follows:

$$PPE_1 = TEF\,(p_{pmax1} - p_{pmin1})$$ 5.11

e. Calculate the *reopening pressure* of the top valve at the depth of the second valve from the formula:

$$p_{o2} = p_{i2} - PPE_1$$ 5.12

f. Find the minimum tubing pressure at the second valve's depth during unloading, p_{pmin2}.

g. Estimate the injection *GLR* required to lift the desired liquid rate from the depth of the second valve and calculate the daily gas injection requirement.

h. Select a choke size that passes the required gas rate with an upstream pressure equal to the reopening pressure of the top valve p_{o2} and a downstream pressure equal to the minimum tubing pressure p_{pmin2}.

6. Find the depth of the third unloading valve and select its port size.

a. The valve is located where the pressure on a line drawn parallel to the unloading gradient and started from the minimum tubing pressure valid at the depth of the second valve p_{pmin2}, is less than the casing pressure by the sum of Δp_{valve}, and the decrease of the top valve's opening pressure, PPE_1.

b. The distance between the second and third valve can also be calculated from:

$$\Delta L_3 = \frac{p_{inj} - p_{pmin2} + L_2\, grad_g - \Delta p_{valve} - PPE_1}{grad_u - grad_g}$$ 5.13

c. Determine the maximum flowing tubing pressure at the second valve p_{pmax2} while lifting from the third valve. This can be done by either using gradient curves or computer calculations.

d. Calculate the decrease, PPE_2, of the second valve's opening pressure caused by the maximum tubing pressure found above and prevailing at the second valve while lifting from the third valve. This decrease is found from the PPEF of the second valve as follows:

$$PPE_2 = TEF \left(p_{pmax2} - p_{pmin2} \right) \qquad\qquad 5.14$$

e. Calculate the *reopening pressure* of the second valve at the depth of the third valve from the formula:

$$p_{o3} = p_{i3} - \left(PPE_1 + PPE_2 \right) \qquad\qquad 5.15$$

f. Find the minimum tubing pressure at the third valve's depth during unloading, p_{pmin3}.

g. Estimate the injection *GLR* required to lift the desired liquid rate from the depth of the third valve and calculate the daily gas injection requirement.

h. Select a choke size that passes the required gas rate with an upstream pressure equal to the reopening pressure of the second valve p_{o3} and a downstream pressure equal to the minimum tubing pressure p_{pmin3}.

7. Find the depth of the fourth unloading valve and select its port size.

   a. The valve is located where the pressure on a line drawn parallel to the unloading gradient and started from the minimum tubing pressure valid at the depth of the third valve p_{pmin3} is less than the casing pressure by the sum of Δp_{valve} and the total of *PPEs* calculated so far.

   b. The distance between the third and fourth valve can also be calculated from

   $$\Delta L_4 = \frac{p_{inj} - p_{pmin3} + L_3\, grad_g - \Delta p_{valve} - \Sigma PPE}{grad_u - grad_g} \qquad\qquad 5.16$$

   c. Determine the maximum flowing tubing pressure at the third valve p_{pmax3} while lifting from the fourth valve. This can be done by either using gradient curves or computer calculations.

   d. Calculate the decrease, PPE_3, of the third valve's opening pressure caused by the maximum tubing pressure found above and prevailing at the third valve while lifting from the fourth valve. This decrease is found from the PPEF of the third valve as follows:

   $$PPE_3 = TEF \left(p_{pmax3} - p_{pmin3} \right) \qquad\qquad 5.17$$

   e. Calculate the *reopening pressure* of the third valve at the depth of the fourth valve from the formula:

   $$p_{o4} = p_{i4} - \Sigma PPE \qquad\qquad 5.18$$

   f. Find the minimum tubing pressure at the fourth valve's depth during unloading, p_{pmin4}.

   g. Estimate the injection *GLR* required to lift the desired liquid rate from the depth of the fourth valve and calculate the daily gas injection requirement.

   h. Select a choke size that passes the required gas rate with an upstream pressure equal to the reopening pressure of the third valve p_{o4} and a downstream pressure equal to the minimum tubing pressure p_{pmin4}.

8. The setting depths of the remaining unloading valves are determined by following the procedure described for the fourth valve. Valves are selected until their spacing becomes too close, called *valve stacking*. If needed, a series of *bracketing valves* may be installed above and below the expected final injection point, as described previously.

After the valve setting depths have been determined, one has to specify the setting of each valve which, for IPO valves, means calculation of their TROs. This is accomplished as follows.

1. The valve temperatures T_i (also required for injection gas rate calculations during the valve spacing procedure) have to be estimated from flowing temperature data.

2. The *reopening pressure* for each valve is calculated. For the top valve, it is defined as the injection pressure at that level p_{i_1}. For the remaining valves, it is calculated by decreasing the injection pressure at valve depth p_i by the sum of the PPEs of the valves above. Expressed for the nth valve, we get

$$p_{on} = p_{in} - \sum_{j=1}^{n-1} PPE_j \qquad\qquad 5.19$$

3. The dome charge pressures at valve setting depths can be expressed from the valves' *opening equations* (Equation 3.17) because injection and production pressures are specified. In order to ensure decreasing closing pressures down the valve string, the operating injection pressure at each valve depth is set to the reopening pressure of the valve above. The production pressure to be used in the valve's opening equation is the minimum tubing pressure at valve depth p_{pmin}. Based on these pressures, the valve's *dome charge pressure* at valve depth is found from Equation 3.20 as given as follows:

$$p_d = p_o\,(1 - R) + p_{pmin}\,R \qquad\qquad 5.20$$

4. Next, the dome charge pressure for surface conditions p_d' of each valve is calculated at the charging temperature of 60 °F. This can be done for a nitrogen gas charge by using the charts in Figure D–1 or D–2 in the Appendix or following the procedure described in connection with Equation 3.9.

5. The *TRO pressure* is found from the surface dome charge pressure, see Equation 3.21, reproduced here:

$$TRO = \frac{p_d'}{1 - R} \qquad\qquad 5.21$$

Example 5–17. Design an unloading valve string for the well given as follows for an unknown injection point with the variable PD procedure using CAMCO J–40 valves..

Tubing Size, in.	2 7/8	Perforation Depth, ft	6,000
Desired Liquid Rate, bpd	800	Formation Temperature, F	170
WHP, psi	100	Wellhead Temperature, F	80
Injection Pressure, psi	950	Load Fluid Gradient, psi/ft	0.468
Injection Gas Gravity	0.65	Gas Gradient, psi/ft	0.015
		Δp Across Valve, psi	50

Solution

Valve spacing results are presented in Figure 5–51, from which it can be seen that the bottom valve can be run to about 4,500 ft, because of the relatively low injection pressure. The calculation of valve port sizes for the top and the second valve is illustrated as follows.

Sizing Valve #1. Injection from the top valve requires a *GLR* of about 400 scf/bbl, and the corresponding gas rate is 320 Mscf/d. At the valve temperature of T_1 = 116 °F, the correction factor from the chart in

Figure C–3 is CF= 1.1. Using the injection and production pressures at valve depth of 977 psi and 390 psi, respectively, Figure C–1 gives a required orifice size of 9/64 in. Using Table F–1, a valve port size of 3/16 in. is selected with $R = 0.094$.

Sizing Valve #2. When injecting through the second valve, an approximate *GLR* of about 700 scf/bbl is required, and the corresponding gas rate is 560 Mscf/d. Gas rate correction factor from the chart in Figure C–3 is found at the valve temperature of $T_2 = 140$ °F as $CF = 1.08$. Injection and production pressures at valve depth being 995 psi and 601 psi, respectively, Figure C–1 gives a required orifice size of 11/64 in. Using Table F–1 a valve port size of 3/16 in. is selected with $R = 0.094$.

The maximum tubing (injection) pressure at the top valve level while injecting through the second valve was found from a pressure traverse as $p_{pmax_1} = 520$ psi and is denoted by a small circle.

Detailed results of the valve string design are presented in Table 5–20. Quite often, an orifice valve is installed at the bottom of the string.

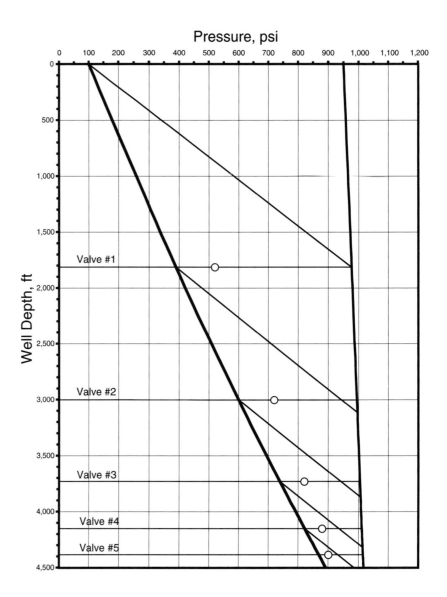

Fig. 5–51 Results of the unloading valve string design for Example 5–17.

Valve No.	Valve Depth	Valve Temp.	Valve Type	Port Size	PPEF	p_i	PPE	Σ PPE	p_o	p_{pmin}	p_d	p'_d	TRO
-	ft	F	-	in	-	psi	psi	psi	psi	psi	psi	psi	psi
1	1,816	107	J-40	3/16	0.104	977	14	0	977	390	922	836	922
2	3,003	125	J-40	3/16	0.104	995	12	14	982	601	946	828	914
3	3,732	136	J-40	3/16	0.104	1,006	8	26	980	740	958	821	907
4	4,152	142	J-40	3/16	0.104	1,012	6	34	978	822	963	817	902
5	4,386	146	J-40	3/16	0.104	1,016	3	40	976	869	966	814	898

Table 5–20 Details of the unloading valve string design calculations for Example 5–17.

5.6.4.1.2 Constant pressure drop per valve. The valve spacing method described in the previous section used a variable pressure drop per valve down the string for ensuring the closing of upper valves which was equal to the cumulative *PPE* of the valves above the actual one. Winkler [33] suggests the use of a constant pressure drop per valve equal to the production pressure effect of the top valve, *PPE1*, while injecting gas from the second valve. The design procedure in API RP 11V6 [34] detailed as follows utilizes a fixed pressure drop per valve of pressure drop = 25 psi as a safety factor. The main steps of the design are detailed as follows, and the calculation of the valve depths being illustrated in Figure 5–52.

1.   Starting from the surface injection pressure p_{inj}, draw the distribution of gas pressure in the annulus with depth.

2.   Establish the flowing pressure traverse above the point of gas injection, starting from the *WHP* for the given tubing size, liquid rate, water cut, and temperature conditions.

3.   Find the depth of the top unloading valve.

a.   By graphical construction, it is found where the pressure on a line parallel to the

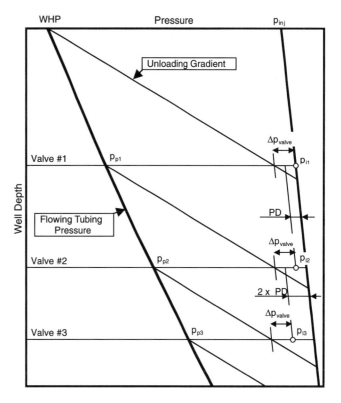

Fig. 5–52 Schematic drawing of the graphical valve spacing procedure for IPO valves using a constant pressure drop per valve.

unloading gradient and started from the *WHP* is less than the casing pressure by Δp_{valve} = 20 psi. This pressure differential is assumed to exist when the valve starts to inject gas and is meant to ensure that gas injection through the valve can occur.

b.   Alternatively, the setting depth of the top valve can be calculated as follows:

$$L_1 = \frac{p_{inj} - WHP - \Delta p_{valve}}{grad_u - grad_g} \tag{5.22}$$

4.   Find the setting depth of the second unloading valve.

a.   Graphically, it is found where the pressure on a line parallel to the unloading gradient and started from the flowing tubing pressure at the top valve p_{p_1} is less than the casing pressure p_{i2}, by the sum of Δp_{valve} and a safety factor of pressure drop = 25 psi.

b.   Mathematically, the depth increment between the top and the second valve is found from

$$\Delta L_2 = \frac{p_{inj} - p_{p1} - L_1\, grad_g - \Delta p_{valve} - PD}{grad_u - grad_g} \tag{5.23}$$

5.   The setting depths of the remaining unloading valves are found similar to the second valve, only an additional pressure drop is taken for each valve. The following universal formula can be applied to find the depth increment for the nth valve:

$$\Delta L_n = \frac{p_{inj} - p_{p(n)} - L_{(n-1)}\, grad_g - \Delta p_{valve} - (n-1)PD}{grad_u - grad_g} \tag{5.24}$$

6.   If the valves become stacked, valve bracketing may be necessary, as described previously.

With the valve setting depths known and the gas lift valve type selected according to local preferences, the size of the valve ports to be used in the unloading valves is determined. This is not a critical step in the design and usually a uniform size is chosen because of the relatively high pressure differentials during unloading. The *operating valve*, however, should be properly sized for the predicted gas injection rate.

Calculation of valve setting data is done as follows.

1.   The temperature at each valve setting depth T_i is found from the flowing temperature data.

2.   The *dome charge pressure* of the valves at their setting depths is calculated from the known injection and production pressures, according to Equation 3.20:

$$p_d = p_i (1 - R) + p_p R \qquad\qquad 5.25$$

where: p_i = injection pressure at valve setting depth, psi

p_p = flowing tubing pressure at valve setting depth

3.   The dome charge pressure for surface conditions p'_d of each valve is calculated at the charging temperature of 60 °F. For the usual nitrogen gas charge, this can be done by using the charts in Figure D–1 or D–2 in the Appendix or following the procedure described in connection with Equation 3.9.

4.   Finally, the TRO pressure is found from the surface dome charge pressure (see Equation 3.21) using the formula as follows:

$$TRO = \frac{p'_d}{1 - R} \qquad 5.26$$

Example 5–18. Design an unloading valve string, using the data of the previous example. Use the constant pressure drop per valve method with the following parameters:

Pressure drop per Valve, psi   25

Δp Across Valve, psi   20

Solution

The results of valve spacing are shown in Figure 5–53; valve setting data are contained in Table 5–21. Comparison with the previous design shows that both procedures gave unloading strings with 5 valves and very similar valve settings. With both design procedures, the well can be lifted from about 4,500 ft because of a relatively low injection pressure.

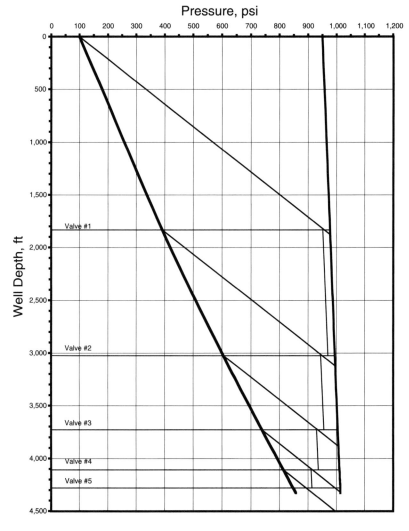

Fig. 5–53 Results of the unloading valve string design for Example 5–18.

Valve No.	Valve Depth	Valve Temp.	Valve Type	Port Size	R	p_i	p_p	p_d	p'_d	TRO
-	ft	F	-	in	-	psi	psi	psi	psi	psi
1	1,832	107	J-40	3/16	0.094	977	392	922	836	922
2	3,025	125	J-40	3/16	0.094	970	605	936	819	904
3	3,731	136	J-40	3/16	0.094	956	740	936	803	886
4	4,110	142	J-40	3/16	0.094	937	814	925	786	867
5	4,281	144	J-40	3/16	0.094	914	848	908	768	848

Table 5–21 Details of the unloading valve string design calculations for Example 5–18.

5.6.4.1.3 Constant surface opening pressure. In contrast to the previous two design methods, the spacing procedure originally suggested by U.S. Industries [36] and called *Optiflow Design* keeps the surface opening pressures of the unloading valves constant and ensures the proper injection transfer by increasing the tubing (production) pressures. The main steps of valve spacing are given as follows for a case when the point of gas injection is known and a kickoff pressure is available (see Fig. 5–54).

1. Starting from the surface injection pressure p_{inj} and the surface kickoff pressure $p_{kickoff}$ draw the distributions of gas pressure in the annulus with depth.

2. Establish the flowing pressure traverse above the point of gas injection, starting from the *WHP* for the given tubing size, liquid rate, injection *GLR*, water cut, and temperature conditions (dashed line).

3. Draw a design tubing pressure line between the point of gas injection (point B) and a design *WHP* at the surface (point A), found as *WHP* + 200 or WHP + 0.2 p_{inj}, whichever is greatest.

4. Find the depth of the top unloading valve.

   a. By graphical construction, it is found where the pressure on a line started from the *WHP* and parallel to the unloading gradient equals the injection pressure at depth.

   b. Alternatively, the setting depth of the top valve can be calculated as follows:

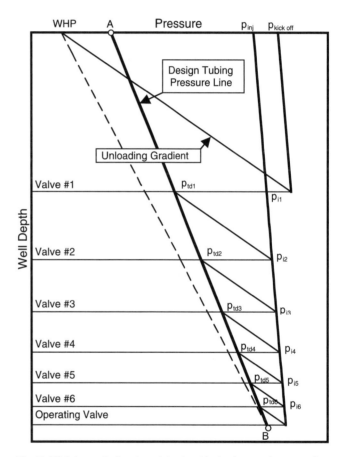

Fig. 5–54 Schematic drawing of the graphical valve spacing procedure for IPO valves using constant surface opening pressures (Optiflow Design).

$$L_1 = \frac{p_{kick\,off} - WHP}{grad_u - grad_g}$$

5.27

5. Determine the *design tubing pressure* at the top valve's setting depth p_{td1}.

6. Find the setting depth of the second unloading valve.

   a. Start a line parallel to the unloading gradient from the design tubing pressure at the setting depth of the top valve p_{td_1} and find the depth where it intersects the casing pressure valid for the normal injection pressure p_{inj}. This determines the depth of the second valve.

   b. Mathematically, the depth increment between the top and the second valve is found from

   $$\Delta L_2 = \frac{p_{inj} - p_{td1} - L_1\, grad_g}{grad_u - grad_g} \tag{5.28}$$

7. Determine the *design tubing pressure* p_{td2} at the setting depth of the second valve.

8. The setting depths of the remaining unloading valves are found similar to the second valve and a universal formula can be applied to find the depth increment for the nth valve:

   $$\Delta L_n = \frac{p_{inj} - p_{td(n-1)} + L_{(n-1)}grad_g}{grad_u - grad_g} \tag{5.29}$$

9. Valves are selected until their spacing becomes too close. If *valve stacking* occurs, installation of a series of *bracketing valves* may be necessary.

After the spacing of the unloading valve string has been determined, the port size for the operating valve is selected, based on the required injection rate and the pressure conditions at the depth of injection. The port sizes of the unloading valves can be smaller or equal to the operating valve's port size; it is a common practice to use uniform sizes. Valve setting calculations can be summarized as follows.

1. Based on flowing temperature data, the temperature at each valve setting depth T_i is found.

2. The *dome charge pressures* of the valves at their setting depth are calculated from the known injection and production pressures, according to Equation 3.20:

   $$p_d = p_i\,(1 - R) + p_{td}\,R \tag{5.30}$$

   where: p_i = injection pressure at valve setting depth, psi

   p_{td} = design tubing pressure at valve setting depth.

3. Dome charge pressures at surface conditions p_d' are calculated at a charging temperature of 60 °F. For the usual nitrogen gas charge, this can be done by using the charts in Figure D–1 or D–2 in the Appendix or by following the procedure described in connection with Equation 3.9.

4. Finally, TRO pressures are found from the surface dome charge pressure (see Equation 3.21) using the formula:

   $$TRO = \frac{p_d'}{1 - R} \tag{5.31}$$

Example 5–19. Use well data given in the previous examples and assume the gas injection point at 4,500 ft, where the tubing pressure, found from the pressure traverse is 893 psi. Design the unloading valve string for a constant surface opening pressure.

Solution

Point A for the design tubing pressure line is found by comparing the values $WHP + 200 = 300$ psi, and $WHP + 0.2\,950 = 290$ psi; and the greater (300 psi) is selected. Valve spacing resulted in five unloading valves to be run as shown in Figure 5–55. Injection gas requirement calculations showed that a valve with a $^3/_{16}$ in. port is sufficient for the operating valve. The unloading valves were selected with the same port size; their required settings are given in Table 5–22.

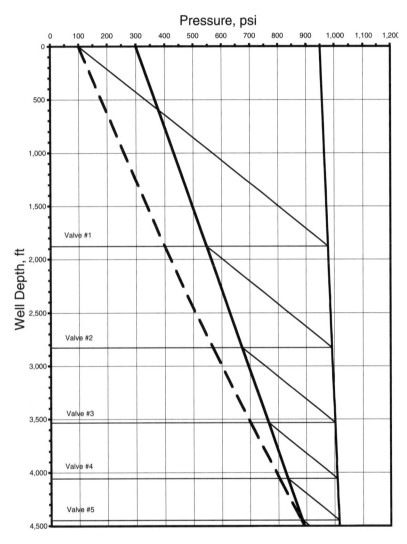

Fig. 5–55 Results of the unloading valve string design for Example 5–19.

Valve No.	Valve Depth	Valve Temp.	Valve Type	Port Size	R	p_i	p_{td}	p_d	p'_d	TRO
-	ft	F	-	in	-	psi	psi	psi	psi	psi
1	1,876	108	J-40	3/16	0.094	978	547	938	848	936
2	2,828	122	J-40	3/16	0.094	992	672	962	847	934
3	3,535	133	J-40	3/16	0.094	1,003	765	981	845	933
4	4,059	141	J-40	3/16	0.094	1,011	834	994	845	932
5	4,449	147	J-40	3/16	0.094	1,017	886	1,004	844	932

Table 5–22 Details of the unloading valve string design calculations for Example 5–19.

5.6.4.2 Balanced valves. Balanced gas lift valves open and close at the same pressure that is why the unloading valve string has to be designed with decreasing injection pressures to ensure that upper valves close successively. The valve spacing procedure described as follows was suggested by the OTIS Engineering Company for use with their flexible sleeve valves. [37] The main steps of the design procedure are described in conjunction with Figure 5–56, for a case when the point of gas injection is known and a kickoff pressure is available.

1. Based on the surface injection pressure p_{inj} and the kickoff pressure $p_{kickoff}$ draw the distributions of gas pressure in the annulus with depth.

2. Establish the flowing pressure traverse above the point of gas injection, starting from the WHP for the given tubing size, liquid rate, injection GLR, water cut, and temperature conditions.

3. The depth of the top unloading valve is based on the kickoff pressure.

    a. By graphical construction, it is found where the pressure on a line started from the WHP and parallel to the unloading gradient equals the injection pressure at depth.

    b. Alternatively, the setting depth of the top valve can be calculated as follows:

$$L_1 = \frac{p_{kick\,off} - WHP}{grad_u - grad_g}$$

5.32

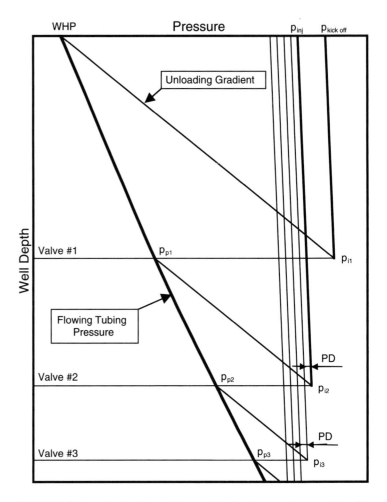

Fig. 5–56 Schematic drawing of the graphical valve spacing procedure for balanced valves.

4. Find the setting depth of the second unloading valve.

    a. From the production (tubing) pressure at the setting depth of the top valve p_{p1}, draw a line parallel to the unloading gradient and find the depth where this line intersects the casing pressure at depth, valid for the normal injection pressure p_{inj}:

    b. Mathematically, the depth increment between the top and the second valve is found from:

$$\Delta L_2 = \frac{p_{inj} - p_{p1} - L_1\,grad_g}{grad_u - grad_g}$$

5.33

5. The setting depths of the remaining unloading valves are found similar to the second valve with the only difference that the injection pressure is decreased by a set *PD* for each successive valve. A universal formula can be applied to find the depth increment for the nth valve:

$$\Delta L_n = \frac{p_{inj} - p_{p(n-1)} - (n-1)PD + L_{(n-1)} grad_g}{grad_u - grad_g}$$

5.34

6. Valves are selected until their spacing becomes too close, when installation of a series of *bracketing valves* may be necessary.

Valve setting calculations are simpler than for IPO valves and can be summarized as follows. Since flexible sleeve valves do not contain a port and can inject any amount of gas entering the casinghead, valve port sizing is not required.

1. Based on flowing temperature data, the temperature at each valve setting depth T_i is found.

2. At its setting depth, the required dome charge pressure p_d of the valve is equal to the opening/closing injection pressure p_i.

3. For balanced valves, the TRO pressure is identical to the dome charge pressure p_d' valid at the surface charging temperature. For the usual nitrogen gas charge, this can be found by using the charts in Figure D–1 or D–2 in the Appendix or by following the procedure described in connection with Equation 3.9. Please note that OTIS used a charging temperature of 80 °F.

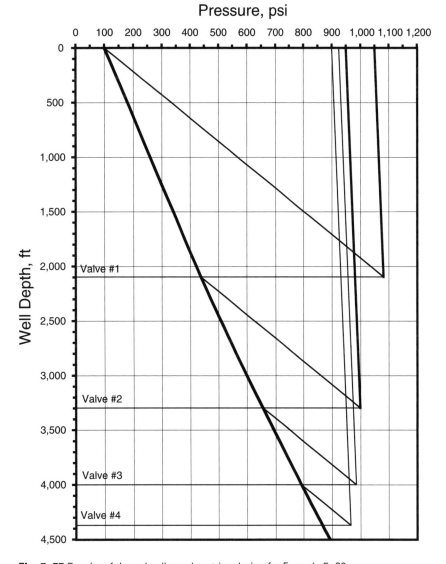

Example 5–20. Use well data given in the previous example and assume a kickoff pressure of 1,050 psi. Design the unloading valve string for OTIS C valves and a pressure drop of pressure drop = 25 psi per valve.

Solution

Valve spacing results are given in Figure 5–57, and an unloading string with four valves was designed. Valve setting data are contained in Table 5–23.

Fig. 5–57 Results of the unloading valve string design for Example 5–20.

5.6.4.3 PPO valves. PPO or fluid operated valves can be normal IPO valves installed in the right type of gas lift mandrel or may have a crossover seat that allows the use of the same mandrel as in IPO service. These valves are opened and closed by production (tubing) pressure that varies substantially during unloading, in contrast to the injection pressure. The valve spacing procedure using the design tubing pressure line (see Section 5.1.4.1.3) was especially developed for this kind of valve. Since the determination of valve setting depths is identical to that detailed before for IPO valves, the calculation steps detailed in conjunction with Figure 5–54 can be used.

Valve No.	Valve Depth	Valve Temp.	Valve Type	p_i	p_d	TRO
-	ft	F	-	psi	psi	psi
1	2,097	111	C	1,081	1,081	1,010
2	3,297	129	C	999	999	901
3	3,999	140	C	985	985	870
4	4,370	146	C	966	966	844

Table 5–23 Details of the unloading valve string design calculations for Example 5–20.

Valve setting calculations are different from that used for IPO valves. The possible minimum port size allowing the injection of the required gas rates must be used to ensure the proper sensitivity of the valves to production pressure. If the port size thus calculated is too big, the valve must be choked upstream or downstream of the port for valves with or without crossover seats, respectively. [38] If unbalanced bellows charged valves in the proper mandrels are used, the following steps describe the required calculations.

1. Based on flowing temperature data, the temperature at each valve setting depth T_i is found.

2. The *dome charge pressure* of the valve at its setting depth is calculated from the opening equation for PPO valves, according to Equation 3.35. The valve's opening production pressure p_{po} is assumed to be equal to the design tubing pressure p_{td}, and the following formula is used.

$$p_d = p_{po} (1 - R) + p_i R \qquad\qquad 5.35$$

   where: p_i = injection pressure at valve setting depth, psi

   p_{po} = valve opening pressure at valve setting depth

3. Dome charge pressures at surface conditions p_d' are calculated at a charging temperature of 60 °F. For the usual nitrogen gas charge, this can be done by using the charts in Figure D–1 or D–2 in the Appendix or by following the procedure described in connection with Equation 3.9.

4. Finally, TRO production pressures are found from the surface dome charge pressure (see Equation 3.39) using the formula:

$$\mathrm{TRO} = \frac{p_d'}{1 - R} \qquad\qquad 5.36$$

Example 5–21. Use well data given in the previous examples and design the unloading valve string if PPO valves are used.

Solution

Valve spacing calculations are identical to that used in Example 5–19 and are contained in Figure 5–55. The same valve (CAMCO J–40) as in the previous example is selected in a different mandrel that allows PPO service. Valve setting data are given in Table 5–24, where the bottom valve is usually replaced with an IPO operating valve.

Valve No.	Valve Depth	Valve Temp.	Valve Type	Port Size	R	p_i	p_{po}	p_d	p'_d	TRO
-	ft	F	-	in	-	psi	psi	psi	psi	psi
1	1,876	108	J-40	3/16	0.094	978	547	588	534	589
2	2,828	122	J-40	3/16	0.094	992	672	702	620	685
3	3,535	133	J-40	3/16	0.094	1,003	765	788	681	752
4	4,059	141	J-40	3/16	0.094	1,011	834	851	725	800
5	4,449	147	J-40	3/16	0.094	1,017	886	898	756	835

Table 5–24 Details of the unloading valve string design calculations for Example 5–21.

5.6.4.4 Throttling valves.

The unbalanced, spring-loaded valves without a bellows charge are frequently called *throttling valves* because they never open fully and always restrict gas injection rates. Due to their many beneficial features (see Section 3.2.2.3.1), these valves are very often used in continuous flow service. Since their gas throughput characteristics were experimentally determined by the manufacturer and made available to users, the design of the unloading string, in contrast to the procedures described so far, can be based on actual injection gas rates instead of the static force balance equations. For a detailed treatment of the gas throughput performance of throttling valves, refer to Section 3.2.3.2.

The design of an unloading valve string using throttling valves was named Proportional Response Design by the original manufacturer of the valves (MERLA) because these valves provide a proper response to changing well conditions by proportionally changing their injection gas rates. In the following, valve spacing and setting procedures are detailed for cases with complete data on well inflow conditions, *i.e.* for a known injection point.

5.6.4.4.1 Injection transfer. In contrast to other IPO valves, throttling valves utilize a constant surface injection pressure and do not require any drop in surface gas pressure to close upper valves during unloading. They are opened and closed by production pressure and are set to completely close before the production (tubing) pressure decreases to the final value valid for injecting from the operating gas lift valve. The basic requirement of a successful unloading operation, *i.e.* an upper valve should close as soon as injection from the valve below starts, is ensured by the operational features of throttling valves. [7]

Figure 5–58 depicts the first two valves in an unloading valve string with the pressure traverses valid during unloading. The two valves are assumed to be properly selected, their gas passage performance curves being displayed at the bottom of the figure. The left-hand part of the figure shows the conditions just before the second valve is *uncovered*, whereas the right-hand part depicts the pressure and injection rate conditions when *injection transfer* from the first to the second valve is almost complete.

After the first (upper) valve is uncovered and starts to inject gas, production (tubing) pressure continuously decreases due to the lightening of the liquid column above the valve. As seen from the left-hand side of the figure, the gas injection rate through this valve decreases according to the upper valve's gas passage performance curve. At the moment shown in the figure,

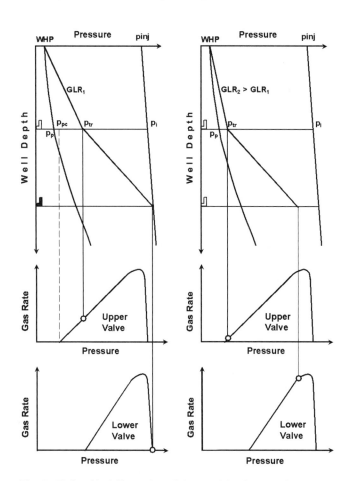

Fig. 5–58 Graphical illustration of the gas injection transfer process when using throttling valves.

tubing pressure at the upper valve equals p_{tr}, called the *transfer pressure* while it is much higher at the lower valve due to the gasless liquid column between the two valves. Because of the continuous gas injection through the upper valve, tubing pressure continuously decreases at this valve level and consequently at the depth of the lower valve, leading to the uncovering of the lower (second) valve.

On the right-hand part of Figure 5–58, both valves are uncovered and admitting gas simultaneously. The tubing pressure opposite the upper valve has substantially dropped due to the high injection gas rate through the two valves, resulting in a diminishing gas injection rate through the valve, as seen from its performance curve. The lower valve, however, injects a gas rate close to its maximum capacity and the transfer of the gas injection point from the upper to the lower valve is almost complete. As shown in the figure, before tubing pressure at the upper valve drops to the design tubing pressure p_b, valid at continuous flow operation from the operating gas lift valve, the upper valve completely closes at a tubing (production) pressure of p_{bc} and gas injection is transferred to the lower valve. The whole process repeats for the second and third valve and unloading continues toward the depth of the operating gas lift valve.

As described previously, during the unloading operation, all valves must supply a sufficient amount of injection gas so as to decrease the tubing pressure to the transfer pressure p_{tr} and they must close after the transfer operation is completed. The proper operation of the unloading, therefore, depends on the proper selection of each valve's transfer pressure p_{tr}.

The proper selection of the valve's transfer pressure p_{tr} is based on the investigation of the effects of the transfer pressure on (a) the gas rate injected through the valve, and on (b) the required gas injection rate for maintaining the multiphase flow in the tubing. It follows from the shape of the valve's gas injection performance curve (see Fig. 5–58) that the greater the transfer pressure, the greater the volume of gas injected through the valve. At the same time, higher transfer pressures require lower GLR values for maintaining the multiphase flow in the tubing string above the valve, see the upper part of Figure 5–58, where $GLR_2 > GLR_1$. The same conclusion can be reached from the study of vertical multiphase gradient curve sheets. Higher $GLRs$ for the same liquid production rate at the given moment of unloading, however, mean higher gas injection demands to ensure continuous multiphase flow.

Since the previous two effects of the transfer pressure work in opposite directions, one can find a value for the transfer pressure p_{tr}, which ensures that the gas volume injected by the gas lift valve matches the gas volume required to continue the unloading operation. The gas rate injected through the valve is readily calculated from the valve's performance curve, according to Equation 3.74. On the other hand, the injection rate required to maintain continuous multiphase flow depends on the unloading liquid rate, found from the well's IPR curve, and the GLR required to flow the actual unloading liquid rate from the transfer to the WHP. Equating these injection rates and utilization of a trial-and-error procedure for hand calculations or a computer solution gives the proper value of the transfer pressure for each valve in the unloading valve string design, as will be shown in the following section.

5.6.4.4.2 Valve spacing and setting calculations. The unloading valve string design procedure developed for throttling valves [7] properly accounts for the gas passage capabilities of the gas lift valves and the varying injection gas requirements during the unloading process. This is why detailed data on the well's inflow properties and an accurate prediction of the point of gas injection is required. Valve spacing calculations, therefore, must be preceded by a basic design of the installation (detailed in Section 5.3.2.) and unloading valves are to be run down to the depth of the operating valve. The main steps of the design are listed as follows, where frequent reference is made to Figure 5–59, a schematic drawing of the graphical valve spacing procedure.

Calculations for the top valve

1. Based on the surface injection pressure p_{inj}, calculate the distribution of gas pressure in the annulus with depth.

2. Establish the flowing pressure traverse above the point of gas injection previously determined, starting from the WHP for the given tubing size, liquid rate, injection GLR, water cut, and temperature conditions.

3. The depth of the top unloading valve is found as follows:

   a. By graphical construction, it is found where the pressure on a line started from the WHP and parallel to the unloading gradient equals the injection pressure at depth.

b. Alternatively, the setting depth of the top valve can be calculated as follows:

$$L_1 = \frac{p_{inj} - WHP}{grad_u - grad_g} \qquad 5.37$$

4. Determine the injection pressure at the depth of the top valve p_i, where the depth of the valve crosses the gas distribution line.

5. Find the design production (tubing) pressure p_p from the flowing pressure traverse curve at the depth of the top valve.

6. Calculate the opening (for throttling valves also the closing) production pressure of the valve. In order to ensure that the valve will close at the design production pressure p_p, this pressure is set as

$$p_{po} = p_{pc} = p_p + 0.1(p_i - p_p) \qquad 5.38$$

7. The next steps involve a trial-and-error selection of the valve's transfer pressure p_{tr}. The objective of the iterative procedure is, as detailed in the previous section, to find a transfer pressure where the gas rate required to continue the unloading operation equals the rate injected by the gas lift valve. For the first guess, a value of 50 psi higher than the valve's opening pressure is normally used.

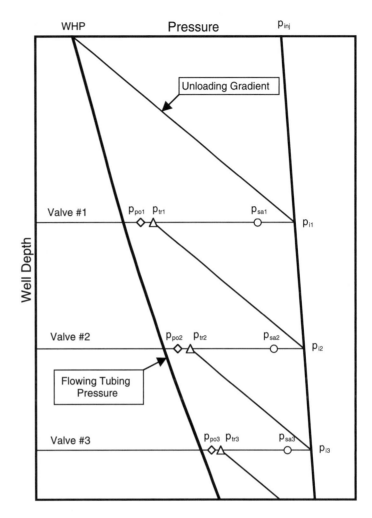

Fig. 5–59 Schematic drawing of the graphical valve spacing procedure for throttling valves.

8. The *FBHP* when reaching the transfer pressure is found by adding the flowing pressure of the fluid column below the top valve to the transfer pressure. The pressure gradient to be used *grad* is the kill fluid gradient as long as there is no inflow from the well and annulus fluid only is produced. After inflow starts, the static gradient of the well fluid is to be used.

$$FBHP = p_{tr} + (L_{perf} - L_1)\, grad \qquad 5.39$$

9. The liquid rate from the well is calculated from the well's IPR curve, when inflow is possible. If not, a minimum rate of 100 bpd is used for unloading the kill fluid from the annulus.

$$q_u = PI\,(SBHP - FBHP) \qquad 5.40$$

where: PI = productivity index, bpd/psi

$SBHP$ = static bottomhole pressure, psi

10. Using gradient curves or computer calculations, find the total gas-liquid ratio GLR_{total} required to produce the unloading rate from the top valve with the given *WHP* and the transfer pressure p_{tr} at the valve setting depth L_1.

11. Find the required injection gas-liquid ratio GLR_{inj} by subtracting the gas produced by the well GLR_p from the GLR_{total} found in the previous step. As long as there is no inflow to the well, $GRL_p = 0$ is used.

12. Determine the required injection gas rate (in Mscf/d units) from the unloading liquid rate and the injected gas-liquid ratio:

$$Q_{inj} = \frac{q_u \, GLR_{inj}}{1,000} \tag{5.41}$$

13. Find the flowing temperature at the top valve level T_1 from the temperature distribution in the well.

14. Calculate the corrected gas rate based on the flowing temperature T_1 at the valve setting depth:

$$Q_{corr} = Q_{inj} \sqrt{\frac{520}{T_1 + 460}} \tag{5.42}$$

15. From manufacturer's data, select the *trim size* (small, medium, or large) for the valve. The increasing trim sizes represent increasing slopes of the gas passage performance curve of the valve, see Figure 3–49. Normally, small trim is selected first.

16. The dynamic A_v/A_b ratio, F_e, as defined by Equation 3.72 is found from the valve's data sheet.

17. The *spring adjustment pressure* p_{sa} of the valve is calculated from Equation 3.46, reproduced here:

$$p_{sa} = p_i - F_e \, (p_i - p_{po}) \tag{5.43}$$

18. Calculate the gas throughput rate of the gas lift valve (in Mscf/d units) under the given conditions from its performance relationship, Equation 3.74. The slope M of the performance curve is found from manufacturer's data (see Fig. 3–49).

$$Q_{valve} = M \, (p_{tr} - p_{po}) \tag{5.44}$$

19. Check if the gas injected by the valve Q_{valve} is sufficient to supply the gas volume Q_{corr} required for the given phase of the unloading process. When doing so, consideration can be given for running two valves to the same setting depth. If the two gas injection rates do not match, the transfer pressure p_{tr} must be properly modified and calculations repeated from Step 7 on. When choosing a better guess for the transfer pressure, consult Section 5.1.4.4.1 on the injection pressure selection for throttling valves.

20. If the injected and required gas rates closely match each other sufficiently, another check is made to ensure that the gas lift valve can pass the required gas volume. If not, a different trim size is selected and calculations are repeated from Step 16. The maximum gas rate through the valve is calculated from the formula as follows (see also Equation 3.75), where factor K is found from manufacturer data (see Fig, 3–49):

$$Q_{max} = K M \, (p_{tr} - p_{po}) \tag{5.45}$$

Calculations for subsequent valves

1. The depths of the other unloading valves are found as follows:

   a. By graphical construction, the depth of the next valve is found where a line started from the *transfer pressure* of the actual valve and parallel to the unloading gradient intersects the injection pressure at depth (see Fig. 5–59).

   b. Alternatively, the depth increment for the nth valve can be calculated as follows:

   $$\Delta L_n = \frac{p_{inj} - p_{tr(n-1)} + L_{(n-1)} \, grad_g}{grad_u - grad_g} \tag{5.46}$$

The rest of the calculations follow Steps 4–20 of the procedure for the top valve.

Calculations for the operating gas lift valve

Since this design procedure is based on the knowledge of the final gas injection point, the spacing of the unloading valves is continued until the depth of the operating gas lift valve is approached. The calculations for the operating valve are considerably different from those applied for the unloading valves because of several reasons:

(a) the operating valve must not close when production (tubing) pressure decreases to flowing tubing pressure valid at the final injection point

(b) there is no transfer pressure for the operating valve since gas injection does not have to be moved deeper down the hole

(c) since the liquid rate and the injection GOR are fixed, the valve must pass a pre-determined amount of lift gas

. Selection of the operating gas lift valve necessitates the determination of the proper valve trim and the right setting of the valve, i.e. the proper selection of its spring adjustment pressure. While selecting these parameters, the objective of the design is to have a valve setting that provides the injection of exactly the same amount of gas that is required for continuous flow operations. The design requires an iterative process because the gas throughput rate of the valve is a function of the M parameter (the slope of the gas passage performance curve) which, in turn, varies with the setting of the valve, i.e. the spring adjustment pressure p_{sa}. The main steps of the design are detailed in the following:

1. Find the operating injection and production pressures at the operating valve's depth, p_i and p_p, respectively.

2. The *FBHP* is calculated from the flowing tubing pressure p_p and the gradient of the well fluid:

$$FBHP = p_p + (L_{perf} - L_1) \, grad \qquad\qquad 5.47$$

3. The liquid rate from the well is calculated from the well's IPR curve.

$$q_l = PI \, (SBHP - FBHP) \qquad\qquad 5.48$$

   where: PI = productivity index, bpd/psi

   $SBHP$ = static bottomhole pressure

4. Using the results of the basic design of the installation, the total gas-liquid ratio GLR_{total} needed for the continuous flow operation from the operating valve is found.

5. The required injection gas-liquid ratio GLR_{inj} is calculated by subtracting the gas produced by the well GLR_p from the GLR_{total} found in the previous step.

6. Determine the required injection gas rate (in Mscf/d units) from the liquid rate and the injected gas-liquid ratio:

$$Q_{inj} = \frac{q_l \, GLR_{inj}}{1,000} \qquad\qquad 5.49$$

7. Find the flowing temperature at the operating valve level T from the temperature distribution in the well.

8. Calculate the corrected gas rate based on the flowing temperature T at the valve setting depth:

$$Q_{corr} = Q_{inj} \sqrt{\frac{520}{T + 460}} \qquad\qquad 5.50$$

9. This is the start of the iterative process for finding the settings of the operating valve and *valve trim* is assumed first. Small trim is preferred since it responds very gently to changes in tubing pressure. If this proves to supply an insufficient gas rate, medium, then large trim has to be selected.

10. By using the assumed valve trim, the slope of the gas passage performance curve M is assumed.

11. By solving the throttling valve's performance equation (Equation 3.74), one can find the valve's required opening production pressure p_{po}:

$$p_{po} = p_p - \frac{Q_{corr}}{M}$$

5.51

12. Based on the valve's dynamic A_v/A_b ratio, F_e, the spring adjustment pressure is calculated as

$$p_{sa} = p_i - F_e \,(p_i - p_{po})$$

5.52

13. Knowledge of the required valve setting makes it possible to find the exact slope M of the valve, using the M vs. p_{sa} chart of the manufacturer's data sheet (see Fig. 3–49). If the calculated and assumed values compare within 5%, the iteration process is finished. Otherwise, repeat the calculation procedure from Step 10.

14. Finally a check is made to ensure that the gas lift valve can pass the required gas volume. If not, a different trim size is selected, and calculations are repeated from Step 10. The valve's maximum gas capacity is calculated from the formula as follows (see also Equation 3.75), where factor K is found from manufacturer data, *e.g.* from Figure 3–49, as a function of the ratio p_i/p_{sa}:

$$Q_{max} = K - M \,(p_i - p_{po})$$

5.53

- - - - - - - - - - - - - - - - - -

Example 5–22. Design an unloading valve string using throttling valves for the well data given in the previous examples. Use MERLA LM–16 type gas lift valves whose performance data is given in Figure 3–49. The injection point was found at 4,500 ft, and the pressure traverse above the point of gas injection is contained in Figure 5–60. Additional well data are given as follows.

SBHP, psi	3,000
Formation GLR, scf/bbl	100
PI, bpd/psi	0.5
Well Fluid Gradient, psi/ft	0.36

Solution

Results of valve spacing calculations are graphically depicted in Figure 5–60, where five unloading valves were designed. Numerical results are contained in Table 5–25, which does not contain the data of iterations for the transfer pressures; final values only are

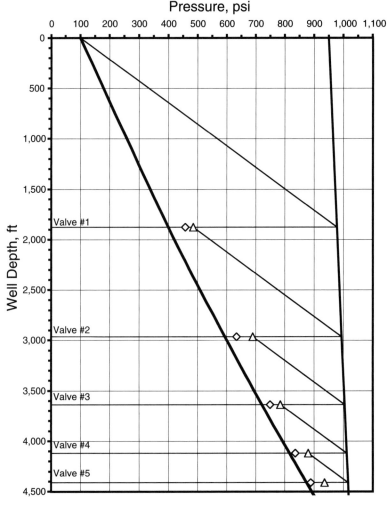

Fig. 5–60 Results of the unloading valve string design for Example 5–22.

given for the first four unloading valves. Since the fifth valve is set to a depth (4,404 ft) quite close to the required point of gas injection, it was selected as the operating valve. The design calculations for the operating valve are given in Table 5–26, which contains the results of the three necessary iterations for finding the valve setting parameters. As shown, the well produces 776 bpd from the depth of the fifth gas lift valve, a rate very close to the desired 800 bpd.

Description	Symbol	Unit	Unloading Valves			
			#1	#2	#3	#4
Valve Type	-	-	LM-16	LM-16	LM-16	LM-16
Valve Depth	L	ft	1,876	2,965	3,637	4,122
Gas Pressure at Valve	p_i	psi	978	994	1,005	1,012
Design Fluid Pressure	p_p	psi	400	594	721	816
Valve Opening Pressure	p_{po}	psi	458	634	750	836
Transfer Pressure	p_{tr}	psi	485	690	787	881
Flowing Bottomhole Pressure	p_{wf}	psi	1,970	1,783	1,638	1,559
Liquid Production Rate	q_l	bpd	515	609	681	721
Required Total GLR	GLR_{total}	scf/bbl	150	200	250	280
Injection GLR	GLR_{inj}	scf/bbl	50	100	150	180
Required Gas Rate	Q_{inj}	Mscf/d	26	61	102	130
Valve Temperature	T	F	108	124	135	142
Corrected Gas Rate	Q_{corr}	Mscf/d	27	65	109	140
Selected Valve Trim	-	-	S	S	M	M
Dynamic Av/Ab Ratio	F_e	-	0.25	0.25	0.5	0.5
Spring Adjustment Pressure	p_{sa}	psi	848	904	877	923
Slope of Throttling Line	M	Mscf/d/psi	1.12	1.18	2.94	3.06
Production Press. Diff.	$p_{tr} - p_{pc}$	psi	27	56	37	46
Gas Thru Valve	q_{valve}	Mscf/d	31	66	110	140
Number of Valves	-	-	1	1	1	1
Pressure Ratio	p_i/p_{sa}	-	1.153	1.100	1.145	1.096
K-factor from Chart	K	-	0.680	0.680	0.468	0.420
Maximum Gas Rate	q_{max}	Mscf/d	396	289	351	227

Table 5–25 Details of the unloading valve string design calculations for Example 5–22.

Description	Symbol	Unit	Iterations		
			#1	#2	#3
Valve Type	-	-	LM-16		
Valve Depth	L	ft	4,404		
Gas Pressure at Valve	p_i	psi	1,016		
Flowing Tubing Pressure	p_p	psi	873		
Flowing Bottomhole Pressure	p_{wf}	psi	1,448		
Liquid Production Rate	q_l	bpd	776		
Required Total GLR	GLR_{total}	scf/bbl	350		
Injection GLR	GLR_{inj}	scf/bbl	250		
Required Gas Rate	Q_{inj}	Mscf/d	194		
Valve Temperature	T	F	146		
Corrected Gas Rate	Q_{corr}	Mscf/d	210		
Selected Valve Trim	-	-	S		
Assumed Slope	M	Mscf/d/psi	1.00	1.15	1.16
Required Gas Rate	Q_{inj}	Mscf/d	210	210	210
Required Pr. Differential	Q_{inj}/M	psi	210	182	181
Valve Opening Pressure	p_{po}	psi	663	691	692
Number of Valves	-	-	1	1	1
Dynamic Av/Ab Ratio	F_e	-	0.25	0.25	0.25
Spring Adjustment Pressure	p_{sa}	psi	928	935	935
Calculated Slope	M	Mscf/d/psi	1.15	1.16	1.16
Pressure Ratio	p_i/p_{sa}	-	1.095	1.087	1.086
K-factor from Chart	K	-	0.68	0.68	0.68
Maximum Gas Rate	q_{max}	Mscf/d	276	256	255

Table 5–26 Details of the design calculations for the operating gas lift valve for Example 5–22.

5.6.5 Practical considerations

During the unloading process, before any valve in the unloading string is uncovered, liquid from the annulus is forced to flow through it. Even if this liquid is a clean kill fluid, the relatively high flow velocities through the small ports of the gas lift valve can cut or *wash out* valve seats, but the presence of any solids, especially sand, in the load fluid exponentially increases the abrasive effect and the possibility of valve damage. This is the reason why more valve damage is done during unloading operations than at any other time of the life of the well. Therefore, it is of utmost importance that well unloading be conducted so as to avoid excessive wear on the gas lift valves.

Preventive measures before the first unloading of the well try to decrease the occurrence of solids in the annulus fluid. Drilling mud should be replaced with a clean *kill fluid* by circulation, so as to completely eliminate or at least minimize the solids content of the annulus fluid. New gas injection lines, before first being connected to the well, must be blown clean of scale, welding slag, and other solids to prevent plugging of surface controls and introduction of solids into the annulus. Only after these precautions can unloading be started.

Liquid damage in gas lift valves mostly depends on the fluid velocities occurring in the valve, which, in turn, are a function of the pace of pressure increase in the annulus. If injection pressure at the casinghead is allowed to increase rapidly, the drop of annulus liquid level will also be rapid, resulting in high liquid rates and velocities across the unloading valves. Valve damage, therefore, can significantly be reduced if injection pressure is gradually applied to the annulus. The procedure described as follows and recommended by the API [40] ensures that damage to the valves in the unloading string is minimized.

1. Before unloading commences, production (tubing) pressure is to be blown down to flowline pressure.

2. Remove any choke installed on the flowline. If after an initial *kickoff*, the well is supposed to flow naturally, an adjustable choke can be left on the wellhead; but it should be fully open this time.

3. Using the surface injection control, introduce lift gas to the annulus so that the rate of increase of the casinghead pressure is about 50 psi for every 8–10 minutes. Continue to build up annulus pressure at this rate until it reaches about 400 psi.

4. Increase the surface gas injection rate so as to maintain an increase rate of the casinghead pressure of about 100 psi for every 8–10 min. Continue until downhole gas injection through the first unloading valve occurs.

5. Adjust the surface gas injection rate to about ½ to ⅔ of the daily gas rate that the installation was originally designed for.

6. After 12–18 hrs of continuous operation with the above reduced injection rate, the surface gas injection control is set to the designed operating gas injection rate.

5.6.6 Conclusions

Out of the many possible unloading valve string designs, the most important ones are those utilizing IPO and throttling valves. The relative advantages and disadvantages of these designs are summarized as follows, after Schmidt et al. [41].

The advantages of IPO valves include the following:

(a) they tolerate errors in reservoir performance data

(b) usually a lower number of valves is needed

Their main drawback comes from the fact that most designs require a drop in injection pressure for every valve in the string to close unloading valves. This is why, especially in deeper wells, the operating valve cannot be run deep enough and, due to the shallower injection point, gas usage may be extensive.

The benefits of using throttling valves are the following:

(a) the full injection pressure can be utilized by the operating gas lift valve ensuring deeper injection points than with IPO valves

(b) the Proportional Response Design properly accounts for the gas injection rates occurring across valves during unloading

(c) the operating valve properly responds to changing inflow conditions

Their disadvantages are the following:

(a) accurate well performance data are needed for design

(b) typically more unloading valves are designed, as compared to the use of IPO valves

(c) the pressure differential through the operating valve is greater than for IPO valves

All of the conventional unloading valve string design procedures, except the Proportional Response Design used for throttling valves, tacitly assume that gas lift valves operate in their *orifice flow* range. This is why they employ the Thornhill-Craver formula (Equation 2.65) to calculate the gas throughput capacity of the unloading valves. During unloading, however, pressure conditions at the given valve's depth considerably change with time and the valve can operate in the *throttling flow* range (consult Section 3.2.3 on dynamic performance of gas lift valves). Although orifice flow may be maintained by installing a small choke downstream of the port in any IPO valve [42], the calculation of gas injection rates during the unloading process still remains a critical phase of the design. As pointed out by Schmidt et al. [41], use of a comprehensive model describing the dynamic performance of gas lift valves is a basic requirement of a successful unloading valve string design.

The basic problem with conventional valve string designs is that they do not consider the *transient nature* of the unloading process and utilize formulas developed for steady-state conditions. Unloading itself is a complex phenomenon that comprises two main phases of transient nature:

(a) first the kill fluid is displaced from the annulus to the tubing string

(b) then gas injected through the unloading valve(s) mingles with tubing and annulus fluids

The performance of the tubing involves concurrent and/or countercurrent multiphase vertical or inclined flow while that of the annulus is governed by the gas injection rates occurring at the surface choke and across the gas lift valve(s).

An ultimate valve string design would be based on the description of the transient processes taking place during the unloading operation and could be developed from the simulation of the unloading process. Several investigators have elaborated numerical simulation models [43–45], and computerized design procedures are also available.

5.7 Surface Gas Injection Control

The efficient gas lifting of any well or a group of wells can only be ensured if lift gas is properly distributed to each well, *i.e.* gas at the proper pressure and of the right volumetric rate is injected to the well(s). Those devices situated at the wellhead and taking care of this requirement constitute the surface gas *injection control* of the given well. Surface injection control, in general, involves the control of the surface injection pressure (measured at the casinghead) as well as of the surface gas injection rate, and for intermittent wells, the duration and frequency of gas injections per day. The different requirements in timing, injection rate, and injection pressure of different wells (continuous, intermittent, etc.) necessitate the use of different surface control devices. Surface injection control, on the other hand, can be manual, semi-, or fully automatic, but since all versions utilize the same general principles, manual controls will be described in the following.

For any surface control to be efficient, continuous measurements of pressures and gas flow rates is required. This can be ensured by installing on the wellhead: (a) a two-pen *pressure recorder* simultaneously recording casinghead and tubinghead pressures, and (b) a *meter run* measuring injection gas rates. It is the information derived from the frequent observation of these recordings that decisions on the proper modifications of surface control settings are made possible. Installation of these devices, therefore, not only ensures an efficient surface gas injection control (and, consequently, high system efficiency) but forms the basis of analyzing and troubleshooting of the well's operation.

Continuous flow gas lift requires the undisturbed injection of a steady rate of lift gas at a steady surface gas injection pressure. This type of operation needs quite simple surface controls, as detailed as follows.

5.7.1 Choke control

The simplest way to inject a constant gas rate into a well placed on continuous flow gas lift is to install a fixed or *positive choke*, or an *adjustable choke* in the injection line at the wellhead (see Fig. 5–61). The choke is sized to pass the required injection gas rate at the designed casinghead pressure by (a) installing the proper size of fixed choke or

(b) setting the right size on the adjustable choke. Use of an adjustable choke or *metering valve* is recommended since adjustment of choke sizes is much easier and does not require the interruption of gas flow, which is the case when using a fixed choke.

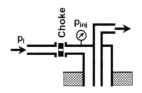

This type of surface injection control, however, can only maintain a constant injection rate if the choke's upstream and downstream pressures are held constant (see chapter 2 on the behavior of fixed chokes). Casinghead (downstream) pressure p_{inj} usually stays constant because it is controlled by the operating gas lift valve in the well, but surface line pressure p_l can fluctuate due to compressor capacity limitations and other causes. If line pressure increases for a shorter or longer period, simple choke control will inject an increased gas rate into the well, thereby increasing the well's instantaneous gas consumption more than the designed value, with an associated drop in system efficiency. A decreased line pressure, on the other hand, results in less lift gas being injected and the well's liquid rate will decrease accordingly. Fluctuating line pressure, therefore, limits the use of fixed or adjustable chokes for surface gas injection control.

Fig. 5–61 Schematic arrangement of choke control of surface gas injection.

A common drawback of all injection control types that employ fixed or adjustable chokes is their vulnerability to *freezing* due to the Joule-Thompson Effect. Injection gas in the choke suffers a sudden decrease of flow area, causing its velocity to increase and, at the same time, its temperature to decrease. Flowing temperature in the choke can become so low that the inherent moisture content of the lift gas freezes and, in extreme cases, may completely block the flow. Out of the preventive measures against freezing problems, the most important is the proper *dehydration* that reduces the moisture content of lift gas to an acceptable level. Injecting methanol in the gas stream, or heating of lift gas can also decrease or eliminate the problems associated with freezing.

5.7.2 Choke and regulator control

When the pressure of the surface gas lift system, available at the wellhead, fluctuates widely, a fixed choke passes fluctuating gas rates into the annulus. Under such conditions, either the injection gas-liquid ratio GLR_{inj} will exceed the designed value, causing a drop in system efficiency; or, due to an insufficient gas injection rate, the well's liquid rate will decrease. The effects of a widely varying line pressure are usually eliminated by a *pressure regulator*, which can be connected parallel or in series with the fixed or adjustable choke.

In a serial configuration (Fig. 5–62), the pressure regulator is installed on the injection line upstream of the choke and is set to regulate its downstream pressure to a value below the *lowest possible* line pressure p_{lmin}. This pressure, in turn, must be higher than the operating valve's surface opening pressure. Gas pressure upstream of the choke being regulated to a constant value, a properly sized choke passes a uniform gas injection rate to the casing annulus. This is due to the fact that for constant upstream and downstream pressures, gas throughput across a fixed choke is constant.

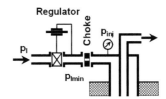

Although choke and regulator control of surface gas injection is usually accomplished with the pressure regulator connected in series to the choke, the parallel configuration may also be used. In this case, the choke is sized to pass the required gas volume with the highest possible surface line pressure. When, due to pressure fluctuations in the gas lift system, line pressure available at the wellhead decreases, the pressure regulator takes over and overrides the choke by passing gas into the annulus, increasing its pressure to the required level.

Fig. 5–62 Schematic arrangement of choke and regulator control of surface gas injection.

5.7.3 Other control methods

There are some special injection control types that can be used in addition to the generally applied methods discussed previously.

In gas lifted wells that intermittently *start to flow*, gas injection is controlled by the tubinghead pressure. When, after an initial kickoff, the well starts to flow, its *WHP* increases, which triggers the shutoff of gas injection to reduce gas usage.

When the danger of freezing of the surface choke is high, a *time cycle controller* (intermitter) is sometimes used to control the gas injection to the well. The intermitter is set for very high cycle frequencies, thereby allowing the use of larger choke sizes that eliminates freezing conditions.

References

1. *Gas Lift*. Book 6 of the Vocational Training Series. Second Edition, Dallas, TX: American Petroleum Institute, 1984.

2. Winkler, H. W. and S. S. Smith. *Gas Lift Manual*. CAMCO Inc., 1962.

3. Raggio, J., "A New Concept of Continuous Flow Gas Lift by Multipoint Injection." Paper SPE 1893 presented at the 42nd Annual Fall Meeting, Houston, TX, October 1–4, 1967.

4. Mach, J. M., E. A. Proano, H. Mukherjee, and K. E. Brown, "A New Concept in Continuous-Flow Gas Lift Design." *JPT*, December 1983: 885–91.

5. McAfee, R. V., "The Evaluation of Vertical-Lift Performance in Producing Wells." *JPT*, April 1961: 390–8.

6. Beadle, G., J. Harlan, and K. E. Brown, "Evaluation of Surface Back-Pressure for Continuous and Intermittent-Flow Gas Lift." *JPT*, March 1963: 243–51.

7. *Gas Lift Manual*. Section 2: Continuous Flow Gas Lift. Teledyne MERLA, Garland, TX, 1970.

8. Beggs, H. D. *Production Optimization Using Nodal Analysis*. Tulsa, OK: OGCI Publications, 1991.

9. Kanu, E. P., "Systems Analysis Hikes Well Performance." *PEI*, May 1981: 96–120.

10. Brown, K. E., J. Mach, and E. A. Proano, "Application of Systems Analysis Techniques in Optimizing Gas Lift Installations." J. Energy Resources Technology, June 1982: 157–61.

11. Gilbert, W. E., "Flowing and Gas-Lift Well Performance." API Drilling and Production Practice, 1954: 126–57.

12. Grupping, A. W., C. W. F. Luca, and F. D. Vermeulen, "Heading Action Analyzed for Stabilization." *OGJ*, July 23, 1984: 47–51.

13. Grupping, A. W., C. W. F. Luca, and F. D. Vermeulen, "These Methods Can Eliminate or Control Annulus Heading." *OGJ*, July 30, 1984: 186–92.

14. Asheim, H., "Criteria for Gas-Lift Stability." *JPT*, November 1988: 1452–6.

15. Blick, E. F., P. N. Enga, and P. C. Lin, "Theoretical Stability Analysis of Flowing Oil Wells and Gas-Lift Wells." *SPE PE*, November 1988: 508–14.

16. Alhanati, F. J. S., Z. Schmidt, D. R. Doty, and D. D. Lagerlef, "Continuous Gas-Lift Instability: Diagnosis, Criteria, and Solutions." Paper SPE 26554 presented at the 68th Annual Technical Conference and Exhibition. Houston, TX, October 3–6, 1993.

17. Jansen, B., M. Dalsmo, L. Nokleberg, K. Havre, V. Kristiansen, and P. Lemetayer, "Automatic Control of Unstable Gas Lifted Wells." Paper SPE 56832 presented at the Annual Technical Conference and Exhibition. Houston, TX, October 3–6, 1999.

18. Tokar, T., Z. Schmidt, and C. Tuckness, "New Gas Lift Valve Design Stabilizes Injection Rates: Case Studies." Paper SPE 36597 presented at the Annual Technical Conference and Exhibition, Denver, CO, October 6–9, 1996.

19. Shaw, S. F. *Gas Lift Principles and Practices*. Houston, TX: Gulf Publishing Co., 1939.

20. Brown, K. E. *The Technology Of Artificial Lift Methods*. Vol. 2a. Tulsa, OK: Petroleum Publishing Co., 1980.

21. Pittman, R. W., "Gas Lift Design and Performance." Paper SPE 9981 presented at the International Petroleum Exhibition and Technical Symposium, held in Beijing, China, March 18–26, 1982.

22. Simmons, W. E., "Optimizing Continuous Flow Gas Lift Wells." Part 1. *PE*, August 1972: 46–8.

23. Simmons, W. E., "Optimizing Continuous Flow Gas Lift Wells." Part 2. *PE*, September 1972, 68–72.

24. Kanu, E. P., J. Mach, and K. E. Brown, "Economic Approach to Oil Production and Gas Allocation in Continuous Gas Lift." *JPT*, October 1981: 1887–92.

25. Tokar, T. J. and B. E. Smith, "Determining an Optimum Gas Injection Rate for a Gas-Lift Well." U.S. Patent 5,871,048, 1999.

26. Blann, J. R. and J. D. Williams, "Determining the Most Profitable Gas Injection Pressure for a Gas Lift Installation." *JPT*, August 1984: 1305–11.

27. Stinson, R., "The Use of Systems Analysis in the Design of a Gas-Lift System." Paper SPE 17584 presented at the International Meeting on Petroleum Engineering, held in Tianjin, China, November 1–4, 1988.

28. Redden, J. D., T. A. G. Sherman, and J. R. Blann, "Optimizing Gas-Lift Systems." Paper SPE 5150 presented at the 49th Annual Fall Meeting held in Houston, TX, October 6–9, 1974.

29. Nishikiori, N., R. A. Redner, D. R. Doty, and Z. Schmidt, "An Improved Method for Gas Lift Allocation Optimization." Paper SPE 19711 presented at the 64th Annual Technical Conference and Exhibition, held in San Antonio, TX, October 8–11, 1989.

30. Dutta-Roy, K., and J. Kattapuram, "A New Approach to Gas-Lift Allocation Optimization." Paper SPE 38333 presented at the Western Regional Meeting held in Long Beach, CA, June 25–27, 1997.

31. Wang, P., M. L. Litvak, and K. Aziz, "Optimization of Production from Mature Fields." Paper B1/F4/04 presented at the 17th World Petroleum Conference, Rio de Janeiro, September 1–5, 2002.

32. Wang, P., M. Litvak, and K. Aziz, "Optimization of Production Operations in Petroleum Fields." Paper SPE 77658 presented at the Annual Technical Conference and Exhibition, held in San Antonio, TX, Sept. 29–Oct. 2, 2002.

33. Bradley, H. B. (Ed.) *Petroleum Engineering Handbook*. Chapter 5. Society of Petroleum Engineers, 1987.

34. "Recommended Practice for Design of Continuous Flow Gas Lift Installations Using Injection Pressure Operated Valves." *API RP 11V6*, 2nd Edition, American Petroleum Institute, 1999.

35. Kanu, E. P., J. M. Mach, and K. E. Brown, "How to Space Gas Lift Mandrels in New Wells." Petroleum Engineer International, November 1981:140–54.

36. *Handbook Of Gas Lift*. USI Garrett Oil Tools. Longview, TX, 1959.

37. Davis, J. B., P. J. Thrash, and C. Canalizo. *Guidelines to Gas Lift Design and Control*. 4th Edition, OTIS Engineering Corp., Dallas, TX, 1970.

38. Winkler, H. W., "Continuous-Flow Gas-Lift Installations Design Utilizing Production-Pressure-Operated Valve Performance." Paper SPE 29451 presented at the Production Operations Symposium held in Oklahoma City, OK. April 2–4, 1995.

39. *Gas Lift Manual*. Section 5: Specifications and Data. Teledyne MERLA, Garland, TX, 1979.

40. "Recommended Practice for Operations, Maintenance, and Trouble-Shooting of Gas Lift Installations." *API RP 11V5*, 2nd Edition, American Petroleum Institute, 1999.

41. Schmidt, Z., D. R. Doty, B. Agena, T. Liao, and K. E. Brown, "New Developments to Improve Continuous-Flow Gas Lift Utilizing Personal Computers." Paper SPE 20677 presented at the 65th Annual Technical Conference and Exhibition, held in New Orleans, LA, September 23–26, 1990.

42. Decker, K., C. Dunham, and B. Waring, "Using Chokes in Unloading Gas-Lift Valves." Proc. 50th Southwestern Petroleum Short Course, Lubbock, TX, 2003: 33–8.

43. Capucci, E. C., and K. V.Serra, "Transient Aspects of Unloading Oil Wells through Gas-Lift Vales." Paper SPE 22791 presented at the 66th Annual Technical Conference and Exhibition, held in Dallas, TX, October 6–9, 1991.

44. Hall, J. W. and K. L. Decker, "Gas-Lift Unloading and Operating Simulation as Applied to Mandrel Spacing and Valve Design." Paper SPE 29450 presented at the Production Operations Symposium held in Oklahoma City, OK, April 2–4, 1995.

45. Tang, Y., Z. Schmidt, and R. Blais, "Transient Dynamic Characteristics of the Gas-Lift Unloading Process." Paper SPE 38814 presented at the Annual Technical Conference and Exhibition held in San Antonio, TX, October 5–8, 1997.

6 | Intermittent Gas Lift

6.1 Introduction

Intermittent gas lift is the default artificial lift method in gas lifted fields where formation pressures have dropped to levels where continuous flow is not sustainable any more. It is also suited for wells with relatively high formation pressures but low productivities. On average, about 10% of gas lifted wells are placed on intermittent lift, mostly in mature fields.

Chapter 6 covers all the topics necessary for the understanding of the intermittent lifting process, its control, operation, and optimization. Discussion is started with the description of the intermittent cycle and its basic mechanism. Since the proper operation of an intermittent installation highly depends on the way of controlling the injection of lift gas into the tubing string, a substantial section deals with the different control arrangements, their operation, relative advantages, and limitations. The calculation of operational parameters is demonstrated by presenting the more important calculation models available. The design of intermittent gas lift installations is detailed as follows, with illustrated examples.

In wells in their latest phase of depletion, conventional intermittent lift installations can turn out to be very inefficient, and the use of an accumulation chamber is advantageous. The application of chamber lift, although only a version of intermittent lift, can substantially increase well rates and decrease gas injection requirements. The features and operational details of chamber lift installations are described in a separate section. Finally, practical recommendations on ensuring an optimum intermittent lift operation are detailed.

6.2 Basic Features

6.2.1 Mechanism of operation.
Intermittent gas lift, although it uses compressed gas from the surface, too, works on a principle completely different from continuous flow gas lift. Lift gas, periodically injected into the flow string at a depth close to the perforations is used to physically *displace* a solid liquid column that was allowed to accumulate above the operating gas lift valve. If the proper amount of lift gas is injected below the accumulated liquid column, the *liquid slug* is propelled

up to the wellhead and into the flowline. Well production, therefore, is done in periodically repeating cycles, with the basic mechanism of fluid lifting being the physical displacement of the liquid slugs by high-pressure lift gas.

The basic operation of a well on intermittent gas lift is illustrated in Figure 6–1. The well has a closed gas lift installation with a standing valve at the bottom of the tubing string, lift gas at a relatively low rate from the surface being continuously injected into the casing-tubing annulus. In *Part A*, the operating (bottommost) valve is closed with formation fluids accumulating above it. Casing and tubing pressures at valve depth continuously increase until the desired liquid slug length is accumulated when the operating valve opens and injects gas below the liquid column, see *Part B*. Lift gas of a relatively high pressure enters the tubing at a great instantaneous rate, creating a large gas bubble below the liquid slug (*Part C*) that propels the slug upward along the tubing string. After the liquid slug enters the flowline, the high-pressure gas column containing dispersed liquid droplets bleeds down to separator pressure (*Part D*), the operating valve closes and well inflow through the now open standing valve commences, and the cycle repeats.

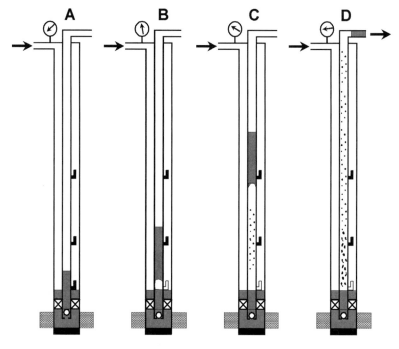

Fig. 6–1 The basic operation of a well on intermittent gas lift with single-point gas injection.

The case just described is valid for *single-point gas injection* when the bottommost valve is usually the operating valve, the upper valves serving only for unloading. *Multipoint injection* is illustrated in Figure 6–2, where upper valves consecutively open and inject gas below the upward-rising liquid slug as it passes. Multipointing is considered if surface injection pressure is low, or in deep wells, and usually requires the use of PPO gas lift valves.

The previous descriptions are valid for cases where a *closed gas lift installation* is used in the well. Intermittent gas lift, however, can be used with other installation types as well. A *chamber installation* contains a downhole chamber of a greater capacity than the tubing where liquid accumulation takes place. The same amount of accumulated liquid, due to the greater chamber capacity, results in a lower backpressure on the formation, and daily production rates can be increased. Another version of intermittent lift uses a plunger between the upward-rising gas and liquid, in order to improve the seal between the two. *Plunger-assisted intermittent lift* is discussed in a separate chapter of this book.

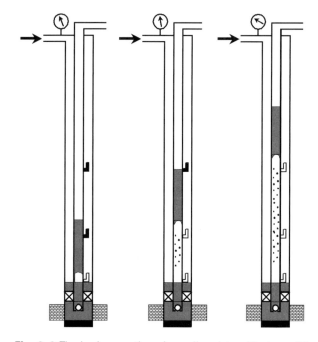

Fig. 6–2 The basic operation of a well on intermittent gas lift with multipoint gas injection.

As shown previously, all variations of intermittent gas lift operate on a completely different principle than continuous flow gas lift whose main mechanism is the reduction of flow resistance of the flow conduit due to a continuous injection of lift gas. Well production is in cycles during which liquid accumulation and fluid lifting follow each other, fluid lifting being done by the physical displacement of a liquid slug by a high-pressure gas stream.

Intermittent gas lift, in contrast to continuous flow operation, is a transient process that is much more difficult to model. This is the reason why even today the design and analysis of intermittent lift relies on theories of limited accuracy.

6.2.2 Applications, advantages, limitations

Intermittent gas lift is the natural choice in gas lifted fields when formation pressures and fluid rates drop to such low levels that continuous flow is inefficient due to the great injection rates required. Approximate minimum liquid rates below which switching from continuous flow to intermittent lift is necessary are 100–150 bpd for 2 $^3/_8$ in., 200–300 bpd for 2 $^7/_8$ in., and 300–400 bpd for 3 $^1/_2$ in. tubing sizes. [1] Since conversion of wells on continuous operation to intermittent lift is easy and involves little additional costs, intermittent lift is usually the preferred method of artificial lift in such cases.

Generally, intermittent gas lift is applied in wells producing low to very low liquid rates, specifically

- In low-productivity wells with relatively high formation pressures

- In wells with low formation pressures and high productivities, where the application of chamber lift is usually recommended

- Instead of pumping (in low producers with relatively high gas rates), the preferred method is plunger-assisted intermittent lift

The basic advantages and limitations of intermittent gas lifting can be summarized as follows:

Advantages

- For low liquid producers, intermittent gas lift is quite flexible to accommodate changes in well inflow parameters.

- It can be used to the well's final abandonment by changing the installation type from the closed to the chamber installation.

- Capital costs, especially for deep, low fluid level wells are lower than for pumping applications.

Limitations

- The energy of formation gas is wasted and is not utilized for fluid lifting.

- Available liquid production rates are limited.

- High fluctuations in the producing bottomhole pressure associated with intermittent lift can present serious sand production problems in unconsolidated formations.

- The high- and low-pressure sides of a closed rotative gas lift system can be overloaded due to the high instantaneous gas flow requirements of the intermittent cycle.

6.3 Surface Gas Injection Control

6.3.1 Introduction

The production efficiency of an intermittent gas lifted well depends largely on the way the injection of lift gas to the well is controlled. If the correct volume of lift gas at the proper timing is injected, liquid production from the well can be maximized with a moderate injection gas requirement. On the other hand, improper timing or injection of too high or too low gas rates lead to low liquid rates and/or excessive usage of lift gas. The selection of the proper surface gas injection control, therefore, is a crucial requirement for achieving a profitable production of wells placed on intermittent gas lift.

Surface *injection control* for intermittent wells involves the control of the following:

(a) the surface injection pressure (measured at the casinghead)

(b) the surface gas injection rate

(c) the duration and frequency of gas injections per day

Compared to wells placed on continuous flow, it is not sufficient to regulate injection pressure and volume. In addition to those, it must also be ensured that downhole injection into the production conduit follows a *cyclic pattern*. This cyclic operation occurs automatically when certain types of gas lift valves are run in the well; in other cases, the gas injection line to the well is intermittently opened and closed.

In continuous flow operations, any surface injection control method can be used with any kind of operating valve. This is not the case in intermittent gas lifting, because some types of gas lift valves will only work with certain surface control methods. Accordingly, in the following discussion, the applicable operating valve types will also be given. It should be noted here that operating valves for intermittent lift must be of the *snap-action* type, except when a surface intermitter is used to control the gas injection to the well.

For an efficient control of the well's operation, continuous measurements of the pressures and gas flow rates are required at the wellsite. This requirement can be met by installing the following on the wellhead:

(a) a two-pen *pressure recorder* that simultaneously records casinghead and THPs

(b) a *meter run* to continuously measure injection gas rates

The regular filing and frequent analysis of these recordings allows one to decide on the required modification of surface control settings. Installation of these devices, therefore, forms the basis of analyzing and troubleshooting of the well's operation. [2, 3]

In the following, intermittent cycles will be illustrated by presenting the schematic variation of the casinghead pressure (CHP) and tubinghead pressure (THP) in the function of time, as well as by indicating the operating valve's actions In these figures, the open or closed state of surface gas injection devices and the operating valve are shown as horizontal bars of white or solid black color, respectively. The fluids produced at the wellhead are indicated by black bars (solid liquid slug), black bars containing white circles (liquid foam), and white circles (mist).

6.3.2 Choke control

Choke control of surface gas injection for intermittent wells, just like in continuous flow operations, means the installation of a *fixed* or an *adjustable choke* on the injection line at the wellhead, see Figure 6–3. The intermittent cycle is controlled by the well's inflow performance and the operating gas lift valve's operational characteristics, as will be detailed as follows.

This type of control is regularly used under such conditions when the injection gas flow rate available at the wellhead is less than the maximum instantaneous gas injection rate required for an efficient lifting of the liquid slug. Because of the insufficient gas supply from the surface

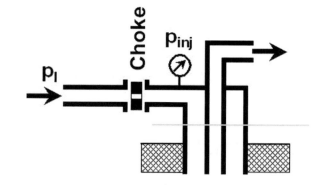

Fig. 6–3 Wellhead arrangement for choke control of surface gas injection.

gas lift system, an additional gas volume has to be provided to meet the gas injection requirement of the intermittent cycle. This is ensured by the well's casing-tubing annulus that, due to the continuous gas injection through the surface choke, stores lift gas during the accumulation period. The annulus, therefore, acts as a buffer that releases the previously stored gas volume when the gas lift valve opens.

The operating valve for choke control can be either an IPO or a PPO gas lift valve. Most frequently, an IPO valve with the proper *spread* (determined from the well's annular volume) is used. The ideal solution, however, is to run a pilot-operated IPO valve which ensures a sufficiently large gas throughput area during the injection period while allowing the operator to change the valve's spread according to the actual annular volume.

For choke control, the intermittent cycle is illustrated in Figure 6–4. After completion of the previous cycle, CHP gradually increases due to the continuous gas injection through the choke at the surface, while the operating valve is closed. Tubing pressure at valve depth also increases, because of the inflow and the accumulation of a liquid column above the gas lift valve's setting depth. The operating valve will open when casing and tubing pressures satisfy the conditions prescribed by the valve's *opening equation*. Lift gas is then injected below the accumulated liquid slug, which starts its journey to the surface.

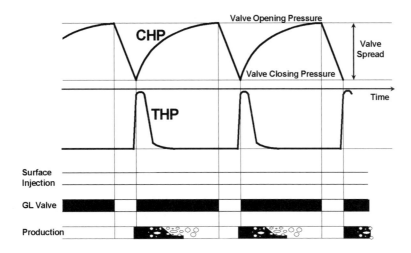

Fig. 6–4 Description of the intermittent cycle for choke control.

The gas rate injected through the operating gas lift valve into the tubing string is composed of the following:

(a) the gas injected into the annulus by the surface choke

(b) the gas previously stored and now released from the annulus.

In a properly designed installation, the total gas volume injected by the operating valve should exactly cover the gas requirement of the intermittent cycle. After this gas volume is injected, the operating valve should close to prevent excessive lift gas usage. This can only happen if gas is injected at a lower rate at the surface than through the gas lift valve, because then casing pressure gradually decreases to the closing pressure of the valve. In order to ensure the necessary drop in casing pressure, the size of the surface choke must always be smaller than the port size of the operating valve. The operating valve closes and the intermittent cycle repeats.

From the previous discussion of the intermittent cycle, one can draw the following conclusions.

1. Most of the injection gas volume required for the efficient lifting of the liquid slug comes from the *gas stored* in the annulus. The volume of gas stored in the annulus depends on the annular volume and the casing pressure reduction occurring during a cycle. The drop in casing pressure, in turn, is governed by the *spread* of the valve

under the actual conditions. Proper design, therefore, requires that the selection of the operating valve's spread is based on the casing annular volume.

2. The number of intermittent *cycles per day* is adjusted by changing the size of the surface choke which affects the time required for the casing pressure to reach the opening pressure of the operating gas lift valve. Increasing the choke size increases the daily number of cycles, and smaller sizes result in fewer cycles.

3. Well *inflow performance*, too, has a direct impact on cycle frequency, because the rate of liquid slug buildup in the tubing string is governed by the inflow to the well. For optimum conditions, the size of the surface choke is selected so that liquid inflow from the well coincides with the annulus pressure buildup, and the operating valve's opening production and injection pressures are reached at the same time.

4. As shown previously, the proper operation of the gas lift valve requires that its port size be greater than the size of the surface gas injection choke. But increasing a valve's port size also increases its *spread*, which can turn out to be too excessive when considering the well's annular volume. In such situations a pilot-operated valve gives the ideal solution, because it has a high gas throughput capacity, but, at the same time, its spread can be set independently.

The *advantages* of using choke control in wells placed on intermittent lift are the following:

1. Surface gas demand of the wells is fairly constant in time, assuring a steady loading of the gas compressors. This is especially important in small *rotative gas lift systems* containing a low number of intermittent wells.

2. Simple choke control provides simple, trouble-free operation.

The main *disadvantages* include the following:

1. Wells with low liquid production rates need low cycle frequencies and the choke size required can be too small to be practical, mainly due to *freezing* problems.

2. The maximum number of daily cycles may be limited by the injection gas rate available at the wellhead.

3. Chokes have a tendency to *freeze*, which can only be prevented by a proper dehydration of the lift gas. Many times, improperly processed gas containing too much water vapor is the reason for the failure of this kind of surface control.

6.3.3 Choke and regulator control

Although a gas *pressure regulator* alone can be used for intermittent gas injection control, its proper selection and adjustment are quite cumbersome. Therefore, a choke is usually installed downstream of the regulator (see Fig. 6–5), which setup works well with all types of operating valves except balanced valves. Choke and regulator control is used to prevent freezing conditions when the required size of the surface gas injection choke is too small. This type of surface injection control is especially well suited for installations where pilot-operated valves are used for the operating valve.

Choke and regulator control is ideally suited to low-rate wells that usually require extremely small choke sizes because of the low injection volumes needed. Small chokes, however, are especially prone to freezing and plugging. This is why the bigger choke sizes made possible by the regulator provide a workable solution. An added advantage is that starting slug length can always be kept at

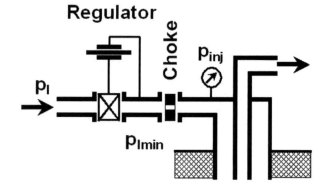

Fig. 6–5 Wellhead arrangement for choke and regulator control of surface gas injection.

its optimum value, in contrast to choke control, where an increase in choke size decreases the starting slug length and causes excessive gas usage per cycle. With choke and regulator control, starting slug length is easily controlled by changing the regulator's set pressure.

The operation of the intermittent cycle under choke and regulator control is explained in Figure 6–6.

The surface gas injection choke is of a relatively large diameter; that is why the casing pressure rises rather quickly, and it reaches the surface opening pressure of the operating valve quite early after the regulator opens. The regulator is set to this pressure, so it shuts down surface gas injection from this moment on, and the casing pressure remains constant. The operating gas lift valve is still closed and will open only after a sufficient head of liquid has accumulated above it. *Starting slug length* can thus be adjusted by adjusting the pressure setting of the regulator on the surface. This is a distinct advantage of this type of surface injection control over simple choke control.

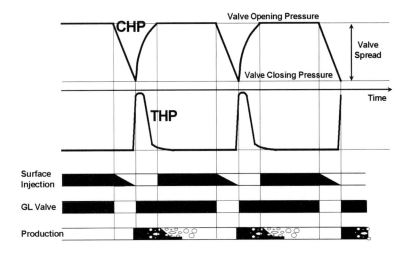

Fig. 6–6 Description of the intermittent cycle for choke and regulator control.

After the operating gas lift valve opens, lift gas injected through the valve starts to lift the liquid slug to the surface. Due to the great amount of gas leaving the annulus, CHP starts to decrease. The *spread* of the regulator should be less than that of the operating gas lift valve, and this causes the regulator to open before the gas lift valve closes. In spite of the gas injection taking place at the surface, casing pressure drops even further because the size of the surface choke is smaller then the operating valve's port size. When casing pressure drops below its closing pressure, the operating valve closes and the intermittent cycle repeats.

The main *advantages* of choke and regulator control are as follows:

1. The well's injection gas demand is more uniform than in the case of using a *time cycle controller* (intermitter), and a more regular loading of the gas compressors is realized.

2. Freezing problems are greatly reduced in comparison to choke control, due to the bigger choke sizes required.

3. Proper adjustment is simpler than for intermitter control, and the required equipment needs less maintenance.

Its *drawbacks* are the relative complexity and the sensitivity to freezing when smaller chokes are used.

6.3.4 Intermitter control

The injection control methods to be discussed in this section all utilize a *time cycle controller*, or *intermitter* for short. The basic feature of any intermitter control is that the cyclic operation is ensured by the programmed opening and closing of the surface gas injection line. Daily cycle number (as well as the duration of gas injection) can be set on the intermitter, independently of any other condition. The intermitter is an adjustable device controlling the operation (opening and closing) of a *motor valve* installed in the gas injection line near the wellhead. Thus, the control of the intermittent lift cycle is, in contrast to choke control, transferred mostly to the surface.

The proper location of the intermitter is at the wellhead rather than at the tank battery or at the gas distribution center. The drawbacks of installing the intermitter at the tank battery are twofold.

(a) The liquid slug's lifting efficiency may be low because of the slow increase in casing pressure due to the need to fill up the injection line.

(b) The high-pressure storage capacity of the gas lift system is reduced by the total volume of the injection lines.

The latter is very detrimental in closed rotative gas lift systems containing only a few wells on intermittent lift and gives rise to high fluctuations in line pressure.

Intermitter control can be used with any type of operating gas lift valve, and it is the only solution when a balanced gas lift valve is used. It is highly recommended for wells with extremely high or very low daily liquid production rates. The *advantages* of this type of injection control are the following:

1. Time cycle controllers provide a reliable operation and are very flexible since the parameters of the intermittent cycle are easy to modify.

2. In gas lift systems containing several intermittent wells, the interference of the individual well's operation can be minimized by synchronizing (*staggering*) the individual intermitters.

3. Intermitters are not sensitive to the liquid content of injection gas and to freezing conditions.

Basic *disadvantages* include the following:

1. Wells on intermitter control require high instantaneous gas injection rates; this is why some provisions for ensuring the steady loading of gas compressors must be provided.

2. Intermitters are more complicated than other control devices and require regular servicing.

6.3.4.1 Simple intermitter control.

The simplest version of intermitter control is the use of only an intermitter on the surface gas injection line, see Figure 6–7. The operating valve can be either an unbalanced or a balanced gas lift valve, depending on the maximum gas flow rate available at the surface. In case the available gas rate from the gas lift system is higher than or equal to the maximum instantaneous gas demand during the intermittent cycle, no gas storage is required in the annulus and a balanced valve with zero spread can be run in the well. If the available surface injection rate is insufficient for the lifting of the accumulated liquid slug, gas storage in the annulus is necessary and an unbalanced valve with the proper spread has to be used.

When using a *balanced valve* for the operating valve, the operation of the intermittent cycle is illustrated in Figure 6–8. At the beginning of the cycle, the surface intermitter opens and injects lift gas into the annulus. CHP increases slightly to reach the surface opening pressure of the operating gas lift valve when the valve opens and admits gas into the tubing. Casing pressure stays constant during the lifting of the liquid slug, because the balanced valve passes all the gas injected at the surface.

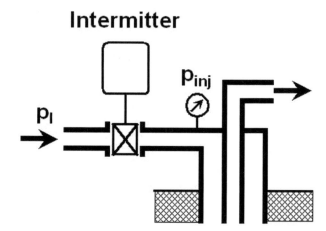

Fig. 6–7 Wellhead arrangement for simple intermitter control of surface gas injection.

After the predetermined injection time has elapsed, the intermitter closes the surface gas injection. Casing pressure then drops slightly and the operating gas lift valve immediately closes. Liquid accumulation in the tubing commences and the cycle is repeated when the intermitter opens again and injects gas into the annulus. Note that the variation of CHP is much less than with unbalanced valves in the well, because balanced gas lift valves open and close at practically the same injection pressure.

If an *unbalanced valve* is used for the operating valve, two cases can occur. In case the operating valve's spread is sufficient for the storage of the required volume of injection gas in the annulus, the intermitter is set to close after the casing pressure has reached the operating valve's opening pressure. Figure 6–9 shows the intermittent cycle in this case.

With a sufficient liquid slug accumulated in the tubing above the operating valve, the intermitter opens gas injection into the annulus. Casing pressure rapidly increases to the valve's surface opening pressure, and the intermitter shuts down the injection of gas. The operating valve opens and the gas volume previously stored in the annulus is injected into the tubing to lift the liquid slug to the surface. During the lifting period, casing pressure drops below the valve's closing pressure, and the gas lift valve closes. Liquid accumulation in the tubing starts again, and the intermittent cycle repeats.

In case the operating valve's *spread* is too low to enable the storage of the total gas volume required for the cycle, the surface intermitter is set to a longer injection period. The proper injection time must be sufficient for the injection of the required amount of gas in the annulus. The cycle follows the pattern shown in Figure 6–10, where the surface and downhole injection times overlap. In the period between the opening of the operating valve and the closing of the surface intermitter, the gas volume required to supplement the volume stored in the annulus is injected into the well.

The advantages and limitations of this type of injection control are identical to those mentioned before in connection with time cycle intermitters. An added *disadvantage* of using simple intermitter control is that it consistently injects the same gas volumes only for a constant gas line pressure. If surface line pressure fluctuates, the gas volumes injected in successive cycles can substantially vary, leading to ineffective gas lift operations. In some cycles, too much gas is injected; whereas in others, the injected gas volume is insufficient to lift the accumulated liquid slug to the surface.

6.3.4.2 Use of an intermitter and a choke.

The high instantaneous gas flow rates associated with intermitter control can be reduced if a choke is installed upstream of the intermitter, as seen in Figure 6–11. The properly sized choke limits the maximum flow rate; the required gas volume is injected in a longer period, and the loading of the surface gas lift system, especially that of the gas compressors can be reduced significantly. The installation of the choke is also advantageous in cases when surface line pressure is too high, as compared to the operating CHP.

The variation of CHP during the intermittent cycle follows the pattern shown in Figure 6–9, but with a lower rate of pressure increase after the intermitter opens. This can cause improper operation and increased gas usage with operating valves that are not of the *snap-acting* type. Such problems are eliminated by the use of snap acting or pilot-operated valves.

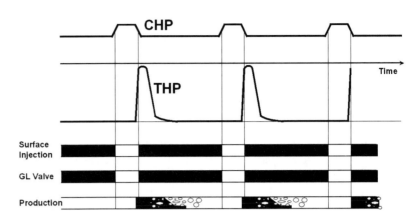

Fig. 6–8 Description of the intermittent cycle for simple intermitter control and a balanced operating valve.

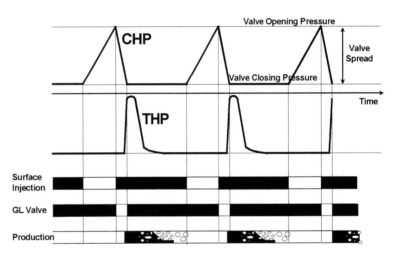

Fig. 6–9 Description of the intermittent cycle for simple intermitter control and an unbalanced operating valve.

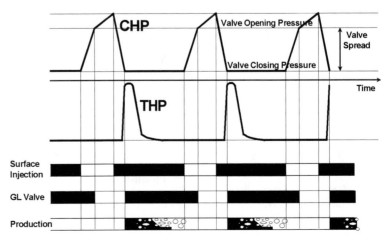

Fig. 6–10 Description of the intermittent cycle for simple intermitter control and an unbalanced operating valve with too-low spread.

Depending on whether the operating valve's spread is sufficient or insufficient for the storage of the required gas volume in the annulus, the intermitter is set to close immediately when the casing pressure reaches the operating valve's surface opening pressure—or later on. By changing the closing time of the intermitter, one can find the proper amount of injection gas volume needed for an efficient intermittent cycle.

6.3.4.3 Use of an intermitter and a regulator.
If gas lift pressure in the surface injection system fluctuates or in case of small annular spaces (macaroni installations), casing pressure is limited by a regulator installed upstream of the intermitter, see Figure 6–12. This type of control is applicable with any type of operating valve and is the ideal solution when using a pilot-operated gas lift valve in the well.

For a pilot-operated operating valve with a sufficient *spread* to supply all the required injection gas volume from the annulus, the CHP and THP vary during the cycle as illustrated in Figure 6–13. After the intermitter opens, CHP quickly reaches the valve's opening pressure set previously on the pressure regulator, and the regulator closes the motor valve on the injection line and gas injection ceases. Since the required starting slug length is not yet available above the operating valve, the valve's production (tubing) pressure is not sufficient to open the pilot section. At one point, however, available slug length, due to the inflow from the formation, reaches the desired value, and the valve's pilot section opens. This forces the power section to open its large injection port and the liquid slug is lifted to the surface.

If the valve's spread is not sufficient for an efficient lifting of the liquid slug, the gas volume stored in the annulus has to be supplemented by gas injected to the annulus. In such cases, the surface intermitter is set to a longer injection period and, with the help of the pressure regulator, will inject gas to the annulus after the operating gas lift valve opens.

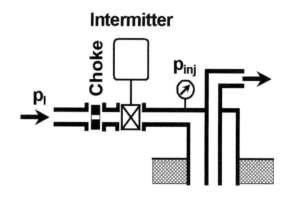

Fig. 6–11 Wellhead arrangement for intermitter and choke control of surface gas injection.

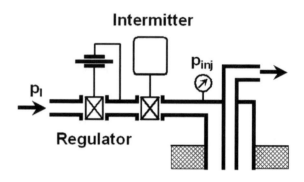

Fig. 6–12 Wellhead arrangement for intermitter and regulator control of surface gas injection.

6.3.4.4 Casing pressure control.
In situations with widely varying surface line pressures, the intermitter's closing may be controlled by the CHP (Fig. 6–14). The intermitter opens at the predetermined time and starts to inject gas into the annulus. Gas injection is continued until a pressure sensor detects that casing pressure reaches a previously set value, at which time, the intermitter shuts down gas injection into the annulus. This installation ensures that gas injection volumes are constant for every cycle, no matter how much the line pressure fluctuates.

Pressures during the cycles vary similar to those shown in Figure 6–9, only the duration of surface gas injection changes from cycle to cycle, depending on the actual line pressure. This type of injection control can be used with all but balanced gas lift valves.

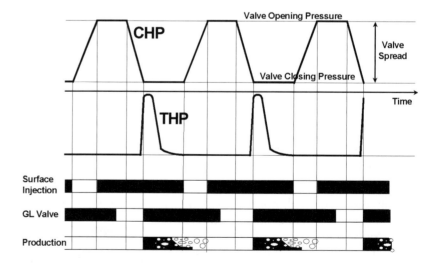

Fig. 6–13 Description of the intermittent cycle for intermitter and regulator control when using a pilot-operated valve.

6.3.4.5 Other controls. In addition to the universally applied surface injection controls detailed previously, many special control methods are known. In the following, tubing pressure control is discussed.

For wells periodically flowing and produced by intermittent lift between their flowing periods, the intermitter is equipped with a pressure sensor connected to the tubinghead. Every time the well starts to flow, its THP increases, triggering a signal to the intermitter that forces it to close. Gas injection ceases until the end of the flowing period when the drop in THP forces the intermitter to start to inject gas and intermittent lift controlled by the intermitter follows.

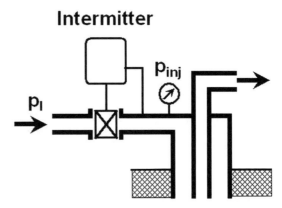

Fig. 6–14 Wellhead arrangement for casing pressure control of surface gas injection.

6.3.5 Control devices

The different devices making up the surface injection control equipment of gas lifted wells are discussed in the following section. They can be used on wells placed on intermittent lift as well as on continuous flow.

6.3.5.1 Fixed and adjustable chokes. Fixed or positive chokes generally used in the petroleum industry for controlling gas or liquid flow can be installed at the wellhead to control gas injection rates to a gas lifted well. Their main drawback is that changing of choke sizes, a common task during the adjustment of well operation, is relatively complicated. Because of the necessity of manual intervention and the bleeding off of gas pressure, changing or replacing a choke consumes valuable production time.

The disadvantages of using fixed chokes are eliminated if *adjustable chokes* or *metering valves* are utilized. These are very easy to adjust since their gas passage area can continuously be set from the fully closed to the fully open position. Actual area open to flow is usually indicated on the housing of the valve (see Fig. 6–15). General features of such valves are the following:

1. The valve can be set at any selected position.

2. For repairs, the valve seat and/or stem can be removed—without the need to remove the valve body from the line.

3. Usually, the valve body (with the valve stem removed) can hold a motor valve, making it possible to quickly change the setup from one kind of surface injection control to the other.

6.3.5.2 Pressure regulators. Pressure regulators used in injection control are composed of a motor valve and a pressure pilot. The pilot can be direct-acting if an increase in the controlled pressure increases, and reverse-acting if an increase in the controlled pressure decreases its output pressure. Motor valves can also be of the direct- or reverse-acting type: a direct-acting (normally open) valve stays open with zero control pressure and starts to close when control pressure is applied; the operation of a reverse-acting one (normally closed) is the opposite. By properly combining the motor valve and pressure pilot types, different injection control functions can be realized.

Figure 6–16 shows a pneumatic pressure regulator where a direct-acting pressure pilot is used to control a direct-acting motor valve. The controlled pressure (the pressure downstream of the valve) enters a *Bourdon tube* that deforms and activates a pneumatic amplifier whose output pressure is connected to the diaphragm of the motor valve.

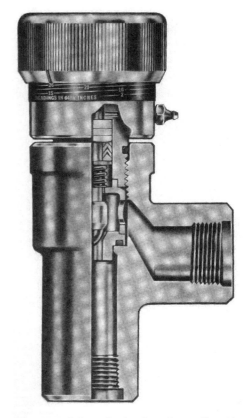

Fig. 6–15 Schematic drawing of an adjustable choke.

Any increase in downstream pressure starts to close the motor valve, whereas a decrease opens it; thus the regulator keeps a constant downstream pressure.

Figure 6–17 depicts the construction of a reverse-acting motor valve.

6.3.5.3 Time cycle controllers.
Intermitters or time cycle controllers open and close a motor valve installed on the gas injection line, according to preset cycle parameters. The motor valves can be actuated by pneumatic pressure or electric current. The controller itself can be mechanical, pneumatic, or electronic.

Figure 6–18 shows the operation and the basic parts of a widely used clockwork-driven pneumatic intermitter. [1] The *timing wheel* is rotated by *clockwork*, at a constant angular velocity. In the position shown, the timing arm is lifted by one of the pins installed on the wheel and does not contact the three-way pneumatic valve. Power gas (usually air), coming from the gas supply system can thus reach the motor valve that keeps the motor valve open and permits gas injection to the well. At the end of the preset duration of gas injection time, the timing wheel has already turned to a position where the timing arm falls to a horizontal position and rests on the three-way valve. Power gas is vented to the atmosphere, thus causing the reverse-acting motor valve to close gas injection.

The duration of gas injection is set by the *adjustment screw*, and the daily number of cycles is set by changing the number of pins on the timing wheel, see Figure 6–18. Clockworks of different rotational time may be used to achieve different cycle frequencies. Instead of the clockwork, timing wheels can also be driven by a small gas engine or electric motor. Gas-driven units are especially attractive since they use the pressure of the lift gas as a power source and do not need periodic maintenance.

Intermitters of fully pneumatic or electronic construction are also available. Setting the daily cycle number as well as the injection period is very simple; their accuracy and reliability is also high. Many times, a single controller drives several timing wheels belonging to different wells allowing the *staggering* of gas injections and preventing well interference. Fully electronic, computer-controlled intermitters with increased reliability and repeatability have completely eliminated the disadvantages of mechanical and pneumatic types. [4]

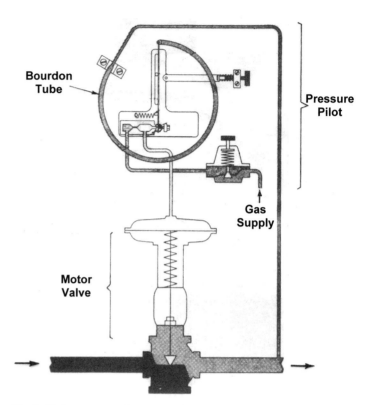

Fig. 6–16 Components of a pneumatic pressure regulator.

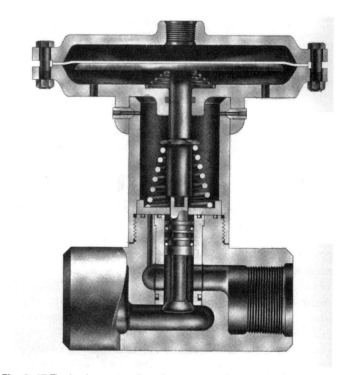

Fig. 6–17 The basic construction of a reverse-acting motor valve.

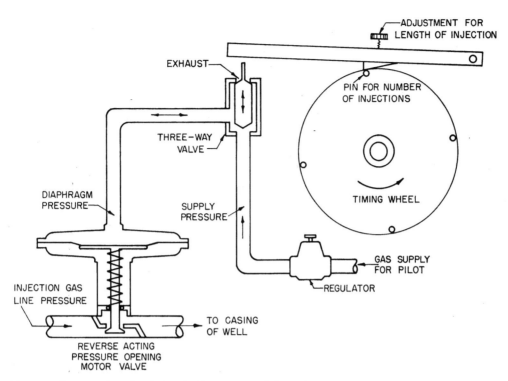

Fig. 6–18 The operation of a clockwork-driven pneumatic intermitter.

6.4 Intermittent Lift Performance

This section describes and analyzes the operation of wells placed on intermittent gas lift having a *closed installation*, the ideal type of installation for this type of artificial lift, and choke control of gas injection. Reference is made here to chapter 4, where the advantages and limitations of the different installations are discussed. Although the performance of intermittent wells with other installations is very similar, *chamber lift* is covered in a different section; whereas a complete chapter is devoted to *plunger lift*.

6.4.1 The intermittent cycle

Wells placed on intermittent gas lift produce in cycles, one cycle covering a time period that lasts until the process first repeats itself. As shown in Figure 6–1, the cycle can be divided into three distinct phases: the periods of liquid accumulation, slug lifting, and pressure blow-down or afterflow. These phases will be discussed in detail as follows, as illustrated in Figure 6–19, for a well with a closed installation and choke control of the surface gas injection. The figure shows (in the function of time) the position of the liquid slug in the well as it moves from the depth of the operating gas lift valve to the wellhead.

Accumulation Period

Liquid accumulation in the tubing above the operating gas lift valve starts as soon the standing valve opens and well fluids start to enter the tubing at time $t = 0$ in Figure 6–19. At this moment, a liquid slug length equivalent to the liquid volume that did not reach the surface in the previous cycle is present above the gas lift valve. Starting from this, the length of the liquid slug continuously increases due to the inflow from the formation and would asymptotically reach the static fluid level corresponding to the well's *SBHP*.

The question arises: should the liquid slug be allowed to reach the static level or should gas injection be started earlier? To solve this problem, one should understand that the well's daily liquid production is the product of the liquid volume per cycle and the number of cycles per day. Accumulation of a long slug requires a long time which, in turn, reduces the number of cycles that can be done in a day, whereas a shorter slug means a greater daily number of cycles. Maximum daily liquid production, therefore, will be found if liquid slugs considerably shorter than that belonging to the static liquid level are produced each cycle. In practice, about 40–50% of the static liquid column is used as a *starting slug length,* i.e. the length of liquid column above the operating valve at the time of gas injection into the tubing.

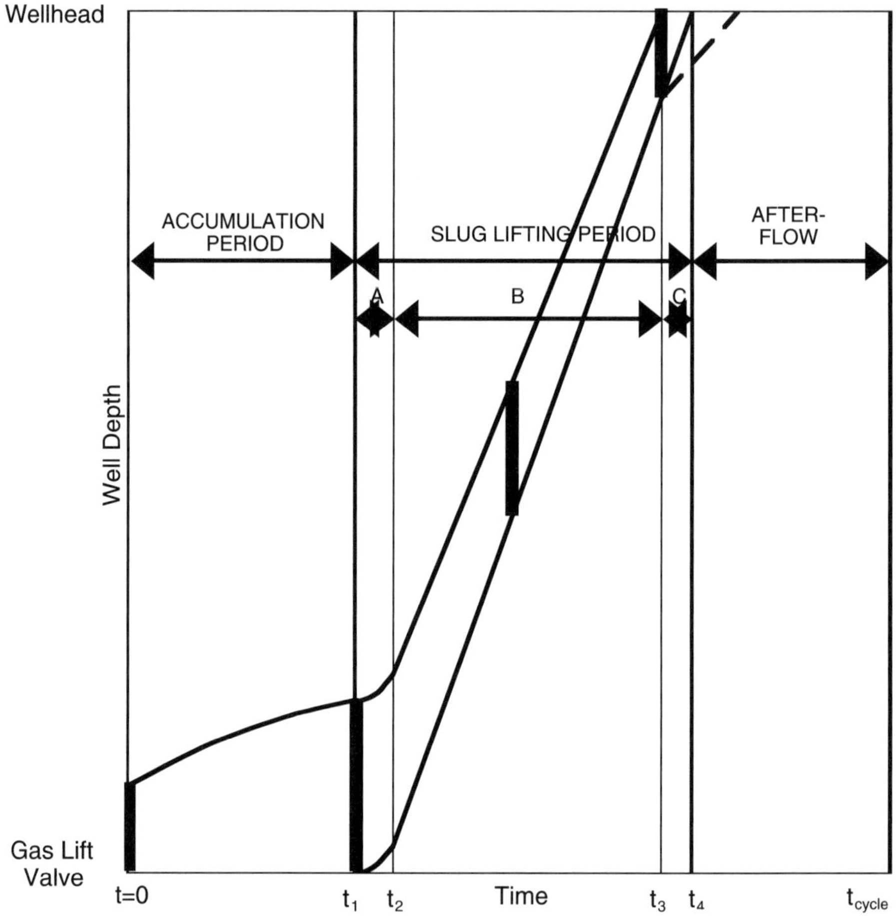

Fig. 6–19 The position of the liquid slug in the well during the intermittent cycle for choke control of gas injection.

Along with the rising of liquid level in the tubing string, casing pressure at valve depth increases due to the continuous gas injection into the casing-tubing annulus at the surface. The operating gas lift valve—either an unbalanced IPO or a pilot valve—will open at a time (t_1 in Fig. 6–19) when actual tubing and casing pressures satisfy its opening equation then the valve starts to inject gas below the liquid slug.

Slug-lifting period

The lifting of the liquid slug starts as soon as the operating valve opens (t_1) and lasts until the slug has completely moved into the flowline, at t_4 in Figure 6–19. This period can further be divided into three distinct phases, defined by the kinetics of the slug's movement.

- *Phase A* where the liquid slug is accelerated

- *Phase B* of a constant rising velocity

- *Phase C* during which the slug enters the flowline

Throughout this period, the standing valve that had closed as soon as gas injection through the operating valve started stays closed, sealing the formation from the high pressure of the lift gas.

During the whole lifting period, the length of the upward-rising liquid slug is continuously reduced due to *gas breakthrough* and *liquid fallback,* two concepts that need clarification. As the liquid slug is lifted by the high-pressure gas bubble, gas continuously penetrates the bottom of the slug due to the high buoyancy force acting on it.

Gas breakthrough, as it is commonly called, results in a loss of slug length because part of the liquid from the gas-penetrated bottom of the slug falls down. At the same time, part of the liquid film present at the tubing inside wall all along the length of the gas bubble may have a downward velocity and will fall back due to gravity. The length received on the surface is less than the starting slug length, due to the combined effect of gas breakthrough and liquid fallback.

The first phase of the slug-lifting period starts after the gas lift valve's opening and lasts until the slug's velocity has reached a constant value (from t_1 to t_2 in Fig. 6–19). In order for the liquid slug to reach the proper rising velocity in a short time, the operating valve must open quickly (*snap action*) and must have a large port because then a sufficiently large gas bubble can develop below the liquid slug. Otherwise, a small valve port or a slow valve action (throttling) results in low gas injection rates into the tubing and gas thus injected will bubble through the liquid column without lifting it. It follows that a large-ported, preferably pilot-operated valve is required for an efficient fluid lifting.

Phase B in Figure 6–19 of the slug-lifting period involves a constant rising velocity of the liquid slug and lasts until the top of the slug reaches the wellhead. During this phase, gas is continuously injected through the open gas lift valve into the tubing and the expansion energy of this gas lifts the liquid column up toward the wellhead. The length of the liquid slug decreases due to gas breakthrough and liquid fallback, the total amount of loss being proportional to the time spent by the slug in the tubing string. Consequently, liquid fallback can be reduced by reducing the time period $t_3–t_2$, which can be achieved by increasing the rising velocity of the slug. Practice has shown that a slug velocity of about 1,000 ft/min gives a minimum of liquid fallback; that's why this value must always be attained.

The final phase of the lifting period lasts until the tail of the surfacing slug leaves the wellhead and is completely transferred to the flowline. At one point during this phase, the gas lift valve can close, provided an adequate amount of lift gas has been injected below the liquid slug, whose expansion is sufficient to finish the lifting process. The length of the slug now very rapidly decreases because an increasing part of it reaches the flowline. Since the liquid slug's length continuously decreases, it exerts less and less pressure on the gas bubble below. If gas pressure decreases (due to the compressibility of the gas), its volume increases, which subsequently increases the gas velocity. The process just described can be very detrimentally affected by the physical arrangement of the wellhead and the Christmas tree assembly that usually include many elbows and other fittings with changes of flow direction. The liquid slug, traveling at a high speed and hitting these restrictions, will slow down, allowing more time for gas breakthrough and liquid fallback. *Phase C* of the slug-lifting period will thus last longer, as shown by the dashed line in Figure 6–19, and recovery of the starting slug length will be low. Streamlining of the Christmas tree (as shown in Fig. 6–20) by removing most or all of the fittings causing rapid changes of flow direction is a very efficient way of increasing the liquid production rate of wells placed on intermittent gas lift. [1]

Afterflow Period

When the tail of the liquid slug leaves the wellhead (t_4 in Fig. 6–19), the tubing string contains a high-pressure gas column with entrained liquid droplets. Because the backpressure presented by the liquid slug is removed from the top of the gas column, gas velocity dramatically increases in the tubing string. The extremely high velocity of the gas stream lifts not only its entrained liquid droplets to the surface but also a part of the liquid film wetting the tubing inside wall. This is the reason why an intermittent well's total liquid production per cycle is always greater than that produced by the solid liquid slug only. The afterflow or blow-down period ends when, due to the rapidly decreasing gas column pressure at the bottom of the tubing string, the standing valve opens; the cycle repeats with the start of the liquid accumulation period.

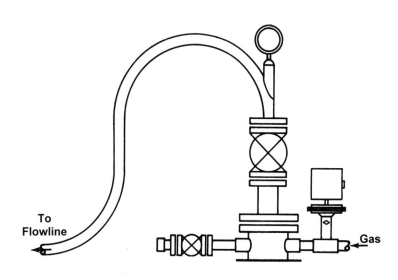

Fig. 6–20 Streamlined Christmas tree for intermittent gas lift wells.

6.4.2 Calculation of operational parameters

6.4.2.1 Rules of thumb. Since the lifting of liquid slugs in intermittent gas lift is a transient process, the description of operational parameters is a complex problem and accurate solutions are still not available. This is the reason why many practical *rules of thumb* may still be in use. The most commonly used ones are listed as follows, according to the *CAMCO Gas Lift Manual.* [5]

- A minimum *slug velocity* of 1,000 ft/min should be attained to minimize liquid fallback.

- For normal conditions, liquid *fallback* is about 5–7% per 1,000 ft of lift.

- Minimum *cycle time* is 3 min per 1,000 ft of lift. The maximum number of daily cycles is easy to find from this approximation.

- To ensure proper operation and a minimum of fallback, the *starting tubing load* (the hydrostatic pressure of the starting slug length plus *WHP*) should be selected as 50–75% of the operating valve's opening pressure at valve depth.

- The injection *gas requirement* of intermittent lift can be approximated as:

  - 200–400 scf/bbl per 1,000 ft of lift for conventional installations

  - 200–300 scf/bbl per 1,000 ft of lift for chamber installations.

Example 6–1. Find the probable daily oil production rate and the gas requirement for the following well on intermittent gas lift.

Depth of Operating Valve = 7,000 ft   Surface Injection Pressure = 700 psi

Tubing Size 2 3/8 in.   Gas Gravity = 0.6

WHP = 50 psi   Oil Gradient = 0.4 psi/ft

Solution

Injection pressure at valve setting depth is calculated from a pressure gradient of 16 psi/1,000 ft, found from Figure B-1 from Appendix B:

$p_i = 700 + 16\ 7,000 / 1,000 = 812$ psi.

Let's assume a starting load of 60% of the valve's opening pressure

Starting Load = 0.6 812 = 487 psi.

Liquid slug pressure is less than this value, the difference coming from the *WHP*:

Slug Pressure = 487– 0 = 437 psi, which equals a slug length of 437 / 0.4 = 1,093 ft

Since tubing capacity for 2 3/8 in. pipe is 3.87 10^{-3} bbl/ft, the volume of the starting slug is:

$V_{slug} = 1,093\ 3.87\ 10^{-3} = 4.2$ bbl.

The volume of the produced slug is less than the previous value by the fallback:

V_{prod} = 4.2 – 4.2 0.07/1,000 7,000 = 2.1 bbl, *i.e.* about 50% of the starting slug falls back.

To find the maximum number of daily cycles, the minimum cycle time has to be found first:

Min. Cycle Time = 3 7,000 / 1,000 = 21 min.

The maximum daily cycle number is thus 1,440 / 21 = 69 cycles/day.

Daily oil production rate is the product of the daily cycle number and the production per cycle:

q_o = 69 2.1 = 145 bpd.

Finally, gas requirement is estimated as 300 scf/bbl 7,000 / 1,000 145 bpd = 305 Mscf/d.

6.4.2.2 Empirical correlations. Although the importance of intermittent gas lift is much less than that of continuous flow gas lifting, many investigations were published on this topic. Because of the transient nature of intermittent lift, however, almost all publications apply empirical methods for the description of production parameters. In the following, an overview of the more important achievements is presented.

Liquid Fallback

Even the first investigations of intermittent gas lift proved that the greater the port of the operating valve, the bigger the portion of the initial liquid slug produced to the surface. [6] Small ports result in low instantaneous gas injection rates that, in turn, simply bubble through the slug without lifting it, or the liquid slug does not reach the surface because all of its length is lost due liquid fallback.

The large-scale investigation of intermittent lift, conducted by White et al. [7] determined the velocities of the liquid slug and the gas bubble following it. Figure 6–21 shows that for a given tubing size and valve port diameter, the rising velocity of the gas bubble in different liquids is constant. Slug velocity, on the other hand, varies with the ratio of injection and production pressures valid at the time the valve opens. In the given case, maximum liquid slug velocity is attained at pressure ratios exceeding 2.2. Total liquid fallback, therefore, can be minimized if the slug moves with the possible greatest velocity.

Based on the results of White et al. [7], the Guiberson Ratiometric System [8] presented a collection of design charts similar to Figure 6–22 that allow the determination of liquid fallback in

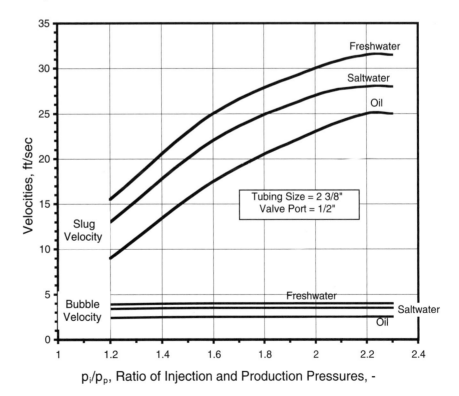

Fig. 6–21 The velocities of the liquid slug and the gas bubble in the function of pressure differential, after White et al. [7].

an intermittent installation. As can be seen on the figure, *liquid recovery* (the fraction of the starting slug received on the surface) very strongly decreases with increasing operating valve depths. On the other hand, liquid recovery can be increased by increasing the injection pressure in relation to the pressure of the starting slug.

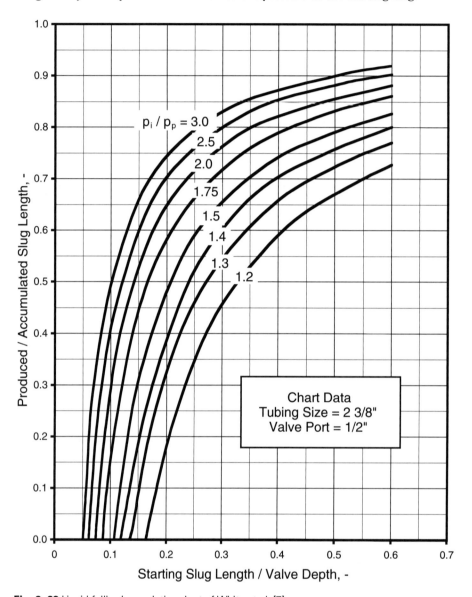

Fig. 6–22 Liquid fallback correlation chart of White et al. [7].

Gas Requirement of the Intermittent Cycle

It follows from the description of the intermittent cycle that the operating gas lift valve closes before the entire liquid slug enters the flowline. At this moment, the tubing string is full of lift gas that later completely blows down to separator pressure before a new cycle can begin. The gas consumption of the cycle, therefore, can be estimated to equal the gas volume present in the tubing string at the time the slug surfaces. This volume depends on the tubing size, depth of the operating valve, and the average tubing pressure at the moment the slug reaches the wellhead. The latter is usually approximated [5, 9] by the arithmetic average of the valve opening pressure at valve depth and the sum of the WHP plus the hydrostatic pressure of the produced liquid slug.

From the many published gas requirement charts, one taken from an OTIS manual [10, 11] is given in Figure 6–23, which shows the injection gas requirement of the intermittent lift cycle in the function of the operating gas lift valve's opening pressure at valve depth and its setting depth.

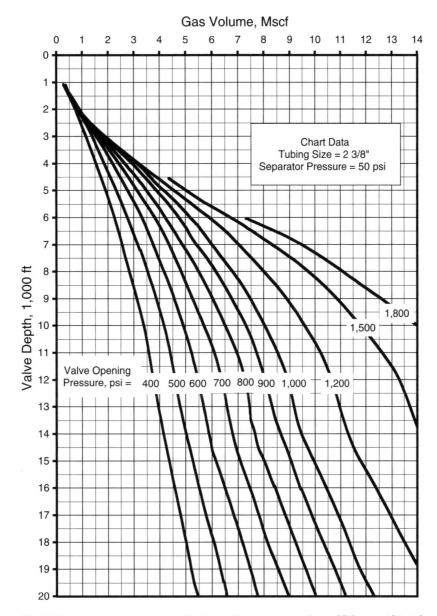

Fig. 6–23 Injection gas requirement of the intermittent cycle according to OTIS manual [10, 11]

Valve Spread Required

If choke control of gas injection is used, then the entire gas volume required to lift the liquid slug to the surface is supplied from the well's annulus. The gas volume released from the annulus in the period when the gas lift valve is open depends on the volume of the annular space and the pressure reduction. As discussed in Section 6.3.2, the pressure reduction of the annulus during the slug-lifting period equals the *spread* of the operating valve. It follows from these that the required valve spread is easily found from the cycle's injection gas requirement and the annular space available.

Figure 6–24 presents a chart [10] that can be used to estimate the required spread of the operating valve from the cycle's gas injection requirement and the annular configuration. As seen, the gas volume stored in a given annulus (a 7 in. casing and any tubing size) increases linearly with an increase in the annular pressure reduction. If the total injection gas volume is known, the chart can be used to find the required spread of the operating valve. In case gas injection at the surface is controlled with an intermitter, the amount of gas used for the cycle depends on the settings of the intermitter, and an operating valve with a lower spread than calculated can be used.

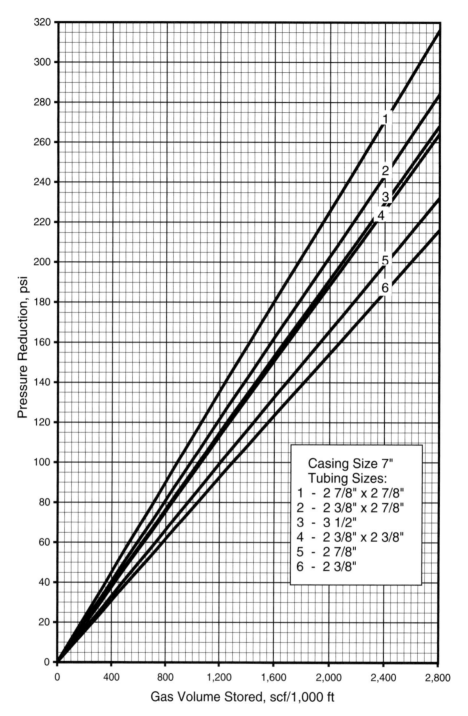

Fig. 6–24 Valve spread required to store a given gas volume in the annulus, after OTIS. [10]

Liquid Production in Afterflow

Several investigators pointed out that a significant portion of the liquid production of the intermittent cycle comes from the entrained liquid droplets produced to the surface during the *afterflow* period. This is due to the fact that gas velocity is very high just after the surfacing of the liquid slug and liquid droplets are carried to the wellhead. Later, as tubing pressure bleeds down, the movement of the remaining droplets slows down and they finally fall to the well bottom. Neely et al. [12] observed (based on their experiments in a 4,884 ft well) that a maximum of 50% of the total liquid production was received after the solid liquid slug entered the flowline. Liao, Schmidt, Doty [13], the developers of the first mechanistic model for intermittent gas lift, also suggest that afterflow can substantially contribute to the well's total liquid production.

6.5 Design of Intermittent Installations

6.5.1 General considerations

As with continuous flow gas lift wells, wells placed on intermittent lift can only be produced if the casing-tubing annulus is free of any liquids down to the operating valve. Since in most cases the operating depth is considerably deeper than the static liquid level, a gas pressure several times higher than that required for normal operations is needed for startup. Therefore, in order to use the normal operating gas lift pressure, installation of an unloading valve string is required at many times. Use of this valve string allows a stepwise transfer of the injection point from the surface down to the operating valve.

Another reason for using an unloading valve string is to allow the changing of the point of gas injection in accordance with changing well conditions. With the depletion of the field, formation pressure can considerably decrease, requiring the lowering of the injection point. This is facilitated by the unloading valve string, and the operating depth can be transferred to a deeper valve to follow the decline in the well's formation pressure. Finally, multipoint gas injection designs also require a valve string to be installed in the well.

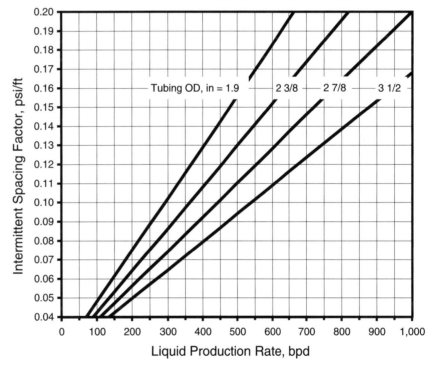

Fig. 6–25 Intermittent spacing factors after CAMCO [5].

Some of the more popular unloading valve string designs for intermittent gas lift are described in the following sections. The spacing of the valve string, just like in continuous flow wells, requires the use of a pressure gradient in the tubing string that represents the minimum tubing pressures during the production cycle. Such pressure gradients are called *spacing factors* and are a function of the well's production rate and tubing size, as shown in Figure 6–25 after [5, 14]. For very low daily production rates, the use of a spacing factor of 0.04 psi/ft is generally accepted; for other rates, the values read from the figure can be used.

6.5.2 Constant surface closing pressure

A common design procedure [1, 9, 15] for installations with *single-point gas injection* uses a constant surface closing pressure for all valves in the unloading valve string. This valve spacing procedure can be applied if unbalanced single element valves or pilot valves are run in the well. Surface gas injection can be controlled either by a choke or an intermitter. The main steps of the design procedure are detailed as follows, the *valve spacing* being illustrated in Figure 6–26.

1. Starting from the surface injection pressure p_{inj}, draw the distribution of gas pressure in the annulus with depth.

2. Establish the *spacing factor* to be used based on the tubing size and the estimated liquid production rate. Figure 6–25 can be used for this.

3. Starting from the *WHP*, draw the spacing pressure line with the gradient established in the previous step.

4. Select the surface closing pressure p_{close} for all valves as 100–200 psi less than the available surface injection pressure.

5. Starting from the surface closing pressure p_{close}, establish the distribution of injection pressure with well depth.

6. Find the depth of the top unloading valve.

   a. It is found where the pressure on a line started from the *WHP* and parallel to the unloading fluid gradient equals the injection pressure at depth.

   b. The setting depth of the top valve can also be calculated analytically as

$$L_1 = \frac{p_{inj} - WHP}{grad_u - grad_g}$$

6.1

   where: p_{inj} = available surface injection pressure, psi

   WHP = design wellhead pressure, psi

   $grad_u$ = unloading fluid gradient, psi/ft

   $grad_g$ = gas gradient in annulus, psi/ft

7. Find the setting depth of the second unloading valve.

   a. Start a line parallel to the unloading gradient from the spacing pressure p_{t1} at the setting depth of the top valve and find the depth where it intersects the closing pressure line. This determines the depth of the second valve.

   b. Mathematically, the depth increment between the top and the second valve is found from

$$\Delta L_1 = \frac{p_{close} - p_{t1} + L_1\, grad_g}{grad_u - grad_g}$$

6.2

   where: p_{close} = surface closing pressure, psi

   p_{t1} = tubing pressure at the first valve's setting depth, psi

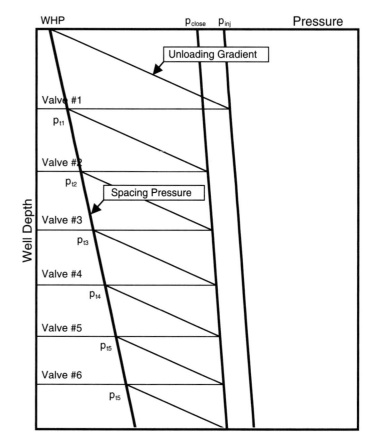

Fig. 6–26 Graphical valve spacing procedure using a constant surface closing pressure for all unloading valves.

8. The setting depths of the remaining unloading valves are found similar to the second valve, and a universal formula can be applied to find the depth increment for the nth valve:

$$\Delta L_n = \frac{p_{close} - p_{t(n-1)} + L_{(n-1)} \, grad_g}{grad_u - grad_g} \qquad 6.3$$

9. Adjust valve setting depths so as to set the bottommost valve as close as possible to the depth of perforations.

10. For *flagging* the operating valve, its surface closing pressure is reduced by about 50 psi below that of the other valves in the string. This helps identify lifting from the operating valve as well as ensures single-point gas injection.

11. Calculate the closing pressures of the valves at valve setting depth p_{ic} from the surface closing pressures. These values are equal to the dome charge pressures p_d of each valve.

After the spacing of the unloading valves has been determined, the port size for the operating valve is selected, based on the required injection gas volume and the pressure conditions at valve setting depth. The ports of the unloading valves can be smaller than the operating valve's port size; it is a common practice to use uniform sizes. The selection of the operating valve's port size follows, assuming that choke control is used at the surface.

1. Based on an assumed surface opening pressure, find the intermittent cycle's gas requirement. After calculating the opening injection pressure p_{io} at valve depth, Figure 6–23 can be used for this purpose.

2. The pressure differential Δp necessary to store the required gas volume in the annulus is found from Figure 6–24. If choke control is used to control the surface gas injection to the well, an operating valve with a *spread* equal to this pressure differential should be selected.

3. The valve's opening pressure is calculated as the sum of its closing pressure at depth and the pressure differential just found:

$$p_{io} = p_{ic} + \Delta p \qquad 6.4$$

4. A *tubing load* p_t is selected that represents the starting slug length's hydrostatic pressure plus the *WHP*. It is customary to set it to half of the *SBHP* of the well.

5. The required port size of the operating valve is found by solving the valve *opening equation* (Equation 3.17) for the ratio of valve areas R:

$$R = \frac{p_{io} - p_{ic}}{p_{io} - p_t} \qquad 6.5$$

6. An unbalanced or pilot-operated gas lift valve with the nearest R ratio is selected from manufacturer's data. For pilot-operated valves, the port size selected is for the *control port* only; the valve's main port can be selected independently so as to ensure a proper instantaneous gas injection rate through the valve.

7. The operating valve's actual opening pressure at depth is found from the valve opening equation (Equation 3.17) as follows:

$$p_{io} = \frac{p_{ic}}{1 - R} - p_t \frac{R}{1 - R} \qquad 6.6$$

8. The *spread* of the valve is calculated as follows and is compared to the required pressure differential Δp calculated in *Step 2*. If the two values differ considerably, adjustment of the assumed valve opening pressure is necessary and the previous calculations are repeated.

$$SPREAD = p_{io} - p_{ic} \qquad 6.7$$

9. If a small difference is found between the assumed and actual spread values, the assumed value is kept and the corresponding tubing load is calculated by solving the valve opening equation (Equation 3.17) for the production (tubing) pressure:

$$p_t = \frac{p_{ic}}{R} - p_{io}\frac{1-R}{R}$$

6.8

In case *intermitter control* is applied at the surface to control the gas injection into the annulus, the spread of the valve is not very critical because most of the gas used for injection is controlled by the surface intermitter. The previous calculations are accordingly modified and the operating valve's opening pressure is found, instead of Equation 6.4, from the following formula that includes only half of the pressure differential determined in *Step* 2. The rest of the calculations are identical.

$$p_{io} = p_{ic} + \frac{\Delta p}{2}$$

6.9

Finally, valve setting data for all valves are calculated as follows.

1. Based on flowing temperature data, the temperature at each valve setting depth T_i is found.

2. The *dome charge pressures* p_d of the valves at their setting depths are equal to their closing pressures at valve depth p_{ic}, calculated from the constant surface closing pressure, taking into account the weight of the gas.

3. The dome charge pressures at surface conditions p_d' are calculated at a charging temperature of 60 °F. For nitrogen gas charge, this can be done by using the charts in Figure D-1 or D–2 in the Appendix or by following the procedure described in connection with Equation 3.9.

4. Finally, TRO pressures are found from the surface dome charge pressures (see Equation 3.21) using the formula:

$$TRO = \frac{p_d'}{1-R}$$

6.10

Example 6–2. Design an intermittent installation lifting off bottom for the well with the data given as follows. Use an unloading valve string with a constant surface closing pressure, CAMCO 1½ in. valves, and evaluate both choke and intermitter control at the wellhead.

Casing Size = 7 in.	Available Injection Pressure = 700 psi
Tubing Size = 2 3/8 in.	Injection Gas Sp.Gr. = 0.6
Depth of Perforations = 7,000 ft	Formation Temperature = 170 °F
SBHP = 800 psi	Wellhead Temperature = 80 °F
WHP = 50 psi	Load Fluid Gradient = 0.42 psi/ft

Solution

The spacing of the unloading valves is given in Figure 6–27, where a *spacing factor* of 0.04 psi/ft was used as the minimum value read from Figure 6–25. The operating valve (Valve #6) had to be moved to the depth of the perforations, as shown in Table 6–1. Its surface closing pressure was reduced to 550 psi from the 600 psi used for the unloading valves.

For finding the port size of the operating valve, a surface opening pressure of 610 psi is assumed, from which the opening pressure at valve depth is p_{io} = 708 psi. Using this value and the valve setting depth of 7,000 ft, the injection gas requirement is found as 4,300 scf, from Figure 6–23. Using Figure 6–24, valid for the casing size of 7 in., the necessary pressure reduction in the annulus is Δp = 48 psi. The operating valve's opening pressure is found from Equation 6.4 as p_{io} = 648 + 48 = 696 psi.

Assuming a tubing load of 400 psi, half of the SBHP, the required R value of the operating valve is, according to Equation 6.5:

$$R = (696 - 648) / (696 - 400) = 0.1622$$

From manufacturers catalog (Appendix F, Table F–2), the nearest R ratio for the pilot-operated valve type RP-6 is 0.168 and the control port size is ⅜ in. The main port size is selected as ½ in., the biggest size available. The valve opening pressure should be modified with the selected R value and using Equation 6.6 we find:

$$p_{io} = 648 / (1 - 0.168) - 400 \cdot 0.168 / (1 - 0.168) = 698 \text{ psi.}$$

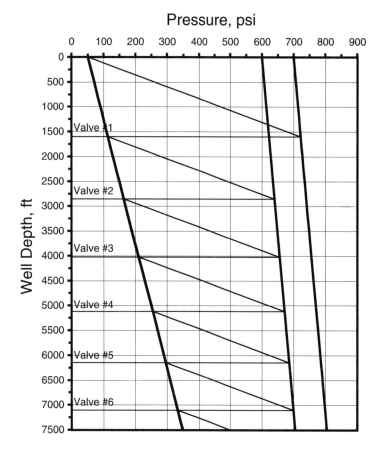

Fig. 6–27 Results of installation design for Example 6–2.

The spread of the operating valve is the difference of its opening and closing pressures, i.e.: 698 – 649 = 50 psi. Since this is very close to the assumed pressure reduction of Δp = 48, there is no need for adjustments. The opening pressure is thus set to 708 psi, the assumed value, when the corresponding tubing load is calculated from Equation 6.8:

$$p_t = 648 / 0.168 - 708 (1 - 0.168) / 0.168 = 351 \text{ psi.}$$

For intermitter control, only half of the pressure reduction is used to find the valve's opening pressure, according to Equation 6.9:

$$p_{io} = 648 + 48 / 2 = 672 \text{ psi.}$$

Valve No.	Valve Depth Chart	Valve Depth Adjusted	Surf. Close Close	Closing at Depth	Tubing Pr. at Depth	Valve Type	Port Size Control	Port Size Main	R	p_{io}	Valve Temp.	p'$_d$	TRO
-	ft	ft	psi	psi	psi	-	in	in	-	psi	F	psi	psi
1	1,601	1,601	600	622	114	R-20		5/16	0.126	696	101	573	656
2	2,853	2,853	600	640	164	R-20		5/16	0.126	709	117	572	654
3	4,025	4,025	600	656	211	R-20		5/16	0.126	721	132	570	652
4	5,122	5,122	600	672	255	R-20		5/16	0.126	732	146	569	651
5	6,149	6,149	600	686	296	R-20		5/16	0.126	742	159	568	650
6	7,110	7,000	550	648	351	RP-6	3/8	1/2	0.168	708	170	527	633

Table 6–1 Details of the intermittent gas lift installation design calculations for Example 6–2.

The required R value is determined from Equation 6.5:

$R = (672 - 648) / (672 - 400) = 0.088$.

For this case, a 5/16 in. port is selected with R = 0.079, and the opening pressure is modified, as given in Equation 6.6:

$p_{io} = 648 / (1 - 0.079) - 400 \ 0.079 / (1 - 0.079) = 669$ psi.

The spread of the valve is now $669 - 648 = 21$ psi, less than the calculated pressure reduction $\Delta p = 48$, required for storing the injection gas volume in the annulus. This is satisfactory, because injection gas is supplied by the intermitter and not from the well's annulus.

This concludes the design of the operating valve. The unloading valves can be standard unbalance valves with spread; CAMCO R–20 valves with 5/16 in. chokes are selected. The installation design is completed with calculating the surface gas charge pressures p_d' and the TRO pressures. Detailed calculation data for choke control are given in Table 6–1.

6.5.3 The Optiflow design procedure

The unloading valve string design developed for IPO valves and introduced by Axelson as *Opti-Flow* [16] is suited for wells with little or no information on well potential. Since the operating point as well as the liquid rate is unknown, the Opti-Flow design provides an unloading valve string that permits injection of gas at the deepest possible point at all times. This is assured by moving the operating point down the well to a point where well inflow is sufficient to prevent operation from the next lower valve. Several variants of this design are known [9, 15, 16] but all assume a certain fraction (about 50%) of the operating gas lift pressure at depth as the tubing pressure at each valve depth.

The main steps of the design procedure are detailed as follows, with the valve spacing calculations being illustrated in Figure 6–28.

1. Set the valves' surface opening pressure equal to $p_{open} = p_{inj} - 100$ and draw the distribution of gas pressure in the annulus with depth.

2. Establish the spacing lines to be used for the determination of valve depths.

   a. For spacing the valves above the static liquid level, use surface gas lift pressures of 30% and 90% of the surface opening pressure p_{open} and establish the distribution of gas pressures with well depth. These are marked and plotted with solid lines in Figure 6–28.

   b. Below the static liquid level, valves are spaced based on 55% and 85% of the surface opening pressure and the distribution of gas pressures with depth are drawn. These are marked and plotted with dashed lines in Figure 6–28.

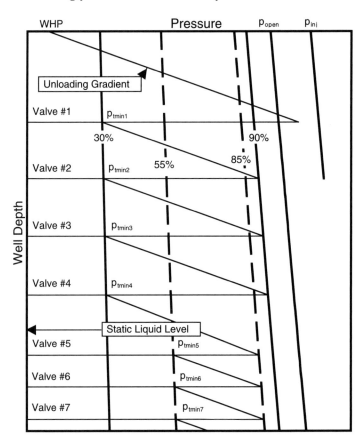

Fig. 6–28 Graphical valve spacing procedure using the Opti-Flow procedure.

3. Find the static liquid level, if the *SBHP* is known.

$$L_{static} = L_{perf} - \frac{SBHP}{grad_u}$$

6.11

where: L_{perf} = depth of perforations, ft

$SBHP$ = static bottomhole pressure, psi

$grad_u$ = unloading fluid gradient, psi/ft

4. Find the depth of the top unloading valve.

   a. It is found where the pressure on a line started from the *WHP* and parallel to the unloading fluid gradient equals the pressure at depth starting from a surface pressure of $p_{inj} - 50$.

   b. The setting depth of the top valve can also be calculated analytically as

$$L_1 = \frac{p_{inj} - 50 - WHP}{grad_u - grad_g}$$

6.12

   where: p_{inj} = available surface injection pressure, psi

   WHP = design wellhead pressure, psi

   $grad_u$ = unloading fluid gradient, psi/ft

   $grad_g$ = gas gradient in annulus, valid for pinj – 50, psi/ft

5. Find the setting depth of the second unloading valve.

   a. Start a line parallel to the unloading gradient from the minimum tubing pressure p_{tmin_1} at the setting depth of the top valve and find the depth where it intersects the maximum tubing pressure line. This determines the depth of the second valve.

   b. The depth increment between the top and the second valve can also be found mathematically as follows:

$$\Delta L_2 = \frac{p_{tmax} - p_{tmin} + L_1 (grad_{gmax} - grad_{gmin})}{grad_u - grad_{gmax}}$$

6.13

   where: p_{tmax} = tubing pressure read from the *maximum* spacing line, psi

   p_{tmin} = tubing pressure read from the *minimum* spacing line, psi

   $grad_{gmax}$ = gas gradient belonging to the *maximum* spacing line, psi/ft

   $grad_{gmin}$ = gas gradient belonging to the *minimum* spacing line, psi/ft

6. The setting depths of the remaining unloading valves are found similar to the second valve, and a universal formula can be applied to find the depth increment for the nth valve. Note that different *maximum* and *minimum* spacing lines with the corresponding pressure and gas gradient values have to be used above and below the static liquid level.

$$\Delta L_2 = \frac{p_{tmax} - p_{tmin} + L_{(n-1)} (grad_{gmax} - grad_{gmin})}{grad_u - grad_{gmax}}$$

6.14

7. If necessary, adjust valve setting depths so as to set the bottommost valve as close as possible to the depth of perforations.

8. Determine for each valve the minimum tubing pressure p_{tmin} read from the proper spacing line. These values are the opening tubing pressures for each valve.

9. Determine each valve's opening casing pressure at depth p_{io} from the distribution of gas pressure in the annulus.

10. Valves above the static liquid level are used for unloading only and must be prevented from opening when the liquid slug passes them. This can be ensured by increasing their surface opening pressures above the value $p_{open} = p_{inj} - 100$.

The selection of the port size for the operating valve is based on the required injection gas volume and the pressure conditions at valve setting depth. For choke control, the selection process is described as follows:

1. Based on the operating valve's opening pressure p_{io} and its setting depth, the intermittent cycle's gas requirement is found from Figure 6–23.

2. The pressure differential Δp necessary to store the required gas volume in the annulus is found from Figure 6–24. An operating valve with a *spread* equal to this pressure differential should be selected, its closing pressure at valve depth is found as:

$$p_{io} = p_{ic} + \Delta p \qquad\qquad 6.15$$

3. The *tubing load* at the operating gas lift valve is the operating valve's opening tubing pressure p_{tmin} determined before.

4. The required port size of the operating valve is found by solving the valve *opening equation* (Equation 3.17) for the ratio of valve areas, R:

$$R = \frac{p_{io} - p_{ic}}{p_{io} - p_{t\,min}} \qquad\qquad 6.16$$

5. From manufacturer data an unbalanced or pilot-operated gas lift valve with the nearest R ratio is selected. For pilot-operated valves, the port size selected is for the *control port* only. The main port is selected independently of the control port. It should be as great as possible in order to ensure a great gas injection rate.

In case *intermitter control* is applied at the surface, the spread of the valve is not very critical because most of the gas used for injection is controlled by the surface intermitter. The previous calculations are accordingly modified, and the operating valve's closing pressure is found by using only the half of the pressure differential determined in *Step* 2. The rest of the calculations are identical.

Finally, valve setting data for all valves are calculated as follows:

1. The port sizes of the remaining valves set below the static liquid level are usually selected so as to be identical to the bottom operating valve. For unloading valves, *i.e.* those set above the static liquid level, standard unbalanced valves are chosen with the same port sizes.

2. Based on flowing temperature data, the temperature at each valve setting depth T_i is found.

3. The *dome charge pressures* p_d of the valves at their setting depths are calculated from the valve *opening equation* (Equation 3.17) as follows:

$$p_d = p_{io} \, (1 - R) + p_{t\,min} \, R \qquad\qquad 6.17$$

4. The dome charge pressures at surface conditions p'_d are calculated at a charging temperature of 60 °F. For nitrogen gas charge, this can be done by using the charts in Figure D–1 or D–2 in the Appendix or by following the procedure described in connection with Equation 3.9.

5. Finally, TRO pressures are found from the surface dome charge pressures (see Equation 3.21) using the formula:

$$TRO = \frac{p'_d}{1 - R}$$

6.18

Example 6–3. Design an intermittent installation using the Opti-Flow procedure for the well with the data given in the previous example assuming choke control at the wellhead.

Solution

The surface opening pressure of the valves is $p_{open} = 700 - 100 = 600$ psi. The spacing lines are depicted in Figure 6–29. The static liquid level is found from Equation 6.11 as

$L_{static} = 7,000 - 800 / 0.42 = 5,095$ psi.

Graphical valve spacing calculations are presented in Figure 6–29; numerical results are contained in Table 6–2.

Table 6–2 contains the adjusted valve setting depths, adjusted surface opening pressures for the unloading valves as well as the minimum tubing pressures at each valve.

The gas requirement of the intermittent cycle is found from Figure 6–23, based on the bottom valve's opening pressure of $p_{io} = 698$ psi and the valve setting depth of $L = 7,000$ ft, as 4,300 scf/cycle. The required casing annulus pressure reduction of $\Delta p = 46$ psi is read off Figure 6–24. The closing pressure of the bottom valve is, according to Equation 6.15:

$p_{ic} = 698 – 46 = 652$ psi.

Using the previous value and the tubing load of $p_{tmin} = 372$ psi, the operating valve's required R value is calculated from Equation 6.16 as:

$R = (698 – 652) / (698 – 372) = 0.141$.

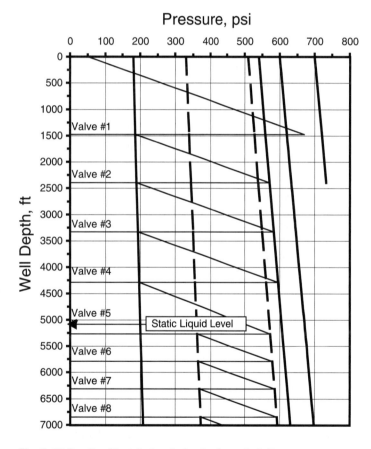

Fig. 6–29 Results of installation design for Example 6–3.

From Appendix F, Table F–2 the nearest RP–6 valve has a control port size of $^5/_{16}$ in. and $R = 0.126$. A valve with this port and a main port size of $^1/_2$ in. is selected. Valves #5 to #7 are selected identically, for unloading valves R–20 are selected with $^5/_{16}$ in. main port sizes.

Calculated valve temperatures, downhole and surface dome charge pressures, and TRO pressures are listed in Table 6–2.

Valve No.	Valve Depth		Surf. Opening	Opening at Depth	Tubing Press.		Valve Type	Port Size		R	p_d	Valve Temp.	p'_d	TRO
	Chart	Adjusted			min.	max.		Contr.	Main					
-	ft	ft	psi	psi	psi	psi	-	in	in	-	psi	F	psi	psi
1	1,478	1,500	680	701	186	-	R-20		5/16	0.126	636	99	587	672
2	2,395	2,400	660	694	190	-	R-20		5/16	0.126	630	111	569	651
3	3,333	3,350	640	687	193	584	R-20		5/16	0.126	625	123	552	631
4	4,291	4,300	620	680	197	596	R-20		5/16	0.126	619	135	535	612
5	5,270	5,300	600	674	362	574	RP-6	5/16	1/2	0.126	635	148	536	613
6	5,789	5,800	600	681	365	580	RP-6	5/16	1/2	0.126	641	155	535	613
7	6,315	6,400	600	690	368	587	RP-6	5/16	1/2	0.126	649	162	535	612
8	6,849	7,000	600	698	372	594	RP-6	5/16	1/2	0.126	657	170	534	611

Table 6–2 Details of the intermittent gas lift installation design calculations for Example 6–3.

6.5.4 Unloading procedure

Unloading an intermittent gas lift well is a manual and closely controlled operation. Its basic objective is to prevent excessive pressure differentials across the valves. This is necessary because liquid from the annulus is forced to flow through the unloading valves. Even a clean kill fluid, given the relatively high flow velocities through the small ports of the valves can cut or damage valve seats. The presence of any solids, especially sand, in the load fluid exponentially increases the abrasive effect and the possibility of valve damage. Therefore, it is of utmost importance that well unloading be conducted so as to avoid excessive wear on the gas lift valves.

Before the first unloading of the well, removal of any solids from the annulus is desirable. Gas injection lines, before first being connected to the well, must be blown clean of scale, welding slag, and other solids to prevent introduction of solids into the annulus. The well's annulus should be circulated with a clean *kill fluid* to minimize or completely eliminate the solids content of the annulus fluid.

Damage of gas lift valves depends on the liquid velocities across the valves that depend on the annular pressures. If injection pressure at the casinghead increases rapidly, the quick drop of annulus liquid level will result in high liquid velocities across the unloading valves. Valve damage, therefore, can significantly be reduced if injection pressure is gradually increased in the annulus. The procedure recommended by the API [17] and described as follows ensures a minimum of damage to the valves in the unloading string.

1. Before unloading, bleed down any pressure greater than flowline pressure from the wellhead.

2. Remove or fully open the flowline choke.

3. Set the intermitter (if the well is on intermitter control) or adjust the gas injection choke (if the well is on choke control) to a gas injection rate that increases the CHP by a rate of 50 psi per 8–10 min until reaching a CHP of 400 psi.

4. After the CHP of 400 psi is reached, increase surface gas injection by adjusting the controls to achieve a pressure buildup rate of 100 psi per 8–10 min. Continue to build up casing pressure until the top valve is uncovered and starts to inject gas into the tubing string.

5. For wells on intermitter control, the first 12 to 24 hrs of the well's operation should be spent with a cycle frequency of 2–3 cycles per hr or less. Gas injection should be stopped as soon as the liquid slug clears the wellhead.

6. For wells on choke control, injection rate should be adjusted to about ⅔ of the design injection rate. Although this rate is usually insufficient to unload the well down to the operating valve, it minimizes the damage to valves. After 12–18 hrs of operation, the gas injection rate can be increased to its full design value.

7. After unloading, the well should be fine-tuned to reach optimum operating conditions.

6.6 Chamber Lift

6.6.1 Basic features

If formation pressure in a well placed on intermittent lift drops to a low level, two unfavorable phenomena occur that can severely decrease the efficiency of fluid lifting. First, the slug volume that accumulates in the tubing string during one cycle considerably decreases and the well's production rate declines accordingly. Second, the injection GLR inevitably increases since the tubing string below the liquid slug must always be filled with injection gas, no matter what the actual starting slug length is. These are the reasons why intermittent gas lifting in a closed or semi-closed installation can turn out to be very inefficient for low to very low formation pressures, *i.e.* in wells near their abandonment.

In a chamber installation, well fluids accumulate in a downhole accumulation chamber of a greater capacity than the tubing string. If the same starting slug length is allowed to accumulate, then the cycle volume is much greater in the chamber installation than in a conventional intermittent installation. For example, if a 2 3/8 in. tubing and a 7 in. casing is used, the capacity of the tubing is 3.87 10^{-3} bbl/ft, whereas that of the casing-tubing annulus is 3.28 10^{-2} bbl/ft. This means that for the same starting slug length of 300 ft, the cycle volume for the tubing string is 1.16 bbl, and for the chamber (tubing plus annulus) is 11 bbl. Since the accumulation time for this 300 ft of liquid is identical in both cases, the cycle volume—and consequently the daily liquid production rate of the well—can be increased drastically if a chamber installation is used.

The operation of the intermittent cycle in a chamber installation of the two-packer type is illustrated in Figure 6–30. After the previous liquid slug has been lifted and the operating valve has closed, fluid inflow to the well starts with the opening of the standing valve. Well fluids simultaneously rise in the casing-tubing annulus as well as in the tubing because of the perforated nipple situated close to the bottom packer, see *Part B*. The bleed valve continuously vents formation gas into the tubing, and thanks to its operation, the chamber fills up fully with liquid. At this moment, gas injection at the surface occurs, forcing the operating gas lift valve to open and to inject gas to the top of the chamber. The bleed valve immediately closes, and the liquid in the casing-tubing annulus is gradually U-tubed (displaced) into the tubing string, as shown in *Part C*. Gas injection below the liquid slug and into the tubing occurs only after all the liquid has left the casing-tubing annulus. Note that the point of gas injection is at the perforated nipple and not at the operating gas lift valve. As seen in *Part D*, the liquid slug starts its upward journey and attains a considerable velocity even before the actual gas injection takes place. This is a crucial feature that distinguishes chamber lifting from simple accumulation chambers sometimes used in intermittent installations. [14]

Applications

Due to its advantageous features, chamber lift is recommended for gas lifting wells with very low formation pressures near their abandonment. It can be used in the latest stages of a well's or a field's productive life and is usually the last kind of gas lifting before the well is finally abandoned. Additionally, chamber lift is ideally suited to produce deep, high productivity wells with low formation pressures. [18]

The selection of the type of chamber lift installation from the many available versions (see chapter 4) is governed by many factors such as well completion type, casing size and condition, production rate, etc. [19, 20] Two-packer chambers (see Fig. 6–30) are more expensive and are justified if the well's production rate is sufficient to cover the additional costs. They are recommended when the dynamic liquid level is above the top of the perforations. The possibility of using the maximum

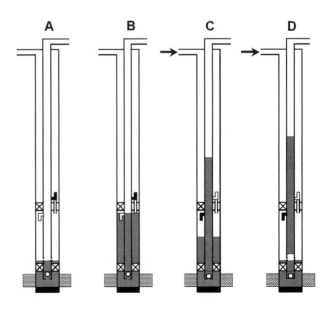

Fig. 6–30 Operation of the two-packer chamber lift installation.

annular capacity of the casing and the utilization of a tubing section as a dip tube allow relatively high production rates to be attained. An added advantage is that all equipment (operating and bleed valves, standing valve) can be of the wireline-retrievable type, ensuring low workover costs.

Insert chambers are required for open hole completions and cases when the dynamic liquid level is below the top of the perforations. The recommended choice for shallow, low capacity or stripper wells is the installation with a simple hookwall packer as shown in Figure 6–31. This inexpensive solution uses a dip tube with a diameter smaller than that of the tubing as well as a standard gas lift mandrel and valve. In deeper wells with more production potential, the use of a more expensive bypass packer can be justified, and the installation given in Figure 6–32 is recommended. Advantages include the use of the tubing as dip tube and the possibility of using wireline retrievable operating and standing valves for easier servicing.

Advantages

Chamber lift installations offer many advantages over conventional intermittent gas lift installations. The more important ones are detailed as follows.

- Chamber lift produces the lowest possible *FBHP* for gas lifting and can thus markedly increase the production rate of most intermittent wells.

- Wells with low formation pressures can be produced until ultimate depletion.

- Liquid fallback is considerably reduced when compared to conventional installations because gas injection takes place only after all of the accumulated liquid is U-tubed to the tubing and the liquid slug is in full motion. Gas breakthrough is thus greatly reduced, leading to a higher liquid recovery than that available with other installation types.

- Injection gas requirements are reduced due to the greater starting slug lengths possible.

- Since the point of gas injection in an insert chamber installation is at the bottom of the dip tube, lift gas can be injected near the total depth in wells with a long perforated interval.

Limitations

The basic limitations of chamber lift are the same as those for intermittent lift in general, *i.e.* lower available liquid rates, wasting of formation gas, etc. Additional disadvantages are as follows:

- Well dimensions like a small casing size or a long perforated interval can severely limit the applicability of chamber installations.

- In wells with high sand production, wireline operations and the pulling of the chamber can be difficult.

6.6.2 Equipment selection considerations

In order to ensure an efficient operation of chamber lift installations, the proper selection of the required equipment is necessary. The following presents a short description of the most important considerations.

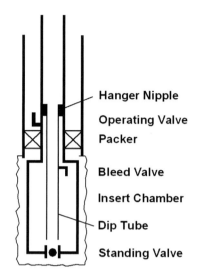

Fig. 6–31 Construction of an insert chamber installation using a hookwall packer.

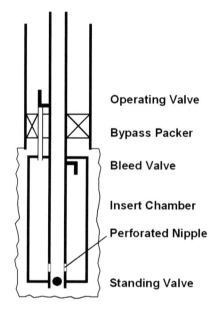

Fig. 6–32 Construction of an insert chamber installation using a bypass packer.

- Preferably, upper packers should be of the bypass type, allowing the tubing to be used as a dip tube. Production packers, on the other hand, are less expensive but require the use of small diameter dip tubes.

- The operating or chamber valve can be any IPO valve suitable for intermittent lift operations. Selection of the valve's spread is very important because tubing pressure at valve depth is extremely low so valve spread is near its maximum value (see Fig. 3–24). Therefore, pilot-operated valves are preferred since their spread can be selected independently of the main port size.

- Standing valves should be mechanically prevented from being blown out from their seating nipples by the great pressure differentials occurring during the afterflow period of the intermittent cycle.

- A properly sized bleed valve or bleed port installed on the dip tube is desirable to vent formation and injection gas from the chamber. Bleed holes of $^3/_{32}$ in. or $^1/_8$ in. size are sufficient for low producers but bleed (differential) valves are required for wells with high formation GLRs and/or high cycle frequencies. For insert chamber installations in gassy wells, an annulus vent valve must also be used to vent formation gas from below the packer into the tubing string. As pointed out by Winkler [21], the accumulation of formation gas due to natural separation below the packer severely restricts the fillage of the chamber with liquid.

- If possible, dip tubes should be the same diameter as the tubing because gas breakthrough can significantly increase when the liquid slug is transferred from the dip tube to the tubing of a greater diameter.

- In order to keep workover costs at a minimum, wireline retrievable equipment should be the first choice.

- The bottom unloading valve should be set immediately above the operating (chamber) valve to ensure proper unloading of the well. Its opening pressure should be at least 50 psi higher than that of the operating valve.

- The top of the chamber should never be placed above the maximum dynamic liquid level. This is because the space between the liquid level and the top of the chamber is filled with injection gas in every cycle, which increases the gas requirements and also the time needed for pressure blow-down. The latter condition can severely limit the cycle frequency and the well's liquid production rate.

6.6.3 Design of chamber installations

Since chamber lifts are intermittent gas lift installations, all the necessary calculations and design procedures follow those required for conventional cases. This is true for the design of the unloading valve string, the selection and operation of surface injection controls, the calculation of gas injection requirements, etc. The most important differences between a conventional and chamber lift installation are that (a) well fluids accumulate in the chamber, and (b) gas is initially injected at the top of the chamber. Implications of these conditions necessitate the knowledge of the length of the chamber, which is discussed as follows.

6.6.3.1 Determination of chamber length. Figure 6–33 illustrates a two-chamber installation immediately before the opening of the operating valve (on the left-hand side) and just before gas injection into the tubing takes place (on the right-hand side). If the chamber has completely filled up, then a liquid column of a length equal to the chamber length (CL) is standing in the chamber and the tubing string. After the chamber's liquid content is completely U-tubed into the tubing string, a liquid column of height H is present in the tubing. This column length is easily found from a balance of the respective volumes and can be expressed using the chamber's geometrical data as

$$H = CL \left(1 + \frac{C_a}{C_t} \right)$$

6.19

where: CL = chamber length, ft

   C_a = capacity of the chamber annulus, bbl/ft

   C_t = capacity of the tubing above the chamber, bbl/ft

The height of the liquid column can also be calculated from a balance of the pressures at the instant the annulus liquid level drops to the bottom of the chamber, if the pressure at the bottom of the liquid slug is known:

$$H = \frac{p_i - p_t}{grad_l} \qquad\qquad 6.20$$

where: p_i = design tubing load, *i.e.* the pressure at the bottom of the liquid column, psi

  p_t = sum of wellhead and gas column pressures at top of liquid column, psi

  $grad_l$ = static gradient of accumulated liquid, psi/ft

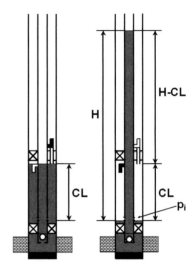

Fig. 6–33 Illustration for CL calculations.

Simultaneous solution of the previous equations yields the length of the chamber:

$$CL = \frac{p_i - p_t}{(1 - R_c)\,grad_l} \qquad\qquad 6.21$$

where: $R_c = C_a / C_t$ = ratio of annular and tubing capacities

As seen, the length of the chamber heavily depends on the design tubing load p_i, which is selected by the designer so as to ensure an efficient lifting of the liquid slug to the surface. As discussed in Section 6.4.2.1, liquid fallback is at a minimum if the tubing load equals 60–75% of the available injection pressure. Since injection pressure is identical to the operating (chamber) valve's opening pressure, the proper value of tubing load is found as follows:

$$0.60\,p_{io} < \mathrm{pi} < 0.75\,p_{io} \qquad\qquad 6.22$$

where: p_{io} = the chamber valve's opening injection pressure at its setting depth, psi.

The previous formula can only be used if the chamber fills up completely and its top is located at the dynamic liquid level. Similar formulas for other possible chamber configurations (insert chamber with a dip tube or a tubing section, etc.) were developed by Winkler and Camp [19, 20].

6.6.3.2 Installation design. As already mentioned, the design of a chamber installation closely follows the procedure for conventional intermittent installations. A complete design for choke control, including the spacing of unloading valves is detailed here, following the procedure required for using a constant surface closing pressure of the unloading valves (see Section 6.5.2 and Fig. 6–26). Although it is developed for a two-packer chamber installation, other installation types can be handled similarly.

Spacing of the unloading valves is identical to the procedure described in *Steps 1–8* of the previous section. The rest of the design calculation is detailed as follows.

1. Select the setting depths of the two packers: the lower packer should be set at the depth of the perforations; the upper packer's setting depth is less by an assumed *CL*.

2. Delete all unloading valves below the assumed depth of the upper packer.

3. Set the operating chamber valve to the depth of the upper packer and select its surface closing pressure as 50 psi less than that of the unloading valves. Calculate the appropriate downhole closing pressure p_{ic}.

4. Calculate the tubing pressure at the chamber valve when the valve is closed:

$$p_t = WHP + L\,grad_g \qquad\qquad 6.23$$

where: WHP = wellhead pressure, psi

   L = setting depth of the chamber valve, ft

   $grad_g$ = gas gradient, psi/ft

5. Assume the opening pressure p_{io} of the chamber valve at depth.

6. Find the intermittent cycle's gas requirement based on the chamber valve's opening pressure p_{io} and its setting depth using Figure 6–23 or a similar chart.

7. The pressure differential Δp necessary to store the required gas volume in the annulus is found. (Fig. 6–24 or a similar chart can be used.) Since a choke is used to control the surface gas injection to the well, an operating valve with a *spread* equal to this pressure differential should be selected.

8. The valve's opening pressure is calculated as the sum of its closing pressure at depth and the pressure differential just found:

$$p_{io} = p_{ic} + \Delta p \qquad\qquad 6.24$$

9. Compare the assumed and calculated opening pressures and repeat *Steps 5–8* until the two values are in agreement.

10. The required port size of the chamber valve is found by solving the valve *opening equation* (Equation 3.17) for the ratio of valve areas, R:

$$R = \frac{p_{io} - p_{ic}}{p_{io} - p_t} \qquad\qquad 6.25$$

11. A pilot-operated gas lift valve with the nearest R ratio is selected from manufacturer's data. The port size selected is for the *control port* only; the main port can be selected independently so as to ensure a proper instantaneous gas injection rate through the valve.

12. The actual opening pressure at depth of the chamber valve is checked with the valve opening equation (Equation 3.17) as follows:

$$p_{io} = \frac{p_{ic}}{1 - R} - p_t\,\frac{R}{1 - R} \qquad\qquad 6.26$$

13. Select the proper tubing load p_i corresponding to the valve opening pressure p_{io} from Equation 6.22.

14. Calculate the capacities of the chamber annulus and that of the tubing; find their ratio by the formulas given as follows:

$$C_a = 9.71 \times 10^{-4}\,(ID_{cb}^2 - OD_t^2) \qquad\qquad 6.27$$

$$C_t = 9.71 \times 10^{-4}\,ID_{cb}^2 \qquad\qquad 6.28$$

$$R_c = \frac{C_a}{C_t}$$

6.29

where: ID_{cb} = chamber inside diameter, in.

ID_t = tubing inside diameter, in.

OD_t = tubing outside diameter, in.

15. Calculate the required length of the chamber by using Equation 6.21.

16. Find the depth of the chamber valve from the *CL* just calculated. If the calculated value is not close enough to the one assumed in *Step 3*, repeat the design with the calculated valve depth. Otherwise, set the bottom unloading valve two tubing joints (about 60 ft) higher than the chamber valve and continue.

17. Make sure that the bottom unloading valve doesn't open when the liquid slug passes. For this, assume the tubing pressure opposite the valve p_t when the bottom of the liquid slug has risen above the valve setting depth.

18. Assume the opening pressure of the bottom unloading valve p_{io} and find the required *R* value from Equation 6.25.

19. Select an unloading gas lift valve with the nearest *R* ratio and calculate the actual opening pressure at depth from Equation 6.26. Compare this value with the assumed one and repeat *Steps 18–19* until the two values agree.

20. Select the type and choke sizes for the unloading valves.

21. The flowing temperature at each valve setting depth T_i is found.

22. The *dome charge pressures* p_d of the valves at their setting depths are calculated from the valve *opening equation* (Equation 3.17) as follows:

$$p_d = p_{io}(1 - R) + p_{t\,min} R$$

6.30

23. The dome charge pressures at surface conditions p'_d are calculated at a charging temperature of 60 °F. For nitrogen gas charge, this can be done by using the charts in Figure D–1 or D–2 in the Appendix or by following the procedure described in connection with Equation 3.9.

24. Finally, TRO pressures are found from the surface dome charge pressures (see Equation 3.21) using the formula:

$$TRO = \frac{p'_d}{1 - R}$$

6.31

Example 6–4. Design a two-packer chamber installation for the well with the data given as follows. Use an unloading valve string with a constant surface closing pressure, CAMCO 1 ½ in. valves, and choke control at the wellhead.

Casing Size = 7 in.	Available Injection Pressure = 1,000 psi
Tubing Size = 2 ⅜ in.	Injection Gas Sp.Gr. = 0.6
Depth of Perforations = 7,000 ft	Formation Temperature = 170 °F
spacing factor = 0.04 psi/ft	Wellhead Temperature = 80 °F
WHP = 50 psi	Load Fluid Gradient = 0.42 psi/ft

Solution

Valve spacing calculations resulted in three unloading valves the depths of which are given in Table 6–3.

Valve	Valve Depth		Surf.	Closing	Tubing Pr.	Valve	Port Size		R	p_{io}	Valve	p'_d	TRO
No.	Chart	Adjusted	Close	at Depth	at Depth	Type	Control	Main			Temp.		
-	ft	ft	psi	psi	psi	-	in	in	-	psi	F	psi	psi
1	2,340	2,340	900	933	144	R-20		5/16	0.126	1,047	110	841	962
2	4,284	4,290	900	960	222	R-20		5/16	0.126	1,067	135	825	944
3	6,103	6,826	900	996	850	R-20		1/2	0.260	1,047	168	806	1,090
4	7,806	6,886	850	946	146	RPB-5	3/8	1/2	0.066	1,003	169	766	820

Table 6–3 Details of the chamber lift installation design calculations for Example 6–4.

For the first iteration, a chamber length of CL = 200 ft was assumed, which gives a chamber valve depth of 6,800 ft. The surface closing pressure of the chamber valve is set to 850 psi, which gives a downhole closing pressure of 946 psi.

The tubing pressure at the chamber valve while the valve is closed is, according to Equation 6.23:

$$p_t = 50 + 6,800 \cdot 0.014 = 146 \text{ psi.}$$

Assume the opening pressure of the chamber valve as p_{io} = 1,000 psi, and find the gas injection requirement from Figure 6–23 as 5,800 scf. The required pressure differential is found as Δp = 63 psi from Figure 6–24.

The chamber valve's opening pressure is found from Equation 6.24 as

$$p_{io} = 946 + 63 = 1,009 \text{ psi, which is sufficiently close to the assumed value.}$$

The port size of the chamber valve is calculated according to Equation 6.25:

$$R = (1,009 - 946) / (1,009 - 146) = 0.073.$$

A CAMCO RPB-5 pilot-operated valve with a ¼ in. port and R = 0.066 is selected. Its opening pressure is determined from Equation 6.26:

$$p_{io} = 946 / (1 - 0.066) - 146 \cdot 0.066 / (1 - 0.066) = 1,013 - 10 = 1,003 \text{ psi, very close to the assumed value.}$$

The tubing load is selected as 60% of the valve's opening pressure: p_i = 0.6 · 1,003 = 602 psi.

Chamber annulus and tubing capacities are calculated from Equations 6.27–8:

$$C_a = 9.71 \cdot 10^{-4} (6.276^2 - 2.375^2) = 0.0328 \text{ bbl/ft}$$

$$C_t = 9.71 \cdot 10^{-4} (1.995^2) = 0.00387 \text{ bbl/ft}$$

From these, the ratio of capacities is found as R_c = 8.48.

The required CL is calculated from Equation 6.21 as

$$CL = (602 - 146) / (8.48 + 1) / 0.42 = 114 \text{ ft, from which the depth of the chamber valve is}$$

7,000 − 114 = 6,886 ft, considered to be close enough to the assumed value of 6,800 ft, so the upper packer and the chamber valve should be run to a depth of 6,886 ft.

The bottom unloading valve (Valve #3) is run 60 ft higher to a depth of 6,826 ft, and valve setting depths in Table 6–3 are corrected accordingly.

To ensure that Valve #3 doesn't open when the liquid slug passes it, a tubing pressure of p_t = 850 psi is assumed. Using an assumed opening pressure of p_{io} = 1,050 psi, the required R value found from Equation 6.25 is

$$R = (1,050 - 996) / (1,050 - 850) = 0.27.$$

An unbalanced R 20 gas lift valve is selected with a choke size of $^1/_2$ in. and an R - 0.26 and its opening pressure is found from Equation 6.26:

$$p_{io} = 996 / (1 - 0.26) - 850\ 0.26 / (1 - 0.26) = 1,345 - 298 = 1,047\ psi,\ very\ close\ to\ the\ assumed\ value.$$

The rest of the unloading valves were selected as R–20 valves with $^5/_{16}$ in. chokes. Valve setting data are listed in Table 6–3.

6.7 Optimization of Intermittent Installations

Optimum conditions for intermittent gas lift wells are often defined as those resulting in the greatest liquid production rate with the minimum of lift gas consumption. When trying to find these conditions, one meets at least two basic problems.

(a) how well inflow under the transient conditions of intermittent lift can be calculated

(b) how lift gas requirement of the transient intermittent cycle can be found

Both of these problems involve flow phenomena of transient nature and of great complexity and, more importantly, require the knowledge of many parameters (inflow performance, gas breakthrough, etc.) only approximately known in most cases. These are the most important reasons why, in contrast to continuous flow gas lift optimization, there are no generally accepted optimization procedures available for intermittent gas lifting in the industry.

The simulation of the intermittent gas lift cycle was done by several investigators [7, 12, 22], and the first mechanistic modeling study was also published [13]. Against all efforts—mainly due to the complexity of the problem—intermittent lift optimization is still based on experience. A comprehensive practical procedure is presented in the API RP 11V5 publication [17], detailed in the following. It is based on the practical observation that liquid fallback is always at a minimum if more than the required gas volume is injected during the cycle. This means that the required cycle number is to be found first by changing the gas injection frequency but always maintaining an over-injection condition. Then, by decreasing the injection volume, the minimum gas requirement will be found.

6.7.1 Intermitter control

The basic objective of efficient intermittent lift (to produce the required liquid rate with a minimum injection gas requirement) can be reached with the proper selection of the cycle frequency and the duration of gas injection. These settings of the intermitter should be checked periodically to ensure efficient operations. Many times, when using an unbalanced operating valve, gas consumption can be excessive due to an improperly selected valve spread. In such situations, before pulling the valve and running another with the proper spread, decreasing the cycle frequency should be considered first. This is recommended because the increased accumulation time results in a higher tubing pressure and a consequently lower spread of the operating valve. The volume of injected gas per cycle is thus reduced and this effect, combined with the well's increased production rate can greatly improve the efficiency of fluid lifting.

The following procedure is recommended for determining the optimum cycle frequency and duration of gas injection for installations where a simple intermitter is used as surface control.

1. First, adjust the duration of gas injection by selecting an injection time that ensures a gas injection volume per cycle that is more than normally required. A good starting value is 500 cu ft per 1,000 ft of lift. Many times, an injection time lasting until the slug reaches the surface gives a more than needed injection volume.

2. Now reduce the daily cycle number until the well does not make the desired liquid rate.

3. The proper cycle frequency is the one immediately before the setting reached in the previous step. The final number of cycles should be set accordingly.

4. The duration of gas injection should be reduced now until the well's production rate decreases. The final setting should be 5–10 seconds longer than the last injection time that provided the desired liquid rate in order to compensate for the effects of injection pressure fluctuations.

The previous procedure works for cases with relatively constant line pressures. If surface line pressure fluctuates significantly, it must be ensured that the required gas volume is injected even with a minimum of line pressure. Since this solution inevitably results in over-injection for those cycles when line pressure is greater than the minimum, the use of a regulator upstream of the intermitter is recommended, see Section 6.3.

6.7.2 Choke control

For choke control of the surface gas injection, the type of the operating valve and its settings should be selected in accordance with the well's annular volume. The valve must have a considerably greater port size than the size of the surface choke to allow injection pressure to decrease to the valve closing pressure.

Setting the daily number of cycles and the injection volume per cycle is more complicated than for intermitter control. The number of *cycles per day* is adjusted by changing the size of the surface choke that affects the time required for the casing pressure to reach the opening pressure of the operating gas lift valve. Well inflow performance, too, has a direct impact on cycle frequency, because the rate of liquid slug buildup in the tubing is governed by the inflow to the well. For optimum conditions, the size of the surface choke is selected so that liquid inflow from the well coincides with the annulus pressure buildup, and the operating valve's opening production and injection pressures are reached at the same time.

The volume of gas injected during the intermittent cycle depends on the spread of the operating gas lift valve. This must be properly selected for the well's annular volume where gas storage during the accumulation period takes place. But valve spread is a function of tubing pressure as well, and it decreases as tubing pressure increases. For optimum operations, therefore, tubing pressure is allowed to increase due to the inflow from the formation to such a level when the valve's spread is sufficient to provide the required injection gas volume.

The recommended procedure for selecting the optimum surface gas injection choke size is given as follows:

1. Use a choke size slightly larger than would be necessary to inject the design injection gas volume.

2. In small increments, reduce the choke size until the well's production rate decreases.

3. The final choke size to be used is the one that gave the optimum fluid production with the least number of cycles, which should also be the least injection gas consumption.

References

1. *Gas Lift.* Book 6 of the Vocational Training Series. Second Edition. Dallas, TX: American Petroleum Institute, 1984.

2. *Field Handbook for Intermitting Gas Lift Systems.* Teledyne MERLA, Garland TX, 1977.

3. *Field Operation Handbook for Gas Lift.* OTIS Engineering Corporation, Dallas TX, 1979.

4. Reis, P. J., "Microprocessors Play a Major Role in Optimizing the Intermittent Gas Lift System in the Ventura Avenue Field." *JPT*, April 1985: 696–700.

5. Winkler, H. W. and S. S. Smith, *Gas Lift Manual.* CAMCO Inc., 1962.

6. Brown, K. E. and F. W. Jessen, "Evaluation of Valve Port Size, Surface Chokes and Fluid Fall-Back in Intermittent Gas-Lift Installations." *JPT*, March 1962: 315–22.

7. White, G. W., B. T. O'Connell, R. C. Davis, R. F. Berry, and L. A. Stacha, "An Analytical Concept of the Static and Dynamic Parameters of Intermittent Gas Lift." *JPT*, March, 1963: 301–8.

8. *Guiberson's Ratiometric Gas Lift Systems.* Dresser-Guiberson, Dallas, TX, 1962.

9. Brown, K. E. *The Technology of Artificial Lift Methods.* Vol. 2a. Tulsa, OK: Petroleum Publishing Co., 1980.

10. Davis, J. B., P. J. Thrash, and C. Canalizo. *Guidelines To Gas Lift Design and Control.* 4th Edition, OTIS Engineering Corp-Dallas, TX, 1970.

11. Brown, K. E. and R. E. Lee, "Easy-to-Use Charts Simplify Intermittent Gas Lift Design." *World Oil*, February 1, 1968: 44–50.

12. Neely, A. B., J. W. Montgomery, and J. V. Vogel, "A Field Test and Analytical Study of Intermittent Gas Lift." *SPEJ*, October 1974: 502–12.

13. Liao, T., Z. Schmidt, and D. R. Doty, "Investigation of Intermittent Gas Lift by Using Mechanistic Modeling." Paper SPE 29454 presented at the Production Operations Symposium held in Oklahoma City, OK, April 2–4, 1995.

14. Bradley, H. B. (Ed.) *Petroleum Engineering Handbook.* Chapter 5. Society of Petroleum Engineers, 1987.

15. *Gas Lift Manual.* Section 3: Intermitting Gas Lift. Teledyne MERLA, Garland TX, 1970.

16. *Handbook of Gas Lift.* Intermittent Opti-Flow Section. Axelson, U.S. Industries, Inc., 1959.

17. "Recommended Practice for Operations, Maintenance, and Trouble-Shooting of Gas Lift Installations." *API RP 11V5*, 2nd Edition, American Petroleum Institute, Washington, D.C., 1999.

18. Davis, J. B., "Deep Chamber Lift." Paper presented at the 8th Annual West Texas Oil Lifting Short Course, held in Lubbock, TX, April 20–21, 1961: 29–33.

19. Winkler, H. W. and G. F. Camp, "Down-Hole Chambers Increase Gas–Lift Efficiency." Part One, *PE*, June, 1956: B-87–100.

20. Winkler, H. W. and G. F. Camp, "Down-Hole Chambers Increase Gas–Lift Efficiency." Part Two, *PE*, August, 1956: B-91–103.

21. Winkler, H. W., "Re-Examine Insert Chamber-Lift for High Rate, Low BHP, Gassy Wells." Paper SPE 52120 presented at the Mid-Continent Operations Symposium held in Oklahoma City, OK, March 29–31, 1999.

22. Schmidt, Z., D. R. Doty, P. B. Lukong, O. F. Fernandez, and J. P. Brill, "Hydrodynamic Model for Intermittent Gas Lifting of Viscous Oil." *JPT*, March 1984: 475–85.

7 | Plunger-assisted Intermittent Lift

7.1 Introduction

As discussed in chapter 6, during the lifting of the liquid slug in intermittent gas lift, gas breakthrough at the bottom of the slug results in fallback, *i.e.* part of the liquid volume falls back and is not produced to the wellhead. Fallback is a natural phenomenon caused by the great difference in density between the liquid and gas phases making the gas bubble below the liquid slug penetrate the liquid. In a deeper well, it can also happen that fallback consumes the total starting slug length and no liquid is produced at the wellhead. Reduction or elimination of fallback by inserting a free piston called *plunger* between the gas and liquid slugs, therefore, can greatly improve the efficiency of intermittent gas lift. Such installations are very similar to plunger lift installations used since the 1920s to produce oil wells or remove liquids from gas wells.

Plunger lift proper (not covered in this book) is a kind of artificial lift utilizing the energy of formation gas to periodically lift liquid slugs accumulated at the bottom of the tubing string. Thanks to the open installation used, formation gas is trapped in the well's annulus until casing pressure builds up to a level sufficient to lift the liquid slug to the surface. Gas from the annulus is injected at the bottom of the tubing string, below the special plunger rising along with the liquid slug, and this prevents an excessive amount of fallback to occur. Intermittent operation is ensured by the periodical opening and closing of a flow valve at the tubinghead. These installations are a low-cost solution to unload gas wells and to produce high GOR, low production rate oil wells. [1]

This chapter deals with the special form of intermittent gas lift utilizing plungers, which may be called plunger-assisted gas lift. Although related to plunger lift, such installations have many distinguishing features but basically are intermittent gas lift installations. The use of plungers, by almost completely eliminating gas breakthrough and liquid fallback, achieves the most efficient form of intermittent gas lift production available. [2]

The main application areas of plunger-assisted intermittent lift are similar to those of conventional intermittent gas lifting (see chapter 6) with the following additions.

- Decreasing the injection gas requirements of intermittent installations with average reductions from 30% to 70% is made possible. [3]

- Application of intermittent lift in deep wells with relatively low surface injection pressures becomes possible. Many times, because of the great amount of fallback in deep wells, no liquid recovery at all is possible without a plunger. In such cases, the use of plungers might be the only way to economically produce the well.

- In wells where the flowing wellhead pressure is excessive after the surfacing of the liquid slug because of a small or too long flowline, etc. the use of plungers increases liquid recovery.

- In wells with emulsion problems, the use of plungers can increase the efficiency of fluid lifting.

- If paraffin or other solids deposit in the tubing, the continuous cleaning action of the plunger provides for the trouble-free operation.

The advantages of using plungers in intermittent gas lift installations are the following:

- Liquid production rates can be increased due to a reduction of liquid fallback.

- An increase in lifting efficiency is observed as compared to conventional installations because a considerable decrease of gas injection requirements is achieved.

- Paraffin or emulsion problems are eliminated.

Limitations include the following:

- In wells requiring high daily cycle numbers, the fall velocity of the plunger may limit the cycle number and the liquid production rate.

- Although special snake plungers are available, well deviation and doglegs may obstruct or prevent the plunger's movement.

- Wells making sand may prohibit the use of plungers.

- If a wireline retrievable unloading valve string is used, special tandem plungers have to be selected.

7.2 Equipment Considerations

7.2.1 Installation types

The most common plunger-assisted intermittent gas lift installation is depicted in Figure 7–1. It is basically a closed gas lift installation with a packer and a standing valve set close to the depth of the perforations. A bumper spring absorbing the kinetic energy of the descending plunger is installed on a tubing stop or in a seating nipple just above the setting depth of the operating gas lift valve. The operation of this installation is very similar to a conventional intermittent installation with the only difference that the plunger rises on top of the lift gas and falls after the pressure in the tubing string is bled down.

A special downhole installation (see Fig. 7–2) is used for extremely low formation pressures where no packer is set in the well and lift gas is supplied to the operating depth by a dedicated conduit. The casinghead is open to the atmosphere or connected to a vacuum pump, thus maintaining a very low flowing bottomhole pressure. This solution allows relatively high production rates to be achieved from great depths and provides a very economic way of producing wells near their abandonment.

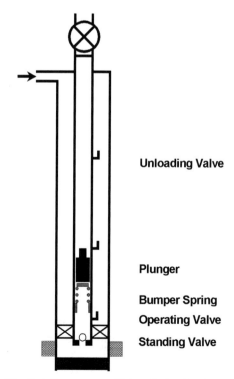

Unloading Valve

Plunger

Bumper Spring
Operating Valve
Standing Valve

Fig. 7–1 Common gas lift installation for plunger-assisted intermittent lift.

Other installations are also possible with plungers; one recent solution combines chamber lift with plunger lift and contains a CT string where the plunger operates. [4]

7.2.2 Surface equipment

Wellhead Arrangement

The usual wellhead arrangement for a plunger-assisted intermittent gas lift installation is shown in Figure 7–3. Lift gas is injected into the casing-tubing annulus through a surface intermitter, as shown, through a simple choke or through a choke and a regulator. The master valve must be sized correctly for the tubing and plunger with a full bore opening equal to, but not greater than, the tubing size. On top of the master valve, a special lubricator should be installed for several reasons discussed as follows. A single or preferably a double flow outlet as shown in the figure allows produced fluids to move into the flowline. [5] The flow outlets must contain properly sized chokes to ensure that the plunger lifts past the lower outlet on arrival to the surface.

Lubricator

Lubricators fulfill three important functions:

1. The high kinetic energy of the plunger arriving at the wellhead is absorbed by a strong bumper spring connected to a strike plate.

2. For periodical inspection, plungers are removed from the well but have to be first caught in the lubricator by a catcher mechanism that allows the plunger to enter the lubricator but prevents it from falling back into the well.

3. The plunger can be inserted into the well or retrieved for service by removing the upper part of the lubricator.

The schematic of a common lubricator with a single flow outlet is given in Figure 7–4. For use with plungers with bypass valves, an actuator rod may be needed; the actuator rod is installed inside the bumper spring to open the plunger's bypass valve on arrival.

7.2.3 Subsurface equipment

Tubing

The tubing is usually run close to perforation depth because of the low formation pressures of the wells and to maximize production. To ensure the free travel of the plunger, the tubing string should be free of bends, tight spots, etc. and must be gauged before the plunger is inserted. The broach should have a length equal to the length of the longest possible plunger. This provides for an unobstructed travel of the plunger.

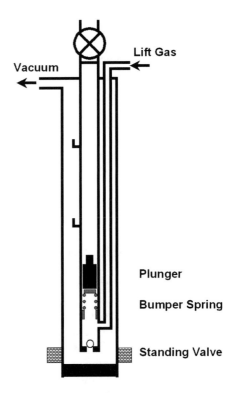

Fig. 7–2 Plunger-assisted intermittent lift installation with a parallel injection string.

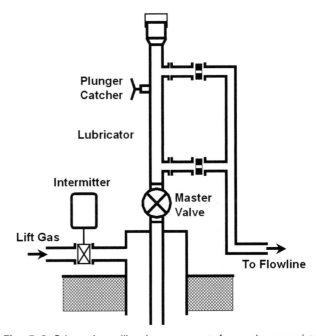

Fig. 7–3 Schematic wellhead arrangement for a plunger-assisted intermittent gas lift installation.

Standing Valve

A standing valve should be run near the tubing bottom to prevent well fluids from being forced by high-pressure injection gas back into the formation. It also helps to keep the average flowing bottomhole pressure at a low level and maximize production.

Bottom Bumper Spring

A strong spring set in the tubing string close above the standing valve absorbs the kinetic energy of the plunger falling at a high speed to the bottom. The bumper spring can be separately set on a tubing stop or seating nipple; or an integral bumper spring-standing valve unit as shown in Figure 7–5 is used.

Plunger

The plunger is the heart of the installation and must meet several criteria, depending on the main requirements of the operation. It should tolerate the high shock loads on arrival to the bottom and surface bumper springs, should be resistant to wear and tear, should provide for the maximum of liquid slug recovery, etc.

Solid plungers are solid steel cylinders with smooth or grooved surfaces and have low fall velocities, especially in a liquid column. In wells operating with greater daily cycle numbers, greater fall velocities are desired and plungers with a bypass valve are used. A bypass plunger (shown in Fig. 7–6) contains a valve that, if open, allows the flow of liquids through the plunger body, thus achieving a high fall velocity. The bypass valve of this plunger is opened at the surface by an actuating rod situated in the lubricator and is closed on hitting

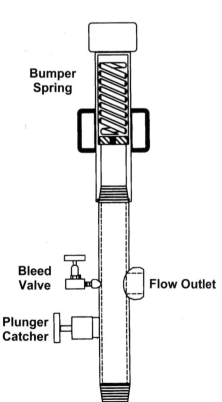

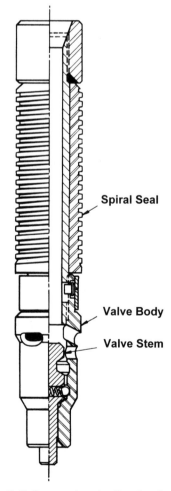

Fig. 7–4 Basic components of a lubricator.

Fig. 7–5 Integral standing valve-bumper spring arrangement.

Fig. 7–6 Construction details of a bypass plunger.

the striker plate as it arrives to the bottom bumper spring. Plungers with integral valve rods are also available; they open and close the bypass valve on arrival to the striker plates without the need of an actuating rod.

Although a perfect seal on the tubing inside wall is not required, the sealing surfaces of plungers are designed to provide a minimum of liquid fallback across the plunger. One of the earliest plunger types uses solid metal pads that are interlocked and spring loaded (see Fig. 7–7) to provide a close mechanical seal on the tubing inside wall. Sealing surface with brush segments, spiral seals, wobble washers are available for different purposes. Tandem plungers, *i.e.* two plungers connected by a rod 2–3 ft long are needed where wireline retrievable gas lift valves are used in the unloading valve string. Special flexible plungers are required in CT applications and in directional wells because they can easily negotiate even sharp tubing bends.

7.3 Operating Conditions

7.3.1 The intermittent cycle

Fig. 7–7 Solid plunger with metal pads.

Being essentially a version of intermittent gas lift, plunger-assisted intermittent lift operates and behaves very similarly to conventional intermittent gas lifting. The intermittent cycle can be divided into three distinct periods, similarly to conventional intermittent gas lift (see Section 6.4.1).

The *Slug Lifting Period* starts with the opening of the operating gas lift valve when lift gas injected below the plunger resting on the bottom bumper spring with a closed bypass valve (if it has one) displaces the liquid slug along with the plunger to the surface. During the slug's upward travel, gas breakthrough and liquid fallback are almost completely eliminated by the action of a properly selected plunger. This period can further be divided into three phases:

1. The slug acceleration phase lasts only a few seconds because the slug very rapidly reaches its maximum velocity.

2. The next phase lasts until the top of the slug reaches the wellhead and is characterized by a constant but small deceleration of the slug. The drop in slug velocity is caused by the interaction of the continuous decrease of gas pressure below the plunger and the largely constant fluid load.

3. After the liquid enters the wellhead, slug velocity starts to increase rapidly due to the decrease of fluid head on the plunger.

The *Gas Blowdown* or *Afterflow Period* starts when the plunger hits the lubricator and lasts until the operating gas lift valve closes. If a plunger with a bypass valve is used, the bypass valve is opened upon impact to the strike plate (in case of plungers with integral valve rods) or to the actuator rod (in case of plungers without integral valve rods). The plunger is kept in the lubricator by the pressure of the tail gas and only falls when gas pressure is insufficient to keep it floating. The high-pressure tail gas, just like in conventional intermittent installations, while bleeding down to the flowline, may carry a significant amount of liquid in the form of entrained liquid droplets.

The *Accumulation Period* starts when the standing valve opens and liquid inflow into the tubing commences. The plunger with its open bypass valve (if it has one) falls to the bottom bumper spring with the impact closing the bypass valve. The cycle repeats when the operating gas lift valve opens again.

7.3.2 Calculation of operating parameters

The operating conditions of plunger lift were studied by a multitude of investigators, and several procedures are available to describe its operation and design including the classical work of Foss-Gaul. [6] The operation, surface control, and design of plunger-assisted intermittent lift installations, however, significantly differ from those of plunger lift proper and are closer to those of conventional intermittent gas lifting. This is the reason why in the following paragraphs only the most significant differences between conventional and plunger-assisted intermittent lifting will be discussed.

The major difference in calculating the operating parameters for conventional and plunger-assisted intermittent lift is the way liquid fallback is treated. Flow conditions are dramatically changed by the presence of the plunger between the liquid slug and the lift gas, and gas breakthrough—responsible for liquid fallback—is greatly reduced. Based on more than 150 tests with different types of plungers, White [7] concluded that an average fallback of 10% can be reached with most types. His results are plotted in Figure 7–8, showing measured fallback values for four different plunger constructions and comparing them to the case of conventional intermittent gas lift. As seen, the use of plungers dramatically decreases the amount of liquid lost due to fallback during the intermittent cycle. His data show that the greatest improvement is reached when using a solid plunger with a capillary hole in its center.

The calculation of fallback during the upward travel of the plunger is made possible by the practical observation that the time rate of liquid fallback is a sole function of plunger rise velocity. Mower et al. [8] experimentally developed these functions for several different types of plungers. The use of their data allowed the development of the first mechanistic model that describes the behavior of plunger-assisted intermittent lift installations. [9]

In wells where a high daily cycle number is desired, the plunger's fall velocity may limit the increase of cycle frequency. In such cases, plungers with a bypass valve are recommended because their fall velocity is extremely high due to the low flow restriction across the open bypass valve. Fall velocities of several different plungers were measured by Rowlan-McCoy-Podio [10], whose data can be used in installation design calculations.

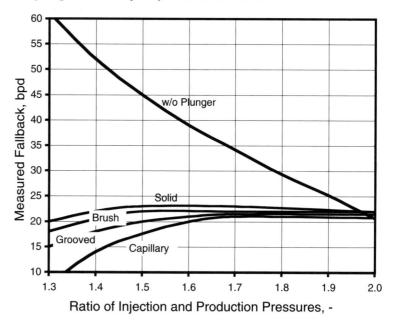

Fig. 7–8 The effect of plunger type on liquid fallback, according to White. [7]

References

1. Brown, K. E. *The Technology of Artificial Lift Methods*. Vol. 2a. Tulsa, OK: Petroleum Publishing Co., 1980.

2. *Gas Lift*. Book 6 of the Vocational Training Series. Second Edition, Dallas, TX: American Petroleum Institute, 1984.

3. Bradley, H. B. (Ed.) *Petroleum Engineering Handbook*. Chapter 5. Society of Petroleum Engineers, 1987.

4. Rogers, J., "Plunger Enhanced Chamber Lift." Proceedings from 50th Southwestern Petroleum Short Course, 2003: 189–96.

5. *Sub-Surface Plunger Handbook*. McMurry Oil Tools, Houston, TX, 1975.

6. Foss, D. L. and R. B. Gaul, "Plunger-lift Performance Criteria with Operating Experience—Ventura Avenue Field." *API Drilling and Production Practice*, 1965: 124–40.

7. White, G. W., "Combine Gas Lift, Plungers to Increase Production Rate." *WO*, November 1982: 69–76.

8. Mower, L. N., J. L. Lea, E. Beauregard, and P. L. Ferguson, "Defining the Characteristics and Performance of Gas-Lift Plungers." Paper SPE 14344 presented at the 60th Annual Technical Conference and Exhibition held in Las Vegas, NV, September 22–25, 1985.

9. Chacin, J., Z. Schmidt, and D. Doty, "Modeling and Optimization of Plunger Lift Assisted Intermittent Gas Lift Installations." *SPE Advanced Technology Series*, Vol. 2, No. 1 (1993): 25–33.

10. Rowlan, O. L., J. N. McCoy, and A. L. Podio, "Determining How Different Plunger Manufacture Features Affect Plunger Fall Velocity." Paper SPE 80891 presented at the Production and Operations Symposium held in Oklahoma City, OK, March 22–25, 2003.

8 | Dual Gas Lift

8.1 Introduction

In order to reduce drilling and completion costs, often several productive formations are produced through a single borehole. This is especially desirable when many formations are located vertically close above one another, and considerable savings on drilling and completion costs can be realized if several zones are produced simultaneously. Although theoretically possible, multiple installations tapping into more than two zones are seldom produced by gas lift. However, dual gas lift, defined as the simultaneous production by gas lifting of two formations opened in the same well, is a time-proven part of gas lift technology. [1–3]

Generally, if several formations are produced from one well, the main rule is that each zone must be produced independently from the others. This is because only through continuous monitoring of the produced fluid amounts can reservoir engineers simulate the behavior of the field. Since commingling of wellstreams (as shown previously) is not recommended, dual gas lift installations have to contain two tubing strings dedicated to the two productive formations. The schematic of the generally used dual installation is shown in Figure 8–1, with two packers isolating the two formations. Both the short string producing the upper zone and the long string producing the lower zone are equipped with their individual operating valves and unloading valve strings. Both production strings share a common annulus, and this is the main source of many problems associated with dual gas lift.

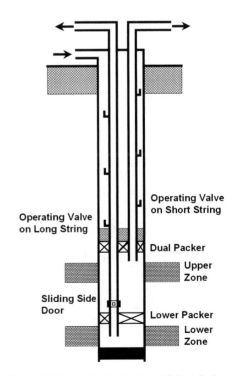

Fig. 8–1 Schematic of a dual gas lift installation.

8.2 General Considerations

Compared to single-zone gas lift installations, there are a host of special problems associated with dual gas lift that need to be properly addressed during installation design and operation. Most of them are related to the space restrictions in the well, as shown in the discussions given here.

Casing Size

The size of the casing string is the greatest constraint on dual gas lift installations. Usually, it is smaller than would be required to accommodate two tubing strings of the proper size. This is why, in most of the cases, tubing strings of less than optimum sizes have to be run, limiting the production rates from both zones.

A common combination of dual installations is two 2 $\frac{3}{8}$-in. tubing strings inside a 7-in. casing column. Because of the tubing size, one has to use 2 $\frac{3}{8}$-in. gas lift mandrels that can only hold 1-in. nominal size gas lift valves. These, however, are generally regarded to be inferior to 1 $\frac{1}{2}$-in. valves, especially in their fluid operated PPO version. All these lead to a less-than-ideal performance of the operating gas lift valve and cause restrictions on the available liquid production rate, especially in high producers.

In addition to the previous problems, too-small casing sizes with a limited cross-sectional area of the casing-tubing annulus can affect the pressure distribution of injection gas in the annulus. In such cases, frictional losses cannot be disregarded as done in larger casings, and the design of unloading strings and the operating valve have to be modified accordingly.

Retrievable Equipment

The use of wireline retrievable equipment (gas lift valves, standing valves, etc.) is highly recommended for dual wells where, because of the more complex well completion, workover operations are more costly. Original installation, replacement, and change of gas lift valves, as well as other pieces of equipment, are easier and less expensive than for conventional equipment. At the same time, special operational practices have to be followed for wireline work because, due to the different formation pressures of the two zones, crossflow between the formations can occur when a valve is pulled. Valve seats can be cut out as a result of the high-velocity liquid jets formed.

Tubing Joints and Valve Mandrels

In casing strings of relatively smaller diameter, tubing running and pulling operations can be cumbersome. Beveling of tubing couplings or the use of integral joints is recommended to prevent tubing running and pulling problems.

If tubing pulling costs are moderate due to a shallow well depth, conventional gas lift mandrels can be run that are designed to pass the tubing collars and mandrels of the other string. In any case, the spacings of unloading valves on the two tubing strings must be checked to avoid problems during running-in of the second tubing string. The principle is that no two mandrels should pass two mandrels on the other string at the same time. Valve spacings should be modified accordingly by moving the given mandrel one tubing joint deeper or higher.

Distance between Zones

As seen in Figure 8–1, the deepest point of gas injection in any of the two zones of a dual gas lift installation can only be situated above the depth of the dual packer. Since the vertical distance between the formations produced by the individual strings is seldom excessive, the lower zone can efficiently be gas lifted with its operating valve above the upper packer. However, if the lower zone is situated much deeper, its operating gas lift valve cannot be run to the desired depth, and the efficiency of lifting that zone can be low. This is why for zones widely spaced apart, special dual installations are required.

Unloading

Unloading a dual gas lift installation is slightly different from unloading a single-zone installation because removal of liquids from the annulus can be accomplished through either of the two unloading valve strings. In practice, the operator should choose that string to remove annulus liquids from the well whose bottom mandrel is at a greater depth; this may be either the short or the long string. During unloading the annulus, the other string should be closed in and can be put back into production after the first has started to produce. This implies that only valve dummies need to be installed in the mandrels of the other unloading string because the valves are not required for unloading. Although this is true, proper unloading design for a dual gas lift well should make it possible to unload the well from either side.

It is generally recommended that a circulating device be installed (usually a sliding side door) on the long string between the two packers for killing, cleaning, or treating of the upper zone (see Fig. 8–1). For similar reasons, a sliding side door above the upper packer can be installed on any of the two tubing strings.

Selection of Operating Valve Type

Because a common annulus is used for gas injection into both zones, any changes in injection pressure can affect the operation of both unloading strings. This is the reason why PPO valves are preferred over IPO valves as operating valves in dual gas lift installations. Since the opening and closing of the valve is controlled by production (tubing) pressure and the valve is insensitive to changes in injection pressure, the use of PPO valves in dual gas lift wells became standard practice.

Application of PPO gas lift valves in dual gas lift installations has its limitations.

- In contrast to IPO valves where the injection pressure distribution with depth is a well-known parameter, information on the flowing tubing pressures is much less accurate and reliable, introducing inaccuracies in valve spacing calculations.

- Gas passage areas in the small gas lift valves (1 in. nominal diameter) are reduced, especially when valves with crossover seats are used.

- The relatively small channels in valves with crossover seats are easily plugged by sand, dirt, or other solid particles.

- PPO valves cannot be used for wells with extremely low formation pressures because the fluid head above the valve could not be enough to open the valve.

In conclusion, for dual gas lift design, the use of other than PPO types of gas lift valves should also be considered.

8.3 Installation Design Principles

For dual gas lift design, the importance of accurate data is more pronounced than for single-zone designs. Knowledge of exact inflow performance data for each zone and the use of an accurate multiphase flow correlation are the most important points to ensure that the installation will work exactly as designed. Otherwise, the point of gas injection, casing, and tubing pressures may change, leading to severe operational problems.

The individual design of the unloading valve string and the operating valve for one string of a dual installation is completely identical to the procedure required for a similar single-zone installation. The result of the two designs for the dual installation, however, must ensure that both valve strings will operate independently of each other without any interference. This objective can be reached only through an analysis of the operation of the individual valve strings, and the conclusions drawn can be used to set up some basic design rules. In the following section, basic principles for dual gas lift design will be discussed [4–6] with references to the unloading valve string design procedures already presented in previous chapters of this book.

8.3.1 Both zones on continuous flow

Two Strings of IPO Valves

Since IPO valves are more sensitive to injection (casing) pressure, the two strings should be designed for a common surface opening pressure with both operating valves passing the desired amounts of gas into the individual zones. The proper distribution of lift gas volume into the two zones can be assured by the proper selection of the valve port sizes. Gas passage characteristics of chokes (see chapter 2) show that critical flow through one of the valves would allow quite big fluctuations in injection pressure without affecting the gas injection rate across the valve. The injection volume to the other zone, therefore, could be controlled by adjusting the surface injection pressure without any influence on the volume of gas injected into the first zone.

In order to come to the ideal solution discussed previously, the operating valve of one zone (usually the one with the shallower point of injection) is selected with its operating point near critical flow. Under such conditions, changes in the injection (casing) or the production (tubing) pressure cause only minor fluctuations in the gas rate injected through the valve. To achieve this behavior, the required port area is selected for the design gas injection rate and a pressure differential of 100–120 psi across the valve. The port size thus selected ensures that despite minor fluctuations in injection pressure, the valve will pass an almost constant volume of gas. The operating valve of the other zone must not be choked, and the injection volume to that zone is easily controlled at the surface without affecting the injection volume of the first zone. The unloading valve strings are usually spaced by assuming a constant surface opening pressure for both valve strings (see chapter 5).

The required surface injection control for this configuration is a simple choke on the injection line near the wellhead. If line pressure fluctuates, installation of a pressure regulator upstream of the choke is recommended.

One String of IPO and One String of PPO Valves

In this case, the two strings of unloading valves can easily operate independently of each other because IPO valves are opened by injection pressure whereas PPO valves open when tubing pressure increases. The IPO string can be designed with a constant surface opening pressure or with a constant drop in surface opening pressures. The PPO string, on the other hand, operates over a limited range of injection (casing) pressures and is spaced by assuming a constant surface opening pressure.

8.3.2 One intermittent and one continuous zone

Two Strings of IPO Valves

For the intermittent zone to operate properly, injection pressure should repeatedly be increased according to the daily cycle number. The operating valve of the zone on continuous flow, however, would inject more gas during the operation of the intermittent zone and an inefficient operation of this zone would occur. For proper operation of both zones, therefore, the operating valve of the formation on continuous flow gas lift must be insensitive to changes in surface injection pressure.

As before, the port size of the operating valve of the zone on continuous flow should be selected so that gas flow across the valve approaches critical conditions. Sizing the port for the required gas rate while maintaining a pressure differential of 100–120 psi across the valve can ensure this. This is the way to make sure that gas injection rates do not increase while the other zone is intermitting. Spacing of the unloading string for this zone should follow the procedure for a constant surface opening pressure (see chapter 5).

The operating valve of the intermittent valve string should have a large port area and a surface closing pressure higher than the operating pressure of the zone on continuous flow. The spacing of unloading valves should be based on a constant surface closing pressure (see chapter 6).

The required surface gas injection control arrangement is shown in Figure 8–2 and consists of an intermitter and a choke and a regulator on a bypass line. The choke and regulator provide the constant injection pressure for the zone on continuous flow, while the intermitter controls the cyclic operation of the other zone on intermittent lift.

One String of IPO and One String of PPO Valves

Installing PPO valves on the string producing the formation on intermittent lift gives proper operation if unloading valves do not require a drop in injection (casing) pressure to close. A simple pressure regulator on the surface is needed to control gas injection.

8.3.3 Both zones on intermittent lift

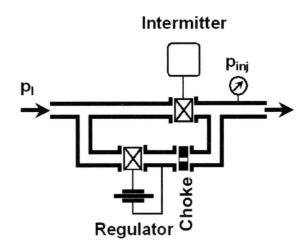

Fig. 8–2 Gas injection control for a dual well—producing one zone by continuous flow and the other by intermittent gas lift.

Two Strings of Pilot Operated Valves

The use of pilot-operated gas lift valves in intermittent lift installations is advantageous because these valves can have very large port sizes while their spread can be kept under control. By selecting the proper control port sizes, the operating valves of the two strings can be adjusted to open when sufficient liquid has accumulated in the tubing string. Valve spacing for the two zones is accomplished by assuming the same surface closing pressures for both strings (see chapter 6).

Surface injection control is simple and conforms to the choke and regulator control discussed in chapter 6. The regulator adjusts the valve opening pressures and the starting slug lengths, while the choke creates the drop in injection (casing) pressure necessary for the pilot valves to close.

Two Strings of PPO Valves

PPO valves can only be used in wells with sufficient *FBHPs* that can support sufficiently large liquid columns to open the operating valve. Both strings are designed for the same surface operating pressure and only a choke is required at the surface to control gas injection.

References

1. Davis, J. B., P. J. Thrash, and C. Canalizo. *Guidelines To Gas Lift Design and Control.* 4th Edition, OTIS Engineering Corp., Dallas, TX, 1970.

2. Brown, K. E. *The Technology of Artificial Lift Methods.* Vol. 2a. Tulsa, OK: Petroleum Publishing Co., 1980.

3. Winkler, H. W. and S. S. Smith. *Gas Lift Manual.* CAMCO Inc., 1962.

4. Winkler, H. W., "More Efficient Dual Gas Lift Installations." *WO,* May 1958: 157–71.

5. Davis, J. B. and K. E. Brown, "Attacking Those Troublesome Dual Gas Lift Installations." Paper SPE 4067 presented at the 47th Annual Fall Meeting held in San Antonio, TX, October 8–11, 1972.

6. Davis, J. B. and K. E. Brown, "Optimum Design for Dual Gas Lift." *PE,* July 1973: 36–9.

9 | The Gas Lift System

9.1 Introduction

Placing all or some of the wells in an oilfield on production by gas lift requires the modification of existing facilities as well as the additional installation of new surface and downhole equipment. The gas lift system comprises all the equipment and facilities necessary for a proper operation of the field by gas lifting. To ensure a profitable oil production over time, the gas lift system should be properly designed and operated.

9.1.1 Functions of the gas lift system

A properly designed and operated gas lift system fulfills the following basic functions:

1. High-pressure lift gas at the proper pressure is provided for gas lifting. Usually a compressor plant is used for this purpose, although high-pressure gas from a pipeline or a gas well can also be utilized.

2. Lift gas of the proper amount and pressure must be made available at every wellsite. A high-pressure gas distribution system with the right control devices is, therefore, an important part of the gas lift system.

3. Gathering of wellstreams containing lift and formation gas coming from the wells, as well as separation of gas from produced liquids is done in the gathering system—another important component of the total gas lift system.

4. The gas lift system should operate under optimum economic conditions and this can be reached if

   • the system was designed for the most economic injection pressure for all the wells involved

   • the compressor plant is fully loaded at all times

- gas volume loss in the system is kept at a minimum

- make-up gas requirements from outside sources involve a minimum of costs

- excess gas produced by the wells and not needed for the proper operation of the system is sold

9.1.2 Types of gas lift systems

Based on field conditions as well as economic considerations, different types of gas lift systems can be designed. The most important condition is the availability of a low-pressure gas source that could provide all the gas necessary for gas lifting. In an open system, as shown in Figure 9–1, low-pressure gas taken from an existing gas line after compression and its use in the gas lifted wells is returned to the gas line at a low pressure. Since gas is always available at the compressor's suction, and all gas leaving the gathering system is led to the gas line, there are no capacity problems in the distribution or the gathering systems.

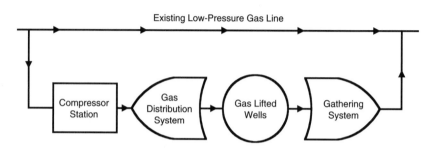

Fig. 9–1 Components and schematic arrangement of an open gas lift system.

A semi-closed system has part of the gas received in the separators led back to the suction of the compressor but a continuous supply of low-pressure gas to the system is still needed. This is because the gathering system's gas storage capacity is not sufficient to keep a steady suction pressure.

Finally, the closed system, called a *closed rotative gas lift system*, returns all gas from the separators back to the suction of the compressors. [1, 2] This system, as shown on the schematic flow diagram in Figure 9–2, requires outside gas for initial make-up only, after which it operates without any outside supplement of gas. After the system is charged with gas at startup, its operational gas requirement equals the sum of the gas rate required for fuel and the gas rate covering leakage losses in system components. These gas volumes, compared to the circulated gas volume, are quite small and are, in the majority of cases, easily covered by the formation gas produced from the wells. If the total amount of formation gas is greater than the gas requirement of the system, gas can be sold. Gas sales can be made at a low pressure from the gathering system or at a high pressure from the compressor discharge.

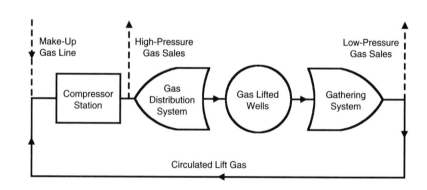

Fig. 9–2 Schematic of a closed rotative gas lift system.

The proper operation of a closed rotative gas lift system requires that both the gas distribution and the gathering systems be designed with enough gas storage capacity to prevent flaring of gas as well as starving of the compressors. The ways to provide proper gas storage capacities and to ensure a continuous supply of gas to the compressors is discussed later in this chapter.

9.1.3 Gas lifting vs. field depletion

It is easy to see, based on the principles of gas lifting, that an oil well can be produced by gas lift from the moment it dies until its total depletion or abandonment. This means that a gas lift system, if properly designed originally, can be used for the entire life of the field with a minimum or no modifications. System design, therefore, should be done with the future production history of the field in mind.

Of the formation parameters changing with the depletion of the field, formation pressure and formation gas production are the most significant ones for gas lifting. Their variation with production time mostly depends on the reservoir depletion mechanism like solution gas drive, water flooding, etc.

Investigating the effects of formation pressure, one can conclude that even a heavily declining reservoir pressure does not pose a problem to the application of gas lifting. With gas lift, FBHP can continuously be reduced to comply with the drop in formation pressure. For the initially greater reservoir pressures, continuous flow gas lift is selected and then conversion to intermittent lift is recommended when formation pressure drops. For extremely low formation pressures, usually experienced before well abandonment, chamber lift or plunger-assisted intermittent lift can be economically applied.

A declining production of formation gas from the field is not a problem as long as it is sufficient to cover fuel requirements (assuming that gas engines drive the compressors) and leakage losses. In the majority of cases, this condition is met and the closed rotative gas lift system operates without any outside gas supply. Fields with low or negligible formation gas production, however, can only be produced by gas lifting if a different source of lift gas is available. This can be accomplished, for example, by converting the exhaust gases of engines to an inert gas and then using that gas for gas lifting. The process described by Kastrop [3] produces an inert, noncorrosive, inflammable gas mixture of high N_2 content that can effectively be used as lift gas in remote areas. Pure nitrogen, produced from air by membrane technology can also be used for gas lifting individual wells or whole fields alike. [4]

9.2 Operation of Gas Lift Systems

9.2.1 System components

For proper system operation, the closed rotative gas lift system schematically depicted in Figure 9–2 must contain several specific components and a host of control devices at the right locations. Selection of the location and the operating pressures of the pressure control devices are extremely important for the efficient operation of the system. They are shown in Figure 9–3, where the most significant components and control devices are indicated for a gas lift system including a single well on intermittent lift. Since such a system represents the greatest challenges in design and operation, understanding of its construction and operation details is essential.

Figure 9–3 illustrates a gas lift system with a compressor station driven by gas engines and specifies, in addition to the necessary piping, the locations of pressure control devices crucial for the proper operation of the system. These pressure regulators are numbered and will be referred to in the following discussion. Two storage vessels, one in the gas distribution and one in the gathering system are necessary to provide proper operating conditions for the compressor. The flow diagram includes all the piping and controls necessary for (a) providing make-up gas, (b) low-pressure gas sales, and (c) high-pressure gas sales.

Controls in the Distribution System

The injection gas distribution system includes all pipes, storages, other vessels, etc. containing lift gas between the discharge flange of the compressor and the surface intermitter of the well. Since the greatest possible storage capacity

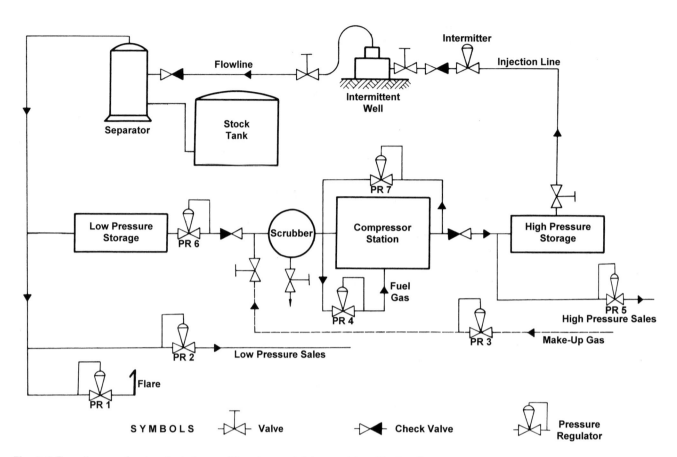

Fig. 9–3 Flow diagram of a closed rotative gas lift system containing one intermittent well.

of the distribution system is desired, the intermitter must always be placed at the wellhead and not at a tank battery. The latter solution not only decreases the storage capacity but may also result in a slow action of the operating gas lift valve and in an increased lift gas usage. The well's annulus must not be considered part of the distribution system except when a choke is used for injection control at the wellsite. In this case, an unbalanced gas lift valve controls the timing of gas injection, and the well's annulus is utilized for gas storage between cycles.

The most important control device in the distribution system is situated on the bypass line connecting the compressor's discharge to its suction. The pressure regulator designated *PR* 7 in Figure 9–3 allows one to set the desired discharge pressure of the compressor station. Normally, this pressure is defined as about 100 psi greater than the injection pressure used in gas lift design calculations, in order to provide the necessary safety and to cover for frictional pressure drops occurring in the gas lines of the distribution system. The purpose of this pressure control is to prevent overloading of the compressors when for some reason (shutdown of wells, etc.) gas pressure in the distribution system increases to dangerous levels.

In the single-well gas lift system discussed here, it is absolutely necessary to include a high-pressure storage, as indicated in Figure 9–3, in the gas distribution system to store gas between intermittent cycles. This storage can be a section of an unused pipeline, the annulus of a well, etc. As the number of wells in the system increases, the capacity of the distribution system proportionally increases and the need for additional gas storage diminishes.

In cases when the well's formation gas production exceeds the operational make-up gas requirements, high-pressure gas can be sold through pressure regulator *PR* 5, whose operational pressure is set below that of *PR* 7. This setting ensures that excess gas is sold as long as there is sufficient pressure in the distribution system.

Controls in the Gathering System

The gathering system includes the separator, all pipes, storages, the scrubber, etc. from the check valve on the well's flowline to the suction flange of the compressor. Since the check valve on the flowline is usually installed near the tank battery, the flowline itself is not part of the gathering system. If gas is sold at low pressure, the necessary pipe is also part of the distribution system. This is why its pressure regulator (*PR* 2 in Fig. 9–3) should be installed as far as possible from the compressor to increase storage capacity. The same holds for the piping used for make-up gas, and pressure regulator *PR* 3 must be placed close to the source of make-up gas.

Gas from the separator enters the low-pressure storage that is always required for a single-well system shown in Figure 9–3 and can be made of a section of an unused pipeline, etc. Its function is to store lift plus formation gas coming from the well at a high rate after the liquid slug surfaces and to prevent the venting of gas. Just like in the distribution system, with an increased number of wells in the system, the gathering system's capacity also increases and less and less additional gas storage is needed.

The maximum pressure in the gathering system is set by the backpressure regulator *PR* 1 on the flare line provided there is no gas sales from the low-pressure system. If excess gas is sold, then the backpressure regulator on the sales line (*PR* 2) controls the separator pressure, and *PR* 1 on the vent line is set to a higher pressure to provide a safety relief should a dangerously high pressure develop in the gathering system.

The compressor's suction regulator is situated downstream of the low-pressure storage (designated as *PR* 6 in Fig. 9–3) and prevents overloading of the compressor. It is set to a pressure below the separator pressure and controls the lowest possible pressure in the gathering system. If system pressure drops below the minimum value thus determined, make-up gas enters through pressure regulator *PR* 3 (set to slightly below the minimum suction pressure). Starving of the compressor is prevented, and by connecting the make-up line downstream of the suction regulator *PR* 6, only a minimum of make-up gas volume is required to keep the suction pressure at a constant level. This is because no make-up gas enters the rest of the gathering system.

To prevent any liquids that might be present in the lift gas to enter the compressor, a gas scrubber is always installed immediately upstream of the compressor's suction flange. Finally, a line with pressure regulator *PR* 4 provides dry gas for the engines driving the compressor.

9.2.2 System operation

The closed rotative gas lift system described before contained only one well on intermittent lift, and this represented the worst scenario for system design. Systems containing several wells operate basically the same way as discussed for the single-well case, with gas storage capacity problems in the distribution and gathering systems becoming less and less pronounced as the total number of wells increases. Figure 9–4 depicts a schematic of a multi-well system where the system's main components along with the pressure regulators are designated identically to that used in the previous figure. The flow diagram includes the possibility of producing any well in the system to a pit, an advantageous feature during well unloading operations.

In order to describe the basic operation of the gas lift system, assume a multi-well system containing several wells on intermittent lift with intermitter control at the wellhead. Assume a time when most of the intermittent wells have just produced their liquid slugs to the surface. Then, since only a few wells are taking gas, the average pressure in the distribution system slowly rises but cannot exceed the set pressure on regulator *PR* 7 or *PR* 5, depending on whether gas is sold or not. Meanwhile, more and more wells start to inject gas and system pressure drops. However, if the gathering system contains the proper storage capacity as a result of an accurate design, injection pressure won't drop below the design injection pressure, and all wells will operate as originally designed. Therefore, the proper design of the gas distribution system ensures that injection pressure will fluctuate between the compressor's discharge pressure and the design injection pressure, which is a basic requirement for efficient operation of the gas lift system.

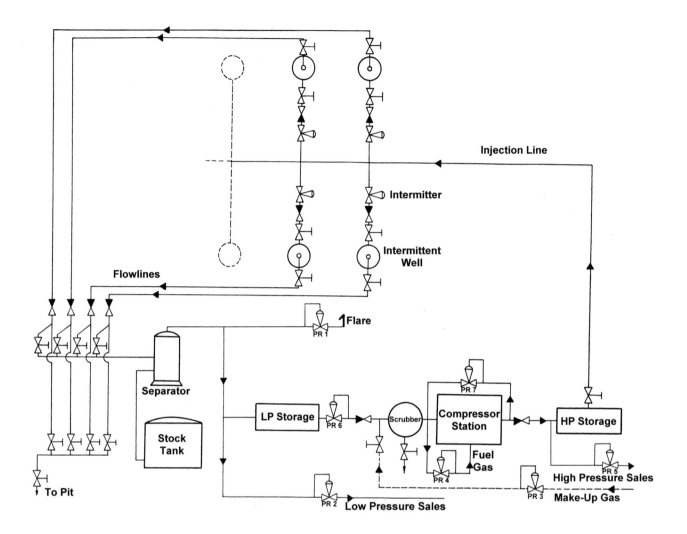

Fig. 9–4 Schematic flow diagram of a multi-well closed rotative gas lift system.

To illustrate the operation of the gathering system, assume that the majority of wells are just starting to produce. The great instantaneous gas rate entering the separator, because it exceeds the compressor's capacity, starts to build up pressure in the gathering system. The increase in gas pressure is limited by the pressure regulators on the sales or the flare line, *PR* 2 or *PR* 1. If the low-pressure storage capacity of the gathering system is properly designed, only excess gas leaves the system through the low-pressure sales line. As more and more wells cease to produce, system pressure drops, but assuming a proper design, it does not drop below the suction pressure of the compressor set on the pressure regulator *PR* 6. If, due to improper design or the simultaneous opening of too many intermitters, the pressure in the gathering system drops below the suction pressure, make-up gas will enter the system through backpressure regulator *PR* 3. By this time, more and more wells start a new intermittent cycle, and the increased flow of lift plus formation gas from the separator starts to build up system pressure again.

As demonstrated previously, an economic operation of the gas lift system is made possible by the proper design of the gas storage capacities of the distribution and gathering systems as well as the proper location and setting of the pressure control devices.

9.3 System Design

9.3.1 Introduction

The closed rotative gas lift system, as discussed in conjunction with Figure 9–2, has the advantage over other types that after being initially charged with gas, no additional make-up gas is needed for its operation. Although more complex than the other types, it is the natural choice if no special conditions (like a gas line nearby, etc.) exist in the field. This is the reason why the design of such systems is detailed in the following section.

If all the wells in the gas lift system are on continuous flow, system design is relatively simple. The compressor is to be sized for a constant injection gas rate calculated as the sum of the gas requirements of the individual wells. The amount of gas returning from the wells is also constant and equals the injected amount plus the formation gas produced. Pressures in the gas distribution and the gathering systems are fairly constant, making the operation of the compressor station ideal with constant suction and discharge pressures. This means that a very efficient gas lift system with no gas make-up requirement can easily be realized.

Wells on intermittent lift pose many problems for the design and operation of a closed rotative gas lift system. The gas demand pattern of these wells follows the intermittent cycle: gas injection is required for short periods of time only but at very high rates, while no gas injection is needed for longer periods. The sizing of compressors is troublesome because a large-capacity unit capable of supplying the peak gas rate would be required. Since this unit would be stalled for any time period when no gas injection is necessary, this is not a feasible solution. In practice, a compressor with a gas capacity equal to the daily average gas requirement is selected and the field's peak gas demand is covered by an additional gas injection coming from the high-pressure gas distribution system having the proper gas storage capacity. This solution is usually economically justified because the cost of increasing the gas distribution system's storage capacity is less than the cost of additional compressor power.

The previous considerations are only valid if injection control of the wells is done with surface intermitters. Choke control, as discussed in chapter 6 on intermittent gas lift, allows the storage of injection gas in the annulus of each well, and no storage in the distribution system is necessary.

Regardless of the surface injection control applied, lift plus formation gas is received in the low-pressure gathering system in short time periods and at relatively high rates following the production of liquid slugs to the surface. If the gathering system does not have the necessary storage capacity, part of this gas must be flared to prevent a dangerous increase of pressure in the system. Gas lost from the system must be supplied from an outside source through the make-up gas line, and this considerably increases operational costs. Proper design of the gas storage capacity of the gathering system, however, prevents the flaring of circulated gas and keeps the operation of the total system at optimum.

It is easy to see that the worst operating conditions for the compressor station occur if the closed rotative gas lift system contains only a single well on intermittent lift controlled by a surface intermitter. Increasing the number of intermittent wells and/or the addition of wells on continuous flow improves the operating conditions of the system, and large systems with several tens or hundreds of wells present smaller design problems. [5, 6]

9.3.2 Factors to consider

Many factors, the features of the field or group of wells, present and future production parameters, etc., must be considered before the actual design of the closed rotative gas lift system is performed. The interaction of the different subsystems—the compressor station, the gas distribution, and gathering systems—and the individual wells have to be properly taken into account if an optimum solution is desired. The following list discusses the most important factors.

1. The location, number, and main features of all existing pieces of lease equipment (wells, flowlines, tank batteries, etc.) must be recorded. This information will form the basis for the decision on the proper location of the required equipment like compressor station, distribution lines, etc.

2. The type of operation for each well (continuous flow or intermittent lift) must be determined. Wells on continuous flow or intermittent wells with choke control have a constant injection gas demand and provide favorable operating conditions for the compressors. Intermittent wells with time cycle intermitter control, however, result in uneven loading of the compressor station, which must be compensated by designing the proper amount of gas storage capacity into the distribution and gathering systems.

3. The separator pressure used on the lease influences the operation of the gas lift system in different ways. Higher separator pressures mean higher bottomhole pressures, and consequently, lower liquid production rates from the wells; but at the same time, they decrease the power needed to compress the required gas volume. Lower separator pressures act the opposite way. It is considered bad practice to save on compressor power and keep a relatively high separator pressure. [2]

4. The use of available gas lines must be fully analyzed.

    • A low-pressure gas line can be used to initially charge the system as well as to provide make-up gas during normal operations. Based on the analysis of operational costs, one can decide whether a closed or a semi-closed gas lift system is more advantageous.

    • Low- or high-pressure sales lines provide an economical way to sell excess gas in systems with a greater formation gas production than the field's make-up gas requirement.

5. Freezing conditions occur when hydrates form in gas distribution lines. Hydrate formation is caused by water and hydrocarbon vapors in the lift gas at specific pressure and temperature conditions. As seen in Figure 9–5, the higher the line pressure, the higher the hydrate formation temperature. [2] Since freezing causes several operational problems in surface lines and control devices, the following preventive actions must be taken:

    • Bleed-off valves at low spots of pipelines have to be installed to remove liquids from gas lines.

    • Injection of chemicals (most frequently methanol) into gas distribution lines decreases the danger of freezing.

    • Heating of injection lines can also help.

    • Gas dehydration removes water vapors from the lift gas.

6. The features of existing gas compressors can have a considerable impact on the design of the gas lift system. The storage capacities of the distribution and the gathering systems must then be designed according to the suction and discharge pressures and the gas capacities of the existing compressors.

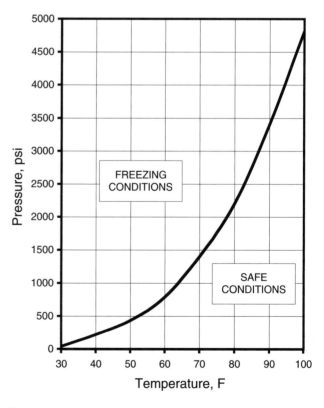

Fig. 9–5 Hydrate formation conditions according to Brown. [2]

9.3.3 Design procedure

The design of the gas lift system, in general, is a very complex task involving the design of each gas lifted well, the gas distribution and gathering systems, and the compressor station. In order to reach an optimum system design, all the mentioned subsystems have to be designed with due respect to their various interactions. The main steps of system design detailed as follows are given for a simplified case where the surface gas injection pressure (used to design the individual gas lift installations) is known or has been assumed previously. The procedure described, therefore, can be considered only as one iteration in the optimum design of the gas lift system.

Installation Design of Gas Lifted Wells

The most important point is to determine whether the given well should be placed on continuous flow or intermittent lift. For continuous flow wells, gas requirements are estimated from vertical multiphase flow calculations. For intermittently produced wells, not only gas requirements but daily cycle numbers and injection periods have to be found. For both cases, unloading valve strings, if required, have to be designed, etc. The details of these calculations were discussed in previous chapters of this book.

Estimation of the System's Gas Requirements

In order to reduce the number of calculations, wells should be grouped into a few representative classes. These can be set up based on the type of operation (continuous, intermittent), production rate, lifting depth, etc. For each class of wells, injection gas requirements are determined and the system's total daily gas injection requirement is calculated with due regard to the number of wells in each class.

Intermittent wells, because their gas demand is not constant over time, necessitate that in addition to their total daily gas requirement, the maximum possible instantaneous injection gas rate required from the system be estimated. This can only be done if the maximum number of wells simultaneously operating is approximated.

To cover unaccountable gas losses in the system due to calculation inaccuracies, leaks, etc., the system's total gas requirement is usually increased by 5–10%.

Selection of the Required Compressors

Compressors are selected on the basis of the necessary suction and discharge pressures and the required gas delivery rate. The suction pressure must be less than the WHP assumed for the calculation of the wells' injection gas requirements. Discharge pressure should be selected 100–200 psi higher than the operating injection pressure used in the field. Finally, compressor capacity is determined by the gas lift system's total daily gas injection requirement. From these data, the necessary brake horsepower of the compressor can be estimated from the formula:

$$BHP = 24.15 \, n \left(\frac{p_d}{p_s} \right)^{\frac{1}{n}} Q_g \qquad\qquad 9.1$$

where: n = number of compressor stages

   p_d = discharge pressure, psia

   p_s = suction pressure, psia

   Q_g = gas volumetric rate, MMscf/d

The necessary number of compressor stages is determined from the total $CR = p_d/p_s$. Single stage compression is sufficient for $CR < 5$, two stages are needed for $CR = 4$–30, three stages are recommended for $CR = 20$–100 values. Final selection of the compressor involves economic considerations that can help with the decision whether portable or stationary units have to be installed.

Calculation of the System's Gas Balance

To maintain the continuous operation of a properly designed closed rotative gas lift system, only a minimum amount of make-up gas is necessary after the initial charge with gas. This should cover the fuel gas requirements of the compressors (if driven by gas engines) as well as any gas losses in the system. Fuel gas requirement is usually approximated as 10–12 scf/BHP-hour, based on the compressor's brake horsepower. If the calculated make-up gas requirement is compared to the total volume of formation gas produced from the wells, two cases can occur. A positive balance means that gas may be sold from the system either to a low- or a high-pressure line, a much more economic solution than flaring the excess gas. If the amount of formation gas is insufficient to cover the gas volume required for fuel and the gas losses, the gas lift system will need a constant source of make-up gas to keep it in operation.

Design of the Distribution System

In bigger fields, the distribution system contains trunk lines leading from the compressor and individual injection lines branching from these and running to the wells. The proper design of the distribution system ensures that (a) frictional pressure losses are kept at a minimum, and (b) the gas storage capacity of the system is sufficient. Trunk lines are sized to keep friction losses to a minimum, and for the same reason, injection lines should have a diameter of at least 2 in.

Gas storage capacity of the distribution system is irrelevant if: (a) only wells on continuous flow are present, or (b) gas injection to the intermittent wells is controlled by surface chokes, or (c) the total number of wells in the system is sufficiently high. For systems with a low number of intermittent wells on time cycle intermitter control, gas must be stored in the distribution system since the compressor delivers only the daily average gas requirement and cannot sustain the great instantaneous gas demand of the intermittent wells. To find the peak instantaneous gas requirement in the system, the possible maximum number of wells producing simultaneously must be estimated first then their total gas requirement is easily found. The difference of the peak gas demand and the capacity of the compressor gives the gas volume to be stored in the distribution system. To provide this amount of gas storage, the capacity of the distribution system should be equal to or exceed the value found from the approximating following formula, derived for ideal gases and a constant temperature:

$$C_{HP} = \frac{p_{sc} \, Q_{sc}}{p_d - p_{inj}}$$

9.2

where: Q_{sc} = gas volume to be stored, cu ft

   p_{sc} = atmospheric pressure, 14.65 psia

   p_d = compressor discharge pressure, psia

   p_{inj} = design injection pressure, psia

In case the total capacity of the distribution system's gas lines is less than the previous value, additional storage volume must be provided by connecting unused pipeline sections, the annulus of abandoned wells, etc. to the distribution system.

Design of the Gathering System

A properly designed gathering system ensures that the compressor is supplied with a sufficient amount of gas at all times. If all the wells are on continuous flow, system design is simple since there is an uninterrupted flow of gas to the separator providing ideal operating conditions for the compressor station. Intermittent wells, however, do not provide a constant supply of gas into the gathering system because they produce in cycles. If more than one well produces its liquid slug to the surface at the same time, the high instantaneous gas rate may overload the separator, causing unwanted amounts of gas to be flared. This may lead to a situation when the gas available in the system is less than the capacity of the compressor, causing the suction pressure to drop below the minimum required pressure and the compressor to be starved of gas.

In order to provide a constant flow of gas at the proper pressure to the compressor's suction, the gathering system must have the right amount of gas storage capacity. To keep this storage requirement from becoming excessive, the total number of intermittent wells producing at the same time should be limited. This is hard to achieve if surface gas injection control is accomplished by a fixed or adjustable choke at the wellhead, because in such cases the timing of well operation highly depends on the well's inflow parameters and cannot be controlled. Therefore, chances are great that at a given time, a significant fraction of the total wells will start to produce simultaneously. On the other hand, the use of intermitters to control the gas injection allows an easy way to program their operation and to limit the number of wells producing simultaneously. Intermitters controlling several wells or central controllers are used for such purposes.

The gathering system's gas storage capacity is properly designed if the gas volume stored in the system (during the period while the gas rate received from the separator is greater than the capacity of the compressor) covers the deliverability of the compressor plus the gas used for fuel and losses. In this case, during the period when gas volume supplied from the separator is less than the capacity of compressor, the gas previously stored in the gathering system is released and ensures the proper loading of the compressor. This way, the only gas leaving the system is the excess formation gas produced by the wells.

If the gas volume to be stored in the gathering system is known, the system's required storage capacity can be found from the change in system pressure. During the liquid production period, gathering system pressure approaches the separator pressure, but it drops to the suction pressure of the compressor between cycles. Based on these data, an approximate formula, similar to Equation 9.2 can be derived to find the gathering system's required gas storage capacity:

$$C_{LP} = \frac{p_{sc}\, Q_{sc}}{p_{sep} - p_s}$$

9.3

where: Q_{sc} = gas volume to be stored, cu ft

   p_{sc} = atmospheric pressure, 14.65 psia

   p_s = compressor suction pressure, psia

   p_{sep} = separator pressure, psia

If the actual capacity of the gathering system is insufficient, connection of various storage vessels, and/or additional gas storages (like wells, pipeline sections) is required. In case such solutions are difficult to accomplish, choking of the wells is usually recommended in order to reduce the high instantaneous gas rates entering the system. Chokes are always installed on the flowlines close to the separator (and never on the wellhead) to prevent the increase of liquid fallback in the wells. This solution allows the produced liquid slugs to enter the flowlines and prevents a rapid inrush of gas to the separator. By considerably reducing the instantaneous gas flow rate entering the separator, the required gas storage capacity of the gathering system is greatly reduced.

Example 9–1. Perform the main design steps for a gas lift system containing two intermittent wells. The wells have identical production parameters: 50 bpd oil production with a formation GOR of 500 scf/bbl. They are intermitted with 50 cycles/d, and 2,000 scf of gas is injected in 3 minutes every cycle. The injection gas lines are 2,000 ft long and of 2 in. nominal diameter. Injection and separator pressures are 700 psi and 100 psi, respectively, and a compressor with a suction and discharge pressure of 50 psi and 800 psi is used.

Solution

First find the required compressor delivery rate. The daily injection gas volume is

22,000 scf/cycle 50 cycles/d = 200 Mscf/d, and corrected for a 5% volume loss in the system:

Q_g = 200 1.05 = 210 Mscf/d.

To find the required brake horsepower of the compressor, use Equation 9.1 and select a two-stage compressor since the compression ratio is $CR = (800 + 14.65) / (50 = 14.65) = 12.6$:

$$BHP = 24.15 \ 2 \ (800 + 14.65) / (50 = 14.65)^{1/2} \ 0.21 = 36 \ HP.$$

The gas balance in the system is calculated as follows. The daily production of formation gas from the two wells equals:

$$Q_{form} = 2 \ 50 \ bbl/d \ 500 \ scf/bbl = 50 \ Mscf/d.$$

Fuel gas requirement is estimated as 10 scf/BHP-hour, which gives:

$$Q_{fuel} = 10 \ scf/BHP/hour \ 35 \ BHP \ 24 \ hour/d = 8.64 \ Mscf/d.$$

System losses are estimated as 5% of the total gas requirement:

$$Q_{loss} = 200 \ Mscf/d \ 0.05 = 10 \ Mscf/d.$$

The balance of daily gas volumes gives:

$$\Delta Q = 50 - 8.64 - 10 = 31.4 \ Mscf/d,$$ a positive balance meaning this gas volume can be sold from the gas lift system.

For designing the gas distribution system's storage capacity, assume the worst case when both wells start to inject gas at the same time. The compressor's deliverability for the 3-minute injection period is

$$Q = 3 \ mins \ 210 \ Mscf/d / 1,440 \ mins/d = 440 \ scf.$$

Since the two wells require the injection of 4,000 scf of gas during the same 3-minute interval, gas stored in the distribution system should supply the following amount of gas:

$$Q_{sc} = 4,000 - 440 = 3,560 \ scf.$$

The required capacity of the high-pressure system is found from Equation 9.2 as

$$C_{HP} = (3,560 \ 14.65) / (800 - 700) = 522 \ scf.$$

The combined internal volume of the two 2-in. injection lines being only about 93 scf, an additional storage volume should be connected to the high-pressure gas distribution system.

For the design of the gathering system, further assumptions need to be made. Let us assume that the gas volume entering the gathering system exceeds the capacity of the compressor for 15 min, during which period about 70% of the total gas production is received from the wells.

Total gas volume received from the two wells equals the sum of the injected and the formation gas volumes, *i.e.* 2 2,500 = 5,000 scf/cycle, 70% of which equals 3.5 Mscf/cycle.

The gas volume needed to fully charge the compressor and to cover the fuel requirements during the 15-min period is

$$(210 + 8.64) \ Mscf/d \ 15 \ min/cycle / 1440 \ min/d = 2.28 \ Mscf/cycle.$$

The gas volume sold during the cycle equals

$$31.4 \ Mscf/d / 50 \ cycles/d = 0.63 \ Mscf/cycle.$$

The gas volume to be stored in the gathering system is the difference between the volume entering the system and the two previous terms:

Q_{sc} = 3.5 – 2.28 – 0.63 = 0.59 Mscf/cycle = 590 scf/cycle.

The required storage capacity of the gathering system, according to Equation 9.3:

C_{LP} = (590 14.65) / (100 – 50) = 173 scf.

If the combined inside volume of the separator, the scrubber, and the piping in the gathering system is less than this value, additional gas storages should be provided for the smooth operation of the gas lift system.

References

1. Winkler, H. W., and S. S. Smith. *Gas Lift Manual.* CAMCO Inc., 1962.

2. Brown, K. E. *The Technology of Artificial Lift Methods.* Vol. 2a. Tulsa, OK: Petroleum Publishing Co., 1980.

3. Kastrop, J. E., "Converted Engine Exhaust Supplies Make-Up Gas." *PE,* May 1961: B-21–26.

4. Aguilar, M. A. L. and M. R. A. Monarrez, "Gas Lift with Nitrogen Injection Generated In Situ." Paper SPE 59028 presented at the International Petroleum Conference and Exhibition in Mexico, held in Villahermosa, February 1–3, 2000.

5. Winkler, H. W., "How to Design a Closed Rotative Gas Lift System." Part 1, *WO,* July 1960: 116–9.

6. Winkler, H. W., "How to Design a Closed Rotative Gas Lift System." Part 2, *WO,* August 1, 1960: 103–14.

10 | Analysis and Troubleshooting

10.1 Introduction

In order to maintain proper operation of a gas lift installation, its operating conditions must frequently be evaluated. In case of deviations from design conditions, appropriate corrective actions have to be taken. Usually, the analysis is started by calculation of operational parameters valid for the original design conditions. Then, using available diagnostic techniques, actual parameters are measured or calculated and compared to those under design conditions. After diagnosing the deviations from ideal conditions, the possible sources of operational problems can be detected and corrective actions can be recommended. This kind of troubleshooting is described in this chapter.

The analysis of operational conditions must always be based on sound and recent data. General data to be determined—and usually available for each installation without any specific diagnostic measurements—can be grouped as given as follows:

1. Fluid properties include the basic parameters (densities, viscosities, etc.) of oil, water and gas as well as bubblepoint pressure, formation GOR, water cut, etc.

2. Reservoir data cover the description of inflow performance of the given well and a recent set of parameters (static reservoir pressure, PI, etc.) must be available.

3. Data of a recent well test (preferably a 24-hour production test) should be available and should include volumetric rates of oil, water, and total (injection plus formation) gas produced.

4. Well completion data include casing, tubing sizes, running depths, well deviation, perforation data as well as surface geometrical data on the flowline, gas injection line, and the wellhead.

5. Gas lift equipment data are related to the running depths of the packer and the gas lift valves, the properties of gas lift valves: type, port size, TRO pressure, etc.

Specific operational data of the gas lift installation are measured or calculated with the diagnostic tools available for the gas lift analyst. The description of these tools and the information derived from their application in troubleshooting are the subject of the next sections.

10.2 Troubleshooting Tools and Their Use

There are many diagnostic tools developed specifically for gas lifted installations to evaluate the performance of these wells. They can be used separately or in combination with other method(s), but final decisions should always be made after consulting all available results. These tools can determine operational conditions at the surface (two-pen pressure recordings, injection gas rate measurements) or in the well (acoustic, pressure, and temperature surveys). Perhaps the most complete information on well performance is provided by a flowing pressure survey run in the tubing string but at a premium cost. Therefore, setting a priority to the available analysis methods can save valuable operational costs.

10.2.1 Two-pen pressure recordings

10.2.1.1 Pressure recorders. A continuous recording of CHP and THP is a basic requirement for the proper operation and analysis of gas lifted installations. The reason for this is that these are the pressures from which downhole pressures at any valve depth can be calculated and compared to design conditions. Such calculations enable one to find out whether the given valve is in the open or closed position. In case of intermittent wells, CHP and THP recordings can be used to evaluate the settings of surface injection controls and locate surface problems.

Casing and tubinghead pressures vs. time are measured and recorded by two-pen pressure recorders on circular preprinted charts. The pressure elements of the recorder should be calibrated before use because of the importance of accurate readings and should have the proper measuring ranges for ease of evaluation. The rotational time of the chart is usually 24 hours, but shorter times are needed to study single intermittent cycles. The installation of the two-pen recorders is crucial because the right pressures must be measured. [1–4] As shown in Figure 10–1, the recorder should be installed at the wellhead with the pressure connections situated as follows:

- The CHP (injection pressure) line must be connected downstream of the injection control device so that the recorder measures the injection pressure at the casing head.

- The THP (production pressure) line must be connected upstream of the production choke body or other restriction at the tubinghead. This position ensures that production pressure at the top of the tubing string is measured.

10.2.1.2 Chart interpretation. The following discussions present schematic two-pen pressure recorder charts and pressure recording shapes typically found on continuous flow and intermittent lift wells. They represent some commonly encountered problems and use the notation *CHP* and *THP* to refer to casinghead and tubinghead pressures, respectively.

Continuous Flow Wells

Figure 10–2 presents typical charts taken on wells placed on continuous flow gas lift. Interpretation of each chart follows.

A   Under ideal conditions, CHP and THP are uniform, and the well produces the design fluid rate with a constant low WHP.

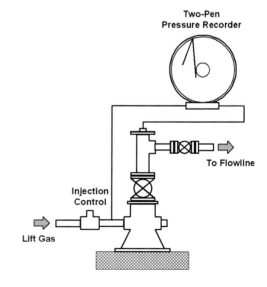

Fig. 10–1 Recommended connections for a two-pen pressure recorder.

B A steady THP curve well above the separator pressure indicates that excessive backpressure is held on the well. Remedies include the removal of the production choke from the wellhead as well as paraffin or scale removal from the flowline. Replacing the flowline with a larger diameter pipe should also be considered for extremely high backpressures.

C Fluctuating surface injection pressure results in changing WHPs and less than optimum operating conditions. Decreasing the operating valve's opening pressure or increasing the gas distribution system's gas storage capacity can help.

D A freezing gas injection choke at the wellhead slowly decreases the CHP below the closing pressure of the operating gas lift valve and the well ceases to produce. THP drops to separator pressure.

E The operating valve throttles because its closing pressure is close to the injection pressure at the casinghead. This can happen if the port of the gas lift valve is oversized and can be eliminated by installing a slightly larger choke at the casinghead.

F A hole in the tubing or a tubing string parted below the depth of the operating valve forces the well to produce intermittently. While CHP is constant, the well produces with a constant THP; but when the hole is uncovered, CHP suddenly drops and a great amount of gas is produced as indicated by the bigger peak in THP. Production ceases and THP drops to separator pressure. Then casing pressure builds up, opening the operating valve again and producing another liquid slug, indicated by the second peak on the THP recording.

Intermittent Wells

Two-pen pressure recorder charts taken on intermittent wells look more complicated than those found on wells on continuous flow since THP and CHP values repeatedly change even in ideal conditions. Also, their patterns vary with the different combinations of surface control and operating valve type used. Ideal pressure recorder patterns are illustrated on the chart given in Figure 10–3 [5] for: A intermitter, B choke, C choke and regulator, D intermitter and choke control of gas injection. The same recordings were discussed in Section 6.3.

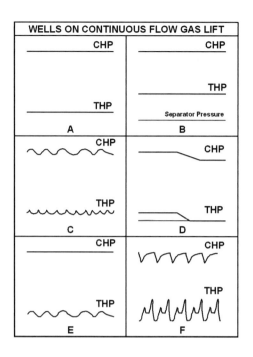

Fig. 10–2 Schematic two-pen pressure recordings made on a continuous flow well.

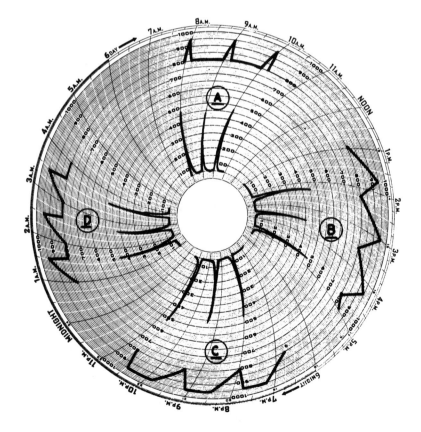

Fig. 10–3 Ideal two-pen pressure recorder charts for different surface injection control types.

The liquid lifting efficiency of the intermittent cycle is clearly indicated by the THP recording. The magnitude of the maximum pressure is closely related to the length and velocity of the liquid slug reaching the wellhead as well as the presence of any restrictions near the wellhead. The time required for the THP to return to separator pressure is also a good indicator: restrictions far from the wellhead tend to increase it. Under ideal conditions, a high but not excessive kick in THP rapidly returning to separator pressure is recorded. It must be noted that although a lot of information can be gained from the study of the THP line alone, a final evaluation should take into account other measurements as well.

THP recordings typical for intermittent wells are schematically shown in Figure 10–4; their interpretation follows.

A   In an ideal case, solid liquid slugs are produced as shown by the thin sharp kicks on the THP line that diminish rapidly, indicating no flow restrictions at the wellhead or in the flowline.

B   If the number of daily cycles is less than required, the longer liquid slugs, while traveling upward, do not reach the proper velocity, and the THP line indicates thick kicks at relatively low pressures.

C   On the other hand, for too high cycle frequencies, THP does not return to separator pressure after the surfacing of the liquid slugs, as indicated by the kicks. This is caused by too little time left between the cycles for the THP to bleed down to separator pressure.

D   Rounded, sluggish kicks on the THP line indicate that short liquid slugs are produced to the surface. This can happen if: (a) not enough gas is injected during the cycle or (b) starting slug lengths are short due to the well's low productivity. To find out the exact cause, check the spread of the operating gas lift valve on the CHP recording.

E   When too much gas is injected during the cycle, tubing pressure kicks are too high and too thick, indicating that too much tail gas is produced at the surface.

F   A wellhead choke or other flow restriction near the wellhead is indicated by abnormally high kicks and a slow pressure reduction in the THP line.

G   Flow restrictions in the flowline, far from the wellhead, behave similarly to a wellhead choke; only the changes in THP are smaller and more gradual.

H   This shape is typical for too much tail gas production or a severe flow restriction at the separator.

The CHP recording carries a lot of information on the performance of the gas injection control (choke, intermitter, etc.), the operation of the gas lift valve, and the capacity of the gas lift system. As with THP, proper evaluation of the installation must not be based on CHP measurements only.

Figure 10–5 displays CHP recordings typical for wells with surface intermitter control and an unbalanced operating gas lift valve. Interpretation of the problems schematically depicted is given as follows.

A   As previously shown in Figure 6–10, this is the ideal change of CHP with time if an unbalanced valve with insufficient spread is used in the well. As seen, gas injection through the surface intermitter continues even after the operating valve has opened.

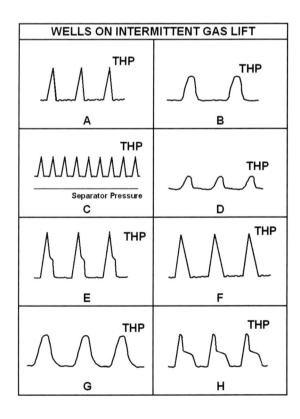

Fig. 10–4 Schematic THP recordings on continuous flow wells.

B   If the gas volume injected to the casing annulus during the intermittent cycle is too low, CHP cannot increase to valve opening pressure in one cycle and the valve may only open during the second cycle.

C   In cases when the instantaneous gas rate injected through the operating valve is much greater than the gas rate provided by the gas lift system at the surface, the valve opens and closes several times during the cycle.

D   This is very similar to case C but with a greater surface injection rate, making the CHP only drop once the operating valve opens.

E   This situation occurs when the injection rates through the operating valve and the surface intermitter are about the same, and a constant CHP is observed after the valve has opened. It may be caused by either (a) an insufficient instantaneous gas rate from the gas distribution system or (b) the opening pressure of the gas lift valve being too close to the available injection pressure.

F   When the surface injection rate is too high, an excessive amount of gas is injected during the cycle and the CHP, instead of dropping instantly, only declines slowly.

G   A leaking surface intermitter admits gas to the annulus at a low rate during the liquid accumulation period.

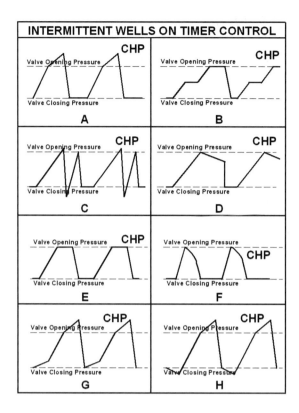

Fig. 10–5 Schematic CHP recordings on intermittent wells with intermitter control.

H   A moderate-sized tubing leak does not prevent the completion of the intermittent cycle and is indicated by the continued decrease of CHP after the gas lift valve closes. To determine the place of the leak, shut down the intermitter and monitor the variation of CHP with time. If the final CHP value is around the separator pressure, the leak is above the liquid level in the tubing. If CHP stabilizes at a pressure greater than separator pressure, the leak should be found below the tubing liquid level.

CHP recordings for wells on choke control of gas injection are shown in Figure 10–6.

A   An ideal CHP variation (see also Fig. 6–4) is characterized by a steady but slow increase followed by a fast decrease of the injection pressure. The two extremes of the CHP are the operating valve's surface closing and opening pressures.

B   If CHP decreases relatively slowly after the gas lift valve opens, then the instantaneous gas injection rate from the annulus may be improperly low due to a small port area of the operating valve or a partially plugged valve. Other possible causes include an annulus volume too great for the actual spread of the operating valve or an injection line of too high capacity.

C   A less than normal difference between the minimum and maximum values of CHP indicates either: (a) that the spread of the operating gas lift valve is too low or (b) an increased flowing tubing pressure reducing the opening injection pressure of the valve.

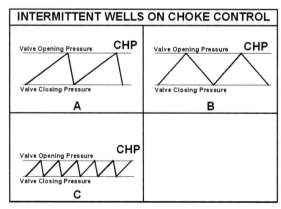

Fig. 10–6 Schematic CHP recordings on intermittent wells with choke control.

10.2.2 Gas volume measurements

Every well on gas lift should be equipped with the proper device to continuously measure the gas volumes injected into the annulus. Orifice, turbine, or mass flow meters can be used with the necessary equipment installed on the gas injection line close to the wellhead. An accurate measurement of gas volumes allows one to calculate the well's injection GLR, a very sensitive indicator of the efficiency of fluid lifting by gas lift. By periodically checking the gas consumption, many malfunctions occurring in the well or at the surface can be detected, and corrective action can be taken in due time.

The measurement of injection gas rates is easier for wells on continuous flow than for those on intermittent lift because injection pressure is practically constant. Intermittent wells take widely fluctuating lift gas volumes during the gas injection period and no gas at all during the rest of the cycle. Under such conditions, conventional measurements with orifice meters are quite difficult to evaluate. Contemporary measurements use flow computers that can easily handle the metering of fluctuating gas rates.

The measurement of the gas rate produced by the gas lifted well (including the injected and the formation gas) is also necessary. This is usually accomplished at the test separator, and many different kinds of measuring devices can be used. Since the accurate knowledge of injected and formation gas volumetric rates is crucial for the analysis and troubleshooting of gas lifted wells, gas volume measurements are a vital part of the diagnostic tools available for the gas lift analyst.

10.2.3 Downhole pressure and temperature surveys

10.2.3.1 Introduction. A complete analysis of a gas lifted well's operating conditions necessitates the measurement of downhole pressures and temperatures. Such pressure and temperature surveys provide the most accurate way to determine the well's performance and form the base of any troubleshooting operations. These surveys involve the running of a pressure and/or temperature recording instrument (pressure/temperature bomb) down the tubing string and making appropriate measurements at specific depths while the well is shut-in or producing under normal conditions. Measurements in shut-in wells provide $SBHPs$, as well as pressure buildup curves, whereas pressures/temperatures measured during normal operation characterize the performance of gas lift valves, downhole malfunctions, etc.

In order to detect downhole pressure and/or temperature conditions representative of the performance of the gas lift installation, these surveys are usually conducted under normal operating conditions. The measuring instrument(s), therefore, is run into the well under pressure, with the help of a lubricator attached to the wellhead. This arrangement requires the use of a permanent swab valve above the main valve in the Christmas tree that facilitates the connection and removal of the lubricator without the need of killing the well. Since the running of the pressure/temperature measuring instrument occurs under normal flow conditions, all subsequent measurements will be representative of the operating mode of the well.

Downhole surveys can be accomplished with separate runs of individual pressure and temperature instruments, but the use of tandem instruments recording pressure and temperature data simultaneously saves time and running costs. Present-day electronic instruments record both pressure and temperature at the same time. Interpretation of pressure/temperature survey data can be used to determine the following main operating parameters of a gas lifted well:

- The depth of gas injection(s)

- The locations of tubing leaks, and/or leaking gas lift valves

- Flowing gradients above and below the injection point

- The static and $FBHPs$

- In intermittent wells, the variation of the $FBHP$ during the cycle

- Inflow performance parameters: PI, pressure buildup curve

10.2.3.2 Running procedures for downhole surveys. Since the pressure/temperature recording instruments are run while the well is producing, several safety precautions must be observed.

- In order to minimize the restriction to flow in the tubing string and to prevent the blowing of the bomb from the well, an instrument with the smallest possible OD must be run. These instruments are usually available in ¾ in., 1 in., and 1¼ in. sizes.

- Because of the high flow velocities in the tubing string (especially close to the wellhead), the danger of the instrument being blown up the hole is high. This can be prevented by adding heavy rod sections *stems* above the instrument as well as by the use of *no-blow latches* with slips that are activated by a sudden upward movement to stop any further movement.

- The greatest caution should be exercised for the first 100 ft of the instrument's travel into the well because flow velocities are at a maximum in this part of the well.

Specific running procedures of pressure/temperature recording instruments into wells placed on continuous flow gas lift are these, according to API RP 11V5 [4]:

1. Ensure that the well has been flowing to the test separator for a sufficient time to reach stable flow conditions.

2. Test the well before running the survey and record the liquid and gas rates, THP and CHP, measured gas rates (injection and total).

3. The pressure/temperature recording instrument should be equipped with one or two *no-blow latches*.

4. Install lubricator containing the recording instrument(s) above the swab valve. Record WHP for 15 minutes and use a dead-weight tester to calibrate the pressure readings.

5. Run instrument(s) into well and make stops about 15 ft below (optionally below and above) each gas lift valve for 15 minutes. If temperature is also measured, make stops at each valve setting depth. Do not shut in well during the entire measuring process.

6. If a tubing leak is suspected, make one or more stops between valve positions.

7. Leave instrument(s) on well bottom for at least 30 minutes.

8. Casing pressure must be measured with a dead-weight tester or recently calibrated pressure gauge.

9. If measurement of a SBHP is required, leave instrument(s) on bottom, shut off gas injection, close wing valve, and wait until pressures stabilize.

Instructions for running pressure/temperature surveys in intermittent wells are given as follows:

1. Ensure that the well has been producing to the test separator for 24 hours to reach stable flow conditions.

2. Test the well before running the survey for at least 6 hours. Record the liquid and gas rates, THP and CHP, measured gas rates (injection and total).

3. The pressure/temperature recording instrument should be equipped with one or two *no-blow latches*.

4. Install lubricator containing the recording instrument(s) above the swab valve. Record WHP during at least one complete cycle and use a dead-weight tester to calibrate the pressure readings.

5. Run instrument(s) into well and make stops about 15 ft below (optionally below and above) each gas lift valve for at least one complete cycle, so that maximum and minimum pressures are recorded. If temperature is also measured, make stops at each valve setting depth. Do not shut in well during the entire measuring process.

6. Leave instrument(s) on well bottom for at least two complete cycles.

7. Maximum and minimum tubinghead and CHPs must be measured with a dead-weight tester or recently calibrated pressure gauge.

8. If measurement of a SBHP is required, leave instrument(s) on bottom, shut off gas injection, let the well bleed down to the separator, and wait for a minimum of 12 hours.

10.2.3.3 Flowing pressure surveys. Flowing pressure surveys allow one to experimentally determine the pressure distribution in the tubing string during normal operating conditions. Measurements below the point of gas injection represent the flowing pressure traverse valid for the formation *GLR*. Those above the injection point correspond to the total (formation plus injected) *GLR*. During the design of the installation, these pressures were either approximated from gradient curve sheets or calculated from multiphase pressure drop calculations of mostly unknown accuracy. This implies that the depth of the design and actual injection points may differ. The pressure survey is an ideal tool to check the original design as well as the effect of changes occurred since the original conditions.

Interpretation of pressure survey data starts with the plotting of the measured pressures in the function of well depth with the setting depths of the gas lift valves superimposed on the depth scale. Such charts are then evaluated to determine average pressure gradients below and above the gas injection, with the shape of the flowing pressure traverse being a source of valuable information. The main point to remember is that any changes in the flowing pressure gradient indicate a change in the gas content of the wellstream that may occur through the operating valve, a leaking valve, or through a tubing leak. The following schematic charts are presented to illustrate the evaluation of pressure survey data in wells placed on continuous flow.

Figure 10–7 presents measured tubing pressures in a well with five unloading valves. The two individual curves plotted in addition to the annular gas pressure traverse clearly indicate two different average pressure gradients—the bottom one being constant and representing single-phase flow. The evaluation of the measured pressures at the individual valve setting depths shows that injection takes place through valve #4; and valve #5, although designed to be the operating valve, was not uncovered. This is a typical valve spacing mistake that can be corrected by re-spacing of the valves.

Observation of the pressure traverse curves given in Figure 10–8 shows that the gradient of the tubing pressure changes at two depths. The deeper point should be the operating valve because it is at the depth of valve #4. The other change in gradient occurs between valves #1 and #2 and most probably indicates a tubing leak.

Pressure survey charts taken in wells on intermittent lift have two traverses above the gas injection, one for the maximum and one for the minimum tubing pressures, because tubing pressure changes during the cycle. Determination of the operating valve(s) usually cannot be accomplished without the evaluation of the temperature survey data.

10.2.3.4 Flowing temperature surveys. Flowing temperature surveys (usually conducted along with pressure surveys) are used to plot the distribution of flowing temperature along the well depth for normal operating conditions. This allows the determination of actual valve temperatures and the comparison of those to the conditions assumed during the design of the installation. Valve temperatures are extremely important for gas lift valves

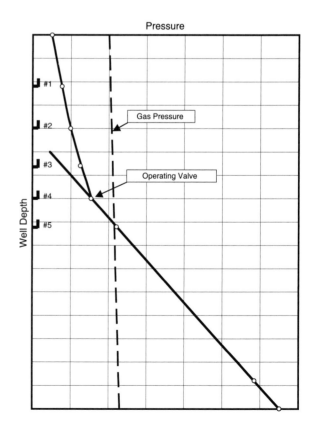

Fig. 10–7 Schematic pressure survey plot in a continuous flow gas lift well.

containing a gas-charged bellows, whereas they are irrelevant when spring-loaded valves are used. An incorrectly assumed temperature distribution may stymie unloading operations and may prevent some or all valves from opening.

In addition to the determination of valve temperatures, the temperature chart can also be used to find all points where gas injection into the tubing string takes place. The reason for this is the Joule-Thompson Effect that significantly reduces the temperature of any gas flowing through a restriction. Gas injection points in the well, therefore, are indicated by local drops in flowing temperature. Figure 10–9 depicts temperature conditions in the previous sample well and clearly indicates injection through valve #4 as well as a tubing leak between valves #1 and #2, supporting the conclusions drawn from the evaluation of the pressure survey data.

10.2.4 Acoustic surveys

Acoustic survey instruments (well sounders) operate on the principles defining the propagation and reflection of pressure waves in gases. During measurement, an acoustic impulse is produced at the top of the casing-tubing annulus, which travels in the form of pressure waves along the length of the annular gas column. The pressure waves are reflected (echoed) from every depth where a change of cross-sectional area occurs, caused by tubing collars, gas lift valves, the fluid level, etc. The reflected waves are picked up and converted to electrical signals by a microphone and are recorded electronically or on a strip chart. Since the annulus liquid level produces the biggest reflection, an evaluation of the reflected signals permits the determination of its depth.

The well sounder consists of two basic components, *i.e.* the wellhead assembly, and the recording and processing unit. The wellhead assembly is easily connected to the casing annulus by means of a threaded nipple and contains a mechanism creating the sound wave and a microphone for picking up the signals. Older well sounders utilize blank cartridges, fired either manually or by remote control. Modern units employ so-called *gas guns*, which provide the required impulse by suddenly discharging a small amount of high-pressure gas (CO_2 or N_2) into the annulus. The recording unit processes the electric signals created by the microphone by filtering and amplification, processed signals are then recorded on a chart recorder in the function of time. The depth of the liquid level in the annulus is found by the proper interpretation of the acoustic chart.

An example acoustic chart is given in Figure 10–10, [4] showing analog recordings of reflected sound signals on a strip-chart. As seen, every tubing collar is identified by a local peak, gas lift valve mandrels produce bigger peaks, and the liquid level is clearly marked by a much larger reflection.

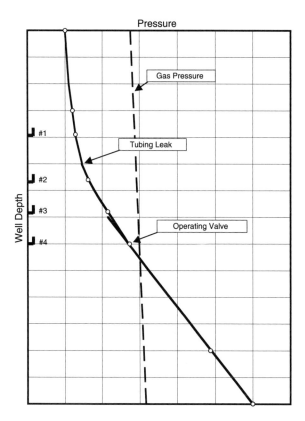

Fig. 10–8 Schematic pressure survey plot indicating a tubing leak.

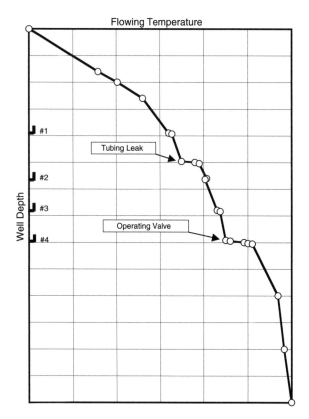

Fig. 10–9 Schematic temperature survey plot in a continuous flow gas lift well.

Determination of the liquid level depth is possible either by counting the number of collar signals and comparing these data to well records, or by finding the reflection time from the liquid level. In addition to finding the annular liquid level, acoustic surveys can be used to estimate SBHPs, locate tubing leaks, and the depth gas lift mandrels.

The use of well sounding has its limitations in detecting the operating valve of an installation. Although the annular liquid level, in most cases, can be quite accurately determined, it does not necessarily indicate the point of gas injection, only the deepest point to which the well has been unloaded. This is why, as already mentioned before, meaningful conclusions can only be drawn if data of several diagnostic methods are also considered.

Although only similar to conventional well sounding, the use of pressure pulses generated by closing a quick-acting valve at the wellhead must be mentioned as a novel technique to find the depth of the operating valve. [6] The background of this new method is that the propagation and reflection of

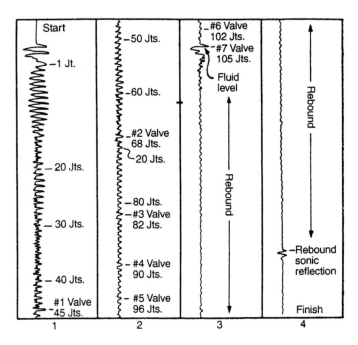

Fig. 10–10 Typical results of an acoustic survey after. [4]

pressure pulses changes with the composition of the multiphase mixture they are traveling in. Since at points where gas injection takes place the gas content of the mixture flowing in the tubing changes, the depth of gas injection can be detected.

10.3 Common Gas Lift Malfunctions

Common problems associated with gas lifting can be classified according to the place they develop, a method followed in many publications on troubleshooting of gas lift installations. [4, 7, 8] The three areas frequently defined are (1) the downhole system, (2) the gas distribution or inlet system, and (3) the fluid gathering or outlet system. Downhole problems include all mechanical problems associated with equipment run in the well: gas lift valves, tubing, casing, etc., as well as unloading valve string design problems. Many times, the gas lift malfunctions stem from problems arising in the gas distribution system: compressors, distribution lines, surface controls, etc. Finally, improper pipe dimensions, pressures, flow restrictions, etc., in the gathering system can prevent individual wells or whole fields alike to achieve optimum operating conditions. The following discussion describes the most important operational problems that may arise in gas lift systems and details the ways to avoid them.

10.3.1 Downhole problems

Well Unloading Problems

Unloading of a gas lifted well can be *stymied* due to an improper design, the usual causes being an excessive backpressure on the well or a too heavy kill fluid in the annulus. Improper spacing of the valves may also prevent unloading to the operating valve depth. Remedial actions include the use of a *kickoff* pressure, if available, removal of tubing liquids by swabbing, or *rocking* the well. Rocking means alternating application of injection pressure to the casinghead and tubinghead. This operation can push some of the liquid in the tubing back to the formation, thereby allowing the unloading process to commence with the available gas lift pressure.

Valve Staying Open

If injection pressure at the casinghead decreases below the closing pressure of any valve in the string, then one of the valves may not close, provided no hole is present in the tubing string. Closing of a valve may be prevented by trash, any kind of deposition, etc. between the valve seat and the stem. Shutting the wing valve until CHP builds up as high as possible and then rapidly opening the valve may create the necessary differential pressure to remove any particles holding the gas lift valve open. If this does not help, the valve should be pulled for repairs.

Holes in Tubing

Excessive injection gas usage and low CHP are the symptoms of a probable tubing leak. The procedure to be followed to confirm a hole in the tubing is as follows:

- Close the wing valve while injecting gas at the wellhead and wait for the tubinghead and CHPs to equalize.

- Shut off gas injection and rapidly bleed off the casing pressure.

- If the tubinghead pressure also drops, a hole is indicated.

- With no hole in the tubing string, the THP stays constant because the gas lift valves and their check valves are closed.

Heading Production

Wells placed on continuous flow gas lift may produce in heads for several reasons. This kind of instable operation can be corrected by redesigning the installation. Possible causes include the following:

- The gas lift valve's port size is too large with an associated large tubing effect, and the valve closes when the flowing tubing pressure decreases below the value necessary to keep it open.

- If valve temperatures temporarily increase due to a greater than assumed liquid production rate, dome charge pressures could increase, forcing the valves to close. After the production has stopped, the well cools down, dome pressures return to their original values, and well production resumes again.

- Limited liquid inflow to the well may also cause production by heads.

Finding the Operating Valve

If available information is not sufficient to pinpoint the operating valve, the following method can help in finding the valve actually injecting gas.

- Shut off gas injection at the casinghead and observe the decrease of CHP.

- The stabilized value of CHP gives the valve's surface closing pressure.

- Based on the TROs of the individual valves in the unloading string, the valve with the same surface closing pressure can be found.

The accuracy of this method is limited because it assumes the tubing pressure to be zero at the valve, therefore it must be used in combination with other diagnostic tools.

Well Producing Dry Gas

The causes of a well producing mostly dry gas can be different.

- If the surface injection pressure is greater than the design pressure, one of the upper unloading valves my admit gas.

- There could be a hole in the tubing. Verify this as recommended previously.

- Gas injection can occur through the bottom valve if the previous two conditions prove to be false. In this case, there is not enough inflow at the well bottom. The perforations and the standing valve (if exists) should be checked.

No Gas Injection

First it must be checked that the injection choke is not frozen, and there are no closed valves on the injection line. In case everything is right on the surface, check the following:

- If fluid-operated valves are used, then insufficient inflow from the well may prevent them from opening because of the low tubing pressures against the valves.

- For IPO valves, check the well's liquid rate. If this is greater than designed, the increased well temperature may cause the valves to close and the well to operate intermittently. For wells producing at the designed rate, check the valves' opening pressures and make sure they are not too high for the surface injection pressure.

10.3.2 Problems in the distribution system

Improper Choke Sizes

The size of the gas injection choke has a great impact on well operation because it controls the gas injection rate. A too-large choke is indicated by a CHP above the design pressure and can cause reopening of upper unloading valves and excessive lift gas usage. In contrast, smaller than required chokes limit gas injection rates and can reduce the liquid production of the well. Also, unloading of the well may not complete because of a small choke size.

Improper CHPs

Lower than design CHPs can be caused by (1) a small, plugged, or frozen choke, (2) a hole in the tubing, or (3) a cut-out gas lift valve. To find the actual cause, check the injection gas volume that must be low for problems with the surface choke. Freezing can be eliminated by line heating or injection of chemicals into the lift gas. A CHP higher than expected can be attributed to the following:

- A gas injection choke of unnecessary large size.

- The reopening of upper unloading valves. This can be checked by measuring the gas injection rate, and an excessive lift gas usage is usually observed.

- A partially plugged operating valve.

- If high CHP is accompanied by low gas injection rates, it is possible that

  - the operating valve is partially plugged

  - downhole tubing pressure may be excessive due to a surface restriction

  - increased valve temperatures raise the dome charge pressures of the gas lift valves and thereby reduce their gas throughput capacity

Improper Gas Injection Rates

Lower than design gas injection rates may indicate partly closed valve(s) on the injection line, a plugged, frozen, or small injection choke. CHP may be lower than required or the available gas rate from the distribution system may be low. If well temperatures were underestimated during valve string design calculations, the higher actual dome charge pressure will limit the throughput capacity of the gas lift valve. Improperly high injection rates can be caused by too-large injection chokes or excessive CHPs. In case CHP is higher than expected, upper unloading valves may be open during normal operation. Although a leak in the tubing or a cut-out valve also increases the gas injection rate, they usually cause lower CHPs.

Improper Intermitter Settings

The basic rules for setting the intermitter are (a) cycle frequency should be set to get the maximum liquid production with the minimum number of cycles, and (b) injection period should then be adjusted to minimize gas consumption. The procedure to be followed to properly set time cycle intermitters is discussed in chapter 6.

10.3.3 Problems in the gathering system

High Separator Pressure

For gas lifted wells, the lowest possible separator pressure should be maintained. In gathering systems with different pressure levels, ensure that wells on gas lift are switched to the low pressure system. Check the separators for flow restrictions.

Wellhead and Manifold Valves

All valves on the wellhead and on the manifold must be completely open and of the proper size. Check all valves on the flowline as well as the condition of the flowline.

High Backpressure

An excessive WHP has very detrimental effects on the well's liquid rate because it directly increases the FBHP. In order to keep a minimum backpressure on the well, do the following:

- Remove any production choke from the wellhead.

- Even the choke body without a choke bean restricts the cross-sectional area open to flow and should be removed.

- Remove as many 90° turns as possible; a streamlined flowline connection is ideal for intermittent lift.

High backpressure can be caused by a flowline of a relatively small diameter or great length. Paraffin or scale buildup in the flowline can also increase backpressure and should be removed by scraping, hot oiling, or other methods.

References

1. *Gas Lift*. Book 6 of the Vocational Training Series. 2nd Ed. Dallas, TX: American Petroleum Institute, 1984.

2. Winkler, H. W. and S. S. Smith. *Gas Lift Manual*. CAMCO Inc., 1962.

3. Brown, K. E. *The Technology of Artificial Lift Methods*. Vol. 2a. Tulsa, OK: Petroleum Publishing Co., 1980.

4. "Recommended Practice for Operations, Maintenance, and Trouble-Shooting of Gas Lift Installations." *API RP 11V5*, 2nd Ed. American Petroleum Institute, Washington D.C., June, 1999.

5. *Field Handbook for Intermitting Gas Lift Systems*. Teledyne MERLA, 1977.

6. Gudmundsson, J. S., I. Durgut, J. Ronnevig, H. K. Korsan, and H. K. Celius, "Pressure Pulse Analysis of Gas Lift Wells." Paper presented at the ASME/API Gas Lift Workshop, Aberdeen, November 12–13, 2001.

7. *Troubleshooting Procedures*. Baker Oil Tools, Houston, TX, 1988.

8. Ortiz, J. L., "Gas-Lift Troubleshooting Engineering: An Improved Approach." Paper SPE 20674 presented at the 65th Annual Technical Conference and Exhibition, held in New Orleans, LA, September 22–26, 1990.

Appendices

Appendix A

Description

Figure A–1 contains the Moody diagram for the calculation of friction factors for pipe flow.

Example problem

Find the friction factor if the value of the Reynolds number is 30,000 and the relative roughness of the pipe is 0.01.

Solution

At a Reynolds number of 30,000 and a relative roughness of 0.01, the friction factor is read as 0.04.

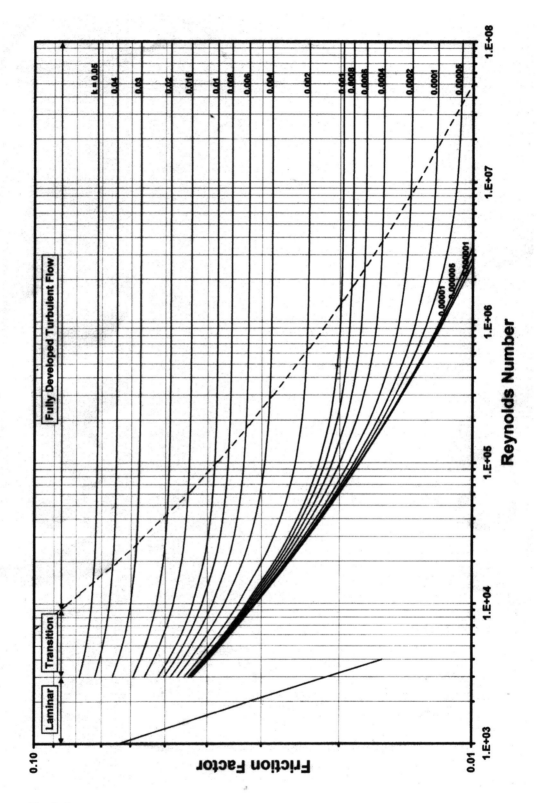

Fig. A–1

Appendix B

Description

Figure B–1 is a chart of calculated pressure gradients in a static gas column vs. surface pressure. The annulus temperature is assumed to change with well depth as shown in the figure. Gas deviation is properly accounted for.

Example problem

Find the gas column pressure at the depth of 4,500 ft if gas specific gravity equals 0.75 and the surface pressure is 800 psia.

Solution

From Figure B–1, gas gradient in the annulus is 25 psi/1,000 ft. Gas column pressure is thus 800 + 4.5 25 = 811.3 psia.

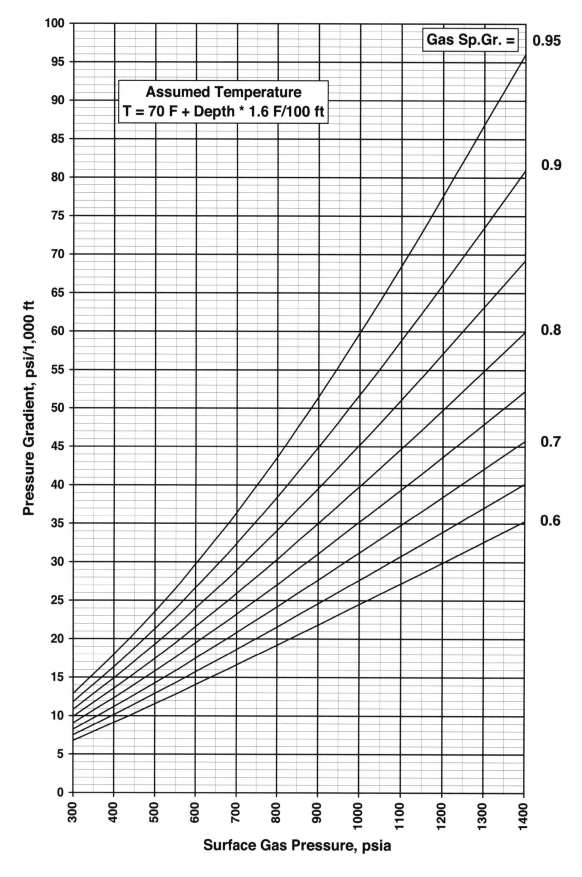

Fig. B–1

Appendix C

Description

Figures C–1 and C–2 present gas capacity charts for square-edge orifices of different sizes and allow the calculation of gas flow rates for different combinations of upstream and downstream pressures. Calculations are based on the following assumptions: Gas Sp.Gr. = 0.65; C_d = 0.865; T_1 = 60 °F. For other gas specific gravities, discharge coefficients and flowing temperatures, the gas rates found from these figures must be corrected according to Figure C–3.

Example problem

Find the gas passage capacity of an 30/64″ choke, if upstream and downstream pressures are 800 psig and 600 psig, respectively. Gas specific gravity is 0.7, flowing temperature is 100 °F, and a discharge coefficient of 0.9 is to be used.

Solution

Based on the choke size, Figure C–2 is to be used. Start at an upstream pressure of 800 psig and go vertically until crossing the curve valid for a downstream pressure of 600 psia. From the intersection, draw a horizontal to the left to the proper choke size (30/64″). Drop a vertical from here to the upper scale to find the gas flow rate of 3,432 Mscf/d. Since actual flow conditions differ from chart base values, the correction as given in Figure C–3 must be applied. At a temperature of 100 °F and a gas specific gravity of 0.7 the correction factor is read to be equal CF = 1.072. Corrected gas volume is according to the formula in Figure C–3 equals:

$$q_{corrected} = 3,432 \; 0.9 \; 1.072 = 3,311 \text{ Mscfd.}$$

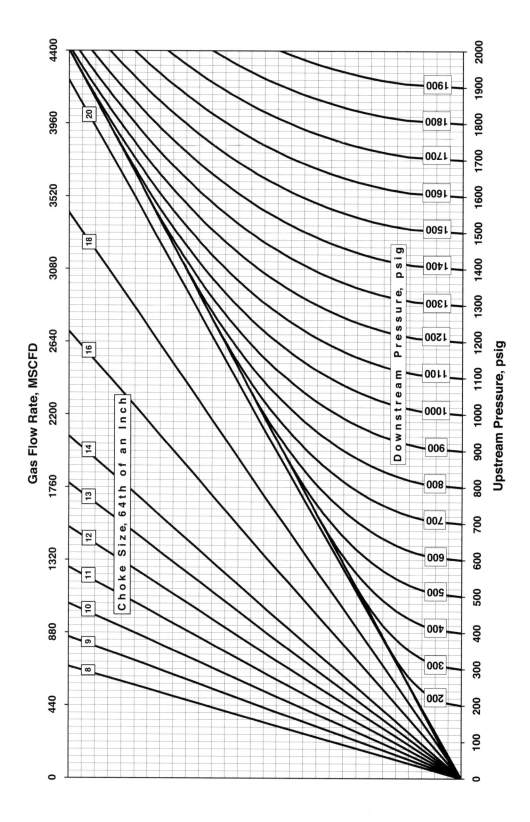

Fig. C–1

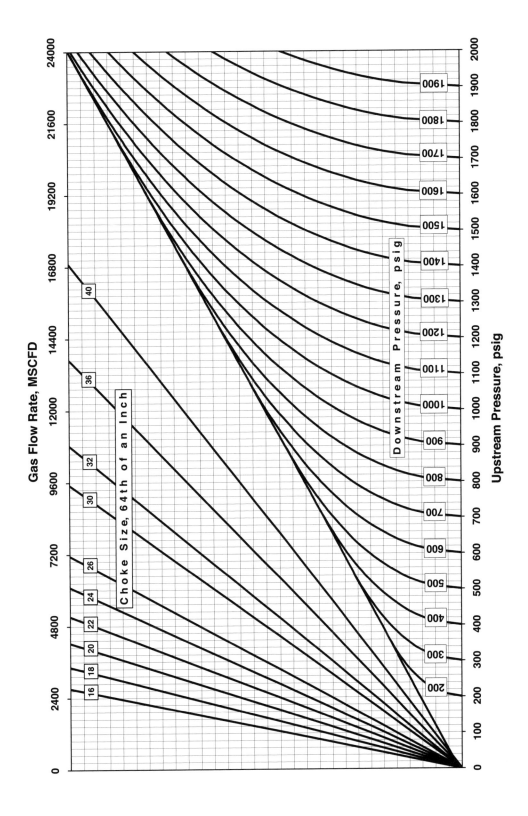

Fig. C–2

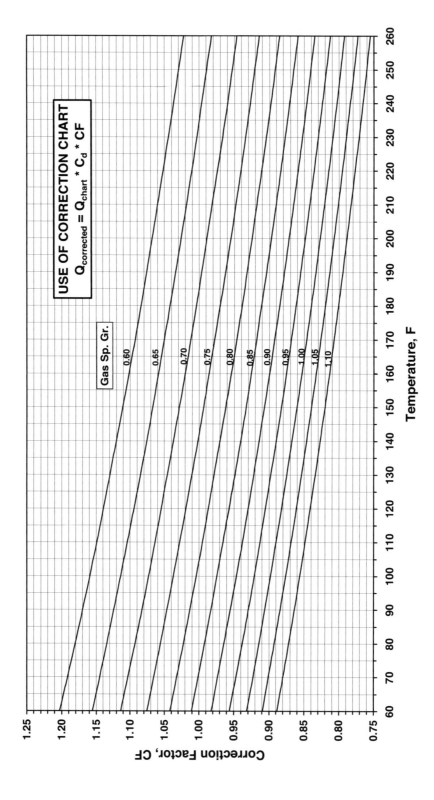

Fig. C–3

Appendix D

Description

Figures D–1 and D–2 show the variation of dome charge pressure with valve temperature for nitrogen gas charged gas lift valves.

Example problem

Find the gas lift valve's dome charge pressure at a valve temperature of 200 °F if the valve was charged with nitrogen gas to a pressure of 800 psig. Charging temperature was 60 °F.

Solution

Using Figure D–1, and reading the vertical pressure scale at the intersection of a vertical line drawn at 800 psig dome pressure and the valve temperature of 200 °F, the actual dome pressure is found as 1,040 psig.

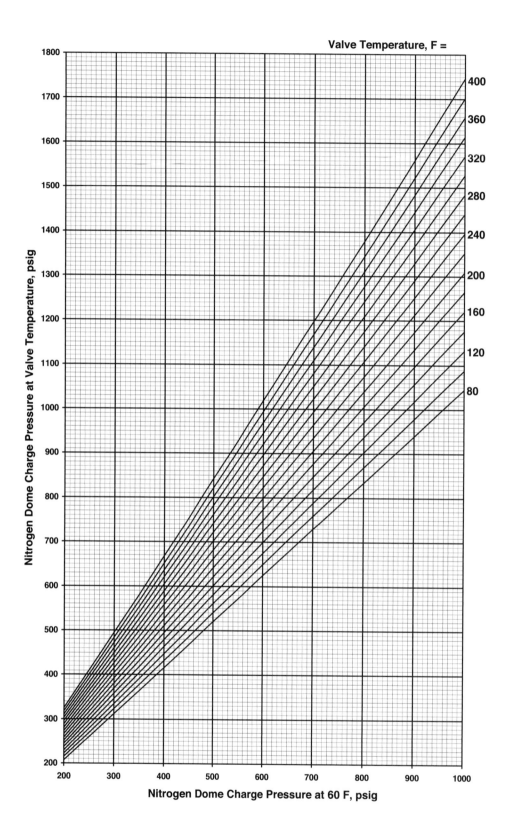

Fig. D–1

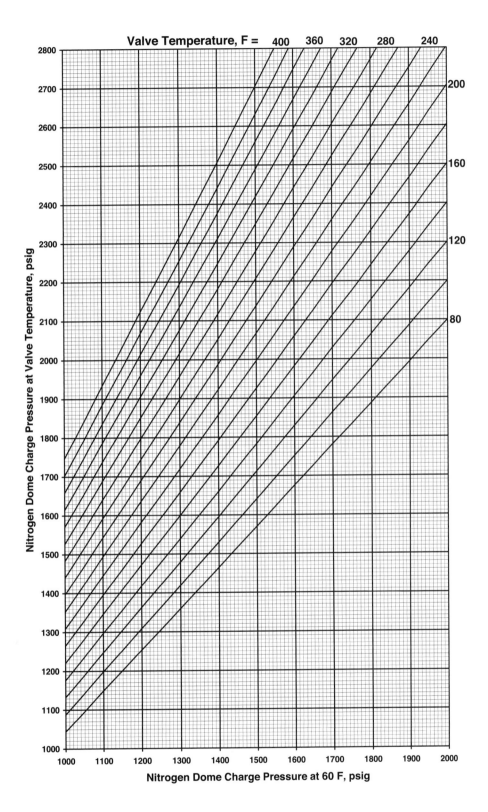

Fig. D–2

Appendix E

Description

Figures E–1 to E–6 allow the calculation of static gas column pressures for different injection gas gravities.

Example problem

Find the gas lift valve's surface opening pressure if the downhole opening pressure is 1,100 psig at the depth of 6,500 ft. The injection gas specific gravity is 0.7.

Solution

Using Figure E–3 and starting from a downhole pressure of 1,100 psig, a horizontal line is drawn to the intersection with the 6,500 ft depth line. The surface pressure is read as 918 psig.

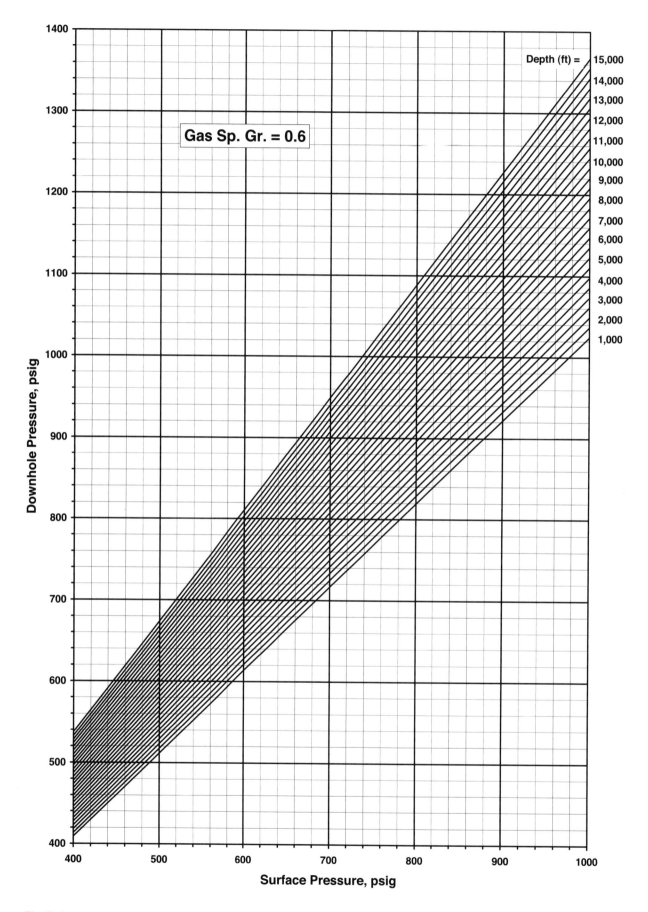

Fig. E–1

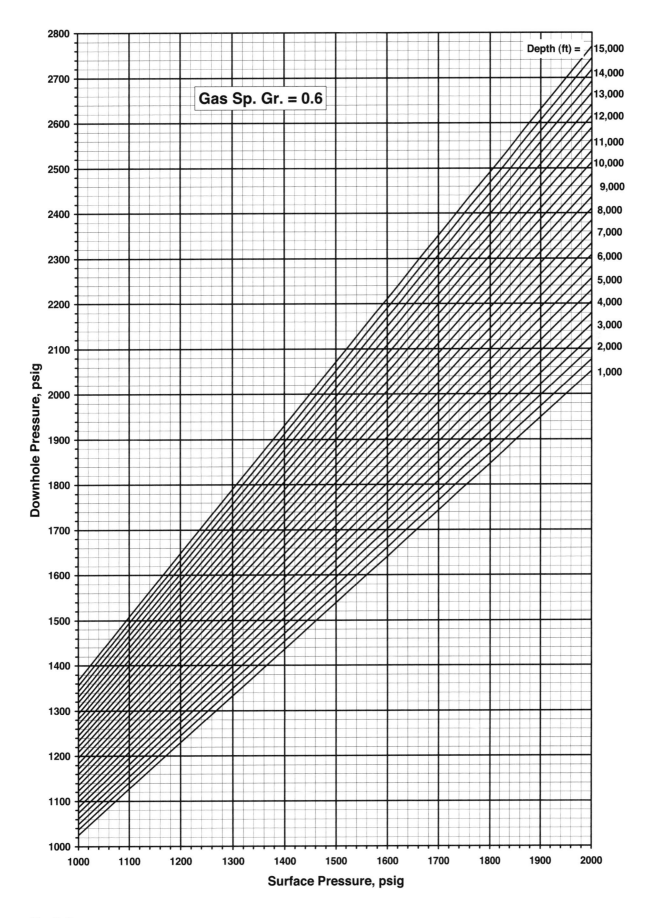

Fig. E–2

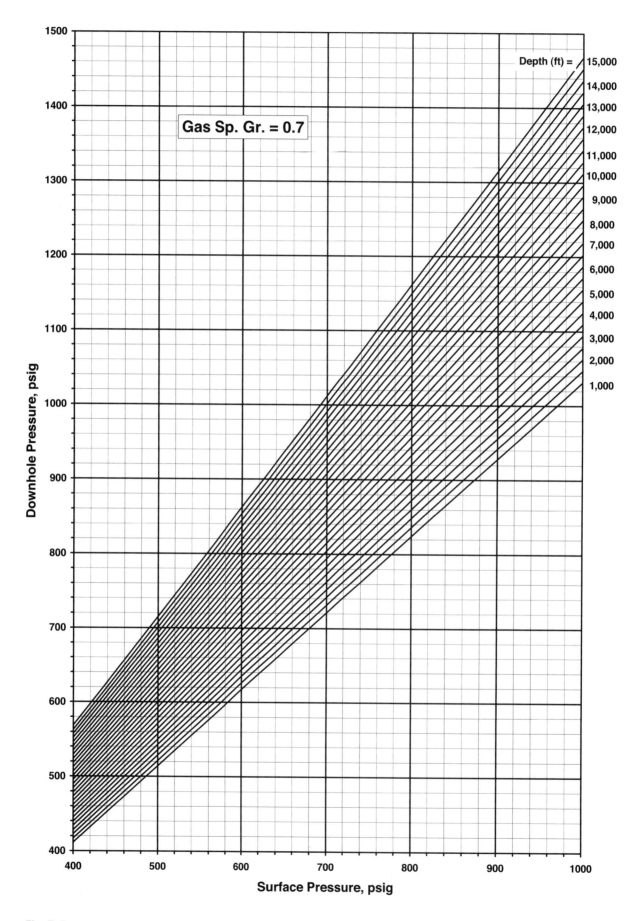

Fig. E–3

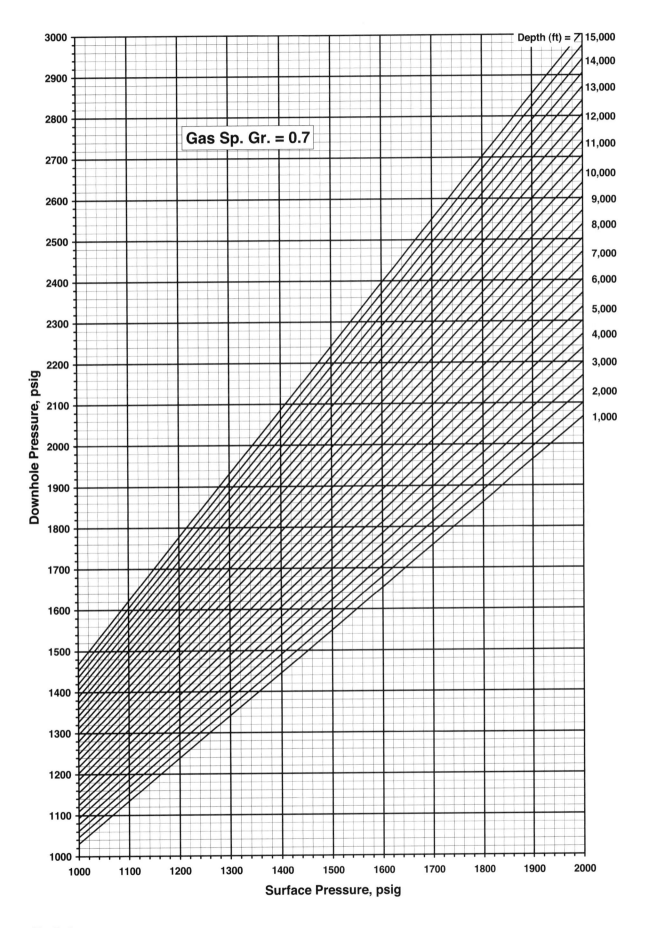

Fig. E–4

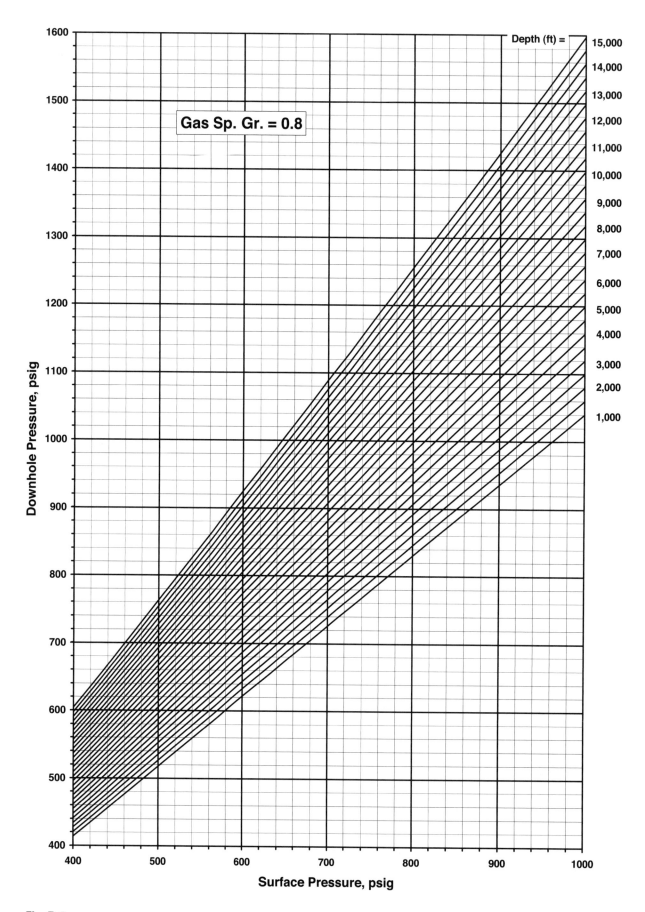

Fig. E–5

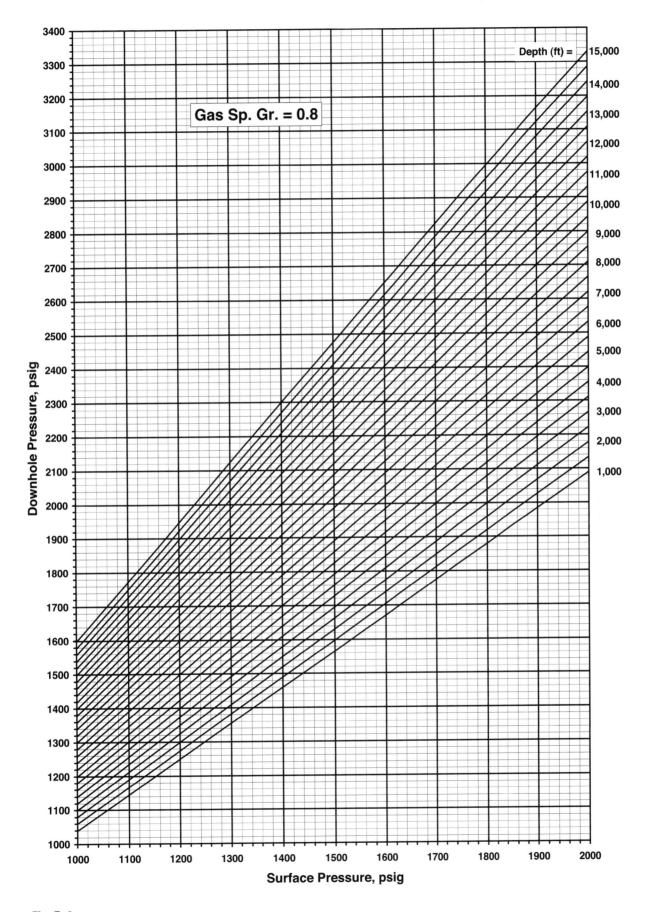

Fig. E–6

Appendix F

Description
Tables F–1 to F–4 contain mechanical data of gas lift valves available from two leading manufacturers.

Valve Type	Valve Description	OD	A_b	Port Size	A_v	A_v/A_b	PPEF
		in	sq in	in	sq in	-	-
J-20	unbalanced, bellows charged	1 ½"	0.77	3/16	0.029	0.038	0.040
				1/4	0.051	0.066	0.071
				5/16	0.079	0.103	0.115
				3/8	0.113	0.147	0.172
				7/16	0.154	0.200	0.250
				1/2	0.200	0.260	0.351
JR-20	unbalanced bellows charged w. crossover	1 1/2"	0.77	1/8	0.013	0.017	0.017
				3/16	0.029	0.038	0.040
				1/4	0.051	0.066	0.071
CP-2*	bellows charged, pilot operated, tubing flow only	1 1/2"	0.77	1/4	0.051	0.066	0.071
				5/16	0.079	0.103	0.115
				3/8	0.113	0.147	0.172
				7/16	0.154	0.200	0.250
				1/2	0.200	0.260	0.351
J-40	unbalanced, bellows charged	1"	0.31	1/8	0.013	0.042	0.044
				3/16	0.029	0.094	0.104
				1/4	0.051	0.165	0.198
				5/16	0.079	0.255	0.342
				3/8	0.113	0.365	0.575
JR-40	unb. bellows w. crossover	1"	0.31	1/8	0.013	0.042	0.044
				3/16	0.029	0.094	0.104
J-46-0	throttling, spring loaded only	1"	0.31	1/8	0.013	0.042	0.044
				3/16	0.029	0.094	0.104
				1/4	0.051	0.165	0.198
BP-2*	bellows charged, pilot operated, tubing flow only	1"	0.31	5/16	0.029	0.094	0.104
				1/4	0.051	0.165	0.198
				5/16	0.079	0.255	0.342
				3/8	0.113	0.365	0.575
J-50	unbalanced, bellows charged	5/8"	0.12	1/8	0.013	0.108	0.121
				5/32	0.020	0.167	0.200
				13/64	0.037	0.308	0.445
				1/4	0.051	0.425	0.739
JR-50	unb. bellows w. crossover	5/8"	0.12	3/32	0.008	0.067	0.072

* Data given for pilot section only.

Table F–1 CAMCO conventional gas lift valves

Valve Type	Valve Description	OD	A_b	Port Size	A_v Monel	A_v Carbide	A_v/A_b Monel	A_v/A_b Carbide	PPEF Monel	PPEF Carbide
		in	sq in	in	sq in	sq in	-	-	-	-
R-20	unbalanced, bellows charged	1 1/2"	0.77	1/8	0.013	0.021	0.017	0.027	0.017	0.028
				3/16	0.029	0.037	0.038	0.048	0.040	0.050
				1/4	0.051	0.058	0.066	0.075	0.071	0.081
				5/16	0.079	0.097	0.103	0.126	0.115	0.144
				3/8	0.113	0.129	0.147	0.168	0.172	0.202
				7/16	0.154	0.166	0.200	0.216	0.250	0.276
				1/2	0.200	0.230	0.260	0.299	0.351	0.427
R-28	unb. bellows w. crossover	1 1/2"	0.77	1/4	0.051	0.058	0.066	0.075	0.071	0.081
				5/16	0.079	0.097	0.103	0.126	0.115	0.144
R-25	unb. bellows w. crossover	1 1/2"	0.77	3/16	0.029	0.037	0.038	0.048	0.040	0.050
				1/4	0.051	0.058	0.066	0.075	0.071	0.081
				5/16	0.079	0.097	0.103	0.0126	0.115	0.144
R-25P	unb. bellows w. crossover	1 1/2"	0.77	3/8	0.113	0.129	0.147	0.168	0.172	0.202
RP-6*	bellows charged, pilot operated	1 1/2"	0.77	1/4	0.051	0.058	0.066	0.075	0.071	0.081
				5/16	0.079	0.097	0.103	0.126	0.115	0.144
				3/8	0.113	0.129	0.147	0.168	0.172	0.202
				7/16	0.154	0.166	0.200	0.216	0.250	0.276
				1/2	0.200	0.230	0.260	0.299	0.351	0.427
RPB-5*	bellows charged, pilot operated, w. chamber vent	1 1/2"	0.77	1/4	0.051	0.058	0.066	0.075	0.071	0.081
				5/16	0.079	0.097	0.103	0.126	0.115	0.144
				3/8	0.113	0.129	0.147	0.168	0.172	0.202
				7/16	0.154	0.166	0.200	0.216	0.250	0.276
RMI	throttling, spring loaded only	1 1/2"	0.65	1/4	0.051	0.058	0.066	0.075	0.071	0.081
				5/16	0.079	0.097	0.122	0.149	0.139	0.175
				3/8	0.113	0.129	0.147	0.168	0.172	0.202
				7/16	0.154	0.166	0.200	0.216	0.250	0.276
				1/2	0.200	0.230	0.308	0.354	0.445	0.548
BK PK-1*	bellows charged, pilot operated	1"	0.31	1/8	0.013	0.021	0.042	0.068	0.044	0.073
				3/16	0.029	0.037	0.094	0.119	0.104	0.135
				1/4	0.051	0.058	0.165	0.187	0.198	0.230
				5/16	0.079	0.097	0.255	0.313	0.342	0.456
BK-1	unbalanced, bellows charged	1"	0.31	1/8	0.013	0.021	0.042	0.068	0.044	0.073
				3/16	0.029	0.037	0.094	0.119	0.104	0.135
				1/4	0.051	0.058	0.165	0.187	0.198	0.230
				5/16	0.079	0.097	0.255	0.313	0.342	0.456
				3/8	0.113	0.129	0.365	0.416	0.575	0.712
BKF-10 BKR-5	unbalanced, bellows charged	1"	0.31	1/8	0.013	0.021	0.042	0.068	0.044	0.073
				3/16	0.029	0.037	0.094	0.119	0.104	0.135
				1/4	0.051	0.058	0.165	0.187	0.198	0.230
BKF-6	throttling, spring loaded only	1"	0.31	1/8	0.013	0.021	0.042	0.068	0.044	0.073
				3/16	0.029	0.037	0.094	0.119	0.104	0.135
				1/4	0.051	0.058	0.165	0.187	0.198	0.230
BKT BKT-1	unbalanced, bellows charged	1"	0.31	1/8	N/A	0.021	N/A	0.068	N/A	0.073
				3/16	N/A	0.037	N/A	0.119	N/A	0.135
				¼	N/A	0.058	N/A	0.187	N/A	0.230
				9/32	N/A	0.083	N/A	0.268	N/A	0.366
				5/16	N/A	0.097	N/A	0.313	N/A	0.456
				3/8	N/A	0.129	N/A	0.416	N/A	0.712

* Data given for pilot section only.

Table F–2 CAMCO wireline retrievable gas lift valves.

Valve Type	Valve Description	OD	A_b	Port Size	A_v	A_v/A_b	PPEF
		in	sq in	in	sq in	-	-
C-2	unbalanced, bellows charged	1 1/2"	0.77	1/8"	0.013	0.017	0.017
				5/32"	0.020	0.027	0.027
				3/16"	0.029	0.038	0.040
				1/4"	0.051	0.066	0.071
				3/8"	0.113	0.147	0.172
				7/16"	0.154	0.201	0.252
CF-2	throttling, spring loaded only	1 1/2"	0.77	3/16"	0.029	0.038	0.040
				1/4"	0.051	0.066	0.071
				5/16"	0.079	0.103	0.115
				3/8"	0.113	0.147	0.172
C-1	unbalanced, bellows charged	1"	0.31	1/8"	0.013	0.042	0.044
				5/32"	0.020	0.066	0.071
				3/16"	0.029	0.094	0.104
				1/4"	0.051	0.165	0.198
				5/16"	0.079	0.255	0.342
CF-1	throttling, spring loaded only	1"	0.31	1/8"	0.013	0.042	0.043
				5/32"	0.020	0.066	0.071
				3/16"	0.029	0.094	0.104
				1/4"	0.051	0.165	0.197
C-3	unbalanced, bellows charged	5/8"	0.12	1/8"	0.013	0.111	0.124
				5/32"	0.020	0.170	0.205
				3/16"	0.029	0.243	0.320

Table F–3 McMurry-Macco conventional gas lift valves.

Table F–4 McMurry-Macco wireline retrievable gas lift valves.

Valve Type	Valve Description	OD	A_b	Port Size	A_v	A_v/A_b	PPEF
		In	sq in	in	sq in	-	-
R-2		1 1/2"	0.77	3/16"	0.029	0.038	0.040
	unbalanced, bellows charged			1/4"	0.051	0.066	0.071
				5/16"	0.079	0.103	0.115
				3/8"	0.113	0.147	0.172
				7/16"	0.154	0.201	0.252
R-2D		1 1/2"	0.77	3/16"	0.029	0.038	0.040
	unbalanced, bellows charged			1/4"	0.051	0.066	0.071
				5/16"	0.079	0.103	0.115
				3/8"	0.113	0.147	0.172
				7/16"	0.154	0.201	0.252
R-2CF		1 1/2"	0.77	3/8"	0.113	0.147	0.172
RF-2		1 1/2"	0.77	3/16"	0.029	0.038	0.040
	throttling, spring loaded only			1/4"	0.051	0.066	0.071
				5/16"	0.079	0.103	0.115
				3/8"	0.113	0.147	0.172
RPV-2*		1 1/2"	0.77	3/16"	0.029	0.038	0.040
	bellows charged, pilot operated			1/4"	0.051	0.066	0.071
				5/16"	0.079	0.103	0.115
				3/8"	0.113	0.147	0.172
RPV-2S*		1"	0.31	3/16"	0.029	0.094	0.104
	spring loaded, pilot operated			1/4"	0.051	0.165	0.198
				5/16"	0.079	0.255	0.342
R-1		1"	0.31	1/8"	0.013	0.042	0.044
	unbalanced, bellows charged			5/32"	0.020	0.066	0.071
				3/16"	0.029	0.094	0.104
				1/4"	0.051	0.165	0.198
				5/16"	0.079	0.255	0.342
R-1BL		1"	0.31	1/8"	0.013	0.042	0.044
	unbalanced, bellows charged			5/32"	0.020	0.066	0.071
				3/16"	0.029	0.094	0.104
				1/4"	0.051	0.165	0.198
				5/16"	0.079	0.255	0.342
R-1D		1"	0.31	1/8"	0.013	0.042	0.043
	unbalanced, bellows charged			5/32"	0.020	0.066	0.071
				3/16"	0.029	0.094	0.104
				1/4"	0.051	0.165	0.197
				5/16"	0.079	0.255	0.342
R-1CF		1"	0.31	5/16"	0.079	0.255	0.342
RF-1		1"	0.31	1/8"	0.013	0.042	0.043
	throttling, spring loaded only			5/32"	0.020	0.066	0.071
				3/16"	0.029	0.094	0.104
				1/4"	0.051	0.165	0.197
RPV-1*		1"	0.31	3/16"	0.029	0.094	0.104
	bellows charged, pilot operated			1/4"	0.051	0.165	0.198
				5/16"	0.079	0.255	0.342
RPV-1S*		5/8"	0.12	3/16"	0.029	0.242	0.319
	spring loaded, pilot operated			1/4"	0.051	0.425	0.575
				5/16"	0.079	0.658	1.924

* Data given for pilot section only.

Index

A

Absolute open flow potential (AOFP), 23, 25

Acceleration losses, 284. *See also* Kinetic losses

Accumulation period
in intermittent cycle, 363–64
in plunger-assisted intermittent cycle, 397

Accuracy/selection, of pressure drop models, 114–21, 119–20

Acoustic surveys, 428–29

Adiabatic power, for gas compression, 289–91

Afterflow period
in intermittent cycle, 364, 365
liquid production in, 370–71
in plunger-assisted intermittent cycle, 397

Air-for-hire, 4

Annular flow
in horizontal flow patterns, 129
pressure drop prediction in, 60
solids in, minimizing, 342

Ansari et al. model, 110–11
accuracy of, 118, 119, 120

AOFP, *See* Absolute open flow potential

API RP 11V2 (API Recommended Practice for Gas-Lift Valve Performance Testing)
procedure, 220–26
orifice flow model, 220–21
throttling flow model, 221–26
valve performance curves, 226–27

API Specification for Gas Lift Equipment, 240–42

Artificial lifting methods, 1–3

Available formation pressure, 257

Average reservoir pressure, 22

Aziz-Govier-Fogarsi model
accuracy of, 118, 119, 120
basic equation, 98
bubble flow in, 99–101
bubble rise velocity in, 100, 102
Drift-flux model in, 98
Duns-Ros model and, 101
flow pattern map, 98–99, 102
friction factor in, 103
mist flow in, 101
modifications to, 101–2
slug flow in, 100–101, 102
summary of, 98
transition boundaries in, 102
transition flow in, 101

B

Balanced valves
  bellows type, 204–5, 233
  flexible sleeve, 204–5, 234
  force balance equations for, 205
  in intermittent gas lift, 358
  intermitter control and, 358
  IPO, 204–6
  PPO, 206
  setting depths for, 332–33

Beggs-Brill correlation, 85–91
  accuracy of, 118, 119, 120
  basic equation, 86
  compared to other methods, 114, 115
  flow pattern in, 85–86, 86–87, 88
  friction factor in, 88–89
  in horizontal/inclined flow calculations, 128, 130
  liquid holdup in, 86, 87–88
  modifications in, 89
  Mukherjee-Brill correlation and, 95
  in multiphase flow calculations, 291
  pressure gradient calculations, 89–91
  summary of, 85–86

Bellows-charged gas lift valves
  setting/testing, 228
  in unloading of continuous flow installations, 319

Black oil model
  in engineering calculations, 45

Bottom dumper springs, in plunger-assisted
  intermittent gas lift, 396

Bottomhole pressure
  bubblepoint pressure and, 24
  in installation design, 259
  variations in open/semi-closed gas lift
    installations, 248

Bubble flow
  in Aziz et al. model, 99–101
  frictional pressure gradient, 68
  in Hagedorn-Brown correlation, 76
  in Hasan-Kabir model, 104–5
  kinetic factors in, 60
  in Orkiszewski correlation, 79
  slip velocity formula for, 67
  transition parameters, 66

Bubblepoint pressure
  bottomhole pressure and, 24
  crude oil, estimating, 13
  errors in, 116, 117
  hydrocarbon fluid systems and, 11
  importance of, 116
  Standing's correlation applied to, 13

Bubble rise velocity
  in Aziz et al. model, 100, 102

in Hasan-Kabir model, 106
  in liquid holdup calculations, 58
  in Orkiszewski correlation, 80

Bubble-slug boundary, 71
  in Hasan-Kabir model, 109

C

Calculation direction, 116–17

Calculation errors, 118
  in pressure drop models, 119

CAMCO Gas Lift Manual, 366

Casing flow installations, 253

Casinghead pressure (CHP)
  in intermittent lift cycle, 354, 355, 357, 358, 359, 360, 424–25
  pressure recorders and, 422–25

Casing pressure control, in intermittent gas lift, 360

CEF (Casing effect factor), See IPEF

Chamber and plunger lift, 5

Chamber installation, 248–50
  hookwall packer in, 382
  intermittent cycle in, 381
  for intermittent gas lift operations, 248–49, 381–88

Chamber lift
  applications/basic features of, 381–82
  calculations for, 383–88
  equipment selection for, 382–83
  gas requirement in, 382
  in intermittent gas lift design, 381–88
  IPO valves in, 383
  liquid fallback in, 382, 384
  pilot valves in, 385, 387
  pros/cons of, 381–82
  valve spacing in, 384, 387–88
  valve spread in, 383
  valve/packer depth settings in, 384
  wireline retrievable equipment in, 383

Chart base values, correcting, 42, 43

Chierici et al. correlation, 85
  accuracy of, 118, 119, 120

Choke control
  advantages of, 357
  calculation data for, 375, 376
  in continuous flow gas lift, 344–45
  gas volume and, 369
  in intermittent gas lift, 354–57, 389
  intermitters as, 359–60
  in Optiflow Design, 378
  water flow rates in, 141

Choke flow, 39–42
  calculating, 40–42, 144–46
  capacity chart, 43
  critical/sub-critical, 40, 41, 42, 139
  Equation of Continuity and, 40
  flow v. pressure chart, 40
  gas capacity charts, 42–43
  single phase, 144

Chokes, 344–46
  adjustable, 361
  continuous flow gas lift and, 264
  fixed, 361
  as flow through restrictions, 139
  freezing in, 345, 356
  in gas injection control, 344–46, 361
  metering valve, 345
  sizes, determining, 140
  for surface gas injection control, 369
  throughput capacity of, 43
  types/functions of, 139, 264

Chokshi, Schmidt, and Doty model, 111
  accuracy of, 118, 119, 120

CHP, *See* Casinghead pressure

Churn flow, in Hasan-Kabir model, 104, 105, 107

Closed gas lift installation, 248
  for intermittent gas lift, 363, 381
  plunger-assisted, 394
  schematic of, 408

Closed rotative gas lift installation, 5
  calculating gas balance for, 416
  compression selection for, 415
  design of, 413–19
  distribution system design, 416
  efficiency of, 413
  freezing in, 414
  gas requirement for, 418–19
  gathering system design, 416–17
  intermitters in, 413
  schematic of, 408
  subsystems of, 413–14

Coiled tubing (CT), 173, 253

Colebrook equation, 29

Compressibility/supercompressibility factor, *See* Deviation factors

Compressor discharge pressure, injection pressure v., 290

Compressor selection, 293–97
  calculated operating points for, 297
  costs and, 293
  for optimizing continuous flow installations, 293–97, 303–8

performance curves for, 295, 296
  pressure calculations for, 293–94
  systems analysis of producing wells and, 293

Constant slip model, 134

Constant surface closing pressure, in intermittent installation design, 371–76

Continuous flow gas lift, *See also* Optimizing continuous flow installations
  basic operational parameters for, 258
  calculating gas requirements for, 263–64
  calculating minimum energy requirement for, 289–91
  choke controls in, 344–45
  compressor selection for, 293–97, 303–8
  cost calculation in, 289
  dehydration for, 345
  design calculations for, 258–60
  design considerations in, 260–61
  design sample for, 260
  efficiency concerns in, 263, 265
  equipment selection for, 245, 246
  flexibility of, 256
  flowline performance in, 295, 296
  gas injection control in, 345–46
  gas injection pressure in, 264–65
  gas liquid ratio in, 259
  injection depth in, 261–62
  installation design basics, 258–61
  intermittent lift v., 5–6, 231, 256, 351, 352–53
  IPO valves in, 343
  mechanism of, 256
  minimum operating costs in, 288
  multipoint injection and, 262–63
  open installation and, 245, 246
  Optiflow design in, 329–31
  optimizing, in single well, 289–308
  optimum conditions in, 288–315
  origins of, 4
  predicted well data in, 289
  pressure recorders in, 422–24
  principles of, 257–65
  problems, common, 431–33
  regulator controls in, 345–46
  semi-closed installation for, 247
  static liquid level in, 318
  surface controls in, 344–46
  system parameters in, 288–89
  systems analysis for, 266
  time cycle control in, 346
  troubleshooting, 422–23, 424, 426, 427, 428
  tubing size in, 265, 293–94
  unloading, 315–44
  valve damage in, 342
  well death v., 258

Cornish correlation, 92–93
  accuracy of, 118, 119, 120

Costs, minimizing, 288
   compressor selection and, 293

Critical flow correlations, 139–42
   gas content affecting, 142

Gilbert's equation for, 139–40
   pressure ratio calculations, 142–44

Critical flow criteria, 142–44

Crude oil, basic properties of, 13–15

CT installations, 253

CT, *See* Coiled tubing

D

Darcy's equation, 23

Darcy's Law, 21

Darcy-Weisbach equation, 28
   frictional pressure gradient and, 68
   for pressure loss in fluid friction, 53

Dead wells, gas lifting v., 257–58

Density, fluid, 10

Deviated wells, flow pattern in, 60

Deviation factors
   calculating, 34, 36
   in gas column pressure calculations, 34, 37
   for natural gases, 16–17

Differential gas lift valve, 238–39

Distribution system problems, common, 432–33

Dome charge pressure
   calculations, 174–77, 184
   charts, 176, 177, 178, 179, 444, 445
   setting, 228
   temperature and, 228

Downhole equipment, evolution of, 5

Downhole pressure
   calculating, 174
   injection pressure v., 179

Downhole problems, common, 430–32

Downhole surveys
   running procedures for, 427–28

Drainage radius, 22

Drift-flux model
   in Aziz et al, 98
   for gas slippage, 48–49
   liquid holdup and, 51

Dual gas lift installations, 251, 401–5
   casing size in, 402
   costs and, 401, 402
   definition of, 401
   design principles for, 403–5
   distance between zones in, 402
   problems with, 401–2
   schematic of, 401
   unloading valve strings for, 403–5
   valve selection for, 403
   valve spacing in, 402
   wireline retrievable equipment in, 402

Dukler correlation, 134–37
   basic equation, 134–35
   traverse calculation flowchart, 138

Duns-Ros correlation, 64–72
   accuracy of, 118, 119, 120
   Aziz-Govier-Fogarsi model and, 101
   basic equation for pressure drop prediction, 64–65
   compared to other methods, 114
   flow patterns in, 65
   gradient curves and, 124
   Orkiszewski correlation and, 78
   in Orkiszewski's method, 82
   slug flow in, 69
   summary of, 64–65
   in systems analysis of producing wells, 288

Duns-Ros flow pattern map, 56–57, 64

Duns-Ros friction factor parameters, 67–69

Dynamic viscosity, *See* Viscosity, fluid

E

Elevation gradient, *See* Hydrostatic gradient

Empirical correlations, 116
   mechanistic models v., 116

Empirical models
   mechanistic models v., 97
   for multiphase flow, 60–61, 114–15

Energy loss factor correlations, 61, 62–63

Engineering Equation of State, 15
   in gas lift valve operation, 174
   gas volume factors and, 17

Equation of Continuity, choke flow and, 40

Equilibrium Curve
   calculation of, 272–77
   in continuous flow gas lift design, 272–76
   in systems performance calculations, 272–76

Errors
    in bubblepoint pressure, 116, 117
    in fluid property data, 116
    gradient curves and, 124
    in input data, 117
    in measured data, 117
    in pressure drop models, 119
    in pressure drop predictions, 115, 260
    in pressure gradient charts, 177
    in reservoir performance data, 343
    sources of, 115
    in valve capacity calculations, 215

Experimental data, lack of, 116

F

Fallback, *See* Liquid fallback

Fetkovich's method, 26

Flag valve, 319, 373

Flexible sleeve pilot valves, 214–15

Flow coefficient data, 219

Flow conditions
    adopted assumptions on, 144
    chart values v., 42, 43
    gas v. liquid, 48
    steady-state, 284

Flow pattern
    in Beggs-Brill correlation, 85–88
    boundary equations for, 99
    determination of, 46
    differential equations for, 53
    in gas lift valves, 219, 220, 224, 225
    in Mukherjee-Brill correlation, 94
    in Orkiszewski's method, 78, 82
    pipe inclination and, 60
    prediction of, 55–57
    vertical, 57

Flow pattern map, 66
    in Aziz-Govier-Fogarsi model, 98–99, 102
    in Beggs-Brill correlation, 86–87
    Duns-Ros, 57
    in Hasan-Kabir model, 104–5
    in Mukherjee-Brill correlation, 94
    in Orkiszewski's correlation, 78–79

Flow pattern maps, 46, 56–57
    empirical method failure and, 97
    for horizontal flow, 129

Flow patterns
    annular (mist) flow, 56
    in annuli, 60
    bubble flow, 55
    calculation models and, 65
    dispersed bubble flow, 55, 57
    in empirical models, 61
    first use of, 64–65
    in Hagedorn-Brown correlation, 76
    horizontal v. vertical, 45
    pressure drop and, 45
    slug flow, 46, 55, 57
    stratification in, 45
    transition (churn) flow, 56
    vertical, 54–55

Flow rate
    basic principles of, 160–61
    calculation models, 144–45
    in gas lifted wells, 155
    in systems analysis of producing wells, 160–61
    in temperature calculations, 146, 147, 149, 150, 151, 152, 153, 154, 155

Flow regimes, *See* Flow patterns

Flow resistance, 1

Flow temperature
    Ramey's formula for, 149

Flow through restrictions, 139–46
    chokes as, 139

Flow velocity
    calculating, 27
    kinetic factors in, 54
    in multi-phase flow, 53

Flowing bottomhole pressure (FBHP), 22, 23

Flowing pressure surveys, 428–29

Flowing temperature surveys, 428–29

Flowline performance
    calculating, 279
    in continuous flow gas lift, 277, 295, 296

Fluctuating line pressure, 260

Fluid differential valves, 210–11
    opening/closing equations for, 210–11

Fluid operated valves, 210–14

Fluid pilot valves
    opening/closing equations for, 212–14
    TRO pressure in, 213

Fluid properties, calculation errors and, 116

Fluid-operated valve, *See* PPO valves

Fluids, oilfield
  flow problems with, 26
  methods for calculating properties of, 10–15
  phase relations of, 9
  physical parameters of, 9
  properties of, 9–15, 10–11, 52
  property correlations for, 11
  water, as comparison standard, 12

Foot-pieces, 5

Freezing
  in chokes, 345, 356
  in closed rotative gas lift system, 414

Frictional losses, 59

Frictional pressure gradient
  calculation, 72, 73
  in Hagedorn-Brown correlation, 76
  in mist flow, 70

Friction factor, 284, *See also* Viscosity, fluid
  in Aziz et al. model, 103
  in Beggs-Brill correlation, 88–89
  calculating, 28, 29–30, 34
  in Cornish correlation, 92–93
  in Dukler correlation, 135
  energy loss factor for, 62–63
  in gas pressure calculations, 32
  in Hasan-Kabir model, 106, 108, 110
  in Mukherjee-Brill correlation, 95–97
  in multi-phase flow, 53
  in Orkiszewski correlation, 79, 84
  parameters determining, 59

G

Gas, *See* Natural gas

Gas allocation, optimum, 309–10, 309–15
  calculations for, 310–14
  for complete network, 314–15

Gas blowdown, in plunger-assisted intermittent cycle, 397

Gas breakthrough, in intermittent gas lift, 364–65

Gas capacity chart, 440, 441

Gas column pressure, calculating, 35–40

Gas compression, calculating costs, 290

Gas density
  evaluating, 35
  in two-phase flows, 50

Gas flow, *See also* Single-phase gas flow

Gas flow metering, pressure drop in, 39–40

Gas flow rate, gas-liquid ratio v., 284

Gas flow rates
  controlling,, 40
  graphical solutions and, 42–43

Gas flow velocity, 33

Gas gradient in annulus, 35–39
  chart, 37
  computer solution, 37
  theoretical background, 35–37

Gas injection
  depth of, 261–62
  multipoint v. single-point, 262–63
  point of, 315
  point/requirement, finding, 259
  pressure, 264–65
  problems with, 432
  rate, 40, 280, 282, 344, 433

Gas injection control
  casing pressure, 360
  choke, 354–57
  devices for, 361–62
  in dual gas lift installations, 404–5
  in intermittent gas lift, 408–9
  intermitter, 357–60
  metering valves in, 361
  pressure regulators in, 361–62
  regulator, 356–57
  tubing pressure, 360

Gas injection rate
  improper, 433
  liquid flow rate and, 280, 282
  in unloading, 344

Gas injection subsystem, stability of, 286

Gas lift
  injection pressure v. depth, 177–80
  malfunctions, common, 430–33
  nitrogen v. natural gas, 177

Gas lift conditions, optimum, 310, 312–13

Gas lift equipment
  evolution of, 170–71
  patents issued for, 172

Gas lift installation
  data types for evaluation, 421–22
  design v. actual parameters, 421–22
  troubleshooting/analysis, 421–33

Gas lift installation types
  casing flow, 253
  chamber installations, 248–50
  closed, 248
  CT installations, 253
  macaroni, 251–52

multiple, 251
open, 245–47
pack-off, 251–52
semi-closed, 247
tubing flow, 245–50

Gas lift performance curves, 287, 307
for multiple wells, 311
typical, 298

Gas lift system, 407–19
closed, 408
field depletion v., 408
functions of, 407–8
modifying existing facilities for, 407
open, 408
semi-closed, 408
types of, 408

Gas lift valve mandrel types, 234–37

Gas lift valve/mandrel designation, in API Specification
for Gas Lift Equipment, 240–42

Gas lift valve performance, 215–27
balance equations and, 216
bellows valves, 216
data sample, 218
deviation factor in, 216
flow rate calculations, 216
general models of, 219–27
injection capacity, 215–16, 217–19
introduction, 215–16
orifice/throttling flow curve, 219, 220
spring loaded/throttling valve, 216–17
testing procedures, 216
throttling valves, 217–19

Gas lift valves, *See also* Pressure operated valves; Valve
mechanics
agers for, 228
API designations for, 240–41
application of, 230–34
bellows operated, 228
bellows-operated, 172, 173
bracketing of, 260
calculating behavior of, 174–80
casing flow valves, 174
classes of, 172–74
coiled tubing (CT), 173
concentric, 172, 173
for continuous flow operations, 230–31
for continuous v. intermittent lifting, 173
differential, 173, 238–39
evolution of, 171–72
flexible sleeve valves, 173
flow pattern in, 219, 220
injection controls, 170–71
for intermittent lift operations, 231–32

introduction, 169
King valve, 172
mechanically controlled, 171, 172
nozzle-Venturi valve, 239–40
operational conditions of, 174
orifice valve, 239
overview of types, 172–74
pack-off, 173
pilot valves, 173
pressure-operated, 172, 173, 174, 180–234
proportional response in, 231
pros/cons of various types, 232–34
pump-down, 173
requirements for different operations, 230–34
retrievable, 173
running/retrieving, 234–38
setting of, 227–30, 230, 260
specific gravity differential, 172
spring-loaded differential, 171, 173, 231
temperature and, 174
temperature/injection pressure and, 174
throttling valves, 231
tubing flow valves, 174
velocity controlled, 172
velocity-controlled, 171
water baths for, 228, 230

Gas lifted wells
calculating temperature in, 155–58
heading in, 284
instability in, 285
peculiarities of, 155
systems analysis of, 161–62
temperature distribution chart, 158

Gas lifting
advantages/limitations of, 6–7
applications of, 6
continuous flow, 2, 3, 4
continuous v. intermittent operations, 231
control methods, early, 170–71
history of, 4–5
intermittent, 2–3
methods compared, 3–4
operational mechanisms, 5–6
plunger-assisted, 3
pressure drop calculation and, 54

Gas-liquid ratio (GLR)
in calculating optimum operation, 290–97
for continuous flow gas lifting, 259
gas flow rate v., 284
in multipoint injection calculations, 262–63

Gas-liquid ratio (GLR), optimum
well performance and, 267–69

Gas phase correlations, 58

Gas property correlations
   deviation factors in, 19
   pseudocritical parameters in, 18–19

Gas requirement
   in chamber lift installation, 382
   in Optiflow Design, 379
   plunger-assisted lift and, 393

Gas requirement chart, 369

Gas slippage
   drift-flux model for, 48–49
   importance of, 115
   liquid holdup and, 50
   in two-phase flow, 48

Gas throughput of chokes, calculating, 40–42

Gas velocity, buoyancy affecting, 48

Gas volume
   calculation of, 47
   choke control and, 369

Gas well delivery equation, 26

Gathering system problems, common, 433

General Energy Equation, 27
   for multi-phase flow patterns, 53
   for static gas column, 35
   in two-phase flow problems, 45

GLR, *See* Gas-liquid ratio

Gomez et al. model, 111
   accuracy of, 118, 119

Gradient curves, 121–37
   computer applications and, 123
   Duns-Ros correlation in, 124
   errors and, 124
   features of various, 123
   Gilbert's, 121–22
   Hagedorn-Brown correlation in, 123
   Poettmann-Carpenter correlation in, 123
   samples of, 121, 124, 125, 126
   steps for use of, 122

Griffith-Wallace model, Orkiszewski correlation and, 78, 79, 80, 81

Guiberson Ratiometric System, 367–68

H

Hagedorn and Brown correlation, in Systems analysis of producing wells, 288

Hagedorn-Brown correlation, 73
   accuracy of, 118, 119, 120
   basic equation, 74

bubble flow in, 76
compared to other methods, 114, 115
flow patterns and, 73, 74, 76
frictional pressure gradient in, 76
gradient curves and, 123
imperfections/modifications in, 76
kinetic component in, 74
liquid holdup in, 73, 74–75
mixture densities in, 74–75
Orkiszewski's use of, 78, 81
pressure gradient calculations in, 76–77
summary, 73–74

Hasan-Kabir model, 103–10
   accuracy of, 118, 119, 120
   basic equation, 104
   bubble flow in, 104–5
   bubble rise velocity in, 106
   Bubble-slug boundary in, 109
   churn flow in, 104, 105, 107
   flow pattern map, 104–5
   friction factor in, 106, 108, 110
   in horizontal/inclined flow calculations, 128
   mist (annular) flow in, 104, 105, 106, 107–9
   pressure gradient calculations in, 104
   slug flow in, 104, 105, 106–7
   summary of, 103
   transition flow in, 104

Hasan-Kabir, modifications to Ramey's model by, 151

Heading in gas lifted wells, 284, 285, 431
   elimination of, 285

Heat loss/transfer
   in gas lifted wells, 155
   in Ramey's model, 147
   in Sagar et al. model, 153–55

Heat transfer coefficient, in Ramey's temperature model, 148

Heat transfer in rock media, 148

Heat transfer in well materials, 147–48

High reservoir pressure, continuous flow gas lifting and, 257

Hookwall packer, in chamber installation, 382

Horizontal flow
   patterns of, 128–29
   pressure traverse calculation in, 137–38

Horizontal gas flow, calculating pressure drop in, 34

Horizontal gas flow in pipes, 32

Horizontal pressure drop, empirical correlations for, 130–37

Horizontal two-phase flow, pressure gradient calculations in, 130–31

Horizontal/inclined flow
  calculating pressure drop in, 127–28
  pressure drop in, 127–38

Horizontal/inclined flow calculations
  Beggs-Brill correlation in, 128

Hasan-Kabir model in, 128

Hydraulic concepts, basic, 27–30
  hydraulic diameter concept, in annuli, 60

Hydrostatic gradient, 27, 284

I

Ideal Gas Law, 15

In dual gas lift installations
  mandrels in, 402
  tubing in, 402
  valves in, 402

Inclined/annulus flow, 60

Inflow performance, 9, 13, 14, 21–26
  curved shape of, 24
  description procedures for, 21
  pressure drawdown and, 23
  productivity index (PI) and, 23
  relationships in, 24–26
  sandface/reservoir pressure v., 21

Inflow performance relationship (IPR), 21, 24–26

Injection conditions, in gas lift valves, evaluating,
  See API RP 11V2 procedure

Injection control, 344–46
  See also Surface gas injection control

Injection controls, See also Gas lift valves
  evolution of, 170–71
  foot-piece/oil injector, 170
  foot-pieces, 170–71
  jet collars/orific inserts, 170–71
  kickoff valves, 171
  open tubing, 170

Injection depth, in continuous flow gas lift design,
  261–62

Injection pressure
  calculating effect of, 265
  compressor discharge pressure v., 290
  depth v., 177–80
  system performance and, 267–72
  valve spread and, 193
  valve stem travel v., 184–85

Injection pressure operated values, See IPO valves

Injection pressure v. depth, 177–80

Injection pressure/rate
  in unloading of continuous flow installations, 318
  in unloading of intermittent gas lift installations, 372

Injection rates, 221–26, See also API RP 11V2 procedure
  fixed orifice v. throttling valve, 216–17
  performance curves for, 306

Insert chamber installations, 249–50

Installation types, gas lift
  introduction, 24
  tubing flow installations, 245–50

Intermittent cycle
  accumulation period, 363–64
  afterflow period, 364, 365
  in chamber installation, 381
  description of, 359, 360
  evaluating efficiency of, 424
  gas requirement of, 368–69
  phases of, 363–65
  in plunger-assisted intermittent gas lift, 397
  slug-lifting period, 364–65
  valve spread required in, 369–70

Intermittent gas lift, 351–89
  accumulation chambers in, 351
  applications of, 353
  basic features of, 351–53
  calculating parameters for, 366–71
  casinghead pressure in, 354, 355
  chamber installations in, 352
  choke controls in, 356
  closed installation for, 363
  in closed/semi-closed installation,, 381
  continuous flow v., 231, 256, 351, 352–53
  cycle in, 356, 357, 363–71
  design of, 371–80
  empirical correlations for, 367–71
  equipment selection for, 245
  fallback in, 393
  gas breakthrough in, 364–65
  gas injection control in, 353–62
  gas requirement for, 366, 368–69
  gas volume in, 355
  inflow performance in, 356
  injection/production pressures in, 367
  intermitter control in, 374
  intermitter/regulator use in, 360
  liquid fallback in, 364–65, 367–68
  liquid recovery in, 368
  low productivity wells and, 351
  mechanism of, 351–53
  multipoint gas injection in, 352
  multipoint injection in, 262–63
  open installation and, 246–47
  performance/analysis of, 363–71
  pilot valves in, 355, 356

plunger-assisted, 352, 393–98
pressure regulator controls in, 356–57
pressure regulators in, 410
pros/cons of, 353
rules of thumb in, 366
single-point gas injection in, 352
starting slug length adjustment in, 357
streamlined Christmas tree for, 365
time cycle control in, 357–61
tubinghead pressure in, 354, 355
unbalanced valves in, 358
valve requirements for, 231–32
valve spread in, 357, 375, 376

Intermittent gas lift design
calculations detailed, 380
chamber lift in, 381–88
spacing factor in, 371, 372, 374
valve setting in, 374

Intermittent gas lift installation
components of, 410
controls in, 411
flow diagram of, 410

Intermittent gas lift well, unloading procedure, 380

Intermittent installation design, 371–80
constant surface closing pressure in, 371–76
for single point gas injection, 371–76
valve spacing in, 371–76

Intermittent installations
choke control in, 389
intermitter control in, 388–89
optimizing, 388–89

Intermittent lift, continuous flow v., 5–6

Intermitter
as choke, 359–60
for surface gas injection control, 369

Intermitter control
in intermittent gas lifting, 357–61, 374
in intermittent installations, 388–89
in Optiflow Design, 378
pros/cons of, 358, 359

Intermitters, *See also* Time cycle controllers
clockwork-driven, 362, 363
in closed rotative gas lift installation, 413
electronic, 362
improper settings in, 433
pneumatic, 362, 363

IPEF (injection pressure effect factor), 197

IPO (injection pressure operated) valves, 189–96
advantages of, for unloading valve string, 343
in chamber lift installation, 383
in continuous flow gas lifting, 343
disadvantages of, for unloading valve string, 343
double element, 194–96
in dual gas lift installations, 403–5, 404–5
in intermittent gas lifting, 355
in multiple lift installations, 251
opening/closing injection pressure of, 321
in Optiflow Design, 329–31, 376
with pilot port, 209–10
pros/cons of, 232–33
single element, 189–94
sizing for unloading valve string, 225–26
in unloading valve strings, 321–31

J

Joule-Thompson effect, 154, 155

K

Kaya, Sarica, and Brill model, 111
accuracy of, 118, 119, 120

Kickoff valves, 5

Kinetic factor, in Aziz et al., 98

Kinetic factors
in flow velocity, 54, 65
in mist flow pattern, 60, 64

Kinetic losses, 59, 60

King valve, 172

L

Liquid density, calculation of, 53

Liquid distribution coefficient
mixture density and, 81
in Orkiszewski correlation, 81–82, 84

Liquid fallback
in chamber lift installation, 382, 384
correlation chart, 368
in intermittent gas lift, 364–65, 367–68, 393
in plunger-assisted intermittent gas lift, 393, 397, 398

Liquid flow rate
gas injection rate and, 280
injection pressure and, 267–69, 312
pressure drop (PD) and, 284
stabilizing, 287
in systems analysis of producing wells, 268–72

Liquid flow rates
  gas injection rates and, 282
  gas liquid ratio and, 284

Liquid holdup
  in Beggs-Brill correlation, 86, 87–88
  bubble rise velocity and, 58
  calculation basics, 66–67
  calculation of, 49–50, 51, 58, 72
  drift-flux model and, 51
  in Dukler correlation, 135–36
  first measurement of, 64
  gas slippage and, 50
  in Hagedorn-Brown correlation, 74–75
  mixture density and, 58
  in Mukherjee-Brill correlation, 94–95
  in Orkiszewski correlation, 83l
  pressure drop and, 58

Liquid production, in afterflow period, 370–71

Liquid production rate
  calculated from well bottom, 277–79
  calculated from well head, 280–84
  injection rate v., 288, 305, 306, 307
  performance curves for, 305

Liquid recovery, in intermittent gas lifting, 368

Lockhart-Martinelli correction factors, 131–32

Lubricators, in plunger-assisted intermittent gas lift, 395, 396

M

Macaroni installations, 252

Mandrels
  API designations for, 241–42
  conventional/outside mounted, 235
  in dual gas lift installations, 402
  types of, 234–37
  in unloading of continuous flow installations, 319
  wireline retrievable/side pocket, 236–37

Mechanistic models, 97–110
  accuracy of, 110
  Ansari et al., 110–11
  Aziz-Govier-Fogorasi, 98–103
  Chokshi, Schmidt, and Doty, 110–11, 111
  empirical models v., 97, 116
  Gomez et al., 111
  Hasan-Kabir, 103–10
  introduction, 97–110
  Kaya, Sarica, and Brill, 111
  Metering valves, in gas injection, 361
  Mist (annular) flow
  in Aziz et al. model, 101

Duns-Ros concept and, 70
  in Hasan-Kabir model, 104, 105, 106, 107–9
  kinetic factor in, 60
  in Orkiszewski's method, 82
  Mixture densities, 59
  calculation of, 51
  in Hagedorn-Brown correlation, 74–75
  liquid holdup and, 58

Mixture density
  liquid distribution coefficient and, 81
  liquid holdup and, 58
  no slip, 50
  in Orkiszewski correlation, 80

Mixture properties
  in multiphase flow, 50–52
  in two-phase flow, 50

Mixture velocity, 59
  pressure gradient and, 82

Moody diagram, 28–29

Motor valve, 361–62
  in intermittent gas lift operations, 357

Mukherjee-Brill correlation, 93–97
  accuracy of, 118, 119, 120
  basic equation, 93
  Beggs-Brill correlation and, 95
  compared to other methods, 114, 115
  flow pattern in, 93
  friction factor in, 95–97
  in horizontal/inclined flow calculations, 130
  liquid holdup in, 94–95
  pressure drop in, 95–97
  stratified flow in, 96
  summary of, 93

Multiphase correlations, for pressure drop, 287–88

Multiphase flow
  in Beggs-Brill correlation, 86
  calculation direction in, 116–17
  comparison of models for, 114–21
  complexity of, 45
  correlations, 27, 44–54
  Duns-Ros dimensional analysis of, 64–65
  empirical models for, 60–61, 114–15
  gas v. liquid phase, 52
  mechanistic models for, 97–110
  Orkiszewski approach to, 78
  pressure drop in, 44
  pressure-drop models for, 52
  in restriction/choke, 144
  thermodynamic parameters in, 52
  variables in, 44–45, 45

Multiple gas lift installations, 251

Multipoint gas injection, 262–63

N

Natural gas
  density calculation, 17–18
  deviation factor for, 15, 16, 19
  flow velocity of, 32
  injection volume v. cost of, 288
  nitrogen v., 177, 182
  property correlations, 18–20
  viscosity, 19
  volume factors, 17

Natural gases
  ideal, 15
  kinetic theory of, 15
  properties of, 15–20

Nodal analysis, *See* Systems analysis of producing wells

No-slip mixture density, calculation of, 50

Nozzle-Venturi valve, 239–40

O

Open gas lift installation, 245–47
  drawbacks in, 246–47
  fluid seal in, 246
  schematic of, 408

Open-hole chamber installation, 250

Open installation
  continuous flow gas lifting and, 245, 246
  intermittent gas lifting and, 246–47

Operating valve, finding, 431

Optiflow Design
  choke control in, 378
  in continuous flow gas lift, 329–31
  gas requirement in, 379
  in intermittent gas lift, 376–80
  intermitter control in, 378
  IPO valves in, 376
  pilot valves in, 378
  steps of, 376–78
  unbalanced valves in, 378
  for unloading valve strings, 376–80
  valve setting in, 376–78
  valve spacing in, 376–78

Optimizing continuous flow installations, 288–315
  balance of costs, 298–99
  compressor selection for, 293–97, 303–8
  constraints in, 315
  cost of gas compression in, 289–90
  economic considerations in, 298
  with existing compressor, 290–92
  field-wide, 314–15
  gas allocation in, 312, 313, 314
  gas flow allocation in, 309–10
  gas injection rates in, 315
  in group of wells, 308–15
  increment of costs v. increment of revenue, 298–99
  introduction, 288–89
  with limited gas availability, 308
  with limited lift gas availability, 312–14
  performance curves for, 298, 301, 302–3
  with prescribed liquid rate, 289–97
  sequential quadratic programming (SQL)
    algorithm in, 315
  in single well, 289–97, 289–308
  steps for, 290–91, 293–94, 300, 304–5
  system parameters for, 288–89, 304–5
  tubing/flowline performance curves for, 295–96
  with unlimited gas availability, 308
  with unlimited lift gas availability, 310–12
  with unlimited liquid rate, 298–308

Orifice valve, 239

Orkiszewski correlation, 78–84
  accuracy of, 118, 119, 120
  basic equation, 78
  bubble rise velocity in, 80
  compared to other methods, 114, 115
  as composite calculation model, 78
  Duns-Ros correlation and, 78
  flow pattern map, 79
  Griffith-Wallace model and, 78, 79, 80, 81
  Hagedorn-Brown model and, 76, 78, 81
  improvements of, 82–83
  liquid distribution coefficient in, 81–82
  mixture density in, 80
  in multiphase flow calculations, 291
  Poettmann-Carpenter in, 78
  slug flow in, 80
  summary of, 78, 84
  in Systems analysis of producing wells, 288
  Taylor bubbles in, 80

Outflow pressure, calculating, 33–35

P–Q

Pack-off installations, 251–52

Papay's equation, 19, 34

PD, *See* Pressure drop

Performance curves, 240–42, 287–88
  computer applications and, 287–88

Physical models, characteristics of, 115

PI, *See* Productivity index

Pilot valves, 206–15
in chamber lift installation, 385, 387
combination valves, 210–14
in dual gas lift installations, 405
flexible sleeve, 214–15
for intermittent gas lift, 206
in intermittent gas lift, 355, 360, 371
IPO valves, 206–9
opening/closing equations for, 207
in Optiflow Design, 378
pros-cons of, 234
setting depth for, 372–73
spread characteristics of, 207–8
TRO pressure, calculating, 208–9

Pilot-operated gas lift valves
for intermittent lift operations, 231–32

Plunger-assisted intermittent cycle
accumulation period in, 397
gas blowdown period in, 397
slug lifting period in, 397

Plunger-assisted intermittent gas lift, 393–98
advantages of, 394
applications, 393–94
bottom dumper spring in, 396
fallback and, 393
installation types, 394–95
liquid fallback in, 397, 398
lubricator in, 395, 396
operating conditions in, 397–98
operating parameter calculations for, 398
plunger in, 396–97
plunger lift v., 393
pros/cons of, 394
standing valve in, 396
subsurface equipment in, 395–97
tubing in, 395
wellhead arrangement in, 395

Plunger, in plunger-assisted intermittent gas lift, 396–97

Plunger lift, 393

Poettmann-Carpenter correlations, 62–64, 114, 115
accuracy of, 118, 119, 120
gradient curves and, 123
Orkiszewski's use of, 78

PPE (Production pressure effect)
calculating, 192–93
in single-element valves, 192

PPEF (Production pressure effect factor)
in single-element valves, 192
in unbalanced valves without spread, 200

PPO (production pressure operated) valves, 189,
197–200
closing equation for, 197–98

with crossover seats, 198–99
in dual gas lift installations, 403, 405
in intermittent gas lifting, 355
in multiple lift installations, 251
with normal seats, 197–98
opening equation for, 197
opening/closing equations for, 206
pros/cons of, 233
setting calculations for, 334
in unloading valve strings, 334

Prediction errors, sources of, 115

Prescribed liquid rate
in single well gas lift production, 289–97
wellhead pressure and, 290

Pressure differential, in high-capacity wells, 260

Pressure drawdown, 22

Pressure drop (PD)
accuracy of models for, 114–21
in Beggs-Brill correlation, 86
calculating, 20, 26, 29, 30–35, 44
calculation models for, 51, 58
components of, 284
computation methods, 44
in continuous flow gas lift, 256
in Cornish correlation, 92–93
in deviated wells, 60
in Duns-Ros correlation, 60, 64–72
empirical correlations for, 116
errors, 115, 260
factors affecting, 1, 2
friction v. elevation in, 59
Hagedorn-Brown model of, 74
in horizontal flowlines, 127
kinetic factor in, 54, 59
liquid flow rate and, 284
liquid holdup and, 58
mechanistic models for, 97–110
mixture weight and, 58
models, accuracy of, 117–21
models compared, 114–21
in Mukherjee-Brill correlation, 95–97
in multiphase flow, 45
in oil well flow, 54
Orkiszewski's approach to, 78
prediction models for, 44, 45
selection of models for, 114–21
slippage loss in, 45
special conditions and, 117
system components and, 160
tubing size and, 265
two-phase, 44
valve settings for, 322, 327
well death and, 257–58

Pressure drop correlations, evaluations of, published, 117–20

Pressure drop models
published evaluations of, 117–21

Pressure gradient calculations
in Beggs-Brill correlation, 89–91
in Hagedorn-Brown correlation, 76–77
in Hasan-Kabir model, 104
in horizontal two-phase flow, 130–31
mixture velocity and, 82
in Orkiszewski correlation, 78, 79, 84

Pressure gradient equations, 53–54, 71–73
solving, 111–13

Pressure gradients, chart, 438

Pressure operated valves
bellows assembly, 182–85
bellows stacking in, 182
check valves, 187–88
construction details, 181–88
core valve/tail plug, 181
damage/protection in, 182–83
Dill valve, 181
dome charge pressure in, 183–84
failures in, 182
gas charge, 182
injection pressure v. valve stem travel, 185
introduction, 180
IPO v. PPO, 189
knife edging in, 182, 183
load rates in, 184–85, 186
nitrogen v. natural gas in, 182
operation of, 180–81
parts of, 180
parts terminology, 180–81
valve charging, 181
valve stem chatter in, 182
valve stem travel in, 185, 186

Pressure recorders
in continuous flow gas lift, 422–24
in intermittent gas lift, 422–25
in troubleshooting/analysis, 422

Pressure regulators
in gas injection control, 361–62
in intermittent gas lifting, 410

Pressure traverse calculation, 111–13
in horizontal flow, 137–38

Pressure traverses, calculating, 111–13

Pressure v. rate curves, in systems analysis of producing wells, 161

Pressure-volume-temperature (pVT) measurements, 10

Production pressure effect factor, *See* PPEF

Production pressure effect, *See* PPE

Production pressure, pressure recorders and, 422

Production rates
bottomhole pressure and, 21
flow restrictions and, 139
gas lift valves and, 169
lift design and, 21
optimizing continuous flow installations, 289–97
for various calculation models, 114–15
various lift methods compared for, 3–4
various methods compared, 21–26
Vogel's IPR correlation and, 24

Productivity index (PI), 23–24

Proportional response valve, 217

Pumping, 2

R

Ramey's temperature model, 147–51
formula for flowing temperature, 149–50
heat loss/transfer in, 147
heat transfer coefficient in, 148
modifications to, 151
production time in, 148, 150

Regulator control
advantages of, 357
in intermittent gas lifting, 356–57

Reverse flow check valves, 187–88

Reynolds number, chart, 436

Reynolds numbers, in pressure drop calculations, 28, 29, 31

S

Sagar-Doty-Smith temperature model, 153–55
formula for, 153
time function in, 154

Semi-closed installation
for continuous flow gas lift operations, 247
for intermittent flow gas lift operations, 247
intermittent gas lifting in, 381
unloading, 316–17

Sequential quadratic programming (SQL) algorithm, for optimizing continuous flow installations, 315

Shiu-Beggs temperature model, 151–52

Single-element valves
  force equation for, 189, 191
  injection pressure in, 190–91
  opening equation for, 190
  PPE in, 192
  PPEF in, 192
  production pressure in, 191
  production v. injection pressure in, 190
  TRO pressure value, 191

Single-phase flow, multi-phase flow v., 44, 45

Single-phase gas flow
  calculating, 32–43
  liquid flow v., 32
  in surface-injection chokes, 40

Single-phase liquid flow
  complexity of, 28
  friction factor in, 28
  General Energy Equation in, 30
  pressure drop in, 26, 30–35
  velocity calculation in, 27

Single point gas injection, in intermittent installation
  design, 371–76

Slip velocity, 49

Slippage loss
  flow rate instability, 286
  in multiphase flow, 45, 60–61
  in two-phase flow models, 61

Slippage velocity correlations, 58

Slug flow, 46
  in Aziz et al. model, 100–101, 102
  Duns-Ros parameters for, 69
  in Hasan-Kabir model, 104, 105, 106–7
  heading and, 285
  in horizontal flow patterns, 129
  kinetic factors in, 60
  in Orkiszewski correlation, 80
  transition parameters, 66

Slug lifting period
  in intermittent cycle, 364–65
  in plunger-assisted intermittent cycle, 397

Solution gas-oil ratio, 11

Solution nodes in systems analysis, 160, 161, 162, 266,
  277–79, 280–83

Spacing factor, in intermittent gas lift design, 371, 372,
  374

Spacing procedure, for IPO valves
  with constant PD per valve, 327–28
  with variable PD per valve, 322–26

Specific gravity, liquid's, 10

Spring-loaded valves
  advantages of, 201
  force balance equations for, 202
  performance of, 202–3

Stability in gas-lifted wells, systems analysis of, 286

Standing valves, in plunger-assisted intermittent gas lift,
  396, 397

Standing-Katz chart, 19

Standing's bubblepoint pressure correlation, 13

Static liquid level, in unloading of continuous flow
  installations, 318

Stratified flow
  in horizontal flow patterns, 129
  in Mukherjee-Brill correlation, 96

Superficial velocities
  calculating, 46–47
  in two-phase flow, 46

Surface dome pressure, calculating, 174

Surface gas injection control, 344–45
  chokes in, 369
  in continuous flow operations, 345–46
  in intermittent gas lift operations, 353–62
  meter run in, 344
  pressure recorder in, 344

System performance calculations
  for limited gas supply, 269–72, 275–76
  for solution node at well bottom, 277–79
  for solution node at wellhead, 280–83
  for unlimited gas supply, 267–69, 272–75

System performance, describing, 266–88
  Constant WHP cases, 267–88
  introduction, 266

System stability, 284–87

Systems analysis, of producing wells, 158–62, 287
  See also System performance, describing
  applicability of, 160
  basic principles of, 160–61
  basic steps for, 266
  component performance, 159
  compressor selection and, 293
  computer applications in, 279, 281, 283, 315
  continuous flow gas lifting and, 266
  Duns and Ros correlation in, 288
  Equilibrium Curve in, 272–76
  evaluating stability via, 286
  example of, 161
  flow rate calculation in, 160–61
  flowline in, 277, 279
  in gas lifting, 161–62
  Hagedorn and Brown correlation in, 288

liquid flow rate calculations in, 268–72
node points for, 160, 266, 277, 280
Orkiszewski correlation in, 288
performance curves in, 277–78
pressure drop between components, 160
pressure v. rate in, 161
production system components, 158, 159–60
solution nodes in, 161, 162, 266, 277, 280
steps for, 278
variable WHP in, 277–84
Wellhead pressure (WHP) in, 279

T

Taylor bubbles, in Orkiszewski correlation, 80

TEF (tubing effect factor), *See* PPEF

Temperature, well, *See* Well temperature

Theorem of Corresponding States, 16

Thermal conductivity of well materials, 147–48

Thornhill-Craver formula for throughput velocity in unloading valves, 344

THP, *See* Tubinghead pressure

Throttling valves, *See also* Unbalanced valves
advantages of, for unloading valve string, 343
beneficial features of, 335
calculating flow in, 221–26
disadvantages of, for unloading valve string, 34
injection transfer in, 335–36
opening/closing equations for, 217
operating valve calculations, 339–40
Proportional Response Design and, 335
pros/cons of, 233
spacing/setting of, 336–42
throughput performance of, 335
in unloading valve string design, 335–42, 343

Time cycle control, in intermittent gas lifting, 357–61

Time cycle controllers. *See also* Intermitters
in intermittent gas lifting, 362–63

Transition boundaries
in Aziz et al. model, 102
in orifice flow modeling, 224

Transition flow
in Aziz et al. model, 101, 102
in Hasan-Kabir model, 104–5
in Orkiszewski's method, 82

Transition flow/region,
calculation of, 71

Transition pressure, calculating, in valves, 220

Traverse calculation flowchart, 138

TRO (test track opening) pressure value
for IPO valves, 207, 325
for PPO valves, 198
in single-element valves, 191

Troubleshooting/analysis
Acoustic surveys in, 428–29
for continuous flow wells, 422–23, 424, 426, 427, 428
downhole pressure surveys in, 426–28
flowing pressure surveys in, 428–29
flowing temperature surveys in, 428–29
gas volume measurements in, 426
for intermittent wells, 423–25
pressure recorders in, 422
temperature surveys in, 426–28
tools for, 421–30
two-pen pressure recordings for, 422–25

TUALP (Tulsa University Artificial Lift Projects), 219
lift valve performance parameter chart, 224

Tubing effect factor, *See* TEF

Tubing flow installations
chamber installations, 248–50
closed, 248
open, 245–47
performance curves, 292
semi-closed, 247

Tubinghead pressure (THP)
in intermittent lift cycle, 354, 355, 357, 359, 360, 361, 424
pressure recorders and, 422–25

Tubing, holes in, 431

Tubing performance curves, 267, 268, 269, 278, 295, 296

Tubing pressure control, in intermittent gas lifting, 360

Tubing-pressure valve. *See* PPO valves

Tulsa University Fluid Flow Projects (TUFFP), 119

Two-packer chamber installations, 249, 381–82

Two-pen pressure recorders
connection recommendations for, 422
in troubleshooting/analysis, 422–25

Two-phase flow, 59
basic principles of, 44–54
calculating phase densities in, 50
concepts of, 45–52
empirical models and, 45
first calculation model for, 60–61
friction factor in, 59
gas slippage in, 48
horizontal/inclined, 127–38
inclined/annulus flow, 60

mechanistic approach to, 45
mixture properties and, 50
in oil wells, 54–127
pressure variations in, 44–45
problems, importance of, 44
superficial velocities in, 46
viscosity in, 51

Two-phase flow in oil wells, background theories, 54–60

U

Unbalanced IPO valves with spread, pros/cons of, 232

Unbalanced valves
for continuous flow gas lift, 200
dynamic performance of, 227
force balance equations for, 200–201
in intermittent gas lifting, 358, 371, 376
IPO valves, 200–203
in Optiflow Design, 378
with pilot port, 209–10
PPO valves, 203–4
setting depth for, 372–73
with spread, 189–200, 207
spring-loaded valves without bellows charge, 201–3
valves with a choke upstream of the port, 200–201
without spread, 200–204

Unlimited liquid rate in single well gas lift, 298–308

Unloading of continuous flow installations, 315–44
basic mechanism of, 316–17
bellows-charged gas lift valves in, 319
injection pressure/rate in, 318, 344
IPO valves in, 343
mandrels in, 319
to a pit, 318
practical considerations for, 342–43
semi-closed installation, 316–17
static liquid level in, 318
throttling valves in, 343
valve damage, preventing, 342–43
valve spacing for, 318, 319
valve string design for, 318, 320–42
well temperature in, 319

Unloading of intermittent gas lift installations, 372, 380

Unloading problems, common, 430

Unloading valve strings
depths/settings, 320, 322–25
design procedures for, 320–42
for dual gas lift installations, 403–5
IPO valves in, 225–26, 321–31, 343
Optiflow Design for, 376–80
PPO valves in, 334
proper results of, 321

requirements of, 321
spacing in, 321–32
throttling valves in, 335, 343

V

Valve(s), *See also* specific values
in dual gas lift installations, 402
problems, common, 431–33

Valve bracketing, 319–20, 324, 327, 330

Valve construction details, 181–88
ball and seat, 186–87
bellows assembly, 182–85
check valve, 187–88
core valve/tail plug, 181
gas charge, 182
spring, 185–86

Valve damage
API recommendations for preventing, 343, 380
in unloading of continuous flow installations, 342
in unloading of intermittent gas lift installations, 380

Valve mechanical data, manufacturers', 456–59

Valve mechanics, 188–234
introduction, 188–89

Valve performance curves, 226–27

Valve problems, common, 431

Valve selection, for dual gas lift installations, 403

Valve setting
in intermittent gas lift design, 374
in Optiflow Design, 376–78

Valve setting depths
for balanced valves, 332–33
calculations for, 322–25, 327–28, 329–30, 372–73

Valve spacing
for balanced valves, 332–33
in chamber lift installation, 384, 387–88
with constant injection pressure drop, 327–29
with constant PD per valve, 327–28
constant surface opening pressure method, 329–31
in dual gas lift installations, 402
in intermittent installation design, 371–76
objectives of, 318
in Optiflow Design, 376–78
procedures per type, 318
safety factors in, 319
with variable injection pressure drop, 322–26
with variable PD per valve, 322–26
well temperature and, 319

Valve spread
  calculating, 193–94
  in chamber lift installation, 383
  defined, 192
  importance of, 192–93
  injection pressure and, 193
  in intermittent gas lift, 355–56, 357, 359, 360, 373–74, 375, 376

Valve stacking, 324, 330

Valve surface opening pressure, charts, 448–53

Valve testers
  encapsulated, 229–30
  sleeve/ring type, 228–29

Velocity numbers, 56–57

Venturi device, 216, 219, 239–40

Vertical flow patterns, 54–55

Viscosity
  crude oil, 14–15
  factors affecting, 14
  fluid, 10, 13, 14
  gas, calculating, 20
  in two-phase flow, 51

Vogel's IPR correction, 24–25

Volume factors
  crude oil, 13–14
  fluid, 11, 13

gas lift valves and, 147, 174, 201, 204, 228,
geothermal gradient and, 146
heat loss/transfer and, 147
lift valve settings and, 147, 228
in pressure drop calculations, 147
Ramey's model, 147–51
Sagar-Doty-Smith model, 153–55
Shiu-Beggs correlation for, 151–52
in startup conditions, 146
steady state conditions in, 146
in unloading of continuous flow installations, 319
valve spacing procedures and, 319

Well unloading, 215–16, 232

WHP, *See* Wellhead pressure (WHP)

Wireline retrievable equipment, in dual gas lift installations, 402

Wireline retrievable mandrels/valves, 236–38
  in chamber lift installation, 383

Wireline tools, 237–38

Z

Z-factor correlations, 16

W–Y

Water
  basic properties of, 12
  volume factor of, 12

Well blower installation, 4

Well death
  continuous flow gas lift v., 258

Wellhead pressure (WHP)
  in continuous flow gas lifting, 263–64
  prescribed liquid rate and, 290
  in systems analysis of producing wells, 279
  variable, well performance analysis and, 277–84

Well performance
  calculating, 267–72
  optimizing continuous flow installations, 289–97

Wells, single, optimizing, 289–308

Well temperature, 146–58
  basics of, 146–47
  calculation models for, 147–54
  distribution, 146